THE HUMAN
RETROVIRUSES

THE HUMAN RETROVIRUSES

EDITED BY

Robert C. Gallo
Laboratory of Tumor Cell Biology
National Cancer Institute
National Institutes of Health
Bethesda, Maryland

Gilbert Jay
Laboratory of Virology
Jerome H. Holland Laboratory
American Red Cross
Rockville, Maryland

ACADEMIC PRESS, INC.
Harcourt Brace Jovanovich, Publishers
San Diego   New York   Boston   London
Sydney   Tokyo   Toronto

Academic Press, Inc.
San Diego, California 92101

United Kingdom Edition published by
Academic Press Limited
24–28 Oval Road, London NW1 7DX

Library of Congress Cataloging-in-Publication Data

The human retroviruses / edited by Robert C. Gallo, Gilbert Jay.
    p. cm.
   Includes index.
   ISBN 0-12-274055-6
   1. Retrovirus infections. 2. Retroviruses. 3. HIV (Viruses)
4. HTLV (Viruses) I. Gallo, Robert C. II. Jay, Gilbert.
   [DNLM: 1. Acquired Immunodeficiency Syndrome. 2. HIV Infections.
3. HTLV Viruses. 4. Retroviridae. 5. Retrovirus Infections.   QW
166 H91835]
QR201.R47H84   1991
616'.0194--dc20
DNLM/DLC
for Library of Congress   90-14480
                                          CIP

PRINTED IN THE UNITED STATES OF AMERICA
91 92 93 94   9 8 7 6 5 4 3 2 1

Contents

3. Human Immunodeficiency Virus (HIV) Gene Structure and Genetic Diversity 35
Mary E. Klotman and Flossie Wong-Staal

4. Human Immunodeficiency Virus (HIV) Gene Expression and Function 69
William A. Haseltine

Part II
Biology 107

5. Biology of Human T-cell Leukemia Virus (HTLV) Infection . 109
Isao Miyoshi

6. Receptors for Human Retroviruses 127
Robin A. Weiss

7. Immunopathologic Mechanisms of Human Immunodeficiency Virus (HIV) Infection 141
Zeda F. Rosenberg and Anthony S. Fauci

Contributors

Numbers in parentheses indicate the pages on which the authors' contributions begin.

Edward A. Berger (379), Laboratory of Viral Diseases, National Institute of Allergy and Infectious Diseases, National Institutes of Health, Bethesda, Maryland 20892

Ruth L. Berkelman (193), Division of HIV/AIDS, Center for Infectious Diseases, Centers for Disease Control, Atlanta, Georgia 30333

William A. Blattner (175), Viral Epidemiology Section, Environmental Epidemiology Branch, National Cancer Institute, National Institutes of Health, Bethesda, Maryland 20892

Dani P. Bolognesi (389), Center for AIDS Research, Duke University Medical Center, Durham, North Carolina 27710

Samuel Broder (335), National Cancer Institute, National Institutes of Health, Bethesda, Maryland 20892

Vijay K. Chaudhary (379), Laboratory of Molecular Biology, National Cancer Institute, National Institutes of Health, Bethesda, Maryland 20892

Irvin S. Y. Chen (21), Departments of Medicine, Microbiology, and Immunology, University of California, Los Angeles, California 90024

Lian-sheng Chen (227), Laboratory of Virology, Jerome H. Holland Laboratory, American Red Cross, Rockville, Maryland 20855

James Chin (213), Surveillance, Forecasting and Impact Assessment Unit, Global Programme on AIDS, World Health Organization, 1211 Geneva 27, Switzerland

Christopher H. Contag (245), Department of Microbiology and Immunology, Stanford University School of Medicine, Stanford, California 94305

James W. Curran (193), Division of HIV/AIDS, Center for Infectious Diseases, Centers for Disease Control, Atlanta, Georgia 30333

Stephen Dewhurst (245), Department of Cancer Biology, Harvard University School of Public Health, Boston, Massachusetts 02115

Timothy J. Dondero, Jr. (193), Division of HIV/AIDS, Center for Infectious Diseases, Centers for Disease Control, Atlanta, Georgia 30333

Anthony S. Fauci (141), National Institute of Allergy and Infectious Diseases, National Institutes of Health, Bethesda, Maryland 20892

David J. FitzGerald (379), Laboratory of Molecular Biology, National Cancer Institute, National Institutes of Health, Bethesda, Maryland 20892

Joseph Fontes (227), Department of Medical Pathology, School of Medicine, University of California, Davis, California 95616

Jun-ichi Fujisawa (3), Department of Cellular and Molecular Biology, Institute of Medical Science, University of Tokyo, Tokyo 108, Japan

Allison L. Greenspan (193), Division of HIV/AIDS, Center for Infectious Diseases, Centers for Disease Control, Atlanta, Georgia 30333

William A. Haseltine (69), Division of Human Retrovirology, Dana-Farber Cancer Institute, Harvard Medical School, Boston, Massachusetts 02115

Steven H. Hinrichs (227), Department of Medical Pathology, School of Medicine, University of California, Davis, California 95616

Jun-ichiro Inoue (3), Department of Cellular and Molecular Biology, Institute of Medical Science, University of Tokyo, Tokyo 108, Japan

Desmond B. Jay (277), Laboratory of Virology, Jerome H. Holland Laboratory, American Red Cross, Rockville, Maryland 20855

Gilbert Jay (227, 277), Laboratory of Virology, Jerome H. Holland Laboratory, American Red Cross, Rockville, Maryland 20855

Mary E. Klotman (35), Laboratory of Tumor Cell Biology, National Cancer Institute, National Institutes of Health, Bethesda, Maryland 20892

Jonathan M. Mann (213), Harvard University School of Public Health, Boston, Massachusetts 02115

J. M. Mc Cune (295), SyStemix, Inc., Palo Alto, California 94303

Hiroaki Mitsuya (335), Clinical Oncology Program, National Cancer Institute, National Institutes of Health, Bethesda, Maryland 20892

Isao Miyoshi (109), Department of Medicine, Kochi Medical School, Kochi 783, Japan

Bernard Moss (379), Laboratory of Viral Diseases, National Institute of Allergy and Infectious Diseases, National Institutes of Health, Bethesda, Maryland 20892

James I. Mullins (245), Department of Microbiology and Immunology, Stanford University School of Medicine, Stanford, California 94305

Ira Pastan (379), Laboratory of Molecular Biology, National Cancer Institute, National Institutes of Health, Bethesda, Maryland 20892

Jonathan A. Rhim (277), Laboratory of Virology, Jerome H. Holland Laboratory, American Red Cross, Rockville, Maryland 20855

Zeda F. Rosenberg (141), National Institute of Allergy and Infectious Diseases, National Institutes of Health, Bethesda, Maryland 20892

Joseph D. Rosenblatt (21), Departments of Medicine, Microbiology, and Immunology, University of California, Los Angeles, California 90024

Paul A. Sato (213), Surveillance, Forecasting and Impact Assessment Unit, Global Programme on AIDS, World Health Organization, 1211 Geneva 27, Switzerland

Kiyoshi Takatsuki (163), Second Department of Internal Medicine, Kumamoto University Medical School, Kumamoto 860, Japan

Gregory A. Viglianti (245), Program in Molecular Medicine and Department of Molecular Genetics and Microbiology, University of Massachusetts Medical Center, Worcester, Massachusetts 01605

Jonathan Vogel (277), Laboratory of Virology, Jerome H. Holland Laboratory, American Red Cross, Rockville, Maryland 20855

Thomas A. Waldmann (319), Metabolism Branch, National Cancer Institute, National Institutes of Health, Bethesda, Maryland 20892

Robin A. Weiss (127), Chester Beatty Laboratories, Institute of Cancer Research, Royal Cancer Hospital, London SW3 6JB, England

Stefan Z. Wiktor (175), Viral Epidemiology Section, Environmental Epidemiology Branch, National Cancer Institute, National Institutes of Health, Bethesda, Maryland 20892

Flossie Wong-Staal (35), Departments of Biology and Medicine, School of Medicine, University of California, San Diego, La Jolla, California 92093

Mitsuaki Yoshida (3), Department of Cellular and Molecular Biology, Institute of Medical Science, University of Tokyo, Tokyo 108, Japan

Foreword

Until about 1950 RNA tumor viruses, as retroviruses were then called, were only known in chickens. Although the mouse mammary tumor virus had been discovered in the 1930s, it was still called the mouse mammary tumor agent at that time and was not considered a true RNA tumor virus. The discovery by Gross of mouse (murine) leukemia virus established that RNA tumor viruses also existed in mammals. Soon other strains of murine leukemia virus were found, and it was shown that the chicken Rous sarcoma virus could cause tumors in mammals. However, in spite of much effort, it still took a long time to find human retroviruses. Now we know of HTLV-1 and -2, HIV-1 and -2, and human spumaretroviruses. Furthermore, although hepatitis B virus is not classified as a retrovirus, its mode of replication has important homologies to that of retroviruses, including reverse transcription. I call it a pararetrovirus.

Retroviruses are important to humans. Some 200 million people are infected with hepatitis B virus, over 10 million are infected with HIV-1, and an unknown total number of others are infected with HIV-2 and HTLV-I and -II. In addition, we know that the human genome is full of retrovirus-related proviruses (endogenous retroviruses) and other sequences related to retroviruses and that over 10% of the human genome is the result of reverse transcription. Furthermore, modified murine retroviruses have been used as vectors to mark human tumor-infiltrating lymphocytes that were injected into human cancer patients, and proposals are being considered to use murine retrovirus vectors to treat human adenosine deaminase deficiency. Therefore, retroviruses can cause human disease and can also be used to treat human disease.

Recently, Simon Wain-Hobson of the Institut Pasteur reviewed in *Nature* (**343,** 706, 1990) *Retrovirus Biology and Human Disease* edited by R. C. Gallo and F. Wong-Staal. Wain-Hobson wrote that "this one (volume) is about on target but is bound to be *depassé* within a year or so. . . .

Perhaps the only solution is to produce frequent reviews and volumes like this one." The present volume provides a response to this suggestion. It certainly should provide more evidence for Wain-Hobson's assertion that "this is a fascinating field." Furthermore, based on previous history, I expect that retrovirology will continue to be fascinating and to need further volumes updating the field.

The chapter titles of this book indicate wide coverage of the traditional human retroviruses and their closest primate relatives. The authors are primarily from the United States; there are also some Japanese authors and one English author. This book should be useful to researchers in the field. It also records the status of the field as of Spring 1990 and the tremendous amount of work and new knowledge that has been gained in the past decade.

In appreciating this knowledge, we must remember that investigation of human retroviruses was based on previous results of basic research with animal retroviruses and with other nonhuman materials. We need to keep the entire field of biomedical and life sciences strong and healthy. We cannot tell what results will be most immediately relevant to future human problems. The only thing of which we can be sure is that we will need all the information and wisdom we can secure to deal with the problems that we know will face us in the future.

Howard M. Temin

Part I

Molecular Biology

Positive and Negative Regulation of Human T-cell Leukemia Virus Type I (HTLV-I) Gene Expression and Replication: Function of the *rex* Gene

Mitsuaki Yoshida, Jun-ichiro Inoue, and Jun-ichi Fujisawa

I. Introduction

The molecular abnormalities found in human cancers seem to be heterogeneous collections of genetic lesions and suggest that each case of human cancer is likely to represent a molecularly unique illness. These abnormalities might be due to numerous carcinogenic factors and genes involved in random human cancers. Therefore, any human cancers in which etiologic agents are identified should be valuable for systematic studies of the molecular mechanisms of tumorigenesis in humans. Adult T-cell leukemia (ATL) (Uchiyama *et al.*, 1977) is one of the best examples from this point of view, because it is caused by infection with a retrovirus, human T-cell leukemia virus type I (HTLV-I) (Yoshida *et al.*, 1984a).

HTLV-I (Poiesz *et al.*, 1980; Hinuma *et al.*, 1981; Yoshida *et al.*, 1982) is a member of the retrovirus family, but is classified in a unique group distinct from general onco-retroviruses (Seiki *et al.*, 1983). The genome has an extra

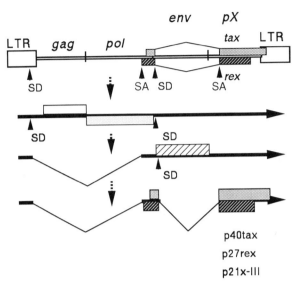

Figure 1. Genome structure and expression of HTLV-I. The boxes represent open reading frames or LTR as specified. SD and SA are the sites for splicing donor and acceptor. Thick horizontal lines represent viral RNA.

sequence named "pX" in addition to the *gag*, *pol*, and *env* genes (Seiki *et al.*, 1983; Fig. 1). The pX region contains three overlapping genes, *tax, rex,* and a third one which codes for p21, whose function is not yet known (Nagashima *et al.*, 1986; Sagata *et al.*, 1985a). HTLV-I shows poor replication competence. Especially *in vivo*, almost all proviral genomes are latent (Hinuma *et al.*, 1982; Clarke *et al.*, 1984; Kitamura *et al.*, 1985), although most proviral genomes are intact (Yoshida *et al.*, 1984a). Some cell lines produce viral progenies rather efficiently, but all these cell lines contain about ten copies of defective proviruses (Watanabe *et al.*, 1984). Another unique feature of HTLV-I is that its infection frequently immortalizes CD4[+] T cells (Miyoshi *et al.*, 1981; Yamamoto *et al.*, 1982; Popovic *et al.*, 1983). This phenomenon has never been observed with ordinary retroviruses which do not carry a typical oncogene. These unique properties are partly explained by the functions of the pX genes, *tax* and *rex*.

Expression of the HTLV-I genes is regulated at two different levels, transcription and RNA processing by a combination of *tax* and *rex* functions.

1. Tax protein (40 kDa) *trans*-activates transcription from the long terminal repeat (LTR) (Sodroski *et al.*, 1984; Chen *et al.*, 1985; Felber *et al.*, 1985; Fujisawa *et al.*, 1985; Seiki *et al.*, 1985, 1986). In addition to the viral genome, transcription of some cellular genes such as IL-2Rα (Inoue *et al.*, 1986a; Cross *et al.*, 1987; Maruyama *et al.*, 1987; Siekevitz *et al.*, 1987), IL-2

(Inoue *et al.*, 1986a) and GM-CSF (Miyatake *et al.*, 1988; Nimer *et al.*, 1988; Shannon *et al.*, 1988) are also *trans*-activated by Tax. In particular, *trans*-activation of the IL-2Rα gene has been proposed to be a molecular mechanism involved in the early stage of ATL development (Inoue *et al.*, 1986a; Yoshida, 1987; Yoshida and Seiki, 1987), since leukemic cells of ATL are always T cells expressing an abnormally high level of IL-2Rα (Hattori *et al.*, 1981; Popovic *et al.*, 1983; Yodoi *et al.*, 1983; Depper *et al.*, 1984). These findings have previously been reviewed (Yoshida, 1987), and so are not covered in this chapter. Cellular factors seem to be involved in the mechanism of Tax-mediated *trans*-activation. NFκB was proposed to be required for *trans*-activation of the IL-2Rα gene (Leung and Nabel, 1988; Lowenthal *et al.*, 1988; Ruben *et al.*, 1988), whereas other cellular proteins have been proposed to be required for *trans*-activation of the LTR (Fujisawa *et al.*, 1989; Yoshimura *et al.*, 1990).

2. The other protein, Rex (27 kDa) (Kiyokawa *et al.*, 1985), is a *trans*-modulator of RNA processing to express unspliced viral RNA (Inoue *et al.*, 1986b, 1987) that is Gag/Pol and Env mRNAs (Fig. 1). These RNA species are otherwise completely spliced out into Tax/Rex mRNA (Hidaka *et al.*, 1988). Therefore, the *rex* gene is essential for the expression of virion proteins. The *rex* function, on the other hand, reduces the Tax/Rex mRNA level and results in reduction of the viral gene transcription (Hidaka *et al.*, 1988).

Thus, the *tax* and *rex* genes in combination regulate HTLV-I gene expression and replication in both positive and negative ways. In this chapter, we review our recent findings on *rex* gene function and discuss the significance of this function in viral replication and ATL development.

II. Discovery of the *rex* Gene Function

In order to study *trans*-activation of the transcription by Tax protein, we and others used the plasmid construct LTR-CAT (gene for chloramphenicol acetyl transferase). The CAT gene expression directed by the LTR is strongly enhanced by Tax protein. In analogy, we tried to express the Gag protein in a human cell line using a defective provirus, LTR-*gag-pol*-LTR in which the *tax/rex* genes were missing, but could not detect any viral protein expression even in the presence of a high concentration of the Tax protein. Surprisingly, we found that the wild-type pX expression plasmid induced expression of the Gag protein (Inoue *et al.*, 1986b, 1987). This observation was unexpected, since the Tax protein should *trans*-activate transcription from the LTR, and the viral mRNA should be expressed efficiently. Therefore, additional function supplied by the pX sequence is essential for Gag protein expression. Testing three open reading frames in the pX sequence, we found that *rex* is required for the induction of Gag protein expression (Inoue *et al.*, 1987).

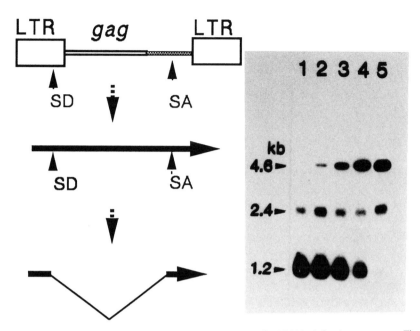

Figure 2. Effect of Rex protein on RNA expression of HTLV-I defective genome. The defective genome and *tax* expression plasmid were transfected into a human cell line in the presence or absence of a *rex* expression plasmid and the viral RNA expressed was analyzed by Northern blotting. The bands of 2.4 kb were derived from *tax* and *rex* expression plasmids. Lanes 1–5, ratios of *rex* DNA to *tax* DNA of 0/1; 0.1/1; 0.3/1; 1/1; and 3/1, respectively. SD and SA, sites for splicing donor and acceptor.

Analysis of mRNA expressed from the LTR-*gag-pol*-LTR construct showed that, in the absence of the Rex protein, all viral mRNAs were completely spliced out into the smallest sized RNA which no longer carried the *gag* sequence (Fig. 2). Upon expression of the Rex protein, unspliced RNA that carried the *gag* sequence was accumulated. This induction of gag mRNA is consistent with expression of the Gag protein. Therefore, Tax protein was clearly effective in *trans*-activation of transcription, but Rex protein was required for expression of unspliced RNA. That is, Rex protein acts at the RNA level. This effect of the Rex protein was dependent on its concentration: the level of unspliced RNA increased with an increase in its concentration (Fig. 2). As a consequence of this modulation, expression of spliced RNA species decreased in a Rex-dependent fashion and was completely shut off when the Rex protein was overexpressed (Inoue *et al.*, 1987). These results clearly indicate that the Rex protein is required for expression of genes located in the intron such as the *gag/pol* and *env* genes, but it represses the expression of genes that are mediated through spliced RNA.

III. Regulation of Expression of *gag*, *pol*, and *env* Genes

To confirm these effects of the Rex protein, the *rex* gene in the full-sized proviral DNA was inactivated by insertion of a frameshift mutation at its 5' region, where the *rex* gene does not overlap the *tax* gene (Hidaka *et al.*, 1988). This mutant proviral DNA was transfected into cells and the RNA expressed was analyzed by Northern blotting. As shown in Fig. 2, the smallest Tax/Rex mRNA was expressed at high level, but no significant expression of Gag and Env RNAs was detected. Upon co-transfection of *rex* expression plasmid, these defects were perfectly complemented. Thus, the essential function of Rex protein in the expression of the Gag/Pol and Env mRNA was directly confirmed. The repressive effect of the Rex protein and its significance were probed by demonstrating time-dependent expression of subgenomic viral RNAs (Hidaka *et al.*, 1988). For this, the wild-type proviral DNA was transfected into cells and the mRNAs expressed were analyzed. As shown in Fig. 3, 10 h after transfection, only spliced Tax/Rex RNA was detected, that is, all the viral RNA was completely spliced out. At 16 h, Gag and Env mRNAs were expressed efficiently and the total level of viral RNA was also increased. Similar levels of expression were observed after

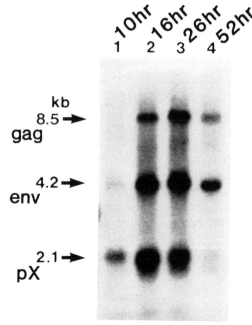

Figure 3. Time-dependent expression of the HTLV-I gene expression. Wild-type HTLV-I DNA was transfected and the RNA expressed was analyzed at the indicated times after transfection.

26 h, but the level of spliced RNA had decreased. Reduction of the RNA level was marked at 52 h: expression of spliced Tax/Rex RNA was completely shut off and the level of total viral RNA also drastically reduced. The disappearance of Tax/Rex RNA by 2 days after transfection cannot be explained simply by decay of transfected DNA. This time-dependent increase expression followed by rapid reduction of subgenomic RNA also cannot be explained simply by accumulation of intermediate RNA, but must reflect the functions of the Rex and Tax proteins, i.e., induction of Gag and Env mRNAs by accumulated Rex protein and reduction of Tax/Rex mRNA which eventually leads to reduction of the Tax protein level.

IV. Two *cis*-acting Elements of *rex* Function

Regulation by the Rex protein was expected to be virus specific, because otherwise HTLV-I-infected cells would be in chaos. To determine the molecular basis of this viral-specific regulation by the Rex protein, we subjected the provirus to deletion mutagenesis to identify the target sequence. As described in the previous section, the construct LTR-*gag-pol*-LTR expressed its unspliced RNA in a Rex-dependent fashion. To determine the significance of the *gag* sequence, we replaced it by the CAT gene (Seiki *et al.*, 1988). The construct LTR-CAT-LTR showed Rex-dependent expression of the CAT gene, indicating that the *gag* sequence is dispensable for the *rex* function. When the 3′ LTR was replaced by the transcriptional terminator of the thymidine kinase gene (TK), the rex responsiveness was completely abolished. Therefore, the 3′ LTR is essential for the rex regulation (Seiki *et al.*, 1988). On the other hand, replacement of the 5′ LTR by the SV40 promoter drastically reduced, but did not abolish, the rex responsiveness. Replacement of both LTRs resulted, of course, in complete inactivation. Therefore, both LTRs are required for efficient regulation by the Rex protein.

For identification of the Rex-responsive element (RxRE) in the 3′ LTR, deletions were made at both ends of the 3′ LTR of the LTR-CAT-LTR. Studies on the resulting mutants indicated that the RxRE includes a short sequence at the 3′ end of the U3 and most of the R segment (Seiki *et al.*, 1988; Toyoshima *et al.*, 1990). The responsive sequence in the 5′ LTR was concluded to be the splice donor signal (SD), because (1) the LTR has an SD and the promoter/enhancer is not responsive if the construct has an SD, and (2) on deletion of the SD the Rex responsiveness was lost. Thus, we concluded that two *cis*-acting elements are required for efficient regulation by the Rex protein: (a) a RxRE transcribed from the 3′ LTR and (b) a splice donor signal. The RxRE determines the viral specificity (Seiki *et al.*, 1988).

To confirm that the RxRE and SD are sufficient for Rex regulation, we introduced the LTR into a cellular gene, the metallothioneine (MT) gene (Seiki *et al.*, 1988). Insertion of the LTR into the third exon of the MT gene results in termination of the transcription within the LTR, and expression of only the completely spliced RNA as with the original MT gene. When this plasmid was coexpressed with the Rex protein, two additional RNA species were detected that were larger than the spliced form and corresponded in size to unspliced or singly spliced RNA. Furthermore the presence of the intron sequence in these RNA molecules was confirmed by hybridization with an intron probe. Similar results were obtained by insertion of the LTR into the second exon or the first intron, but its insertion into the first exon did not make a construct responsive to the Rex protein. These observations clearly indicated that the 3' LTR and the SD are required and are sufficient for the *rex* regulation and that no other viral sequence is required.

V. Mode of Action of the Rex Protein

Most cellular genes contain introns and their transcripts are spliced in the nucleus and then transported into the cytoplasm. Unexpected cytoplasmic expression of unspliced RNA seems to be prevented by some cellular mechanism. If precursor RNA is not effectively spliced, it will be degraded or stuck to the nuclear structure, and thus will not be expressed in the cytoplasm. Thus studies on the mechanism of rex function that induces the expression of unspliced RNA in the cytoplasm should be useful for understanding RNA processing in the nucleus.

As discussed previously, the LTR-CAT-LTR construct that has no splicing acceptor signal is still regulated by the Rex protein, indicating that the splicing acceptor is not required for Rex regulation (Inoue *et al.*, 1987). Therefore, the splicing process itself is not the direct target of the Rex function. Another possible target is the process of transport into the cytoplasm. To test this possibility, we analyzed nuclear RNA to determine whether unspliced RNA is accumulated in the nucleus by Rex protein.

When LTR-*gag*-LTR was expressed in the absence of the Rex protein, no unspliced RNA was detectable in either the nucleus or cytoplasm. However, in the presence of the Rex protein, unspliced RNA accumulated in the nucleus and cytoplasm to similar extents (Inhoue *et al.*, in preparation; Fig. 4). These results clearly indicated that the Rex protein modulated some processes before splicing and also before transport of the precursor RNA. Thus it may inhibit spliceosome formation or stabilize the precursor RNA, or both. The preferential localization of Rex protein in nucleoli (Siomi *et al.*, 1988) may be significant for its effects. On the other hand, when a CM-*env*-LTR (CM is cytomegalovirus promoter) was used, Rex did not induce accumulation of unspliced RNA in the nucleus, although it induced accumulation of unspliced RNA in the cytoplasm, as repeatedly confirmed. This observation was in contrast to results obtained with the *gag* construct and suggested that Rex may simply activate the transport of precursor RNA into the cytoplasm. Surprisingly, two constructs gave results suggesting that Rex oper-

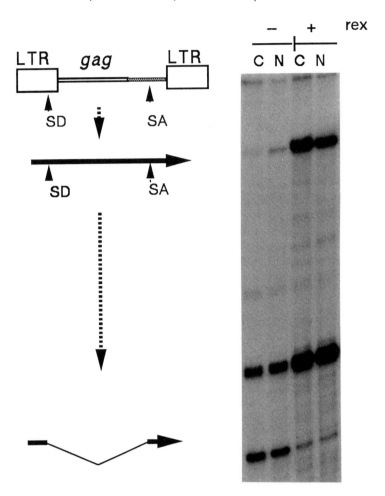

Figure 4. Nuclear accumulation of unspliced RNA induced by Rex protein. Transfected cells were disrupted and the nuclear (N) and cytoplasmic (C) fractions were separated. Spliced and unspliced RNAs in each fraction were analyzed by RNA protection assay. SD and SA, sites for splicing donor and acceptor.

ated in a different way. This curious situation could be explained by postulating the following mechanisms: (a) Rex is bifunctional and alternative functions are apparent with different constructs; (b) Rex has a single function but another process is closely associated with its function in the nucleus, thus affecting it indirectly. For example, Rex rescues the precursor RNA from any events that prevent its cytoplasmic expression, such as splicing, degradation, or retardation on the nuclear structure, and thus the results reflect the properties of the precursor RNA.

VI. Cross-reaction of the Rex Protein on Human Immunodeficiency Virus (HIV) RNA

HIV also has a gene *rev* that regulates expression of unspliced Gag and Env mRNAs (Feinberg *et al.*, 1986; Sodroski *et al.*, 1986; Knight *et al.*, 1987; Malim *et al.*, 1988; Terwilliger *et al.*, 1988; Sadaie *et al.*, 1988a). HIV and HTLV-I are classified in different retroviral groups and their genomic sequences are only distantly related. However, the *rex* gene function of HTLV-I was found to be able to control RNA expression of HIV (Rimsky *et al.*, 1988). This finding was unexpected not only because of the low homology of the sequences of HIV and HTLV-I, but also because of their different requirements for regulation: as discussed above, HTLV-I *rex* requires RxRE derived from the 3' LTR and a splicing donor signal (SD) Seiki *et al.*, 1988), whereas HIV *rev* requires RRE (Rev-responsive element) in the *env* coding sequence and a *cis*-acting repressive element (CRS) also in the *env* gene (Rosen *et al.*, 1988; Fig. 5). HIV *rev* does not require an SD to regulate its own expression (Malim *et al.*, 1989; Felber *et al.*, 1989a; Hadzopoulou-Cladaras *et al.*, 1989). The locations of these elements and the difference in the properties of SD and CRS suggested that these two genes operated in different ways. Furthermore, Rev was proposed to activate nuclear transport of unspliced RNA into the cytoplasm or to stabilize the precursor RNA (Malim *et al.*, 1989; Felber *et al.*, 1989a), whereas Rex seems to act before the transport process. To clarify these apparently contradictory phenomena, we made hybrid constructs between HTLV-I and HIV and tested their responses to Rex and Rev regulations (Itoh *et al.*, 1989).

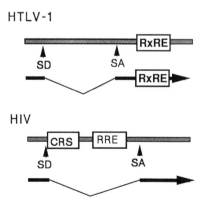

Figure 5. Elements previously thought to be required for HTLV-I *rex* and HIV *rev* regulations. RxRE and RRE stand for the Rex-responsive element and Rev-responsive element, respectively. CRS is a *cis*-acting repressive sequence defined by Rosen *et al.* (1988).

Plasmids containing RRE in the intron or exon of the HTLV-I *env* construct both responded to both the Rex and Rev protein and induced expression of unspliced RNA in the cytoplasm, although the effect of the Rex protein was only weak. These findings indicated that the Rev protein can function on the combination of SD and RRE, but does not necessarily require the negative sequence CRS, because the HTLV-I *env* sequence does not contain any element functionally equivalent to CRS. The location of the target sequence was not significant, as reported previously (Itoh *et al.*, 1989; Malim *et al.*, 1989). Therefore, the differences in their requirements were dependent on the constructs used, and may not be dependent on different mechanisms. In other words, the different requirements for the two regulations reflect differences in the properties of the indicator construct, but not the functions of the Rex and Rev proteins (Fig. 5). Thus, it is possible that the Rex and Rev proteins act through the same mechanism, although the Rev protein could not regulate HTLV-I RNA (Felber *et al.*, 1989b; Itoh *et al.*, 1989). Recently, the Rev protein was clearly demonstrated to bind directly to the RRE sequence (Zapp and Green, 1989; Daly *et al.*, 1989). Thus, similar binding of the Rex protein has been suggested, but no direct information is available so far.

VII. Secondary Structure of the Rex-responsive Element

Previously, we proposed that the sequence containing the R segment in the LTR can form a very stable secondary structure (Seiki *et al.*, 1983). We also suggested that this secondary structure is functionally important for bringing the poly(A) signal (AATAAA) to within consensus distance of the poly(A) site, which is about 300 bases away. Very interestingly, the RxRE identified above completely overlapped this region forming the secondary structure. Therefore, we are very much interested in the significance of this secondary structure in Rex regulation.

We previously proposed that the sequence can be folded into the more stable form shown in Fig. 6 (Toyoshima *et al.*, 1990). This structure was examined by testing the sensitivity of the RNA segment to digestions with nuclease. The sites that were sensitive to the nucleases coincided well with the computer-predicted structure, suggesting that this structure is in fact formed in solution. The RxRE defined by mutagenesis was mapped in the upper part of this secondary structure (Toyoshima *et al.*, 1990).

To determine the significance of this structure in Rex regulation, we introduced mutated RxRE into the 3' side of the LTR-*gag* sequence and tested the effect of the Rex protein on expression of unspliced Gag RNA. The two mutants in which one of the two strands of the long stem was replaced by *Eco*RI linker were completely inactive in Rex regulation (Toyoshima *et al.*, 1990). However, when both strands were replaced with the

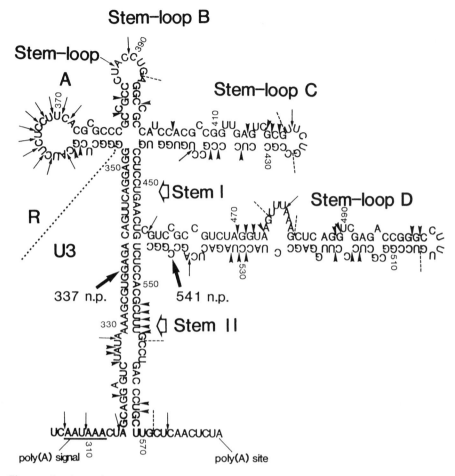

Figure 6. Secondary structure of RxRE. The structure was predicted by computer-assisted analysis and confirmed by nuclease-sensitivity assay. Active RxRE was defined as the sequence from 337 to 541 n.p. (n.p. is the nucleotide number from the 5′ end of the LTR).

linker, the original activity was recovered. These observations clearly indicate that formation of a stem structure, but not the sequence itself, is important in Rex regulation.

Deletions of either stemloops A, B, or C did not abolish the rex responsive activity, however, deletion of all three resulted in the inactivation. That is any one of these stemloops is required to maintain the responsiveness to rex regulation. But, there is no significant sequence homology among these stem-loop structures, again suggesting the importance of the secondary structure. On the other hand, stem-loop D was essential. Deletion of the external half of this stem completely inactivated the Rex responsiveness, but

(a)   **(b)**

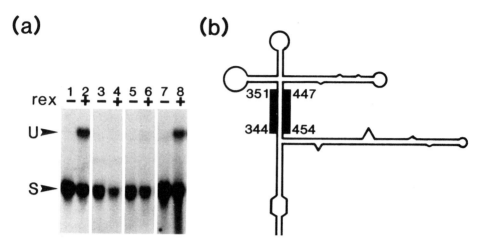

Figure 7. Demonstration of importance of the stem in the secondary structure of RxRE. RxRE and its mutants were linked to the 3' end of the LTR-*gag* construct and tested for ability to induce Rex-dependent expression of unspliced RNA. S and U stand for spliced and unspliced RNA, respectively. Closed bars indicate *Eco*RI linker. Lanes 1 & 2, wild RxRE; 3 & 4, nucleotides 344–351 were replaced with *Eco*RI linker; 5 & 6, nucleotides 447–454 were replaced with *Eco*RI linker; 7 & 8, both 344–351 and 447–454 were replaced with *Eco*RI linker.

deletion of the top loop or bottom half of the stem did not affect the activity significantly (Fig. 7). Thus, the external half of stem D is required for Rex regulation.

These studies on mutants indicated that two separated regions in the secondary structure, around stems A, B, and C, and the external half of stem D, are essential for activity of RxRE. Thus the crucial role of the tertiary structure was proposed (Toyoshima *et al.*, 1990). This idea that the target sequence for Rex function is formed by the secondary or tertiary structure of the RNA explains the following three observations.

1. A part of the U3 sequence is important to maintain this secondary structure (see Fig. 6), explaining why Rex requires the 3' LTR but not the 5' LTR (Seiki *et al.*, 1988), since the RNA from the 5' LTR has no U3 segment.

2. The secondary structure of the sequence covering the U5 and R region has two functions, one for forming an active poly(A) signal (Toyoshima *et al.*, in preparation) and the other for Rex regulation of the expressions of Gag and Env mRNAs. These overlapping functions explain why the R in HTLV-I is such a long segment (228 bases).

3. Rex regulation may not require any primary sequence homology in the responsive element. The same region of HTLV-II (Shimotohno *et al.*, 1984) and bovine leukemia virus (BLV) (Rice *et al.*, 1984; Sagata *et al.*, 1985b) also form secondary structures similar to that of HTLV-I RxRE,

suggesting that each *rex* gene also requires the same secondary structure. HTLV-I Rex protein is able to control BLV RNA (Felber *et al.*, 1989b); however, the sequence has no significant homology to that of HTLV-I RxRE except for a region of 11 nucleotides in the external half of stem D (Toyoshima *et al.*, 1990). More surprisingly, RRE of HIV which can be regulated by HTLV-I Rex has almost no homology with RxRE (Rimsky *et al.*, 1988), but can be folded into a similar secondary structure to that of RxRE (Toyoshima *et al.*, 1990). These observations can be explained by supposing that these RNA segments can form a similar conformation and thus that Rex can control RNAs that have no significant sequence homology (Fig. 8). We found that a sequence was conserved in the external half of the stem D region of HTLV-I, HTLV-II, and BLV (see Fig. 8), suggesting that this is the recognition site for the Rex protein. However, the RRE sequence of the HIV has almost no homology with this sequence, raising the question of the significance of this conserved sequence. Initially, Rex and Rev were thought

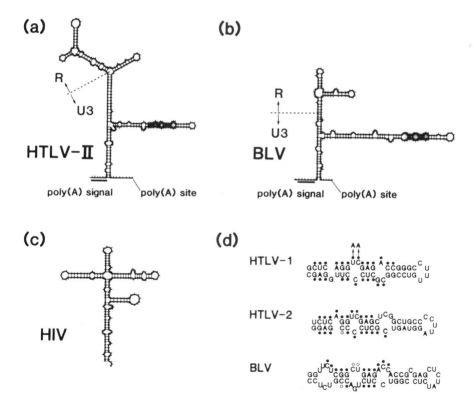

Figure 8. Secondary structures of sequence corresponding to RxRE in HTLV-II (a) and BLV (b) genomes and RRE in the HIV genome (c). In (d) sequences conserved in the HTLV subgroup are compared. Bases marked with dots are conserved.

to be virus specific, because they required viral-specific sequences. However, the involvement of the secondary or tertiary structure of the RNA segment suggests that Rex may affect some cellular gene(s) that have no sequence homology. This would be a very interesting possibility to be tested.

VIII. Significance of Rex Regulation in HTLV-I Replication

As described above, the Rex protein functions in cytoplasmic expression of unspliced RNA, that is, Gag/Pol and Env mRNAs encoding virion proteins for maturation of viral particles. However, this Rex function has more significance in regulation of viral gene expression and replication, and also in survival under the pressure of the host immune response (Fig. 9).

In an early stage of viral gene expression, all viral mRNAs are spliced into Tax/Rex mRNA, because no Rex protein is available. Thus, low levels of Tax and Rex proteins are produced, but virion proteins are not produced (Hidaka et al., 1988). The Tax protein thus produced will further *trans*-activate the transcription of the viral gene producing more viral RNA. At this stage, the Tax protein also activates cellular genes such as IL-2, IL-2Rα, and GM-CSF, thus inducing the proliferation of infected T cells. Then accumulated Rex protein modulates RNA processing and induces the cytoplasmic expression of unspliced Gag/Pol and Env mRNAs producing virion

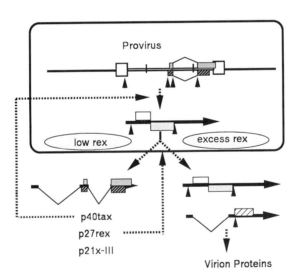

Figure 9. Significance of Tax and Rex functions in HTLV-I gene expression and replication. The left half of the figure represents the early stage of viral gene expression with little or no Rex protein. The right half represents the late stage with excess Rex protein which induces virion protein expression and represses viral gene transcription.

proteins. Thus, Rex protein is an early gene product and a trigger of virion protein expression in the late phase of viral gene expression. Then excess expression of the Rex protein will promote expression of more RNA in an unspliced form, and finally shut down expression of Tax/Rex mRNA resulting in reduction or shut off of viral gene transcription. That is, the *rex* gene exerts a feedback control of HTLV-I gene expression. For this coupled regulation, balanced expressions of the Tax and Rex proteins would be critical. This organized expression of two regulatory proteins is achieved by expression of two proteins by a single species of mRNA (Nagashima *et al.*, 1986). In this respect, it is noteworthy that the *tat and rev* genes of HIV seem to be expressed by different mRNA species (Sadaie *et al.*, 1988b), indicating that the gross regulations of HIV and HTLV-I are probably different.

When viral gene expression is reduced, proliferation of infected T cells will also decrease, with a shift to a resting state. If the HTLV-I genes are expressed continuously, infected T cells will grow continuously, but these cells will be rejected by host immune surveillance. Therefore, transient expression of viral genes resulting in transient production of progeny viruses and promotion of infected cell growth seems to be a very efficient strategy for achieving survival of the virus under the strong pressure of the host immune response. This negative control by the *rex* gene may help some cellular functions to maintain the latent proviral genome, thus explaining why most HTLV-I proviruses are latent *in vivo* and why HTLV-I replicates poorly. For efficient production of viruses, expression of the *trans*-activator *tax* gene should be independent Rex regulation. In some cell lines this separation of *tax* expression from Rex regulation is achieved by the integration of several copies of defective proviruses that express Tax protein. These defective proviruses have lost splicing signals (Watanabe *et al.*, 1984; Yoshida *et al.*, 1984b) and thus Tax protein is expressed by mRNA that is no longer spliced. Therefore, *tax* expression is not repressed by Rex function, and viral gene expression become continuous.

The regulatory system that is activated by enhancement of transcriptional initiation and repressed by modulation of RNA processing seems to be a unique mode of regulation of gene expression. However, most viral functions are similar to certain cellular events. Thus gene regulatoin mediated by Rex protein through RNA processing might have a cellular equivalent. Such a system could provide a new strategy for studying gene regulation at the level of RNA processing.

Acknowledgments

We thank many colleagues for their collaboration and valuable discussion. Work in our laboratory was supported by a Grant-in-Aid for Special Project Research on Cancer Bio-Science from the Ministry of Education, Science, and Culture of Japan.

References

Chen, I. S. Y., Cann, A. J., Shah, N. P., and Gaynor, R. B. (1985). *Science* **230**, 570–573.

Clarke, M. F., Trainor, C. D., Mann, D. L., Gallo, R. C., and Reitz, M. S. (1984). *Virology* **135**, 97–104.

Cross, S. L., Feinberg, M. B., Wolf, J. B., Holbrook, N. J., Wong-Staal, F., and Leonard, W. J. (1987). *Cell* **49**, 47–56.

Daly, T. J., Cook, K. S., Gray, G. S., Maione T. E., and Rusche, J. R. (1989). *Nature* (*London*) **342**, 816–819.

Depper, J. M., Leonard, W. J., Kronke, M., Waldmann, T. A., and Treene, W. C. (1984). *J. Immunol.*, **133**, 1691–1695.

Felber, B. K., Paskalis, H., Keinman-Ewing, C., Wong-Staal, F., and Pavlakis, G. N. (1985). *Science* **229**, 675–679.

Felber, B. R., Derse, D., Athanassopoulos, A., Campbell, M., and Pavlakis, G. N. (1989a). *New Biologist* **1**, 318–328.

Felber, B. K., Hadzopoulou-Cladaras, M., Cladaras, C., Copeland, T., and Pavlakis, G. N. (1989b). *Proc. Natl. Acad. Sci. U.S.A.* **86**, 1495–1499.

Feinberg, M. B., Jarrett, R. F., Aldovini, A., Gallo, R. C., and Wong-Staal, F. (1986). *Cell* **46**, 807–817.

Fujisawa, J., Seiki, M., Kiyokawa, T., and Yoshida, M. (1985). *Proc. Natl. Acad. Sci. U.S.A.* **82**, 2277–2281.

Fujisawa, J., Toita, M., and Yoshida, M. (1989). *J. Virol.* **63**, 3234–3239.

Hattori, T., Uchiyama, T., Tobinai, K., Takatsuki, K., and Uchino, H. (1981). *Blood*, **58**, 645–647.

Hadzopoulou-Cladaras, M., Felber, B. K., Cladaras, C., Athanassopoulos, A., Tse, A., and Pavlakis, G. N. (1989). *J. Virol.* **63**, 1265–1274.

Hidaka, M., Inoue, M., Yoshida, M., and Seiki, M. (1988). *EMBO J.* **7**, 519–523.

Hinuma, Y., Nagata, K., Misaka, M., Nakai, M., Matsumoto, T., Kinoshita, K., Shirakawa, S., and Miyoshi, I. (1981). *Proc. Natl. Acad. Sci. U.S.A.* **78**, 6476–6480.

Hinuma, Y., Gotoh, Y., Sugamura, K., Nagata, K., Goto, T., Nakai, M., Kamada, N., Matsumoto, T., and Kinoshita, K. (1982). *Gann* **73**, 341–344.

Inoue, J., Seiki, M., Taniguchi, T., Tsuru, S., and Yoshida, M. (1986a). *EMBO J.* **5**, 2883–2888.

Inoue, J., Seiki, M., and Yoshida, M. (1986b). *FEBS Lett.* **209**, 187–190.

Inoue, J., Yoshida, M., and Seiki, M. (1987). *Proc. Natl. Acad. Sci. U.S.A.* **84**, 3653–3657.

Itoh, M., Inoue, J., Toyoshima H., Akizawa, T., Higashi, M., and Yoshida, M. (1989). *Oncogene* **4**, 1275–1279.

Kitamura, T., Takano, M., Hoshino, H., Shimotohno, K., Shimoyama, M., Miwa, M., Takaku, F., and Sugimura, T. (1985). *Int. J. Cancer* **35**, 629–635.

Kiyokawa, T., Seiki, M., Iwashita, S., Imagawa, K., Shimizu, F., and Yoshida, M. (1985). *Proc. Natl. Acad. Sci. U.S.A.* **82**, 8359–8363.

Knight, D. M., Flowerfelt, F. A., and Ghrayeb, J. (1987). *Science* **236**, 837–840.

Leung, K., and Nabel, G. J. (1988). *Nature* (*London*) **333**, 776–778.

Lowenthal, J. W., Böhnlein, E., Ballard, D. W., and Greene, W. (1988). *Proc. Natl. Acad. Sci. U.S.A.* **85**, 4468–4472.

Malim, M. H., Hauber, J., Fenrick, R., and Cullen, B. R. (1988). *Nature* (*London*) **335**, 181–183.

Malim, M. H., Bauber, J., Le, S-Y., Maizel, J. V., and Cullen, H. R. (1989). *Nature* (*London*) **338**, 254–257.

Maruyama, M., Shibuya, H., Harada, H., Hatakeyama, M., Seiki, M., Fujita, T., Inoue, J., Yoshida, M., and Taniguchi, T. (1987). *Cell* **48**, 343–350.

Miyatake, S., Seiki, M., Malefijt, R. D., Heike, T., Fujisawa, J., Takebe, Y., Nishida, J., Shlomai, J., Yokota, T., Yoshida, M., Arai, K., and Arai, N. (1988). *Nucleic Acid Res.* **16**, 6547–6566.

Miyoshi, I., Kubonishi, I., Yoshimoto, S., Akagi, T., Ohtsuki, Y., Shiraishi, Y., Nagata, K., and Hinuma, Y. (1981). *Nature (London)* **294**, 770–771.

Nagashima, K., Yoshida, M., and Seiki, M. (1986). *J. Virol.* **60**, 394–399.

Nimer, S. D., Morita, E. A., Martis, M. J., Wachsman, W., and Gasson, J. C. (1988). *Mol. Cell. Biol.* **8**, 1979–1983.

Poiesz, B. J., Ruscetti, F. W., Gazdar, A. F., Bunn, P. A., Minna, J. D., and Gallo, R. C. (1980). *Proc. Natl. Acad. Sci. U.S.A.* **77**, 7415–7419.

Popovic, M., Lange-Wantzin, G., Sarin, P. S., Mann, D., and Gallo, R. C. (1983). *Proc. Natl. Acad. Sci. U.S.A.* **80**, 5402–5406.

Rimsky, L., Hauber, J., Dukovich, M., Malim, M. H., Langlois, A., Cullen, B. K., and Greene, W. C. (1988). *Nature (London)* **335**, 738–740.

Rice, N. R., Stephens, R. M., Couez, D., Deschamp, J., Kattmann, R., Burny, A., and Gilden, R. (1984). *Virology* **138**, 82–93.

Rosen, C. A., Tervilliger, E. F., Dayton, A. I., Sodroski, J. G., and Haseltine, W. A. (1988). *Proc. Natl. Acad. Sci. U.S.A.* **85**, 2071–2075.

Ruben, S., Poteat, H., Tan, T.-H., Kawakami, K., Roeder, R., Haseltine, W., and Rosen, C. A. (1988). *Science* **241**, 89–92.

Sadaie, R. M., Benter, T., and Wong-Staal, F. (1988a). *Science* **239**, 910–914.

Sadaie, M. R., Rappaport, J., Benter, T., Josephs, S. F., Willis, R., and Wong-Staal, F. (1988b). *Proc. Natl. Acad. Sci. U.S.A.* **85**, 9224–9228.

Sagata, N., Yasunaga, T., and Ikawa, Y. (1985a). *FEBS lett.* **192**, 37–42.

Sagata, N., Yasunaga, T., Tsuzuku-Kawamura, J., Ohishi, K., Ogawa, Y., and Ikawa, Y. (1985b). *Proc. Natl. Acad. Sci. U.S.A.* **82**, 677–681.

Seiki, M., Hattori, S., Hirayama, Y., and Yoshida, M. (1983). *Proc. Natl. Acad. Sci. U.S.A.* **80**, 3618–3622.

Seiki, M., Inoue, J., Takeda, T., Hikikoshi, A., Sato, M., and Yoshida, M. (1985). *Gann* **76**, 1127–1131.

Seiki, M., Inoue, J., Takeda, T., and Yoshida, M. (1986). *EMBO J.* **5**, 561–565.

Seiki, M., Inoue, J., Hidaka, M., and and Yoshida, M. (1988). *Proc. Natl. Acad. Sci. U.S.A.* **85**, 7124–7128.

Shannon, M. F., Gamble, J. R., and Vadas, M. A. (1988). *Proc. Natl. Acad. Sci. U.S.A.* **85**, 674–678.

Shimotohno, K., Golde, D. W., Miwa, M., Sugimura, T., and Chen, I. S. Y. (1984). *Proc. Natl. Acad. Sci. U.S.A.* **81**, 1079–1083.

Siekevitz, M., Feinber, M. B., Holbrook, N., Wong-Staal, F., and Greene, W. C. (1987). *Proc. Natl. Acad. Sci. U.S.A.* **84**, 5389–5393.

Siomi, H., Shida, H., Nam, S. H., Nosaka, T., Maki, M., and Hatanaka, M. (1988). *Cell* **55**, 197–209.

Sodroski, J. G., Rosen, C. A., and Haseltine, W. A. (1984). *Science* **225**, 381–385.

Sodroski, J. G., Goh, W. C., Rosen, C., Dayton, A., Terwilliger, E., and Haseltine, W. A. (1986). *Nature (London)* **321**, 412–417.

Terwilliger, E., Burghoff, R., Sia, R., Sodroski, J. G., Haseltine, W. A., and Rosen, C. (1988). *J. Virol.* **62**, 655–658.

Toyoshima H., Itoh, M., Inoue, J., Seiki, M., Takaku, F., and Yoshida, M. (1990). *J. Virol.* **64**, 2825–2832.

Uchiyama, T., Yodoi, J., Sagawa, K., Takatsuki, K., and Uchino, H. (1977). *Blood* **50**, 481–491.

Watanabe, T., Seiki, M., and Yoshida, M. (1984). *Virology* **133**, 238–241.

Yamamoto, N., Okada, M., Koyanagi, Y., Kannagi, M., and Hinuma, Y. (1982). *Science* **217**, 737–739.

Yodoi, J., Uchiyama, T., and Maeda, M. (1983). *Blood* **62**, 509–511.

Yoshida M. (1987). *Biochim. Biophys. Acta* **970**, 145–161.

Yoshida, M., and Seiki, M. (1987). *Annu. Rev. Immunol.* **5**, 541–559.

Yoshida, M., Miyoshi, I., and Hinuma, Y. (1982). *Proc. Natl. Acad. Sci. U.S.A.* **79,** 2031–2035.

Yoshida, M., Seiki, M., Yamaguchi K., and Takatsuki, K. (1984a). *Proc. Natl. Acad. Sci. U.S.A.* **81,** 2534–2537.

Yoshida, M., Seiki, M., Hattori, S., and Watanabe, T. (1984b). *In* "Human T-cell Leukemia/Lymphoma Virus," pp. 141–148. Cold Spring Harbor, New York.

Yoshimura, T., Fujisawa, J., and Yoshida, M. (1990). EMBO J. **9,** 2537–2542.

Zapp, M. L., and Green, M. R. (1989). *Nature (London)* **342,** 714–716.

Human T-cell Leukemia Virus Type II

Irvin S. Y. Chen and Joseph D. Rosenblatt

I. Introduction

Since its first discovery in 1982, human T-cell leukemia virus type II (HTLV-II) has been a virus for which both diseases and its origin have been sought. We now know that it is associated with rare forms of human leukemia, and may be the etiologic agent for a disease related to hairy cell leukemia. Recently, HTLV-II has been recognized to be far more prevalent in the population than originally believed, although the origins of the virus are still obscure. This chapter addresses what is currently understood about the biology, molecular biology, pathogenesis, and epidemiology of HTLV-II, and in particular, traces the central role molecular biology has played in understanding this virus.

II. History of HTLV-II: Discovery and Epidemiology

HTLV-II was first identified in 1982 in a T-cell line known as the Mo-T cell line. The cell line was derived by culturing splenic tissue from a patient who had undergone therapeutic splenectomy following diagnosis of hairy cell leukemia in 1976 (Saxon *et al.*, 1978a,b). The cells were cultured in the absence of any exogenous growth factors, including interleukin 2 (IL-2), growing more rapidly with continued adaptation to culture. The virus identified to be produced from this cell line cross-reacted serologically with human T-cell leukemia virus type I (HTLV-I) isolates. However, competitive radioimmune analysis demonstrated that this virus was related to but distinct from isolates of HTLV-I (Kalynaraman et al., 1982). In retrospect, it is now clear that the Mo-T cell line is a transformed cell line derived from the individual infected with HTLV-II, having many of the characteristics of other HTLV-I- and HTLV-II-transformed cells; namely, T-cell phenotype and characteristics of activated T-cells. In fact, the first use of this cell line prior to isolation of HTLV-II was for production of lymphokines expressed by activated T-cells such as granulocyte-macrophage colony-stimulating factor (GM-CSF) (Golde et al., 1980).

The individual from whom the Mo-T cell line was established had a disease originally described as a T-cell-variant hairy cell leukemia. Hairy cell leukemia is a rare form of chronic leukemia characterized by pancytopenia, splenomegaly, and the presence of morphologically distinct cells with cytoplasmic protrusions in the peripheral blood. Variants of hairy cell leukemia with T-cell phenotype are much less common than B-cell phenotype. The patient was serologically positive for HTLV, and HTLV-II could be reisolated from his peripheral blood by establishment of T-cell and B-cell lines seven years subsequent to the isolation of the Mo-T cell line. Association of a rare virus with an equally rare disease suggested a possible causal relation between HTLV-II and some forms of T-cell hairy cell leukemia, but further evidence awaits subsequent isolates of HTLV-II in association with the same disease.

III. Additional Cases of HTLV-II and Evidence for an Etiologic Role in a Chronic T-cell Leukemia Related to Hairy Cell Leukemia

A second individual with leukemia associated with HTLV-II infection was described in 1986 (Rosenblatt et al., 1986). This individual, termed NRA, also had an atypical form of hairy cell leukemia, having been first described clinically in 1978 as having a CD8$^+$ T-cell proliferation. It is perhaps of interest that he was one of the few individuals enrolled in a β-interferon clinical trial for hairy cell leukemia who showed no response to the treatment regimen. The patient possessed antibodies to HTLV-II, and virus was repeatedly isolated from his peripheral blood.

Through molecular analysis of the viral genome in the peripheral blood mononuclear cells of NRA, molecular evidence for an etiologic role was provided (Rosenblatt et al., 1988a). Studies of HTLV-I and bovine leukemia virus (BLV) have demonstrated two invariant molecular characteristics of their pathogenesis. One of these characteristics is the presence of a clonal tumor cell population in which only one or a few integrated proviruses are present (Burny *et al.*, 1984; Seiki *et al.*, 1984; Yoshida *et al.*, 1984). Secondly, although the tumor cells contain detectable levels of proviral DNA, no viral RNA can be detected by standard hybridization means (Franchini *et al.*, 1984; Yoshida *et al.*, 1984), and polymerase chain reaction (PCR) analysis indicates that although some transcripts are present, they are of extremely low abundance (Kinoshita *et al.*, 1989). Examination of the peripheral blood of this individual indicated that he had a monoclonal tumor cell population, as indicated by the presence of two dominant proviral bands which remained constant in several analyses of DNA from peripheral blood cells up until death. However, RNA extracted from these same cells did not reveal the presence of any viral RNA transcripts. Thus, these molecular characteristics provided strong support for an etiologic role for HTLV-II in the development of malignancy.

There were several other noteworthy features about this individual (Rosenblatt *et al.*, 1988a). First, given the patient's elevated CD8 cell count, we sought to ascertain the infected cells in his peripheral blood by cell fractionation. Such studies revealed that the CD8+ rather than the CD4+ cells, as is typically the case in HTLV-I-associated leukemia, harbored the HTLV-II provirus, consistent with elevated CD8 cell count. There have also been descriptions of CD8+ T-cell lymphoproliferative disorders in association with HTLV-I, but they are usually in the context of other infections such as in dual infection with HIV-I (Harper *et al.*, 1986). Secondly, at autopsy, it was found that NRA harbored massive infiltration of B-cells in his major internal organs. These B-cells did not harbor HTLV-II or Epstein-Barr virus (EBV), but were monoclonal, as determined by analysis of immunoglobulin gene rearrangement. Thus, this individual harbored two coexisting lymphoproliferative cell populations: one a CD8+ population harboring HTLV-II, and the second, a B-cell population which did not harbor HTLV-II or EBV, but which was probably the major cause of death.

This patient represents the first and only documented case of HTLV-II infection in which the molecular properties point to an etiologic role in the malignancy. Other individuals infected with HTLV-II, including the original Mo case, have not been examined for these molecular traits.

Screening of numerous cases of hairy cell leukemia of B-cell type reveals no other cases associated with HTLV-II to date (Rosenblatt *et al.*, 1987). Thus, it is clear that the majority of hairy cell leukemia cases are not associated with HTLV-II. Further screening for HTLV-II in association with leukemia should focus on those forms of hairy cell leukemia that are atypical, having properties of T-cell lymphoproliferations and/or resistance to treatment with interferon, and that occur in patients at risk for viral infection, such as transfusion recipients or intravenous drug abusers (IVDA) (see below). We refer to the two cases of leukemia linked to HTLV-II as HTLV-II-associated chronic T-cell leukemia.

IV. HTLV-II Appears on the World Map

Prior to 1989, in addition to the two cases of HTLV-II associated with malignancy, other rare cases of HTLV-II infection had been identified which were not associated with malignancy (Table I). One isolate of HTLV-II was obtained from an AIDS patient (Hahn *et al.*, 1984). Given what is known about the recent epidemiology of HTLV-II (see below), it is likely that this individual acquired the virus infection by intravenous drug use. An additional isolate of HTLV-II was obtained from a hemophiliac who had previously had blood transfusions (Kalyanaraman *et al.*, 1985); it is likely that this individual acquired HTLV-II through transfusion. Other reports of HTLV-II infection were made in various regions, primarily in IVDA populations of New York (Robert-Guroff *et al.*, 1986) and Great Britain (Tedder *et al.*, 1984). Infection was determined by serologic means, and HTLV-II infection was distinguished from HTLV-I by a relatively crude competition ELISA assay. In none of these individuals was infection confirmed by molecular means or by virus isolation.

Recently, two developments have enabled confirmed detection of HTLV-II in the U.S. population. One was the more widespread use of screening for HTLV-I in the American blood supply. In 1988, the FDA licensed the use of commercial kits for screening blood in blood banks in this country. These tests are designed to be used with disrupted whole HTLV-I virion preparations which will also react with sera from known cases of HTLV-II infection. The second technical advance was the development of

Table I

Confirmed Cases of HTLV Infection

Date of diagnosis	Case or population	Etiologic role of HTLV-II	Reference
1976	Atypical hairy cell leukemia	?	Kalyanaraman *et al.* (1982)
1979	Atypical hairy cell leukemia	Yes	Rosenblatt *et al.* (1986)
1984	AIDS	No	Hahn *et al.* (1984)
1985	Hemophiliac with pancytopenia	?	Kalyanaraman *et al.* (1985)
1989	IV drug abusers	None yet	Lee *et al.* (1989)
1989	American volunteer blood donors	None yet	Unpublished

the polymerase chain reaction (PCR) technology (Saiki *et al.*, 1985). This highly sensitive molecular assay allows detection of HTLV-II even in carriers where the proportion of infected cells was very low. More importantly, PCR allows one to easily distinguish between HTLV-I and HTLV-II at the genetic level. The first population to be systematically screened for presence of HTLV-II was an IVDA population in New Orleans (Lee *et al.*, 1989). About 20% of these individuals harbored HTLV. Surprisingly, when PCR was performed on DNA extracted from their peripheral blood, it was clear that the majority of these individuals (about 90%) were infected with HTLV-II and not HTLV-I. This result was unexpected, since the southeastern United States had previously been recognized as a region where HTLV-I infection was present, probably as a result of proximity to the Caribbean basin, an endemic region for HTLV-I. Other IVDA populations that have been screened by PCR and/or virus isolation and have revealed the presence of HTLV-II are in Miami (G. Shaw and W. Blattner, personal communication) and New York (B. Poiesz, personal communication; M. Kaplan and W. Hall, personal communication). Clinical characterization of these infected individuals is currently ongoing. In the New Orleans population, we have so far been unable to detect any clinical symptoms that could be attributed solely to HTLV-II infection, although about 20% of patients manifested low-grade lymphocytosis (Rosenblatt *et al.*, 1990). In one group of New York IVDA also coinfected with HIV-I, some unusual symptoms involving the skin and at least one additional hairy cell leukemia were identified (W. Hall and M. Kaplan, personal communication).

V. The Role of HTLV-II in Human Disease

Unlike for HTLV-I, the numbers of HTLV-II-infected individuals, until just recently, have been too low for clear epidemiological linkage to any specific disease in man. The linkage between a relatively rarely recognized virus and an equally rare disease suggests that HTLV-II may be the etiologic agent for some forms of HTLV-II-associated chronic T-cell leukemia. Furthermore, in the second case of HTLV-II-associated chronic T-cell leukemia, the molecular basis of HTLV-II pathogenesis is consistent with an etiologic role for HTLV-II in that individual's malignancy. As indicated above, two molecular features also characteristic of HTLV-I infection are noteworthy: first, the monoclonal pattern of viral integration in the tumor cells, and second, the lack of viral RNA expression in the tumor cells. This individual with atypical hairy cell leukemia developed a second malignancy in which HTLV-II was not directly involved. This is reminiscent of some cases of B-cell chronic lymphocytic leukemia (CLL) associated with HTLV-I infection in the Caribbean (Blattner *et al.*, 1983).

Given the low frequency of HTLV-I-related cancer or neurological disorder among infected individuals (1% over the course of a lifetime), the small numbers of HTLV-II-infected individuals currently being followed are unlikely to yield significant new information about disease caused by HTLV-II. Clearly a more in-depth, larger scale follow-up of infected individuals will be necessary to determine the role and extent of HTLV-II in cancer or any neurological disorders similar to HTLV-I-associated myelopathy (HAM/ TSP).

Some projections regarding potential clinical consequences of HTLV-II infection can be made from *in vitro* studies of HTLV-II. HTLV-II immortalizes T-cells *in vitro*, similarly to HTLV-I (Chen *et al.*, 1983a), and thus, it possesses transforming potential for T-cells. In addition, an observation has been made that both HTLV-I and HTLV-II can induce T-cell mitogenesis in the absence of infection (Gazzolo and Duc Dudon, 1987; Zack *et al.*, 1988; and our unpublished observations). Although the mechanism has not yet been defined at the molecular level, it presumably involves interaction between a viral component, most likely the envelope glycoprotein, and a receptor on the surface of the T-cells. The consequences of this mitogenic effect are unknown, but may be relevant in cases of dual infection with HTLV-I/HTLV-II and HIV. Here, mitogenic stimulation of cells harboring HIV by HTLV virions could increase the likelihood of active HIV replication. *In vitro*, this prediction is borne out, since HTLV-I or HTLV-II stimulation of T-cells previously nonproductively infected with HIV leads to activation of HIV viral production (Zack *et al.*, 1988).

For future studies, it is important, in examining populations for any diseases associated with HTLV-II infection, that both direct and indirect processes of pathogenesis be considered.

VI. *In Vitro* Biological Studies of HTLV-II

HTLV-II, like HTLV-I, was first discovered to be produced from a T-cell line transformed by the virus and producing infectious virions. Similarly, cell lines derived from the patient NRA are of T-cell phenotype, and are transformed by HTLV-II. As in the case of HTLV-I, most cell lines obtained from patients do not represent the same T-cell clone as the malignancy, but rather result from *de novo* infection and transformation of previously uninfected cells once the blood sample is placed in culture (Hoshino *et al.*, 1983). Thus, the integration sites for $HTLV-II_{NRA}$ in the cell lines differ from integration sites in the original tumor cell clone (unpublished observations).

T-cell transformation by HTLV is defined as indefinite proliferation of T-cells at a time when normal T-cells would have ceased replicating. This can usually be achieved in the absence of exogenously added IL-2 or other growth factors, although the early addition of growth factors during the initial period of culturing may assist in development of transformed lines. HTLV-II transforms normal human peripheral blood T-cells, usually by cocultivation of irradiated or mitomycin-C-treated infected cells with normal human or cord blood mononuclear cells. The actual efficiency of transformation relative to HTLV-I is unclear, since the requirement for cocultivation precludes good quantitative means to assess HTLV transformation efficiency. However, when parallel cocultivations of HTLV-I and HTLV-II are set up, they appear to be fairly comparable in terms of ease of development of transformed T-cell lines.

The phenotypes of cells transformed by HTLV-II *in vitro* are similar to those of HTLV-I, having the phenotype of an activated mature T-cell, CD3$^+$ and CD4$^+$ and producing a number of lymphokines characteristic of activated T-cells such as GM-CSF, γ-interferon (γ-IFN), and IL-3. These cells also invariably express IL-2 receptors on their surface and produce IL-2 early during culture, but generally do not persist in secretion of IL-2.

A simpler, more reproducible transformation assay is required for HTLV-I/II in order to facilitate study of its properties. Development of such a transformation assay has been hampered by requirements for primary T-cell cocultivation to obtain efficient HTLV infection and relatively low efficiencies of transformation. Since an infectious molecular clone of HTLV-II exists, molecular genetic studies of HTLV-II transformation would be feasible if an assay were developed.

HTLV-II can also infect some B-cell lines. EBV-transformed B-cell lines harboring HTLV-II have been isolated from patients Mo and NRA. The ability to infect B-cells with HTLV-II has been exploited experimentally by using certain B-cell lines for selective isolation of HTLV-II from infected individuals (W. Hall, personal communication), and as targets for *in vitro* infection and transfection by HTLV-II and molecular clones, respectively (Chen *et al.*, 1985a).

VII. The Molecular Genetics of HTLV-II

In general, insights gained from investigating the molecular genetics of HTLV-II have shown it to be similar to HTLV-I in terms of the overall number and arrangement of viral genes and their function in regards to viral replication (see Cann and Chen, 1989, and other chapters of this text for review). In addition, the study of two related viruses, HTLV-I and HTLV-II, has complemented the understanding of the biology of this group of retroviruses. At present, study of the molecular genetics of HTLV-II has the additional advantage of the availability of a molecular clone of an HTLV-II Mo provirus which will give rise to infectious and transforming HTLV-II (Chen *et al.*, 1983b, 1985a) when transfected into B-cell lines. This molecular clone with the full complement of necessary HTLV genes has allowed elucidation of a number of important features. For example, the infectious clone allowed the first demonstration that the *tax* and *rex* genes were necessary for replication of HTLV, with mutants having these genes deleted being unable to synthesize high levels of viral RNA (Chen *et al.*, 1985a). Similarly, examination of the complete sequence of HTLV-II allowed determination of the position of the protease open reading frame (Shimotohno *et al.*, 1985), which was closed in molecular clones of HTLV-I. Comparison of the sequences also assisted in determination of the boundaries of the *tax* and *rex* open reading frames, and the lack of reading frames in a region between *env* and the *tax/rex* genes.

Below is a description of the molecular genetics of HTLV-II, with particular emphasis on those features of HTLV-II that are distinct from those of HTLV-I (a more detailed review of HTLV-II molecular biology can be found in Cann and Chen, 1989). HTLV-II, in comparison to HTLV-I, has the same complement of genes encoding Gag proteins, reverse transcriptase, protease, and envelope, which are components of the virion, as well as two regulatory genes, *tax* and *rex*, whose products are synthesized in infected cells and are not present in the virion (Rosenblatt *et al.*, 1988b; Slamon *et al.*, 1984; Wachsman *et al.*, 1984, 1985). The HTLV-I and HTLV-II viruses share about 65% overall nucleic acid sequence homology, with the homology being the greatest in the *tax/rex* regions and the least in the long terminal repeats (LTR) and the "nontranslated region" between *env* and the *tax/rex* genes.

A. Regulatory Genes of HTLV-II

The products of both the *tax* and *rex* genes of HTLV-II act to increase the overall levels of viral gene expression. However, their modes of action are quite distinct. HTLV-II Tax acts at the level of transcriptional activation, increasing the overall levels of viral RNA in cells (Cann *et al.*, 1985, 1989; Chen *et al.*, 1983b, 1985a). HTLV-II Rex acts at a post-transcriptional level (Ohta *et al.*, 1988; Rosenblatt *et al.*, 1988b). Both proteins are apparently expressed from the same doubly spliced HTLV-II mRNA (Wachsman *et al.*, 1984, 1985).

B. *tax* Gene

The general function and mechanism of action of the HTLV-II *tax* gene is similar to that of the HTLV-I *tax* gene in regards to transcriptional activation; however, some quantitative differences have been noted in their actions. In certain cell types, the HTLV-I and HTLV-II Tax proteins differ in their *trans*-activation phenotypes when tested on the heterologous LTR (Shah *et al.*, 1986). These differences suggest that the sequence differences between these proteins result in quantitative effects on levels of activity. Given the current idea that Tax interacts with cellular proteins which bind to the LTR, it is possible that these sequence differences result in differing rates or affinities of interaction with these cellular proteins. The genetic differences between HTLV-I and HTLV-II Tax proteins have been exploited to further map the functional regions critical for Tax activity. Formation of chimeric molecules between HTLV-I and HTLV-II Tax maps regions critical for Tax activity to regions within the first 59 amino acids (Cann *et al.*, 1989). Single amino acid or small deletions within this region can have dramatic effects on the function of the HTLV-II Tax protein (Cann *et al.*, 1989; Wachsman *et al.*, 1987). Of particular interest, a single amino acid change at position 5 which converts a proline to a leucine residue results in an HTLV-II Tax protein which has a "transdominant" phenotype suppressing the *trans*-activation capabilities of the wild-type protein when cotransfected in *trans*. These mutagenesis and recombination studies indicate that the N-terminal regions of the protein are critical for Tax protein function, and possibly crucial for interactions with cellular proteins.

The sequences in the LTR which respond to the effects of Tax are located within the U3 region of the LTR, and encompass a region in which there are three 21-nucleotide repeats. These repeats are present in both HTLV-I and HTLV-II, and were first recognized after comparison of the sequences of HTLV-I and HTLV-II (Shimotohno *et al.*, 1984). They represent the most conserved portion of the U3 region of the LTR. As in the case of HTLV-I, these elements are absolutely critical for responsiveness to the Tax protein (Kitado *et al.*, 1987; Shimotohno *et al.*, 1986). These elements are likely to be binding sites for specific cellular transcription factors, and are not likely to be bound directly by the Tax protein.

C. *rex* Gene

Transient cotransfection experiments with *rex* indicate that it will alter overall LTR-directed expression without altering the absolute level of total RNA within the cell. Cell fractionation studies indicate that the role of Rex is to redistribute those RNAs containing a Rex-responsive element (RxRE) from the nucleus to the cytoplasm. For HTLV-II, the RxRE has been mapped to the LTR, positioned downstream of the splice donor site. The precise location of the RxRE has been mapped differently by two groups: one group has mapped it exclusively to the R region of the LTR (Ohta *et al.*, 1988), whereas we have mapped it, at least partially, to the U5 region of the LTR (Black *et al.*, 1990). The discrepancy might be related to the different assays used by the two laboratories. These results also differ from those for HTLV-I, where the 3' LTR has been implicated in the response to Rex (Inoue *et al.*, 1987; Rimsky *et al.*, 1989). The location of the RxRE, downstream of the splice donor site in the 5' LTR, would predict that only *gag* RNA should be regulated by Rex, whereas *tax/rex* RNA would not be regulated, since the RxRE would be spliced out of these species. Since *env* RNA also appears to be regulated, other RxRE may be present elsewhere in the genome. In contrast, the presence of the RxRE of HTLV-I solely in the 3'

LTR does not explain how HTLV-I *rex* would differentially regulate distribution of different RNAs, since all RNAs would be expected to harbor the same 3' LTR sequences.

Removal of part of the RxRE sequences increases the basal level of HTLV-II LTR activity in the absence of Rex, indicating that these sequences contain intrinsic negative *cis*-acting regulatory elements which maintain the RNA in the nucleus. Thus, the apparent function of Rex is to relieve the effect of the negative sequences of this region of the LTR and allow RNA to be redistributed to the cytoplasm.

The Rex protein of HTLV-II also has an additional function (Rosenblatt *et al.*, 1988b) which has not been described for the Rex protein of HTLV-I. This is a negative regulatory effect at concentrations of Rex greater than that required for the positive effect described above. In contrast to the positive regulatory effects of Rex which do not affect total RNA levels, the negative regulatory effects of Rex act to decrease the steady state levels of mRNA synthesized by the LTR. It is unclear whether this negative regulation is a result of decreases in the rate of transcription initiation or decreases in the stability of HTLV-II RNA.

In summary, the comparative molecular biology of HTLV-I and HTLV-II has been extremely informative, both in highlighting important sequence and functional similarities that these viruses share, and in bringing to light differences that may ultimately be relevant to their apparently different pathogenesis.

VIII. Molecular Mechanisms of HTLV-II-Induced T-cell Transformation

The mechanism of *in vitro* T-cell transformation by HTLV-II is unknown, although a number of hypotheses have emerged which parallel those proposed for mechanisms of HTLV-I transformation. The first hypothesis involves the HTLV-II regulatory genes, *tax* and *rex*. It is proposed that, since these genes are capable of regulating viral gene expression, they may also be able to aberrantly regulate cellular gene expression, resulting in immortalization of infected T-cells (Chen *et al.*, 1986). The HTLV-II Tax protein was the first human retrovirus *trans*-activating protein shown to activate a heterologous promoter, the adenovirus E3 promoter, which is normally *trans*-activated by an adenovirus *trans*-activating gene known as Ela (Chen *et al.*, 1985b). Subsequent studies have shown that the HTLV-II Tax protein, like the HTLV-I Tax protein, can activate cellular genes which may be more relevant to T-cell transformation, including the IL-2 receptor and the GM-CSF promoter (Nimer *et al.*, 1989). Both of these gene products may be important in maintaining T-cell proliferation. IL-2 and IL-2 receptor participate in autocrine T-cell proliferation; GM-CSF could participate by activating macrophages which in turn could release cytokines such as IL-1 to feed back upon T-cells to maintain their proliferation.

An additional hypothesis on HTLV transformation has emerged from studies demonstrating that HTLV-I and HTLV-II can mitogenically stimulate quiescent T-cells to divide in the absence of viral infection (Gazzolo and Duc Dodon, 1987; Zack *et al.*, 1988). This property (discussed above) may involve the HTLV envelope glycoprotein and a receptor on the surface of T-cells which may ordinarily be involved in regulating normal T-cell proliferative responses. An HTLV-infected T-cell could then respond to autocrine proliferation signals induced by either HTLV virions or an HTLV envelope glycoprotein interacting with this putative receptor. Future studies will focus on the specific virion components and on the cellular receptor involved in the mitogenic response.

Overall, these models suggest that HTLV-II transformation of T-cells may be similar to processes of transformation in systems involving oncogenes (Land *et al.*, 1983; Ruley, 1983), where factors in the nucleus, in this case, Tax and possibly Rex, and factors at the cell membrane (envelope) may act in concert to transduce signals, resulting in the final transformed state of the T-cells.

IX. Conclusions

Until recently, HTLV-II was an extremely rare virus for which much more was known about the molecular biology than about its origins, epidemiology, or associated clinical diseases. Precise molecular characterization of the virus has resulted in a greater understanding of how it replicates *in vitro*, and some of the processes that might be involved in T-cell transformation. This molecular understanding has in turn led to the identification of more cases of HTLV-II infection. In the future, it will also lead to reagents which are specific for HTLV-II diagnosis and screening. Although the origins of the virus are as yet undetermined, it is clear that HTLV-II is spreading among IVDA in this country; the consequences of that spread are uncertain. In tissue culture, HTLV-II clearly has the same potential as HTLV-I to immortalize T-cells. Whether or not this will be reflected in a larger number of HTLV-II associated chronic leukemias in man has yet to be determined.

References

Black, A. C., Chen, I. S. Y., Arrigo, S. J., Ruland, C. T., Allogiamento, T., Chin, E., and Rosenblatt, J. D. (1991). *Virology* **181,** 433–444.

Blattner, W. A., Saxinger, C., Clark, J., Hanchard, B., Gibbs, W. N., Robert-Guroff, M., Lofters, W., Campbell, M., and Gallo, R. C. (1983). *Lancet* **ii,** 61–64.

Burny, A., Bruck, C., Cleuter, Y., Dekegel, D., Deschamps, J., Ghysdael, J., Gilden, R. V., Kettmann, R., Marbaix, G., Mammerickx, M., and Portetelle, D. (1984). *In* "Mechanisms of Viral Leukaemogenesis'' (J. M. Goldman and O. Jarrett, eds.), pp. 229–260. Churchill Livingstone, London.

Cann, A. J., and Chen, I. S. Y. (1989). *In* ''Virology'' (B. N. Fields, D. M. Knipe *et al.*, eds.), 2nd edn. Raven Press, New York.

Cann, A. J., Rosenblatt, J. D., Wachsman, W., Shah, N. P., and Chen, I. S. Y. (1985). *Nature (London)* **318**, 571–574.

Cann, A. J., Rosenblatt, J. D., Wachsman, W., and Chen, I. S. Y. (1989). *J. Virol.* **63**, 1474–1479.

Chen, I. S. Y., Quan, S. G., and Golde, D. W. (1983a). *Proc. Natl. Acad. Sci. U.S.A.* **80**, 7006–7009.

Chen, I. S. Y., McLaughlin, J., Gasson, J. C., Clark, S. C., and Golde, D. W. (1983b). *Nature (London)* **305**, 502–505.

Chen, I. S. Y., Slamon, D. J., Rosenblatt, J. D., Shah, N. P., Quan, S. G., and Wachsman, W. (1985a). *Science* **229**, 54–58.

Chen, I. S. Y., Cann, A. J., Shah, N. P., and Gaynor, R. B. (1985b). *Science* **230**, 570–573.

Chen, I. S. Y., Wachsman, W., Rosenblatt, J. D., and Cann, A. J. (1986). *In* ''Cancer Surveys'', Vol. 5, pp. 329–342. Oxford University Press, London.

Franchini, G., Wong-Staal, F., and Gallo, R. C. (1984). *Proc. Natl. Acad. Sci. U.S.A.* **81**, 6207–6211.

Gazzolo, L., and Duc Dodon, M. (1987). *Nature (London)* **326**, 714–717.

Golde, D. W., Bersch, N., Quan, S. G., and Lusis, A. J. (1980). *Proc. Natl. Acad. Sci. U.S.A.* **77**, 593–596.

Hahn, B. H., Popovic, M., Kalyanaraman, V. S., Shaw, G. M., LoMonico, A., Weiss, S. H., Wong-Staal, F., and Gallo, R. C. (1984) *In* ''Acquired Immune Deficiency Syndrome'' (M. S. Gottlieb and J. E. Groopman, eds.), pp. 73–81. Alan R. Liss, New York.

Harper, M. E., Kaplan, M. H., Marselle, L. M., Pahwa, S. G., Chayt, K. J., Sarngadharan, M. G., Wong-Staal, F., and Gallo, R. C. (1986). *N. Engl. J. Med.* **315**, 1073–1078.

Hoshino, H., Esumi, H., Miwa, M., Shimoyama, M., Minato, K., Tobinai, K., Hirose, M., Watanabe, S., Inada, N., Kinoshita, K., Kamihira, S., Ichimaru, M., and Sugimura, T. (1983). *Proc. Natl. Acad. Sci. U.S.A.* **80**, 6061–6065.

Inoue, J. I., Yoshida, M., and Seiki, M. (1987). *Proc. Natl. Acad. Sci. U.S.A.* **84**, 3653–3657.

Kalyanaraman, V. S., Sarngadharan, M. G., Robert-Guroff, M., Miyoshi, I., Blayney, D., Golde, D., and Gallo, R. C. (1982). *Science* **218**, 571–573.

Kalyanaraman, V. S., Narayanan, P., Feorino, P., Ramsey, R. B., Palmer, E. L., Chorba, T., McDougal, S., Getchell, J. P., Holloway, B., Harrison, A. K., Cabradilla, C. D., Telfer, M., and Evatt, B. (1985). *EMBO J.* **4**, 1455–1460.

Kinoshita, T., Shimoyama, M., Tobinai, K., Oto, M., Ito, S.-I., Ikeda, S., Tajima, K., Shimotohno, K., and Sugimura, T. (1989). *Proc. Natl. Acad. Sci. U.S.A.* **86**, 5620–5624.

Kitado, H., Chen, I. S. Y., Shah, N. P., Cann, A. J., Shimotohno, K., and Fan, H. (1987). *Science* **235**, 901–904.

Land, H., Parada, L. F., and Weinberg, R. A. (1983). *Nature (London)* **304**, 596–602.

Lee, H., Swanson, P., Shorty, V. S., Zack, J. A., Rosenblatt, J. D., and Chen, I. S. Y. (1989). *Science* **244**, 471–475.

Nimer, S. D., Gasson, J. C., Hu, K., Smalberg, I., Chen, I. S. Y., and Rosenblatt, J. D. (1989). *Oncogene* **4**, 671–676.

Ohta, M., Nyunoya, H., Tanako, H., Okamoto, T., Akagi, T., and Shimotohno, K. (1988). *J. Virol.* **62**, 4445.

Rimsky, L., Duc Dodon, M., Dixon, E. P., and Greene, W. C. (1989). *Nature (London)* **341**, 453–456.

Robert-Guroff, M., Weiss, S. H., Giron, J. A., Jennings, A. M., Ginzburg, H. M., Margolis, I. B., Blattner, W. A., and Gallo, R. C. (1986). *JAMA* **255**, 3133–3137.

Rosenblatt, J. D., Golde, D. W., Wachsman, W., Jacobs, A., Schmidt, G., Quan, S., Gasson, J. C., and Chen, I. S. Y. (1986). *N. Engl. J. Med.* **315**, 372–375.

Rosenblatt, J. D., Gasson, J. C., Glaspy, J., Bhuta, S., Aboud, M., Chen, I. S. Y., and Golde, D. W. (1987). *Leukemia* **1,** 397–401.

Rosenblatt, J. D., Giorgi, J. V., Golde, D. W., Ben Ezra, J., Wu, A., Winberg, C. D., Glaspy, J., Wachsman, W., and Chen, I. S. Y. (1988a). *Blood* **71,** 363–366.

Rosenblatt, J. D., Cann, A. J., Slamon, D. J., Smalberg, I. S., Shah, N. P., Fujii, J., Wachsman, W., and Chen, I. S. Y. (1988b). *Science* **240,** 916–919.

Rosenblatt, J. D., Plaeger-Marshal, S., Giorgi, J. V., Chen, I. S. Y., Chin, E., Wang, H., Canavaggio, M., Swanson, P., Black, A. C., and Lee, H. (1990). *Blood.* In press.

Ruley, H. E. (1983). *Nature (London)* **304,** 602–606.

Saiki, R. K., Scharf, S., Faloona, F., Mullis, K. B., Horn, G. T., Erlich, H. A., and Arnheim, N. (1985). *Science* **230,** 1350–1354.

Saxon, A., Stevens, R. H., and Golde, D. W. (1978a). *Ann. Intern. Med.* **88,** 323–326.

Saxon, A., Stevens, R. H., Quan, S. G., and Golde, D. W. (1978b). *J. Immunol.* **120,** 777–782.

Seiki, M., Eddy, R., Shows, T. B., and Yoshida, M. (1984). *Nature (London)* **309,** 640–642.

Shah, N. P., Wachsman, W., Souza, L., Cann, A. J., Slamon, D. J., and Chen, I. S. Y. (1986). *Mol. Cell. Biol.* **6,** 3626–3631.

Shimotohno, K., Golde, D. W., Miwa, M., Sugimura, T., and Chen, I. S. Y. (1984). *Proc. Natl. Acad. Sci. U.S.A.* **81,** 1079–1083.

Shimotohno, K., Takahashi, Y., Shimizu, N., Gojobori, T., Chen, I. S. Y., Golde, D. W., Miwa, M., and Sugimura, T. (1985). *Proc. Natl. Acad. Sci. U.S.A.* **82,** 3101–3105.

Shimotohno, K., Takano, M., Terunchi, T., and Miwa, M. (1986). *Proc. Natl. Acad. Sci. U.S.A.* **83,** 8112–8116.

Slamon, D. J., Shimotohno, K., Cline, M. J., Golde, D. W., and Chen, I. S. Y. (1984). *Science* **226,** 61–65.

Tedder, R. S., Shanson, D. C., Jeffries, D. J., Cheingsong-Popov, R., Dalgleish, A., Clapham, P., Nagy, K., and Weiss, R. A. (1984). *Lancet* **ii,** 125–128.

Wachsman, W., Shimotohno, K., Clark, S. C., Golde, D. W., and Chen, I. S. Y. (1984). *Science* **226,** 177–179.

Wachsman, W., Golde, D. W., Temple, P. A., Orr, E. C., Clark, S. C., and Chen, I. S. Y. (1985). *Science* **228,** 1534–1537.

Wachsman, W., Cann, A., Williams, J., Slamon, D., Souza, L., Shah, N., and Chen, I. S. Y. (1987). *Science* **235,** 674–677.

Yoshida, M., Seiki, M., Yamaguchi, K., and Takatsuki, K. (1984). *Proc. Natl. Acad. Sci. U.S.A.* **81,** 2534–2537.

Zack, J. A., Cann, A. J., Lugo, J. P., and Chen, I. S. Y. (1988). *Science* **240,** 1026–1029.

Human Immunodeficiency Virus (HIV) Gene Structure and Genetic Diversity

Mary E. Klotman and Flossie Wong-Staal

I. Introduction

Human immunodeficiency viruses type 1 and type 2 (HIV-1 and HIV-2) are human retroviruses associated with the acquired immunodeficiency syndrome (AIDS) (Gallo and Montagnier, 1988). HIV-1 is the opportunistic virus found in most of the world, while HIV-2 is so far restricted to countries in west Africa. HIV-2 also appears to be less pathogenic than HIV-1 in man (Marlink, 1988). These viruses are highly related in their genetic, morphologic, and biologic properties, and must have recently diverged from a common progenitor. Similar viruses have also been identified in a number of simian species, with a virus from wild caught mangabey (SIVmangabey) being most closely related to HIV-2, and to a virus obtained from captive macaques (SIVmac), probably representing relatively recent interspecies transmissions (Hirsch *et al.*, 1989). Although our discussions in this chapter should pertain to both HIV-1 and HIV-2, unless otherwise stated, most of the information has been derived from studies of HIV-1.

The HIVs share many biologic and molecular characteristics with the human T-cell leukemia viruses (HTLV-I and HTLV-II) including routes of transmission, T-cell tropism, syncytia formation, a magnesium-dependent reverse transcriptase, a p24 major core protein, and, most strikingly, the presence of viral-encoded regulatory proteins that function at both transcriptional and post-transcriptional levels. However, despite these similarities with HTLV, HIV is more appropriately placed in the Lentivirinae subfamily on the basis of genetic and morphologic parameters (Gonda et al., 1985, 1986). Like all members of the lentivirus group and unlike the type-C retroviruses, HIV has a distinct bar-shaped central core as shown by electron microscopy. There is also significant homology between HIV-1 and the animal lentiviruses. For example, HIV-1 and visna virus share 66% amino acid sequence homology in their *pol* regions (Gonda et al., 1986) and 43% amino acid homology in their transmembrane proteins (Braun et al., 1987). Other shared genetic features include a large, glycosylated transmembrane envelope protein, a long *env* open reading frame, overlapping *gag* and *pol* regions, and a long separation between the *pol* and *env* genes (Sonigo et al., 1985). Parallels in the biologic properties of HIV and animal lentiviruses (e.g., visna virus, equine infectious anemia virus (EIAV), and caprine arthritis-encephalitis virus) also include their propensity to induce progressive wasting disease and diseases involving the immune and neurologic systems. Both HIV and EIAV demonstrate rapid genetic drift, with the generation of microvariants in the course of a single infection and macrovariants in the course of an epidemic. This feature in HIV has important implications for vaccine development, and will be one of the major focuses of this chapter.

II. Genomic Structure

A. Introduction

The HIV-1 genome is 9.7 kb in length and has the same general organization as other retroviruses (Hahn et al., 1984; Luciw et al., 1984). The integrated proviral genome has long terminal repeat regions flanking the genes coding for the major structural proteins, *gag, pol,* and *env* located 5' to 3' (Ratner et al., 1985; Wain-Hobson et al., 1985; Muesing et al., 1985; Sanchez-Pescador et al., 1985). The presence of at least six (possibly eight) additional open reading frames that code for proteins involved primarily in gene regulation gives the virus an extraordinary level of complexity (Fig. 1).

B. Long Terminal Repeats (LTR)

1. Overview   As with other retroviruses, the 634-bp noncoding sequences at either end of the genome are identical repeated sequences divided into discrete functional units designated U3, R, and U5 going from the 5' to the 3' end. These functional units are critical to integration of the virus

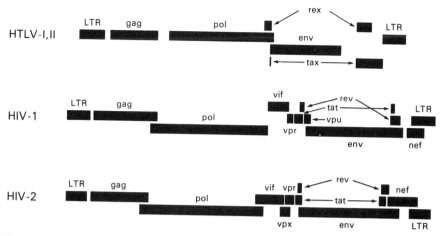

Figure 1. Comparison of the genomic structure of HIV-1 with the other human retroviruses HTLV-I and -II and HIV-2.

into the host genome and contain promoter and enhancer elements with signals recognized by cellular and possibly viral transcriptional factors. The LTRs are flanked by short internal repeats with the conserved dinucleotide TG (Starcich *et al.*, 1985). Immediately downstream from the 5' LTR is the tRNA primer binding site with 18 bp complementary to tRNA Lys which initiates minus strand DNA synthesis. Most other infectious retroviruses have a tRNA Pro as the primer binding site. Immediately upstream from the 3' LTR is a perfect 15–16-bp polypurine tract important in initiation of plus strand DNA synthesis (Ratner *et al.*, 1985; Wain-Hobson *et al.*, 1985; Muesing *et al.*, 1985; Sanchez-Pescador *et al.*, 1985). The U3 region which is 453–456 bases long terminates at the RNA transcription initiation site. The TATAA box common to many eukaryotic and viral genes functions in positioning the initiation of transcription and is located −24 to −27 relative to the RNA cap site. The surrounding TATA region (−42 to −13) contains two direct repeats and is a site for binding a number of cellular factors involved in transcription (Garcia *et al.*, 1988, 1989; Jones *et al.*, 1988). The U3 region contains the viral promoter and enhancer elements that are recognized by a number of cellular factors and will be discussed in further detail in Section II.B.2.

The R region which is 98 nucleotides long extends from the mRNA initiation site to the polyadenylation site with the terminal CA being the site of endonucleolytic cleavage (Bohnlein *et al.*, 1989b). At the junction of the U3 and R regions (-17 to +44) are the sequences required for *trans*-activation by the *Tat* protein. The polyadenylation signal, AATAAA, is located 19 bases from the R–U5 junction. The U5 region is 3' to the transcription unit and is 83–85 bp long. It has recently been demonstrated that G-T clusters within the U5 region are required and are sufficient for proper and efficient polyadenylation within the LTR (Bohnlein *et al.*, 1989b).

2. Promoter/Enhancer Elements The *cis*-acting control elements of the viral genome are located within the viral promoter, enhancer, and negative regulatory regions in U3 and extending into R. At least five regions have been described, including the *trans*-acting responsive region (TAR), TATA, SP1, enhancer, and negative regulatory regions, that have been shown to interact directly or indirectly with cellular and perhaps homologous and heterologous viral factors (Garcia *et al.*, 1987; Wu *et al.*, 1988b). The location of some of these sequences will be discussed relative to the mRNA cap site at +1 (Fig. 2). As noted above, the TATAA box is located at −27 and determines proper inititation of transcription. The promoter region lies further upstream between −45 and −77 where Jones *et al.* (1986) mapped three tandem sites that bind the cellular transcription factor SPI (I is −46 to −55, II is −57 to −66, and III −68 to −77). Site III has the greatest affinity for SPI while sites I and II appear to be primarily responsible for promoter activity *in vitro*. Mutational analysis and oligonucleotide competition studies suggest that the Sp1 binding sites stabilize factors binding downstream in the TATA region as well. *In vivo* mutagenesis studies demonstrate a marked decrease in Tat induction of transcription when all three sites are changed which is associated with decreased binding of factors to the enhancer, TATA, and TAR regions suggesting interactions with other DNA-binding factors (Garcia *et al.*, 1987; Harrich *et al.*, 1989).

Rosen *et al.* (1985) first demonstrated an enhancer element in the HIV-1 LTR located between −17 and −137 that could enhance gene expression from a heterologous promoter irrespective of location and orientation. This region has been more precisely mapped to −82 to −105 (Kaufman *et al.*, 1987) and contains two imperfect 11-bp direct repeats with the consensus sequence GGGACTTTCC. The 3′ repeat is identical to the sequence, κB,

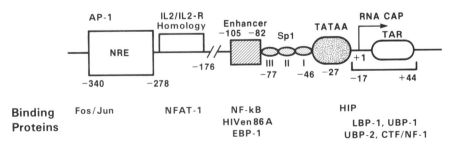

Figure 2. Structure of the long terminal repeat of HIV-1. Cellular proteins that have been suggested to interact with specific promoter and enhancer elements are noted below the appropriate regions.

found in the κ-immunoglobulin enhancer region. The 51-kDa B-cell-specific NFκB factor purified using the κ-enhancer NFκB recognition sequence binds specifically to this distal enhancer sequence of HIV-1 and stimulates *in vitro* transcription (Kawakami *et al.*, 1988). Activation of T cells with phorbol myristate acetate and phytohemagglutinin post-transcriptionally induces expression of the DNA-binding protein NFκB and results in augmentation of HIV promoter activity as well as binding to these sequences (Nabel and Baltimore 1987; Kaufman *et al.*, 1987). Nonlymphoid cells stimulated with 12-*O*-tetradecanoyl phorbol acetate (TPA) also exhibit increased binding of cellular factors to the enhancer region associated with activation of the HIV-1 promoter *in vitro* and *in vivo* which appears to be independent of protein synthesis (Dinter *et al.*, 1987). The pattern of binding to these two sequences is dependent on cell type as well as the state of activation of the cell. Extracts from unstimulated HeLa cells have several binding proteins (EBP-1) as well between 61 and 63 kDa that bind to the proximal enhancer sequence (Wu *et al.*, 1988a). Four enhancer-binding polypeptides between 36 and 42 kDa have been isolated from the human B cell line BALL-1 (Maekawa *et al.*, 1989) as well as an 86-kDa polypeptide (HIVen86A) induced in H9, Jurkat, and two B-lymphoblastoid cell lines. The latter protein specifically binds to a 12-bp enhancer element of the interleukin-2 receptor-alpha gene as well (Franza *et al.*, 1987; Bohnlein *et al.*, 1988). The relationship of these factors to each other and to NFκB is not clear.

A number of factors have been shown to induce or enhance expression of the HIV-1 promoter via the enhancer region. The *trans*-activator *tax* of HTLV-I stimulates the HIV-1 LTR most likely via induction of the enhancer-binding protein HIVen86A (Siekevitz *et al.*, 1987; Bohnlein *et al.*, 1989a). Several DNA viruses have been shown to enhance expression of the HIV-1 LTR as well. Deletion analysis, binding studies, and mutational analysis indicate that both the NFκB enhancer sequences and SP1 sequences are required for full activation of the HIV-1 LTR by herpes simplex virus (HSV) (Mosca *et al.*, 1987; Gimble *et al.*, 1988). However, stimulation of the LTR by the HSV immediate early gene product, ICP0, in T cells in the presence of Tat does not appear to require an intact enhancer (Nabel *et al.*, 1988). Human herpes virus 6 has also been shown to enhance LTR expression associated with increased binding to the enhancer region (Ensoli *et al.*, 1989). While the herpes viruses CMV (Davis *et al.*, 1987) and EBV (Kenney *et al.*, 1988) can also enhance expression of the HIV-1 LTR, the precise mapping of the region responsible for enhancement remains unclear. The X gene of hepatitis B *trans*-activates the HIV-1 LTR via the enhancer κB-like region as well (Twu *et al.*, 1989a,b).

Stimulation of the HIV-1 promoter by the cytokines tumor necrosis factor-alpha (TNFα) and interleukin 1 (IL-1) appears to be dependent on and associated with binding to the κB sites (Okamoto *et al.*, 1989; Osborn *et al.*, 1989). This response is maximal in the presence of *Tat* as well as an intact

TAR sequence. The *cis*-acting element responsible for UV irradiation-induced as well as heat-shock-induced expression of the HIV-1 promoter also appears to be the NFκB-binding sites within the enhancer. The sequence GGACTTC within the κB site shares 100% similarity with a sequence in the soybean heat-shock protein 17 element, suggesting that similar proteins are involved in these responses (Geelen *et al.*, 1988; Stein *et al.*, 1989).

Further upstream from the enhancer elements are two sequences that share significant homology with the IL-2 and IL-2 receptor regulatory regions at −275 and −221 respectively (Starcich *et al.*, 1986). An additional inducible factor found in T cells, NFAT-1, has been found to interact with upstream promoter elements between −216 and −254 and −288 and −303 in the HIV-1 LTR. These sequences compete for binding of the NFAT-1 protein that binds to the IL-2 enhancer. Synthesis of this factor and binding to the IL-2 promoter occurs immediately preceding T-cell activation following antigen exposure (Shaw *et al.*, 1988). Rosen et al. (1985) noted that deletions between −185 and −340 result in an increase in HIV-1 LTR gene expression suggesting the existence of a negative regulatory element (NRE) which has subsequently been mapped between −278 and −340. Within this region are sequences with significant similarity to the AP-1 consensus sequences (−315 to −316 and −291 to −299) that appear to interact with an inducible Fos-containing complex and Fos-related antigens (Franza *et al.*, 1988).

Cis-acting regulatory elements within the promoter/enhancer region of whole infectious molecular clones have been studied by site directed mutagenesis to determine their function during active infection. While mutations in the NFκB binding sites or one or two of the Sp1 sites are tolerated, altering both of the NFκB sites along with all three of the Sp1 sites results in loss of virus replication (Leonard *et al.*, 1989).

3. 5′ Untranslated Leader Sequence   The R region of the LTR contains the 5′ untranslated leader sequence shared by all HIV-1 mRNAs. This region and sequences immediately adjacent have been shown to be binding sites for a number of factors involved in transcription and also to contain the *trans*-acting-responsive region (TAR) critical in determining responsiveness to the potent *trans*-activator Tat. Several cellular proteins involved in transcriptional regulation bind to this region. A RNA polymerase II transcription factor, leader-binding protein-1 (LBP-1), binds in the region −17 to +27 and is dependent on the presence of the reiterated motif 5′TCTGG3′ for optimal binding. A second RNA polymerase II transcriptional factor, CTF/NF1, binds to sequences between +32 and +52 and specifically recognizes 5′AGCCAG3′ (+40+45) (Jones *et al.*, 1988; Jones, 1989). A recently described protein from HeLa nuclear extracts (HIP) binds to the TATA/initiation region (−37−+2) as well (Jones, 1989). The purified factor or complex, untranslated binding protein 1 (UBP-1), binds to the two direct repeated sequences +5 to +12 and +37 to +44 as well as to the TATA region (Wu *et* *al.*, 1988a). The relationship of this protein to LBP-1 is unclear although it has been suggested that they are identical factors (Jones, 1989). UBP-2 binds over the critical open loop sequence in the TAR element (Garcia *et al.*, 1989).

The sequences surrounding transcription initiation, -17 to $+80$, are sufficient to confer responsiveness to the major viral transactivator, Tat, and were initially designated the TAR element (Rosen *et al.*, 1985). Further deletion analysis has defined the sequence -17 to $+44$ as the Tat responsive element which is dependent on orientation and position (Peterlin *et al.*, 1986; Muesing *et al.*, 1987; Jakobovits *et al.*, 1988; Garcia *et al.*, 1989). Within this region are two 14-bp imperfect direct repeats. The first 59 nucleotides of this region can form a very stable stem loop RNA structure with a predicted free energy of 37 kcal/mol (Okamoto and Wong-Staal, 1986; Muesing *et al.*, 1987; Hauber and Cullen, 1988). A second stable stem loop forms between nucleotides 60 and 104. These sequences are downstream from the mRNA cap site and are present in both the DNA and RNA. The sequence CUGGG ($+30$–$+34$) in the open loop of the first RNA hairpin structure appears critical to *trans*-activation whereas limited substitutions within the stem are tolerated as long as compensatory substitutions preserve the stem structure (Feng and Holland, 1988). However, completely artificial stems which alter binding sites with the conserved open loop sequence are not active (Garcia *et al.*, 1989). This specific 5′ structure that is part of every HIV mRNA appears to inhibit *in vitro* protein synthesis which can be relieved by mutations that disrupt this secondary structure (Parkin *et al.*, 1988). In a transient expression assay, transcriptional elongation beyond nucleotide $+59$ (the end of the first stem loop) appears to be blocked in the absence of Tat, suggesting that Tat acts as an anti-terminator of transcription (Kao *et al.*, 1987). However, mutations that would be predicted to relieve the termination effects of the stem loop do not result in increased expression in the absence of Tat. The importance of the stem loop structure at the RNA level in *trans*-activation by Tat is supported by the observation that selective alteration of the structure but not the primary sequence by the placement of specific sequences 5′ to TAR completely eliminate *trans*-activation while 3′ sequences that allow for the transient formation of the stem loop can be completely *trans*-activated by Tat (Berkhout *et al.*, 1989). Microinjection into *Xenopus* oocytes of presynthesized TAR-bearing transcripts results in Tat induced activation if the protein and the transcripts are located in the nucleus (Braddock, *et al*, 1989). This supports the role of TAR in *trans*activation at the RNA level. Although the exact mechanism by which Tat produces its effects remains unclear, the purified protein can bind directly to the RNA TAR sequence in vitro (Dingwall *et al.*, 1989); however, this binding does not appear to be absolutely specific for this stem loop structure (Rappaport *et al.*, personal communications). Although a complete discussion of the mechanisms of *trans*-activation is beyond the scope of this chapter, the evidence strongly suggests that the secondary structure of the TAR sequence is critical to *trans*activation. Either the direct or indirect interaction of Tat and/or cellular factors with TAR results in an increase in rate of initiation of HIV-1 transcription as well as stabilization of elongated messages (Laspia *et al.*, 1989; Jakobovits *et al.*, 1988; Kao *et al.*, 1987).

C. Structural Genes

1. gag The first 5' open reading frame codes for the precursor of the internal structural proteins of the virus and extends from the ATG initiation codon at nucleotide 334–337 to nucleotide 1837 relative to the cap site (unless otherwise stated all nucleotide numbers referred to are relative to the cap site and are according to Ratner *et al.*, 1985). In HIV-1, as with other retroviruses, the region between the 5' LTR and the first ATG *gag* initiation codon contains highly conserved sequences required for efficient viral RNA packaging. Deletion of 19 bp immediately downstream from the first splice donor and upstream of the *gag* ATG results in normal viral protein production and particle formation but markedly reduced levels of virion RNA suggesting the presence of the packaging sequence in this region (Lever *et al.*, 1989).

A 55-kDa precursor Gag peptide, and perhaps a 41-kDa alternative form, are synthesized from the unspliced genomic-length mRNA. The myristoylated 55-kDa Gag protein is processed to a p41 intermediate and ultimately to at least four smaller peptides. The N-terminal protein, p17, is the major matrix protein and is myristoylated as well as phosphorylated. The major phosphorylated core protein, p24, initiates at the carboxy end of p17 at codon 133. A precursor peptide, p15, is cleaved at the C-terminus of p24 at residue 378. This is further processed between amino acid 447 (Phe) and 448 (Leu) to a 7-kDa N-terminal protein and a 6-kDa C-terminal protein (Veronese *et al.*, 1987). The p7 region has two runs of repeated cysteine residues corresponding to nucleic acid binding regions in other retrovirus proteins (Mervis *et al.*, 1988). The processing of the precursor peptide is mediated by the specific virus-coded protease gene in the adjacent and overlapping region. Inclusion of this overlapping protease reading frame has allowed the successful expression and processing of the Gag proteins in yeast, *Escherichia coli,* and vaccinia vectors (Kramer *et al.*, 1986; Gowda *et al.*, 1989).

2. pol The second long open reading frame of the HIV-1 genome overlaps the *gag* reading frame by 241 nucleotides with *pol* in the -1 phase with respect to gag (Ratner *et al.*, 1985). *Pol* is expressed as a 90–92 kDa Gag/Pol fusion protein from a genomic length mRNA. To produce both the Gag and the out of phase Gag/Pol fusion proteins, there is a ribosomal frame shift during translation that occurs at a UUA leucine codon with an efficiency of 11%. This type of frame shifting is also seen in RSV and MMTV (Jacks *et al.*, 1988). A minimum of 26 nucleotides in this region (with a run of six uracil residues) appears to be sufficient for frame shifting (Wilson *et al.*, 1988). This suggests that a predicted stem loop structure immediately downstream from UUA may not be required for frame shifting (Jacks *et al.*, 1988; Wilson *et al.*, 1988).

Early sequence similarities with other retroviruses allowed tentative identification of functional discrete proteins coded within the *pol* open reading frame. At the N-terminus are sequences similar to proteases of a number of retroviruses including BLV, HTLV-I, HTLV-II, Mo-MuLV, RSV, and visna virus. The location of the protease at the 5' end of the *pol* open reading frame is structurally similar to Moloney murine leukemia (Mo-MuLV) and visna virus and in contrast to BLV and HTLV-I where such activity maps between the *pol* and *gag* genes. This gene product responsible for enzymatic cleavage of *gag* and *gag/pol* precursors is released from the *gag/pol* precursor in an *E. coli* system by a two-step cleavage first at the Phe-Pro bond between protease and reverse transcriptase (RT) resulting in an 18kDa product, and a subsequent internal cleavage at a Phe-Pro bond releasing the stable active 10-kDa protein (Mous *et al.*, 1988). Inclusion of the protease region in *cis* or *trans* with *gag* or *gag/pol* precursor sequences in bacterial (*E. coli*) and yeast expression vectors leads to the production of mature, fully processed Gag or reverse transcriptase (Farmerie *et al.*, 1987; Mous *et al.*, 1988; Kramer *et al.*, 1986). Loss of activity associated with an internal mutation of the asparagine to threonine or asparagine to alanine and partial inhibition by pepstatin A suggest that this protease belongs to the class of aspartic proteases found in Mo-MuLV, bovine leukemia virus, and HTLV-I (Seelmeier *et al.*, 1988; Le Grice *et al.*, 1988).

The reverse transcriptase protein is encoded adjacent to the C-terminus of protease and exists in two forms of 66 kDa and 55 kDa respectively. The p66/p55 HIV-1 reverse transcriptase proteins are recognized by antibodies in up to 80% of HIV-1-positive individuals (Veronese *et al.*, 1986). Reverse transcriptase activity is localized to the N-terminus of p66 and p55 which share 156 amino acids downstream of the start of the open reading frame. When the RT gene is expressed in bacterial vectors, two enzymatic activities, DNA polymerase as well as RNaseH activity, can be detected (Hansen *et al.*, 1987, 1988; Larder *et al.*, 1987b). The DNA polymerase activity is localized to the N- terminal region while RNase H activity appears to be localized to the terminus of p66 and can also be detected in a smaller p15 polypeptide which appears to be the C-terminus derived from further processing of p66 to p51 (Prasad and Goff, 1989; Hansen *et al.*, 1988). When the amino and carboxy ends of the Pol region are deleted and the ends of the insert from nucleotide 2129 to 3808 are modified to have initiation and termination codons, fully functioning 66-kDa reverse transcriptase is produced. Further deletion of the N- or C-terminal 23 amino acids results in loss of activity while smaller deletion mutants at the C-terminus retain activity (Hizi *et al.*, 1988). Mutational analysis of the reverse transcriptase gene suggests two areas that might be good candidates for triphosphate-binding sites. Substitution of a glycine for asparagine or serine for an alanine within amino acids 109--116 and substitution of a serine for a tryptophan in region 257–266 result in a significant decline in azidothymidine triphosphate sensitivity (Larder *et al.*, 1987a).

At the 3′ end of the *pol* open reading frame is a region coding for a protein with endonuclease-like activity similar to the integrase of other retroviruses (Ratner *et al.*, 1985). This 34-kDa (integrase) protein at the N-terminus maps to residue 716 of the *pol* open reading frame (Lightfoote *et al.*, 1986). Thus it appears that the long *pol* open reading frame codes for at least four polypeptides with four associated enzymatic functions, the 10-kDa fully processed protease, reverse transcriptase/RNaseH (p66 and p55), and the 34-kDa endonuclease/integrase. The viral protease mediates at least four cleavage events of these proteins (Le Grice *et al.*, 1988).

3. env　The large open reading frame coding for the envelope glycoprotein is situated at the 3′ end of the genome extending from nucleotide 5781 to 8369 with the ATG at nucleotide 5802 (Ratner *et al.*, 1985). The *env* gene codes for a large 854–873 amino acid precursor protein which is processed by endoproteolytic cleavage to form the N-terminal exterior gp120 and the C-terminal transmembrane protein, gp41 (Veronese *et al.*, 1985). The 4.2-kb *env* message is a singly spliced polycistronic message with the splice donor at nucleotide 287 and the splice acceptor at 5358 (Muesing *et al.*, 1985) or 5557 (Sadaie *et al.*, 1988b) (Fig. 3). Sequences coding for a number of regulatory genes (*tat, rev, nef, vpu*) overlap with this RNA and one or more of them may be expressed from the same message. Also located within the *env* gene are *cis*-acting sequences that negatively regulate gene expression as well as the Rev-responsive element (RRE) that is required to relieve this negative effect in the presence of the regulatory protein Rev (Rosen *et al.*, 1988; Dayton *et al.*, 1988).

The precursor Env polypeptide has three stretches of hydrophobic residues corresponding to an N-terminal signal peptide that is cleaved prior to further processing, an internal arginine-rich site of proteolytic cleavage and the membrane-spanning region of the transmembrane protein. After cleavage of the leader signal sequence at amino acid 37, the N-terminus of the precursor polypeptide codes for the 480-amino acid hydrophilic peptide, gp120 (amino acids 38–518 in BH10 clone) (Starcich *et al.*, 1986). Sequence and mutational analyses have confirmed that the 345-amino acid transmembrane protein gp41 (519–863) at the C-terminus of the precursor polypeptide is cleaved from gp120 at a site located in an arginine-rich region adjacent to the hydrophobic stretch of gp41 between the arginine (amino acid 518) and alanine (amino acid 519) in clone BH10 (Veronese *et al.*, 1985; McCune *et al.*, 1988). The 480-amino acid exterior glycoprotein is heavily glycosylated (23–25 potential sites) while there are seven sites on the transmembrane

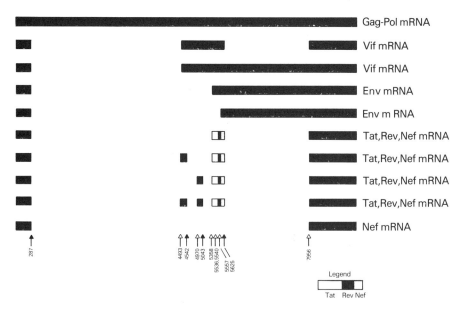

Figure 3. Map of the known spliced messages for HIV-1. The splice donor and acceptors are noted and are numbered according to Ratner *et al.* (1985).

protein. After cleavage of gp120 from gp41, the two molecules continue to be noncovalently linked via a number of sites at the N-termini of both proteins (Kowalski *et al.*, 1987).

The critical role the envelope may play in immune recognition with obvious implications for vaccine development has led to extensive sequence comparisons as well as epitope mapping. Highly conserved, variable and hypervariable regions have been identified within the envelope sequence. Five hypervariable regions with less than 25% conserved amino acids (amino acids 135–154, 163–203, 305–330, 396–414, and 459–468) are dispersed among four conserved regions with at least 75% amino acid homology (38–134, 204–279, 415–458, 470–510) (numbering from Modrow *et al.*, 1987) (Fig. 4). There are two additional highly conserved regions within the transmembrane protein. Virtually all of the cysteine residues (18) found in gp120 are conserved, as are many of the glycosylation sites, suggesting that despite extensive sequence variability within the envelope the structure is well conserved. Computer-assisted analysis initially revealed that many of the hypervariable regions had characteristics predictive of antigenic sites such as hydrophilicity, high B-turn potential, flexibility, and surface probability. A smaller number of antigenic sites were predicted in conserved regions as well (i.e., sequence 470–483) suggesting that these sites may have group rather than type specificity (Starcich *et al.*, 1986; Willey *et al.*, 1986; Alizon *et al.*, 1986; Modrow *et al.*, 1987). Evolving work has confirmed at least 20

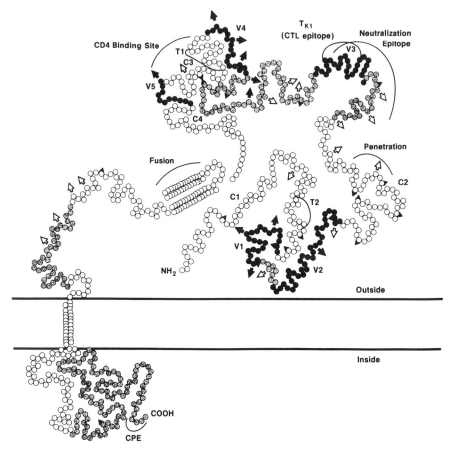

Figure 4. Structure of the HIV-1 envelope proteins gp120 and gp41. The conserved and hypervariable regions are indicated by open and closed circles respectively (☻, less variable region). Major neutralizing and CTL epitopes are indicated, as are regions involved in CD4 binding and cell fusion. ▲, Cysteine residues; ⇕, conserved glycosylation sites; ♦, variable glycoslylation sites; V1–V5, variable regions; C1–C4, conserved regions.

epitopes involved in antibody and cellular (CTL, ADCC) recognition (Bolognesi, 1989).

Neutralizing antibodies can be identified in the majority of patients with AIDS, although they are of relatively low titer (Robert-Guroff *et al.*, 1985; Weiss *et al.*, 1985). Immunization studies using purified envelope preparations or synthetic peptides in a variety of animals have confirmed that the HIV-1 envelope elicits neutralizing antibodies to a number of epitopes within gp120 and gp41 (Matthews *et al.*, 1986; Ho *et al.*, 1987). The primary immunodominant domain for eliciting type-specific neutralizing antibodies appears to be localized in the variable region V3 between amino acids 307 and

330. The site consists of a loop between two highly conserved cysteines which are thought to be connected by a disulfide bond (Putney *et al.*, 1986; Palker *et al.*, 1988; Rusche *et al.*, 1988). Specific antibodies to this region are elicited in mice, chimpanzees, and goats and as few as eight amino acids at the crown of this loop can elicit type-specific antibodies that neutralize cell-free virus and prevent cell fusion (Goudsmit *et al.*, 1988; Matsushita *et al.*, 1988). Despite the overall variability in this loop (30–50%), sequence analysis of this region in field isolates reveals a high prevalence of a specific amino acid sequence at the crown, IHGPGRAF, offering hope for a subunit vaccine. A synthetic peptide derived from within the second conserved region of HIV-1 (amino acids 254–274) has been shown to elicit neutralizing antibodies reactive against a number of HIV-1 isolates. However, the antibodies do not block binding of the virus to CD4$^+$ cells, suggesting that this region may function in a postbinding event during viral entry (Ho *et al.*, 1988). Substitution of an amino acid within this region converting an asparagine to glutamine results in noninfectious provirus even though the resulting envelope protein is properly processed, capable of binding to CD4 and forming syncytia. This supports the role of this region in internalization of the virus (Willey *et al.* 1988; Dowbenko *et al.*, 1988). Additional relatively well conserved epitopes in gp120 and gp41 have been shown to elicit neutralizing antibodies in mice (amino acids 458–532, 616–632, 728–752) (Ho *et al.*, 1987).

In contrast to gp120, the transmembrane protein gp41 has no hypervariable region but at least two predicted antigenic sites. Chanh *et al.* (1986) first demonstrated that synthetic polypeptides from gp41 (735–752) could elicit neutralizing antibodies in rabbits. This site was confirmed and a second epitope between 616 and 632 described by Ho *et al.* (1987). The highly conserved sequence 598–609 detects antibodies in the majority of sera tested (Gnann *et al.*, 1987a). The minimal essential epitope within this region is a seven-amino acid sequence with flanking cysteines suggesting that the formation of a loop via a disulfide bond might be required for antigenicity (Gnann *et al.*, 1987b). This epitope elicits an IgG response which is T-cell-dependent, an IgM response early in infection which is T-cell-independent, and an *in vitro* lymphocyte proliferative response in 24% of patients. Rabbit antibodies generated against this epitope have *in vitro* neutralizing capabilities (Schrier *et al.*, 1988). There have been reports of several antigenic sites that elicit nonneutralizing antibodies as well, including the synthetic peptide-spanning amino acids 583–599, a region homologous to immunosuppressive domains found within the transmembrane proteins of other type C retroviruses (Klasse *et al.*, 1988; Banapour *et al.*, 1987).

T-cell recognition sites have been mapped within the envelope region as well. Two such helper T-cell sites, amino acids 112–124 (T2) and 428–443 (T1), elicit lymph node proliferative responses in F1 mice (Cease *et al.*, 1987). The peptide sequence 308–322 from the hypervariable region V3 appears to be the predominant epitope eliciting murine HIV-specific CTL activity (Takahashi *et al.*, 1988).

A number of other critical functional domains of gp120 and gp41 have been identified. The cell tropism of HIV-1 for CD4$^+$ cells is mediated at least in part through a specific CD4-binding site located in a region of conserved amino acids at the C-terminus of gp120 (between amino acids 397 and 439) (McDougal *et al.*, 1986; Lasky *et al.*, 1987; Dowbenko *et al.*, 1988). However, random mutagenesis studies indicate that mutations in at least three discontinuous sites can interfere with this interaction suggesting the importance of tertiary structure in binding (Kowalski *et al.*, 1987). The domain involved in fusion of infected cells resulting in the formation of the characteristic syncytia is located at the hydrophobic N-terminal region of gp41. This location was initially suggested on the basis of sequence homology with other retroviruses and paramyxoviruses (Gonzalez-Scarano *et al.*, 1987; Gallaher, 1987) and subsequently confirmed by mutational analysis (Kowalski *et al.*, 1987). Since antibodies to the V3 region of gp120 block cell fusion, that region may also contribute to the fusion process (Rusche *et al.*, 1988).

The 350-mino acid membrane protein gp41 contains a hydrophobic membrane-spanning segment (amino acids 666–722 according to Kowalski *et al.*, 1987), hydrophilic region that anchors the protein, and a long cytoplasmic domain at the C-terminus which consists of a hydrophilic region followed by an alternating hydrophilic/hydrophobic region (Starcich *et al.*, 1986). This terminal region is not necessary for syncytia formation but does affect virus replication (Kowalski *et al.*, 1987) and cytopathicity (Fisher *et al.*, 1988). Within gp41 there is a highly conserved region with amino acid sequence homology to the B1 domain of HLADR and DQ. Antibodies generated to synthetic peptides from this region are cross-reactive and may play a role in autoimmunity (Golding *et al.*, 1988). A small six-amino acid sequence at the C-terminus of gp41 is homologous to a region in IL-2 that has been predicted to be the IL-2 receptor binding site. This region of homology can be found in a number of retroviruses and raises the possibility of a role in immunosuppression (Reiher *et al.*, 1986).

D. Regulatory Genes

1. *vif* Unlike those of many other retroviruses, the *pol* and the *env* open reading frames do not overlap in the genomes of lentiviruses including HIV-1 and visna virus. The 5′ end of a short open reading frame (nucleotides 4588–5196) coding for the *vif* (sor, *orf* Q, *orf*-1 and P′) protein overlaps with the 3' end of *pol* and terminates prior to the beginning of *env* (Ratner *et al.*, 1985; Wain-Hobson *et al.*, 1985; Muesing *et al.*, 1985; Sanchez-Pescador *et al.*, 1985). This open reading frame codes for a 23-kDa protein that generally elicits weak antibody responses in some HIV-positive patients (Lee *et al.*, 1986; Kan *et al.*, 1986; Sodroski *et al.*, 1986a). The protein is coded for by a

5.0-kb singly spliced message (Rabson *et al.*, 1985) or a smaller 2.5-kb doubly spliced message. The first splice donor is at 267 and the splice acceptor at 4493 for the singly spliced message. In the shorter message there is a second splice donor at 5625 with the splice acceptor at 7956 (Arya *et al.*, 1985). There is a high degree of amino acid sequence conservation among different HIV-1 strains as well as analogous proteins in other lentiviruses including HIV-2, SIV, visna, and caprine arthritis-encephalitis virus (Strebel *et al.*, 1987).

Mutant proviruses lacking a functional *vif* gene product are capable of both replication and cytopathicity, but with markedly reduced efficiency, especially in cell-free transmission (Sodroski *et al.*, 1986a; Strebel *et al.*, 1987; Fisher *et al.*, 1987).

2. *vpr* An additional open reading frame overlaps the 3' end of the *vif* gene and terminates before the initial methionine of Tat. This highly conserved open reading frame (among HIV-1 and HIV-2 isolates as well as other lentiviruses, e.g., visna) codes for either a 77 or a 95–96-amino acid protein that appears to be immunogenic. At least two variants of the C-terminus have been described. In HIV/HTLV-IIIB-derived clones, there is a termination codon at 77 while in other unrelated isolates, there appears to be a frame shift at amino acid 77 resulting in an additional 18–19 amino acids. The initial ATG begins at nucleotide 5592 and extends to nucleotide 5828 in the truncated 77-amino acid protein (Wong-Staal *et al.*, 1987). Mutational analysis reveals that this gene product is dispensable for replication and cytopathicity. However, at least under low multiplicities of infection, replication is inefficient (Dedera *et al.*, 1989; Ogawa *et al.*, 1989). Recently, the full length Vpr protein has been shown to accelerate virus replication in Jurkat T cells and activate the HIV-1 LTR as well as heterologous promoters in *trans* (Cohen *et al.*, 1990). These data, along with the recent observation of Haseltine (personal communication) that the protein may be incorporated in the virion, support a role for Vpr early in the virus life cycle.

3. *tat* It was first noted that genes linked to the HIV-1 LTR promoter sequences were more highly expressed in HIV-1-infected cells than in uninfected cells suggesting a *trans*-activating function associated with the virus. Such *trans*-activating factors had only been associated with the other human retroviruses HTLV-I and HTLV-II at that time (Sodroski *et al.*, 1984, 1985a). The HIV-1 *trans*-activating factor, Tat, is an 86-amino acid protein coded for by a doubly spliced mRNA. The 5' untranslated exon is found in all viral mRNAs with the splice donor at nucleotide 287. The 268-base middle exon is the first coding exon spanning nucleotides 5358 to 5625 and is necessary and sufficient for activity. The 1258-base 3' exon has a splice acceptor at 7956 and is not necessary for activity (Arya *et al.*, 1985; Sodroski *et al.*, 1985b). Although this spliced message is the predominant message coding for the Tat protein, several alternative doubly spliced messages have been described that could code for functional Tat proteins (Fig. 3). Expression of a functional Tat protein is necessary for viral expression and replication (Fisher *et al.*, 1986; Dayton *et al.*, 1986).

The LTR sequence necessary for *trans*-activation by this protein was originally mapped to between -17 and $+80$ relative to the cap site (Rosen *et al.*, 1985). As noted in Section II.B.3, further mapping of the TAR region defined nucleotides $+18$ to $+44$ as absolutely necessary (Garcia *et al.*, 1989, Jakobovits *et al.*, 1988). The RNA sequence encoded by this region could form a stable stem loop structure which appears to be an important requisite for *trans*-activation by Tat. The accumulating evidence suggests that the effects of Tat can be mediated at transcriptional and post-transcriptional levels although the relative contribution of each remains unclear (Cullen, 1986; Peterlin *et al.*, 1986; Rosen *et al.*, 1986; Wright *et al.*, 1986; Kao *et al.*, 1987; Muesing *et al.*, 1987; Rice and Mathews, 1988; Feinberg *et al.*, 1988; Jakobovits *et al.*, 1988; Sadaie *et al.*, 1988a; Laspia *et al*, 1989; Braddock *et al.*, 1989).

Several functional domains have been identified in the Tat protein by sequence comparisons and mutational analysis, including an acidic domain, a cysteine-rich domain, and a basic domain. The first 13 amino acids of Tat can potentially form an amphipathic α-helix with the appropriate periodicity of the acidic residues. This characteristic is seen in the *trans*-acting domains of other *trans*-activators. Substitutions that alter the acidic or amphipathic nature of this domain generally result in reduced or complete loss of function (Ruben *et al.*, 1989; Rappaport *et al.*, 1989). The highly conserved cysteine-rich domain located from amino acids 22 to 37 has four potential zinc-binding sites reminiscent of DNA-binding domains. Frankel *et al.* (1988) have shown that bacterially expressed Tat binds to such divalent ions to form dimers. Point mutations in six of the seven cysteines in this region result in complete loss of activity which appears to be independent of the ability to bind metals (Garcia *et al.*, 1988; Sadaie *et al.*, 1988b; Ruben *et al.*, 1989). N-terminal and cysteine mutants still display nuclear localization despite loss of activity. A third domain, amino acids 49–57, is a highly conserved basic region rich in arginine and lysine. Mutations in this region markedly reduce but do not eliminate activity. The mutations interfere with nuclear localization as well as RNA stability suggesting that the basic region functions as a nuclear/nucleolar localization signal (Ruben *et al.*, 1989; Hauber *et al.*, 1989).

4. rev The 20-kDa *trans*-regulatory protein Rev of HIV-1 (previously named Art or Trs) located predominantly in the nucleolus (Perkins *et al.*, 1989) is coded for by another multiply spliced mRNA that overlaps the *tat* reading frame extensively. Early mutational analysis of the *tat* gene revealed that certain mutations within the two coding exons of *tat* could not be complemented in *trans* by Tat-producing plasmids and resulted in a decrease in expression of *env* and *gag/pol* gene products with normal expression of multiply spliced genes (Sodroski *et al.*, 1986b; Freinberg *et al.*, 1986; Sadaie *et al.*, 1988a; Malim *et al.*, 1988). This novel regulatory protein, Rev, is expressed from a doubly spliced RNA with the first noncoding exon identical to *tat* extending to the major splice donor at position 287. The splice acceptor for the first coding exon is now known to be downstream from the *tat* splice acceptor. Sadaie *et al.* (1988b) mapped this acceptor to nucleotide 5540 by S1 nuclease assay while more recent data (Robert-Guroff *et al.*, 1990) using polymerase chain reaction amplification of cDNA suggest a splice acceptor at nucleotide 5536 as well. The splice donor sites for the second exons and the splice acceptors for the third exons of both *rev* and *tat* (and *nef*) are identical at nucleotides 5625 and 7956 respectively. Alternative multiply (triple) spliced messages capable of coding for Rev include the addition of small noncoding exons 4493–4542 and/or 4970–5043 (Robert-Guroff *et al.*, 1990). The stop codon for *rev* is at position 8227, thus coding for a protein of 116 amino acids (Sodroski *et al.*, 1986b).

Cis-acting repressive sequences within the *gag/pol* and *env* genes appear to have a negative effect on their own as well as heterologous gene expression (Sodroski *et al.*, 1986b; Rosen *et al.*, 1988). Furthermore, this effect can be overcome by Rev only if an RRE is present (Dayton *et al.*, 1988; Malim *et al.*, 1989a; Hadzopoulou-Cladaras *et al.*, 1989). This highly conserved 234-bp element located immediately 3′ to the start of gp41 (nucleotides 7346–7579) can form a very stable complex stem loop structure suggesting a potential role of secondary structure in Rev responsiveness (Malim *et al.*, 1989a). When this sequence is in proper orientation anywhere within an unspliced gene before the polyadenylation site, expression of the unspliced message is observed in the presence of Rev. This appears to be the result of enhanced nuclear export as well as increased stability of the unspliced message (Malim *et al.*, 1989a; Hadzopoulou-Cladaras *et al.*, 1989; Felber *et al.*, 1989; Hammarskjold *et al.*, 1989). Partially purified Rev protein specifically and stably binds directly to RNA coded for by this RRE (Zapp and Green, 1989; Daly *et al.*, 1989). Further mapping appears to localize the binding of the protein to the first ninety nucleotides of the RRE which can potentially form a stable stem loop as well (Daefler *et al.*, 1990). This binding may be a critical step in the enhanced expression of unspliced mRNAs.

The Rev protein appears to have at least two functional domains. Most single amino acid changes in the N-terminal portion result in loss of function. Mutations in the basic arginine-rich domain toward the N-terminus result in loss of function that is associated with loss of nuclear localization. It is felt that this region contains the nuclear localization signal (Perkins *et al.*, 1989; Malim *et al.*, 1989b; Kubota *et al.*, 1989). A specific sequence required for nucleolar localization, distinct from the NRRRN motif that can confer nuclear localization, has been described as well and appears to reside within the N-terminus of the protein (Cochrane *et al.*, 1990). Specific mutations in the carboxy end of the protein have been identified that have no *rev* activity but are able to inhibit in *trans* normal Rev function. This has led to the suggestion by Malim *et al.* (1989b) that Rev may be analogous to other transcriptional activators with a specific binding domain that is distinct from the activation domain. Mutations in the activating domain would lead to *trans*-dominant phenotypes by competing with the wild-type protein for the appropriate binding site. Deletion of the last 25 amino acids at the C-terminus results in no loss of *rev* function (Perkins *et al.*, 1989).

5. *vpu* A small open reading frame located at the 3' end of the genome downstream of the first coding exons of *tat* and *rev* codes for a small 16-kDa protein. A number of cloned viruses including the prototype HIV/HXB2 lack the initiation codon and therefore produce no functional *vpu*. This protein is unique to HIV-1 with no equivalent reading frame found in closely related viruses HIV-2 or SIV (Strebel *et al.*, 1989). Evidence suggests that the protein is expressed from the same spliced message coding for the Env protein.

The Vpu protein is phosphorylated but not glycosylated (Strebel *et al.*, 1989). A hydrophobicity profile of the protein reveals a 27-amino acid hydrophobic stretch at the N-terminus suggesting membrane association. This is followed by a well-conserved 15-amino acid hydrophilic domain (Strebel *et al.*, 1988). Functional analyses of the Vpu protein by several laboratories have suggested its role in enhancing virus release from the cell. Mutation of *vpu* results in a decrease in supernatant virus and accumulation of intracellular viral proteins with more rapid formation of syncytia and cytopathic effects (Strebel *et al.*, 1988, 1989; Terwilliger *et al.*, 1989; Klimkait *et al.*, 1990).

6. *nef* The conserved *nef* (3' orf) open reading frame is situated at the 3' end of the *env* gene extending into the 3' LTR U3 region (Ratner *et al.*, 1985; Wain-Hobson *et al.*, 1985; Muesing *et al.*, 1986; Sanchez-Pescador *et al.*, 1985). The myristoylated 27-kDa protein (Guy *et al.*, 1987) is associated primarily with the cytoplasm (Franchini *et al.*, 1986; Hammes *et al.*, 1989). The protein is expressed from a multiply spliced mRNA with the ATG start site at nucleotide 8390 and extending to nucleotide 9009. The first exon splice donor and third exon splice acceptor are identical to *tat* and *rev*. The second exon is noncoding and has a splice acceptor at 5557 and donor at 5625 (Arya and Gallo, 1986). *nef*-Coding messages with additional noncoding exons encompassing nucleotides 4493–4542 or 4970–5043 have been reported. Robert-Guroff *et al.* (1990) have demonstrated a number of additional spliced messages capable of coding for *nef*. One alternative is a message containing only two exons with the splice donor at 287 and splice acceptor at 7956. Two other versions included a splice donor at 269 with the acceptor at 7956 and splice donor at 325 with the accelerator at 8013. These two-exon *nef* messages predominate in *in vitro* infected macrophages (with BA-L strain) while the former three- and four-exon versions predominate in H9 cells.

Isolates have been reported with a premature stop codon at amino acid 122–124 (Ratner *et al.*, 1985). While this mutant as well as other deletion mutants fail to produce the 27-kDa Nef protein, replication and cytopathicity are retained. Furthermore, infectivity appears to be enhanced in these mutants, suggesting a negative regulatory role of the protein (Terwilliger *et al.*, 1986; Luciw *et al.*, 1987; Niederman *et al.*, 1989). There are also *in vitro* data which indicate that the protein can suppress in *trans* the HIV-1 LTR possibly via the *cis* negative regulatory element (Ahmad and Venkatesan, 1988). However, recent data from several groups have failed to substantiate a negative regulatory role of the Nef protein (Kim *et al.*, 1989; Hammes *et al.*, 1989). Whether some of the contradictory data may be related to strain variation in the negative effects of Nef (Cheng-Mayer *et al.*, 1989) remains to be determined.

Unique structural characteristics of the Nef protein may shed light on its function. There are amino acid homologies with the Ras family of proteins including regions involved with nucleotide binding (amino acids 93–112). Functional assays have suggested that Nef may have both GTPase and GTP-binding activities (Guy *et al.*, 1987) although this has not been confirmed with the purified protein. A highly conserved and highly charged amino acid sequence between 104 and 114 of the Nef protein shares 41% homology with the intracytoplasmic tail of the human IL-2 receptor which contains two sites that are phosphorylated during T-cell activation (Samuel *et al.*, 1987). These homologies suggest potential functions of this poorly understood protein particularly involving signal transduction and cell regulation.

The majority of human sera contain antibodies that recognize the Nef protein. Mapping of the immunoreactive domains reveals that while deletion of amino acids 35–73 does not result in loss of recognition by human sera, deletions toward the C-terminus (amino acids 73–91) result in loss of immunoreactivity (Arya, 1987).

7. *vpx* HIV-2 and SIV isolates have an additional open reading frame that is highly conserved among these viruses but is not found in HIV-1 (Kappes *et al.*, 1988). This reading frame, *vpx*, is located in the central part of the genome between the *pol* and *env* open reading frames with partial overlap of the *vif* gene at the 5′ end. The 14-kDa protein is immunogenic and its product is not required for infectivity and replication *in vitro* (Henderson *et al.*, 1988; Kappes *et al.*, 1988; Yu *et al.*, 1988; Franchini *et al.*, 1988). Although the function of the Vpx protein remains unknown, the highly conserved nature of the protein within HIV-2 and SIV isolates and the complete absence in HIV-1 suggests an important role, perhaps in regulating pathogenicity.

8. Others

a. tnv* or *tev At least two additional reading frames that code for virus-induced proteins have been identified. A 26-kDa protein reactive with Tat, antisera has been shown to be a fusion protein containing Tat, Env, and Rev sequences. The mRNA coding for this protein contains the initial standard noncoding leader sequence spliced to the small noncoding sequence spanning 4970–5043, the first coding exon of *tat* spanning 5557–5625, a new exon within *env* spanning 6185–6300 and the second coding exon of *tat* and *rev* spanning 7956–9210. The fusion protein produced has Tat but no apparent Rev activity (Salfeld *et al.*, 1990). A second reading frame codes for a 17-kDa fusion protein with the N-terminus from the *tat* reading frame and the C-terminus from the T open reading frame (*vpt*). Such a fusion protein would be produced by a ribosomal frame-shifting event (Haseltine, personal communication).

b. Positive Strand Reading Frames Computer analysis of the DNA plus strand reveals a potential, highly conserved open reading frame in a region complementary to the envelope sequence 7403–7972 of the minus strand (Miller, 1988). Although there has not been confirmation of a protein encoded by this open reading frame, transcription from the plus strand of DNA has recently been described (Bukrinsky and Etkin, 1990).

III. Genetic Diversity

A. Introduction

One of the more striking characteristics of HIV-1 is its extreme genetic variability that is not only manifest in strains isolated from individual to individual but also in those from a single individual. This remarkable heterogeneity has thwarted efforts to develop a universal vaccine. The genetic diversity most likely contributes to many of the unique biologic and pathogenic features of the virus such as cell tropism and the heterogeneous clinical course experienced by those infected by the virus. The rapid genetic change of the virus over time in any given individual host may contribute to the prolonged and progressive nature of the infection by allowing the virus to escape immune destruction (Desai *et al.*, 1986). Understanding the specific relationships between genetic variability and biologic characteristics of the virus in the host will prove critical for determining prognosis, as well as for the development of vaccines and therapeutic programs.

While such genetic variability is a hallmark of retroviruses, the HIV-1 reverse transcriptase has one of the highest error rates described. Like other reverse transcriptases, HIV-1 RT does not have exonucleolytic proofreading capabilities to correct misreadings and readily polymerizes from a mismatched base pair. Calculated error rates from *in vitro* assays range from

1/1700 to 1/2000 per detectable nucleotide incorporated, about ten times higher than AMV RT. Furthermore, the mutations appear to be nonrandom and can include substitutions, additions, and deletions (Roberts *et al.*, 1988; Preston *et al.*, 1988).

Whether genetic recombination plays a role in generating diversity *in vivo* is not known; however, such events occur extensively *in vitro*. Introducing noninfectious proviruses into cell lines either transiently or stably carrying other defective proviruses results in recombination and production of infectious virus as long as there is an overlapping region (Srinivasan *et al.*, 1989). However, there has been no evidence of recombination with cellular DNA as a source of genetic variability.

The considerable genetic diversity of HIV-1 was realized soon after the first virus isolations, based on restriction enzyme analyses of cloned as well as fresh uncultured HIV-1 isolates (Shaw *et al.*, 1984; Luciw *et al.*, 1984; Alizon *et al.*, 1984; Hahn *et al.*, 1985; Wong-Staal *et al.*, 1985; Benn *et al.*, 1985). Restriction enzyme polymorphism ranges from single site variations to changes in >50% of sites. This is in striking contrast to ATL-associated HTLV-I isolates which show very little heterogeneity (Shaw *et al.*, 1984; Wong-Staal *et al.*, 1985; Benn *et al.*, 1985). Although the genetic organization is well conserved among all isolates, sequence polymorphism is found in all isolates compared and is most striking among African strains (Alizon *et al.*, 1986). Overall, sequence comparisons demonstrate nucleic acid sequence divergence varying from as little as 1.5% to as high as 9.3% between isolates with an amino acid variability ranging from 2.2 to 14.25% (Starcich *et al.*, 1986).

Genetic diversity is noted among isolates from the same individual at a single point in time as well as when comparing isolates over time. Restriction enzyme analysis of cloned variants from single isolates as well as direct amplification and sequence determination of parts of the viral genome in fresh tissue has confirmed that individuals are infected with a multitude of closely related but distinct viruses at any time (Saag *et al.*, 1988; Goodenow *et al.*, 1989). It is therefore more appropriate to consider HIV isolates as quasispecies, as seen in other RNA viruses. In general, multiple isolates are more closely related within an individual than among individuals. The population of viruses can also undergo "drift" in the course of infection. Sequence analyses of cloned viruses from sequential isolations reveal point mutations and short deletions and insertions that are greatest in the *env* region but also scattered throughout the genome. The rate of change has been calculated to be 10^{-3} nucleotide substitutions per site per year in *env* and 10^{-4} in *pol* (Hahn *et al.*, 1986). Despite a notable degree of microheterogeneity in the virus genomes within the individual, coexistence of greatly divergent strains is rarely observed, raising the possibility that a single exposure may interfere with subsequent infection with unrelated strains. An alternative possibility is that a new incoming virus may be rapidly "absorbed" into the pool of existing viruses through genetic recombination.

The evaluation of genetic as well as biologic diversity of viral isolates is further complicated by the observation that any *in vitro* manipulation such as passage in tissue culture can alter the dominant genotype as well as phenotype of the virus (Meyerhans *et al.*, 1989).

B. *env*

1. Sequence Comparisons Perhaps the most striking feature of this heterogeneity is the distinct pattern and location of variable regions within the envelope gene. Heteroduplex analysis as well as nucleic acid sequencing indicates that the greatest degree of diversity among HIV-1 isolates from different individuals as well as from the same individual over time is in the envelope region, particularly the N-terminus coding for the signal peptide and exterior glycoprotein gp120 (Hahn *et al.*, 1985; Starcich *et al.*, 1986). The *env* variability is much more striking than that observed within the HTLV/BLV family (Yasunaga *et al.*, 1986). Amino acid sequence divergence in the extracellular region of gp120 is as high as 21% compared to the more conserved gp41 (13%) and highly conserved *gag* and *pol* genes where maximum divergence is 6.5% (Starcich *et al.*, 1986; Desai *et al.*, 1986). While the majority of the nucleotide changes in *gag* and *pol* are silent single point mutations in the third codon, *env* changes are more diverse, often leading to clustered amino acid changes, in-frame deletions, insertions, and duplications. Single point mutations in *env* are most often located in the first and second codon and when in the third position more often result in a change in amino acids when compared to *gag* alterations (Starcich *et al.*, 1986). The variability in the Env protein is not randomly dispersed but discretely distributed between areas of relatively conserved sequences as discussed in Section II.C.3. While the greatest variation is in the exterior glycoprotein, gp120, there are some highly conserved regions including the positions of at least 18 cysteine residues indicating an overall structural conservation. Analysis of African strains reveals the same overall pattern of variable and conserved regions within the envelope (Srinivasan *et al.*, 1987).

Prior to extensive immunologic studies, computer-assisted analyses allowed for the prediction of potential antigenic sites within the envelope sequence based on such characteristics as hydrophilicity, number of B turns/ surface probability, flexibility, and glycosylation sites. The majority of these predicted epitopes fall within the highly variable regions. However, there are some notable exceptions including the highly conserved predicted antigenic site adjacent to gp41 (amino acids 473–518) (Modrow *et al.*, 1987). As discussed in section II.C.3, immunologic data have confirmed many of the computer-generated predictions. As predicted by their localization within variable domains, many of the highly immunogenic epitopes elicit type-specific and not group-specific responses (Palker *et al.*, 1988).

2. Role in Immune Selection This marked genomic heterogeneity, particularly in envelope immune epitopes, has profound implications for vaccine development and may be extremely important in understanding the behavior of the virus in the host. The extensive envelope variability in antigenic sites raises the possibility that immune selection plays a role in the ability of the virus to escape immune recognition and ultimately in determining disease expression in the host. Evidence generated from work with other animal lentiviruses demonstrates that antigenic variants that can no longer be neutralized by previously generated antibodies arise over time (Scott *et al.*, 1979). In models such as EIAV and visna, it has been proposed that this immune selection and antigenic variability might contribute to the chronicity of the disease in the host.

A panel of sera from HIV-1-infected patients and immunized animals shows marked variability in neutralizing activity against closely related cloned viruses derived from a single infected cell line with <1% envelope nucleotide sequence diversity, suggesting that minor variability can profoundly affect susceptibility to neutralization (Looney *et al.*, 1988). This apparent type-specific neutralization may pose a major problem for vaccine development as new variants can continually emerge with immune selection. Such a phenomenon has been demonstrated *in vitro*. Passage of virus in the presence of immune sera readily selects for escape variants (Robert-Guroff *et al.*, 1986). Neutralization-resistance was shown to be localized within the *env* of the new variant and mapped to a single amino acid substitution in the amino end of the gp41 transmembrane protein. Because this mutant existed at a low level in the parental virus stock, immune selection resulted in the enhanced growth of this preexisting variant (Reitz *et al.*, 1988). This provides support for the idea that viral heterogeneity and immune selection can result in the predominance of resistant variants in the host and may contribute to the persistent and chronic nature of infection.

C. Other Regions of Variability

While genetic variability is not as striking in other areas of the HIV genome, even small changes may have implications concerning the biologic activity of the virus. Asjo *et al.* (1988) demonstrated that *in vitro* propagation of HIV-1 isolates that grow slowly and with low reverse transcriptase activity (slow/low) can be improved by using a *tat*-expressing cell line. This observation raises the possibility that variation in the *tat* gene may be responsible for biologic differences between isolates. There is significant variation in the C-terminus of Tat; however this region is generally not required for *trans*-activation. To study further the functional role of genetic variability in this critical viral regulatory gene, Meyerhans *et al.* (1989) analyzed a number of sequential *in vitro* isolates from a male homosexual and compared them to *in vivo* isolates from the same patient. Tat gene sequences were specifically amplified and sequenced using the polymerase chain reaction. A number of critical observations concerning variability in general and *tat* variability specifically were made. As has been observed by others, an individual at any point in time has a number of major and minor *in vivo* and *in vitro* isolates, reinforcing the concept of HIV quasispecies. Although there are changes in sequential isolates in *tat* sequences *in vitro* and *in vivo* over a $2\frac{1}{2}$-year period, there is little correlation between the major isolates *in vitro* and *in vivo*. Major forms of the virus after culture are often minor forms *in vivo*, highlighting the dangers of correlating *in vivo* biologic characteristics with genetic information acquired from *in vitro*-passed virus. Moreover, while there are dramatic and rapid changes in the predominant isolates over time, such changes do not correlate with biologic measurements such as CD4 counts. One third of the *tat* variants analyzed for function were defective and these included *in vivo* as well as *in vitro* isolates. Overlapping *rev* and *vpu* sequences were found to be defective as well. The role these HIV-1-defective genomes play *in vivo* remains unclear.

Other areas of genetic variability include the U3 region of the LTR (8% nucleotide variability), and the *nef* gene where there is up to 10.6% nucleotide variability and 17% amino acid variability (Starcich *et al.*, 1987; Desai *et al.*, 1986). One variant of the *nef* gene has a premature stop codon resulting in a functional protein of 86 rather than 101 amino acids (Desai *et al.*, 1986).

D. Biologic Diversity

As with genetic variability, individuals can have a number of isolates with distinct biologic characteristics as well including cell tropism and cytopathicity (Sakai *et al.*, 1988). When divergent envelope sequences are placed into an otherwise identical provirus, the viruses exhibit marked biologic variation in terms of growth characteristics and cell tropism, suggesting that genomic variability even within the same host can contribute to biologic variability and perhaps ultimately to disease expression (Fisher *et al.*, 1988).

In vitro biologic variability has been observed in isolates obtained directly from patients as well. Isolates have been classified according to their replication characteristics, syncytium-inducing capacities, cytolytic properties, and cell tropisms. Isolates that infect a number of T-lymphoid cell lines with a rapid rise in reverse transcriptase activity (rapid/high) have been shown to have high infectivity and are cytopathic with the production of syncytia. These isolates have been obtained more frequently from patients with advanced disease (Asjo *et al.*, 1987; Fenyo *et al.*, 1988; Cheng-Mayer *et al.*, 1988). The infectivity of these isolates and syncytial formation may be related to the same genotypic property. Isolates with very slow and restricted growth in T-cell lines (slow/low) are more often isolated from asymptomatic patients. In some patients a progression to the more highly cytopathic isolates has been shown to occur with progression of disease in the same host (Cheng-Mayer *et al.*, 1988).

Cell tropism appears to be another biologic variant that may be important to disease manifestations in the host. There are clear differences in the *in vitro* host range of HIV-1 isolates and cytolytic properties of these isolates are dependent on the particular target cell (Evans *et al.*, 1987; Dahl *et al.*, 1987). Of particular interest is the apparent macrophage tropism of some isolates, particularly those isolated from mononuclear phagocytes of the lung and brain. Such isolates show a 10–100-fold more efficient infection in macrophages over T-cell lines *in vitro* with the production of large amounts of virus as well as the production of large syncytia (Gartner *et al.*, 1986). Different isolates with unique cell tropisms can be obtained from different sites within the same patient as demonstrated by the isolation of a genetically distinct macrophage-tropic isolate from the brain and a lymphocyte-tropic variant from the cerebral spinal fluid of a patient with encephalopathy (Koyanagi *et al.*, 1987). This result confirms the existence of genetically and biologically distinct variants of virus within the same host.

Despite the evidence that supports the existence of biologic variants of the virus as measured by cell tropism, replication characteristics, and cytopathicity, it has been difficult to link these characteristics directly with distinct genetic profiles and with the course of the disease in the host. It has been observed that syncytium-inducing isolates are more frequently found in patients with AIDS or AIDS-related complex (ARC) (Tersmette *et al.*, 1988), that replication rate correlates with rate of decline of $CD4^+$ cells in the host, and that prior to the decline in $CD4^+$ cell counts and development of disease in some individuals there is a phenotypic switch from nonsyncytia to syncytia-forming isolates (Tersmette *et al.*, 1989). These observations all suggest that there is a correlation between biologic variability of HIV isolates and the course of disease in the host. It is reasonable to conclude that genetic diversity contributes to these biologic variations and such variations have important implications concerning the behavior of the virus within an individual. A more complete understanding of the role that genetic variation plays in these relationships will be extremely helpful in developing therapeutic options and vaccine candidates.

Acknowledgments

M.E.K is supported by the Veteran's Administration Research Service. The authors would like to thank Steven Josephs for his contribution to Fig. 2, Anna Mazzuca for her assistance in the preparation of this manuscript, and Robert C. Gallo for his continued support.

References

Ahmad, N., and Venkatesan, S. (1988). *Science* **241**, 1481–1485.
Alizon, M., Sonigo, P., Barre-Sinoussi, F., Chermann, J.-C. Tiollais, P., Montagnier, L., and Wain-Hobson, S. (1984). *Nature (London)* **312**, 757–759.

Alizon, M., Wain-Hobson, S., Montagnier, L., and Sonigo, P. (1986). *Cell* **46**, 63–74.

Arya, S. K. (1987). *Proc. Natl. Acad. Sci. U.S.A.* **84**, 5429–5433.

Arya, S. K., and Gallo, R. C. (1986). *Proc. Natl. Acad. Sci. U.S.A.* **83**, 2209–2213.

Arya, S. K., Guo, C., Josephs, S. F., and Wong-Staal, F. (1985). *Science* **229**, 69–73.

Asjo, B., Ivhed, I., Gidlund, M., Fuerstenberg, S., Fenyo, E. M., Nilsson, K., and Wigzell, H. (1987). *Lancet* **ii**, 660–662.

Asjo, B., Albert, J., Chiodi, F., and Fenyo, E. M. (1988). *J. Virol. Methods* **19**, 191–196.

Banapour, B., Rosenthal, K., Rabin, L., Sharma, V., Young, L., Fernandez, J., Engleman, E., McGrath, M., Reyes, G., and Lifson, J. (1987). *J. Immunol.* **139**, 4027–4033.

Benn, S., Rutledge, R., Folks, T., Gold, J., Baker, L., McCormick, J., Feorino, P., Piot, P., Quinn, T., and Martin, M. (1985). *Science* **230**, 949–951.

Berkhout, B., Silverman, R. H., and Jeang, K-T. (1989). *Cell* **59**, 273–282.

Bohnlein, E., Lowenthal, J. W., Siekevitz, M., Ballard, D. W., Franza, B. R., and Greene, W. C. (1988). *Cell* **53**, 827–836.

Bohnlein, E., Siekevitz, M., Ballard, D. W., Lowenthal, J. W., Rimsky, L., Bogerd, H., Hoffman, J., Wano, Y., Franza, B. R., and Greene, W. C. (1989a). *J. Virol.* **63**, 1578–1586.

Bohnlein, S., Hauber, J., and Cullen, B. R. (1989b). *J. Virol.* **63**, 421–424.

Bolognesi, D. P. (1989). *Science* **246**, 1233–1234.

Braddock, M., Chambers, A., Wilson, W., Esnouf, M. P., Adams, S. E., Kingsman, A. J., and Kingsman, S. M. (1989). *Cell* **58**, 269–279.

Braun, M. J., Clements, J. E., and Gonda, M. A. (1987). *J. Virol.* **61**, 4046–4054.

Bukrinsky, M. I., and Etkin, A. F. (1990). *AIDS Res. Hum. Retroviruses* **6**, 425–426.

Cease, K. B., Margalit, H., Cornette, J. L., Putney, S. D., Robey, W. G., Ouyang, C., Streicher, H. Z., Fischinger, P. J., Gallo, R. C., DeLisi, C., and Berzofsky, J. A. (1987). *Proc. Natl. Acad. Sci. U.S.A.* **84**, 4249–4253.

Chanh, T. C., Dreesman, G. R., Kanda, P., Linette, G. P., Sparrow, J. T., Ho, D. D., and Kennedy, R. C. (1986). *EMBO J.* **5**, 3065–3071.

Cheng-Mayer, C., Seto, D., Tateno, M., and Levy, J. A. (1988). *Science* **240**, 80–82.

Cheng-Mayer, C., Iannello, P., Shaw, K., Luciw, P. A., and Levy, J. A. (1989). *Science* **246**, 1629–1632.

Cochrane, A. W., Perkins, A., and Rosen, C. A. (1990). *J. Virol.* **64**, 881–885.

Cohen, E. A., Terwilliger, E. F., Jalinoos, Y., Proulx J., Sodroski, J. G., and Haseltine, W. A. (1990). *J. AIDS* **3**, 8–11.

Cullen, B. (1986). *Cell* **46**, 973–982.

Daefler, S., Klotman, M. E., and Wong-Staal, F. (1990). *Proc. Natl. Acad. Sci. U.S.A.* **87**, 4571–4575.

Dahl, K., Martin, K., and Miller, G. (1987). *J. Virol.* **61**, 1602–1608.

Daly, T. J., Cook, K. S., Gary, G. S., Maione, T. E., and Rusche, J. R. (1989). *Nature (London)* **342**, 816–819.

Davis, M. G., Kenney, S. C., Kamine, J., Pagano, J. S., and Huang, E.-S. (1987). *Proc. Natl. Acad. Sci. U.S.A.* **84**, 8642–8646.

Dayton, A. I., Sodroski, J. G., Rosen, C. A., Goh, W. C., and Haseltine, W. A. (1986). *Cell* **44**, 941.

Dayton, A. I., Terwilliger, E. F., Potz, J., Kowakski, M., Sodroski, J. G., and Haseltine, W. A. (1988). *J. AIDS* **1**, 441–452.

Dedera, D., Hu, W., Heyden, N. V., and Ratner, L. (1989). *J. Virol.* **63**, 3205–3208.

Desai, S. M., Kalyanaraman, V. S., Casey, J. M., Srinivasan, A., Andersen, P. R., and Devare, S. G. (1986). *Proc. Natl. Acad. Sci. U.S.A.* **83**, 8380–8384.

Dingwall, C., Ernberg, I., Gait, M. J., Green, S. M., Heaphy, S., Karn, J., Lowe, A. D., Singh, M., Skinner, M. A., and Valerio, R. (1989). *Proc. Natl. Acad. Sci. U.S.A.* **86**, 6925–6929.

Dinter, H., Chiu, E., Imagawa, M., Karin, M., and Jones, K. A. (1987). *EMBO J.* **6**, 4067–4071.

Dowbenko, D., Nakamura, G., Fennie, C., Shimasaki, C., Riddle, L., Harris, R., Gregory, T., and Lasky, L. (1988). *J. Virol.* **62**, 4703–4711.

Ensoli, E., Lusso, P., Schachter, F., Josephs, S. F., Rappaport, J., Negro, F., Gallo, R. C., and Wong-Staal, F. (1989). *EMBO J.* **8**, 3019–3027.

Evans, L. A., McHugh, T. M., Stites, D. P., and Levy, J. A. (1987). *J. Immunol.* **138**, 3415–3418.

Farmerie, W. G., Loeb, D. D., Casavant, N. C., Hutchison III, C. A., Edgell, M. H., .and Swanstrom, R. (1987). *Science* **236**, 305–308.

Feinberg, M. B., Jarrett, R. F., Aldovini, A., Gallo, R. C., and Wong-Staal, F. (1986). *Cell* **46**, 807–817.

Felber, B. K., Hadzopoulou-Cladaras, M., Cladaras, C., Copeland, T., and Paulakis, G. N. (1989). *Proc. Natl. Acad. Sci. U.S.A.* **86**, 1495–1499.

Feng, S., and Holland, E. C. (1988). *Nature (London)* **334**, 165–170.

Fenyo, E. M., Morfeldt-Manson, L., Chiodi, F., Lind, B., Gegerfelt, A. V., Albert, J., Olausson, E., and Asjo, B. (1988). *J. Virol.* **62**, 4414–4419.

Fisher, A. G., Feinberg, M. B., Josephs, S. F., Harper, M. E., Marselle, L. M., Reyes, G., Gonda, M. A., Aldovini, A., Debouk, C., Gallo, R. C., and Wong-Staal, F. (1986). *Nature (London)* **320**, 367–371.

Fisher, A. G., Ensoli, B., Ivanoff, L., Chamberlain, M., Petteway, S., Ratner, L., Gallo, R. C., and Wong-Staal, F. (1987). *Science* **237**, 888–893.

Fisher, A. G., Ensoli, B., Looney, D., Rose, A., Gallo, R. C., Saag, M. S., Shaw, G. M., Hahn, B. H., and Wong-Staal, F. (1988). *Nature (London)* **334**, 444–447.

Franchini, G., Robert-Guroff, M., Shroyeb, J., Chang, N. T., and Wong-Staal, F. (1986). *Virology* **155**, 593–599.

Franchini, G., Rusche, J. R., O'Keeffe, J., and Wong-Staal, F. (1988). *AIDS. Res. Hum. Retroviruses* **4**, 243–250.

Franza, Jr., B. R., Josephs, S. F., Gilman, M. Z., Ryan, W., and Clarkson, B. (1987). *Nature (London)* **330**, 391–395.

Franza, Jr., B. R., Rauscher III, F. J., Josephs, S. F., and Curran, T. (1988). *Science* **239**, 1150–1153.

Frankel, A. D., Triezenberg, S. J., and McKnight, S. L. (1988). *Nature (London)* **335**, 452–454.

Gallaher, W. R. (1987). *Cell* **50**, 327–328.

Gallo, R. C., and Montagnier, L. (1988). *In* "The Science of AIDS" (Scientific American, Inc.) pp. 1–11. W. H. Freeman and Company, New York.

Garcia, J. A., Wu, F. K., Mitsuyasu, R., and Gaynor, R. B. (1987). *EMBO J.* **6**, 3761–3770.

Garcia, J. A., Harrich, D., Pearson, L., Mitsuyasu, R., and Gaynor, R. B. (1988). *EMBO J.* **7**, 3143–3147.

Garcia, J. A., Harrich, D., Soultanakis, E., Wu, F., Mitsuyasu, R., and Gaynor, R. B. (1989). *EMBO J.* **8**, 765–778.

Gartner, S., Markovits P., Markovitz, D. M., Kaplan, M. H., Gallo, R. C., and Popovic, M. (1986). *Science* **233**, 215–219.

Geelen, J. L. M. C., Minnaar, R. P., Boom, R., Van der Noordaa, J., and Goudsmit, J. (1988). *J. Gen. Virol.* **69**, 2913–2917.

Gimble, J. M., Duh, E., Ostrove, J. M., Gendelman, H. E., Max, E. E., and Rabson, A. B. (1988). *J. Virol.* **62**, 4104–4112.

Gnann Jr., J. W., Schwimmbeck, P. L., Nelson, J. A., Truax, A. B., and Oldstone, M. B. A. (1987a). *J. Infect. Dis.* **156**, 261–267.

Gnann Jr., J. W., Nelson, J. A., and Oldstone, M. B. A. (1987b). *J. Virol.* **61**, 2639–2641.

Golding, H., Robey, F. A., Gates III, F. T., Linder, W., Beining, P. R., Hoffman, T., and Golding, B. (1988). *J. Exp. Med.* **167**, 914–923.

Gonda, M. A., Wong-Staal, F., Gallo, R. C., Clements, J. E., Narayan, O., and Gilden, R. V. (1985). *Science* **227**, 173–177.

Gonda, M. A., Braun, M. J., Clements, J. E., Pyper, J. M., Wong-Staal, F., Gallo, R.C., and Gilden, R.V. (1986). *Proc. Natl. Acad. Sci. U.S.A.* **83**, 4007–4011.

Gonzalez-Scarano. F., Waxham, M. N., Ross, A. M., and Hoxie, J. A. (1987). *AIDS Res. Hum. Retroviruses* **3**, 245–252.

Goodenow, M., Huet, T., Saurin, W., Kwok, S., Sninsky, J., and Wain-Hobson, S. (1989). *J. AIDS* **2**, 344–352.

Goudsmit, J., Debouck, C., Meloen, R. H., Smit, L., Bakker, M., Asher, D. M., Wolff, A. V., Gibbs Jr, C. J., and Gajdusek, D. C. (1988). *Proc. Natl. Acad. Sci. U.S.A.* **85**, 4478–4482.

Gowda, S. D., Stein, B. S., Steimer, K. S., and Engleman, E. G. (1989). *J. Virol.* **63**, 1451–1454.

Guy, B., Kieny, M. P., Riviere, Y., Le Peuch, C., Dott, K., Girard, M., Montagnier, L., and Lecocq, J.-P. (1987). *Nature,* (London) **330**, 266–269.

Guyader, M., Emerman, M., Sonigo, P., Clavel, L., Montagnier, L., and Alizon, M. (1987). *Nature (London)* **326**, 662–669.

Hadzopoulou-Cladores, M., Felber, B. K., Cladaras, C., Athanassoppulos, A., Tse, A., and Pavlakis, G. N. (1989). *J. Virol.* **63**, 1265–1274.

Hahn, B. H., Shaw, G. M., Arya, S. K., Popovic, M., Gallo, R. C., and Wong-Staal, F. (1984). *Nature (London)* **312**, 166–169.

Hahn, B. H., Gonda, M. A., Shaw, G. M., Popovic, M., Hoxie, J. A., Gallo, R. C., and Wong-Staal, F. (1985). *Proc. Natl. Acad. Sci. U.S.A.* **82**, 4813–4817.

Hahn, B. H., Shaw, G. M., Taylor, M. E., Redfield, R. R., Markham, P. D., Salahuddin, S. Z., Wong-Staal, F., Gallo, R. C., Parks, E. S., and Parks, W. P. (1986). *Science* **232**, 1548–1553.

Hammarskjold, M.-L., Heimer, J., Hammarskjold, B., Samgwan, I., Albert, L., and Rekosh, D. (1989). *J. Virol.* **63**, 1959–1966.

Hammes, S. R., Dixon, E. P., Malim, M. H., Cullen, B. R., and Greene, W. C. (1989). *Proc. Natl. Acad. Sci. U.S.A.* **86**, 9549–9553.

Hansen, J., Schulze, T., and Moelling, K. (1987). *J. Biol. Chem.* **262**, 12393–12396.

Hansen, J., Schulze, T., Mellert, W., and Moelling, K. (1988). *EMBO J.* **7**, 239–243.

Harrich, D., Garcia, J., Wu, F., Mitsuyasu, R., Gonzalez, J., and Gaynor, R. (1989). *J. Virol.* **63**, 2585–2591.

Hauber, J., and Cullen, B. R. (1988). *J. Virol.* **62**, 673–679.

Hauber, J., Malim, M. H., and Cullen, B. R. (1989). *J. Virol.* **63**, 1181–1187.

Henderson, L. E., Sowder, R. C., Copeland, T. D., Benreniste, R. E., and Oroszlan, S. (1988). *Science* **241**, 199–201.

Hirsch, V. M., Olmsted, R. A., Murphy-Corb, M., Purcell, R. H., and Johnson, P. R. (1989). *Nature (London)* **339**, 389–392.

Hizi, A., McGill, C., and Hughes, S. H. (1988). *Proc. Natl. Acad. Sci. U.S.A.* **85**, 1218–1222.

Ho, D. D., Sarngadharan, M. G., Hirsch, M. S., Schooley, R. T., Rota, T. R., Kennedy, R. C., Chanh, T. C., and Sato, V. L. (1987). *J. Virol.* **61**, 2024–2028.

Ho, D. D., Kaplan, J. C., Rackauskas, I. E., and Gurney, M. E. (1988). *Science* **239**, 1021–1023.

Jacks, T., Power, M. D., Masiarz, F. R., Luciw, P. A., Barr, P. J., and Varmus, H. E. (1988). *Nature (London)* **331**, 280–283.

Jakobovits, A., Smith, D. H., Jakobovits, E. B., and Capon, D. J. (1988). *Mol. Cell. Biol.* **8**, 2555–2561.

Jones, K. A, (1989). *New Biol.* **2**, 127–135.

Jones, K. A., Kadonaga, J. T., Luciw, P. A., and Tjian, R. (1986). *Science* **232**, 755–760.

Jones, K. A., Luciw, P. A., and Duchange, N. (1988). *Genes Dev.* **2**, 1101–1114.

Kan, N. C., Franchini, G., Wong-Staal, F., DuBois, G. C., Robey, W. G., Lautenberger, J. A., and Papas, T. S. (1986). *Science* **231**, 1553–1555.

Kao, S.-Y., Calman, A. F., Luciw, P. A., and Peterlin, B. M. (1987). *Nature* (*London*) **330**, 489–493.

Kappes, J. C., Morrow, C. D., Lee Shei-Wen, Jameson, B. A., Kent, S. B., Hood, L. E., Shaw, G. M., and Hahn, B. H. (1988). *J. Virol.* **62**, 3501–3505.

Kaufman, J. D., Valandra, G., Roderiquez, G., Bushar, G., Giri, C., and Norcross, M. A. (1987). *Mol. Cell Biol.* **7**, 3759–3766.

Kawakami, K., Scheidereit, C., and Roeder, R. G. (1988). *Proc. Natl. Acad. Sci. U.S.A.* **85**, 4700–4704.

Kenney, S., Kamine, J., Markovitz, D., Fenrick, R., and Pagano, J. (1988). *Proc. Natl. Acad. Sci. U.S.A.* **85**, 1652–1656.

Kim, S., Ikeuchi, K., Byrn, R., Groopman, J., and Baltimore, D. (1989). *Proc. Natl. Acad. Sci. U.S.A.* **86**, 9544–9548.

Klasse, P. J., Pipkorn, R., and Blomberg, J. (1988). *Proc. Natl. Acad. Sci. U.S.A.* **85**, 5225–5229.

Klimkait, T., Strebel, K., Hoggan, M. D., Martin, M. A., and Orenstein, J. M. (1990). *J. Virol.* **64**, 621–629.

Kowalski, M., Potz, J., Basiripour, L., Dorfman, T., Goh, W. C., Terwilliger, E., Dayton, A., Rosen, C., Haseltine, W., and Sodroski, J. (1987). *Science* **237**, 1351–1355.

Koyanagi, Y., Miles, S., Mitsuyasu, R. T., Merrill, J. E., Vinters, H. V., and Chen, I. S. Y. (1987). *Science* **236**, 819–822.

Kramer, R. A., Schaber, M. D., Skalka, A. M., Ganguly, K., Wong-Staal, F., and Reddy, E. P. (1986). *Science* **231**, 1580–1584.

Kubota, S., Siomi, H., Satoh, T., Endo, S-I., Maki, M., and Hatanaka, M. (1989). *Biochem. Biophys. Res. Commun.* **162**, 963–970.

Larder, B. A., Purifoy, D. J. M., Powell, K. L., and Darby, G. (1987a). *Nature* (*London*) **327**, 716–717.

Larder, B., Purifoy, D., Powell, K., and Darby, G. (1987b). *EMBO J.* **6**, 3133–3137.

Lasky, L. A., Nakamura, G., Smith, D. H., Fennie C., Shimasaki, C., Patzer, E., Berman, P., Gregory, T., and Capon, D. J. (1987). *Cell* **50**, 975–985.

Laspia, M. F., Rice, A. P., and Mathews, M. B. (1989). *Cell* **59**, 283–292.

Lee, T.-H., Coligan, J. E., Allan, J. S., McLane, M. F., Groopman, J. E., and Essex, M. (1986). *Science* **231**, 1546–1548.

Le Grice, S. F. J., Mills, J., and Mous, J. (1988). *EMBO J.* **7**, 2547–2553.

Leonard, J., Parrott, C., Buckler-White, A. J., Turner, W., Ross, E. K., Martin, M. A., and Robson, A. B. (1989). *J. Virol.* **63**, 4919–4924.

Lever, A., Gottlinger, H., Haseltine, W., and Sodroski, J. (1989). *J. Virol.* **63**, 4085–4087.

Lightfoote, M. M., Coligan, J. E., Folks, T. M., Fauci, A. S., Martin, M. A., and Venkatesan, S. (1986). *J. Virol.* **60**, 771–775.

Looney, D. J., Fisher, A. G., Putney, S. D., Rusche, J. R., Redfield, R. R., Burke, D. S., Gallo, R. C., and Wong-Staal, F. (1988). *Science* **241**, 357–359.

Luciw, P. A., Potter, S. S., Steimer, K., Dina, D., and Levy, J. A. (1984). *Nature* (*London*) **312**, 760–763.

Luciw, P. A., Cheng-Mayer, C., and Levy, J. A. (1987). *Proc. Natl. Acad. Sci. U.S.A.* **84**, 1434–1438.

Maekawa, T., Itoh, F., Okamoto, T., Kurimoto, M., Imamoto, F., and Ishii, S. (1989). *J. Biol. Chem.* **264**, 2826–2831.

Malim, M. H., Hauber, J., Fenrick, R., and Cullen, B. R. (1988). *Nature* (*London*) **335**, 181–183.

Malim, M. H., Bohnlein, S., Hauber, J., and Cullen, B. R. (1989a). *Cell* **58**, 205–214.

Malim, M. H., Hauber, J., Le, S-Y., Maizel, J. V., and Cullen, B. R. (1989b). *Nature* (*London*) **338**, 254–257.

Marlink, R. G. (1988). *AIDS Res. Hum. Retroviruses* **4**, 137–148.

Matthews, T. J., Langlois, A. J., Robey, W. G., Chang, N. T., Gallo, R. C., Fischinger, P. J., and Bolognesi, D. P. (1986). *Proc. Natl. Acad. Sci.. U.S.A.* **83,** 9709–9713.

Matsushita, S., Robert-Guroff, M., Rusche, J., Koito, A., Hattori, T., Hoshino, H., Javaherian, K., Takatsuki, K., and Putney, S. (1988). *J. Virol.* **62,** 2107–2114.

McCune, J. M., Rabin, L. B., Feinberg, M. B., Lieberman, M., Kosek, J. C., Reyes, G. R., and Weissman, I. L. (1988). *Cell* **53,** 55–67.

McDougal, J. S., Kennedy, M. S., Sligh, J. M., Cort, S. P., Mawle, A., and Nicholson, J. K. A. (1986). *Science* **231,** 382–385.

Mervis, R. J., Ahmad, N., Lillehoj, E. P., Raum, M. G., Salazar, F. H. R., Chan, H. W., and Venkatesan, S. (1988). *J. Virol.* **62,** 3993–4002.

Meyerhans, A., Cheynier, R., Albert, J., Seth, M., Kwok, S., Sninsky, J., Morfeldt-Manson, L., Asjo, B., and Wain-Hobson, S. (1989). *Cell* **58,** 901–910.

Miller, R. H. (1988). *Science* **239,** 1420–1422.

Modrow, S., Hahn, B. H., Shaw, G. M., Gallo, R. C., Wong-Staal, F., and Wolf, H. (1987). *J. Virol.* **61,** 570–578.

Mosca, J. D., Bednarik, D. P., Raj, N. B. K., Rosen, C. A., Sodroski, J. G., Haseltine, W. A., Hayward, G. S., and Pitha, P. M. (1987). *Proc. Natl. Acad. Sci. U.S.A.* **84,** 7408–7412.

Mous, J., Heimer, E. P., and Le Grice, S. F. J. (1988). *J. Virol.* **62,** 1433–1436.

Muesing, M. A., Smith, D. H., Cabradilla, C. D., Benton, C. V., Lasky, L. A., and Capon, D. J. (1985). *Nature (London)* **313,** 450–458.

Muesing, M. A., Smith, D. H., and Capon, D. J. (1987). *Cell* **48,** 691–701.

Nabel, G., and Baltimore, D. (1987). *Nature (London)* **326,** 711–713.

Nabel, G., Rice, S., Knipe, D., and Baltimore, D. (1988). *Science* **239,** 1299–1301.

Niederman, T. M. J., Thielman, B. S., and Ratner, L. (1989). *Proc. Natl. Acad. Sci. U.S.A.* **86,** 1128–1132.

Ogawa, K., Shibata, R., Kiyomasu, T., Higuchi, I., Kishida, Y., Ishimoto, A., and Adachi, A. (1989). *J. Virol.* **63,** 4110–4114.

Okamoto, T., and Wong-Staal, F. (1986). *Cell* **47,** 29–35.

Okamoto, T., Matsuyama, T., Mori, S., Hamamoto, Y., Kobayashi, N., Yamamoto, N., Josephs, S. F., Wong-Staal, F., and Shimotohno, K. (1989). *AIDS Res. Hum. Retroviruses* **5,** 131–137.

Osborn, L., Kunkel, S., and Nabel, G. J. (1989). *Proc. Natl. Acad. Sci. U.S.A.* **86,** 2336–2340.

Palker, T. J., Clark, M. E., Langlois, A. J., Matthews, T. J., Weinhold, K. J., Randall, R. R., Bolognesi, D. P., and Haynes, B. F. (1988). *Proc. Natl. Acad. Sci. U.S.A.* **85,** 1932–1936.

Parkin, N. T., Cohen, E. A., Darveau, A., Rosen, C., Haseltine, W., and Sonenberg, N. (1988). *EMBO J.* **7,** 2831–2837.

Perkins, A., Cochrane, A. W., Ruben, S. M., and Rosen, C. A. (1989). *J. AIDS* **2,** 256–263.

Peterlin, B. M., Luciw, P. A., Barr, J. P., and Walker, M. D. (1986). *Proc.* Natl. Acad. Sci. U.S.A. **83,** 9734–9738.

Prasad, V. R., and Goff, S. P. (1989). *Proc. Natl. Acad. Sci. U.S.A.* **86,** 3104–3108.

Preston, B. D., Poiesz, B. J., and Loeb, L. A. (1988). *Science* **242,** 1168–1173.

Putney, S. D., Matthews, T. J., Robey, W. G., Lynn, D. L., Robert-Guroff, M., Mueller, W. T., Langlois, A. J., Ghrayeb, J., Petteway Jr., S. R., Weinhold, K. J., Fischinger, P. J., Wong-Staal, F., Gallo, R. C., and Bolognesi, D. P. (1986). *Science* **234,** 1392–1395.

Rabson, A. B., Daugherty, D. F., Venkatesan, S., Boulukos, K. E., Benn, S. I., Folks, T. M., Feorino, P., and Martin, M. A. (1985). *Science* **229,** 1388–1390.

Rappaport, J., Lee, S.-J., Khalili, K., and Wong-Staal, F. (1989). *New Biol.* **1** 101–110.

Ratner, L., Haseltine, W., Patarca, R., Livak, K. J., Starcich, B., Josephs, S. F., Doran, E. R., Rafalski, J. A., Whitehorn, E. A., Baumeister, K., Ivanoff, L., Petteway, S. R., Jr., Pearson, M. L., Lautenberger, J. A., Papas, T. S., Ghrayeb, J., Chang, N. T., Gallo, R. C., and Wong-Staal, F. (1985). *Nature (London)* **313,** 277–283.

Reiher III, W. E., Blalock, J. E., and Brunck, T. K. (1986). *Proc. Natl. Acad. Sci. .U.S.A.* **83**, 9188–9192.

Reitz Jr., M. S., Wilson, C., Naugle, C., Gallo, R. C., and Robert-Guroff, M. (1988). *Cell* **54**, 57–63.

Rice, A. P. and Mathews, M. B. (1988). *Nature (London)* **332**, 551–553.

Robert-Guroff, M., Brown, M., and Gallo, R. C. (1985). *Nature (London)* **316**, 72–74.

Robert-Guroff, M., Reitz, M. S., Robey, W. G., and Gallo, R. C. (1986). *J. Immunol.* **137**, 3306–3309.

Robert-Guroff, M., Popovic, M., Gartner, S., Markham, P., Gallo, R. C. and Reitz, M. S. (1990). *J. Virol.* **64**, 3391–3398.

Roberts, J. D., Bebenek, K., and Kunkel, T. A. (1988). *Science* **242**, 1171–1173.

Rosen, C. A., Sodroski, J. G., and Haseltine, W. A. (1985). *Cell* **41**, 813–823.

Rosen, C. A., Sodroski, J. G., Goh, W., Dayton, A. I., Lippe, J., and Haseltine, W. A. (1986). *Nature (London)* **319**, 555–559.

Rosen, C. A., Terwilliger, E. F., Dayton, A. I., Sochoski, J. F., and Haseltine, W. A. (1988). *Proc. Natl. Acad. Sci. U.S.A.* **85**, 2071–2075.

Ruben, S., Perkins, A., Purcell, R., Joung, K., Sia, R., Burghoff, R., Haseltine, W. A., and Rosen, C. A. (1989). *J. Virol.* **63**, 1–8.

Rusche, J. R., Javaherian, K., McDanal, C., Petro, J., Lynn, D. L., Grimaila, R., Langlois, A., Gallo, R. C., Arthur, L. O., Fischinger, P. J., Bolognesi, D. P., Putney, S. D., and Matthews, T. J. (1988). *Proc. Natl. Acad. Sci. U.S.A.* **85**, 3198–3202.

Saag, M. S., Hahn, B. H., Gibbons, J., Li, Y., Parks, E. S., Parks, W. P., and Shaw, G. M. (1988). *Nature (London)* **334**, 440–443.

Sadaie, M. R., Benter, T., and Wong-Staal, F. (1988a). *Science* **239**, 910–913.

Sadaie, M. R., Rappaport, J., Benter, T., Josephs, S. F., Willis, R., and Wong-Staal, F. (1988b). *Proc. Natl. Acad. Sci. U.S.A.* **85**, 9224–9228.

Sakai, K., Dewhurst, S., Ma, X., and Volsky, D. J. (1988). *J. Virol.* **62**, 4078–4085.

Salfeld, J., Gottlinger, H. G., Sia, R. A., Park, R. E., Sodroski, J. G., and Haseltine, W. A. (1990). *EMBO J.* **9**, 965–970.

Samuel, K. P., Seth, A., Konopka, A., Lautenberger, J. A., and Papas, T. S. (1987). *FEBS Lett.* **218**, 81–86.

Sanchez-Pescador, R., Power, M. D., Barr, P. J., Steimer, K. S., Stempien, M. M., Brown-Shimer, S. L., Gee, W. W., Renard, A., Randolph, A., Levy, J. A., Dina, D., and Luciw, P. A. (1985). *Science* **227**, 484–492.

Schrier, R. D., Gnann, J. W., Langlois, A. J., Shriver, K., Nelson, J. A., and Oldstone, M. B. A. (1988). *J. Virol.* **62**, 2531–2536.

Scott, J. V., Stowring, L., Haase, A. T., Narayan, O., and Vigne, R. (1979). *Cell* **18**, 321–327.

Seelmeier, S., Schmidt, H., Turk, V., and von der Helm, K. (1988). *Proc. Natl. Acad. Sci. U.S.A.* **85**, 6612–6616.

Shaw, G. M., Hahn, B. H., Arya, S. K., Groopman, J. E., Gallo, R. C., and Wong-Staal, F. (1984). *Science* **226**, 1165–1171.

Shaw, J.-P., Utz, P. J., Durand, D. B., Toole, J. J., Emmel, E. A., and Crabtree, G. R. (1988). *Science* **241**, 202–206.

Siekevitz, M., Josephs, S. F., Dukovich, M., Peffer, N., Wong-Staal, F., and Greene, W. C. (1987). *Science* **238**, 1575–1578.

Sodroski, J. G., Rosen, C. A., and Haseltine, W. A. (1984). *Science* **225**, 381–385.

Sodroski, J., Patarca, R., Rosen, C., Wong-Staal, F., and Haseltine, W. (1985a). *Science* **229**, 74–77.

Sodroski, J. G., Rosen, C. A., Wong-Staal, F., Salahuddin, S. Z., Popovic, M., Arya, S., Gallo, R. C., and Haseltine, W. A. (1985b). *Science* **227**, 171–173.

Sodroski, J., Goh, W. C., Rosen, C., Dayton, A., Terwilliger, E., and Haseltine, W. (1986a). *Nature (London)* **321**, 412–417.

Sodroski, J., Goh, W., C., Rosen, C., Tartar, A., Portetelle, D., Burny, A., and Haseltine, W. (1986b). *Science* **231**, 1549–1552.

Sonigo, P., Alizon, M., Staskus, K., Klatzmann, D., Cole, S., Danos, O., Retzel, E., Tiollais, P., Haase, A., and Wain-Hobson, S. (1985). *Cell* **42**, 369–382.

Srinivasan, A., Anand, R., York, D., Ranganathan, P., Feorino, P., Schochetman, G., Curran, J., Kalyanaraman, V. S., Luciw, P. A., and Sanchez-Pescador, R. (1987). *Gene* **52**, 71–82.

Srinivasan, A., York, D., Jannoun-Nasr, R., Kalyanaraman, S., Swan, D., Benson, J., Bohan, C., Luciw, P. A., Schnoll, S., Robinson, R. A., Desai, S. M., and Devare, S. G. (1989). *Proc. Natl. Acad. Sci. U.S.A.* **86**, 6388–6392.

Starcich, B., Ratner, L., Josephs, S. F., Okamoto, T., Gallo, R. C., and Wong-Staal, F. (1985). *Science* **227**, 538–540.

Starcich, B. R., Hahn, B. H., Shaw, G. M., McNeely, P. D., Modrow, S., Wolf, H., Parks, E. S., Parks, W. P., Josephs, S. F., Gallo, R. C., and Wong-Staal, F. (1986). *Cell* **45**, 637–648.

Stein, B., Kramer, M., Rahmsdorf, H. J., Ponta, H., and Herrlich, P. (1989). *J. Virol.* **63**, 4540–4544.

Strebel, K., Daugherty, D., Clouse, K., Cohen, D., Folks, T., and Martin, M. A. (1987). *Nature (London)* **328**, 728–730.

Strebel, K., Klimkait, T., and Martin, M. A. (1988). *Science* **241**, 1221–1223.

Strebel, K., Klimkait, T., Maldarelli, F., and Martin, M. A. (1989). *J. Virol.* **63**, 3784–3791.

Takahashi, H., Cohen, J., Hosmalin, A., Cease, K. B., Houghten, R., Cornette, J. L., DeLisi, C., Moss, B., Germain, R. N., and Berzofsky, J. A. (1988). *Proc. Natl. Acad. Sci. U.S.A.* **85**, 3105–3109.

Tersmette, M., de Goede, R. E. Y., Al, B. J. M., Winkel, I. N., Gruters, R. A., Cuypers, H. T., Huisman, H. G., and Miedema, F. (1988). *J. Virol.* **62**, 2026–2032.

Tersmette, M., Gruters, R. A., de Wolf, F., de Goede, R. E. Y., Lange, J. M. A., Schellekens, P. T. A., Goudsmit, J., Huisman, H. G., and Miedema, F. (1989). *J. Virol.* **63**, 2118–2125.

Terwilliger, E., Sodroski, J. G., Rosen, C. A., and Haseltine, W. A. (1986). *J. Virol.* **60**, 754–760.

Terwilliger, E. F., Cohen, E. A., Lu, Y., Sodroski, J. G., and Haseltine, W. A. (1989). *Proc. Natl. Acad. Sci. U.S.A.* **86**, 5163–5167.

Twu, JR-S., Rosen, C. A., Haseltine, W. A., and Robinson, W. S. (1989a). *J. Virol.* **63**, 2857–2860.

Twu, JR-S., Chu, K., and Robinson, W. S. (1989b). *Proc. Natl. Acad. Sci. U.S.A.* **86**, 5168–5172.

Veronese, F. D., DeVico, A. L., Copeland, T. D., Oroszlan, S., Gallo, R. C., and Sarngadharan, M. G. (1985). *Science* **229**, 1402–1405.

Veronese, F. D. M., Copeland, T. D., DeVico, A. L., Rahman, R., Oroszlan, S., Gallo, R. C., and Sarngadharan, M. G. (1986). *Science* **231**, 1289–1291.

Veronese, F. D. M., Rahman, R., Copeland, T. D., Oroszlan, S., Gallo, R. C., and Sarngadharan, M. G. (1987). *AIDS Res. Hum. Retroviruses* **3**, 253–264.

Wain-Hobson, S., Sonigo, P., Danos, O., Cole, S., and Alizon, M. (1985). *Cell* **40**, 9–17.

Weiss, R. A., Clapham, P. R., Cheingsong-Popov, R., Dalgleish, A. G., Carne, C. A., Weller, I. V. D., and Tedder, R. S. (1985). *Nature (London)* **316**, 69–71.

Willey, R. L., Rutledge, R. A., Dias, S., Folks, T., Theodore, T., Buckler, C. E., and Martin, M. A. (1986). *Proc. Natl. Acad. Sci. U.S.A.* **83**, 5038–5042.

Willey, R. L., Smith, D. H., Lasky, L. A., Theodore, T. S., Earl, P. L., Moss, B., Capon, D. J., and Martin, M. A. (1988). *J. Virol.* **62**, 139–147.

Wilson, W., Braddock, M., Adams, S. E., Rathjen, P. D., Kingsman, S. M., and Kingsman, A. J. (1988). *Cell* **55**, 1159–1169.

Wong-Staal, F., Shaw, G. M., Hahn, B. H., Salahuddin, S. Z., Popovic, M., Markham, P., Redfield, R., and Gallo, R. C. (1985). *Science* **229**, 759–762.

Wong-Staal, F., Chanda, P. K., and Ghrayeb, J. (1987). *AIDS Res. Hum. Retroviruses* **3,** 33–39.

Wright, C. M., Felber, B. K., Paskalis, H., and Pavlakis, G. N. (1986). *Science* **234,** 988–992.

Wu, F. K., Garcia, J. A., Harrich, D., and Gaynor, R. B. (1988a). *EMBO J.* **7,** 2117–2129.

Wu, F., Garcia, J., Mitsuyasu, R., and Gaynor, R. (1988b). *J. Virol.* **62,** 218–225.

Yasunaga, T., Sagata, N., and Ikawa, Y. (1986). *FEBS Lett.* **199,** 145–150.

Yu, X-F., Ito, S., Essex, M., and Lee, T. H. (1988). *Nature (London)* **335,** 262–265.

Zapp, M. L., and Green, M. R. (1989). *Nature (London)* **342,** 714–716.

Human Immunodeficiency Virus (HIV) Gene Expression and Function

William A. Haseltine

I. Introduction

The molecular biology of the human immunodeficiency virus type 1 (HIV-1) and related human and nonhuman primate retroviruses provides a basis for understanding the peculiar pathogenesis of acquired immunodeficiency syndrome (AIDS). Knowledge of the molecular biology of these viruses also provides the basis for rational antiviral drug design and vaccine development. Application of principles of viral replication to drug development has already lead to clinical trials of novel drugs. Vaccines based on a detailed knowledge of the viral protein structure are also undergoing clinical trials. Many more vaccine candidates based on fundamental principles are also being developed.

Beyond the immediate practical application of the study of the molecular biology of these viruses, this work has revealed new fundamental principles regarding the mechanisms by which eukaryotic organisms control gene expression. The molecular biology of HIV-1 has provided, for the first time, both the *cis-* and *trans*-acting regulatory components which govern the fate of nascent RNA molecules within the nucleus. Other viral regulatory genes affect fundamental processes of replication including assembly and export as well as the infectivity of the mature virus particle. The complexity of the regulatory network which governs virus replication is such that the functions of at least two recently discovered viral genes are not fully understood. Novel use of the information encoded in the genome of HIV-1 is still being described five years after the nucleotide sequence of the virus was determined. Continued investigation of the molecular biology of these intricate organisms will be practically as well as theoretically rewarding. Advances made in understanding of this organism are likely to translate into understanding of other diseases including cancer and other types of genetically based chronic diseases.

II. The Virus Particle

A. Capsid Proteins

The mature virus particle contains an RNA–protein core surrounded by a lipid membrane into which are imbedded viral proteins (Fig. 1). The duplex RNA genome, approximately 10,000 nucleotides long, is bound to a basic nucleocapsid protein (Coffin, 1985). The nucleocapsid protein contains a region that is rich in histidines and cysteines. This protein can bind zinc and may do so by formation of a zinc finger (Green and Berg, 1989; Schiff *et al.*, 1988). Mutation studies demonstrate that each of the cysteines and histidines of the proposed finger are required for the binding of RNA and the formation of an infectious particle (Gorelick *et al.*, 1988; Jentoft *et al.*, 1988).

A specific region of the viral RNA genome, which is located near the 5' terminus of the molecule, is required for encapsidation of the viral nucleic acid. Deletion of this region prevents entry of viral RNA into the capsid protein (Lever *et al.*, 1989). Together the mutation studies on the nucleocapsid protein and the viral RNA indicate that the RNA is captured into the virus particle via a specific nuclear protein—RNA interaction. The capture

Virion Proteins

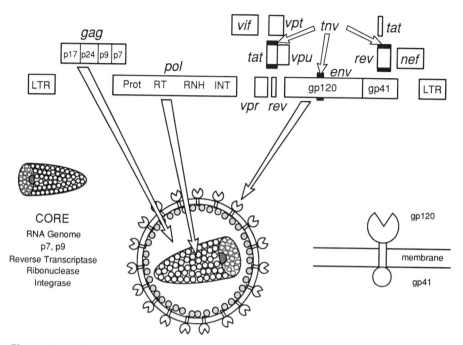

Figure 1. Schematic representation of the genetic organization of the HIV-1 genome and of the virus particle. The genes that encode structural proteins (*gag, pol,* and *env*) are shaded. Regulatory regions and genes including the LTR, *vif, vpr, tat, rev, vpu,* and *nef,* are not shaded. The membrane association of the envelope glycoprotein is indicated as is the organization of the core of the virus particle.

of viral RNA occurs when the nucleocapsid protein is part of a much larger precursor polyprotein. The small nucleocapsid protein is cleaved from the precursor protein only after the virus particle is budded from the cell. The uncleaved nucleocapsid precursor of HIV-1 has been found to be associated with the nucleus of infected cells (Goh, Sodroski, and Haseltine, unpublished observations). Perhaps capture of the full-length viral RNA occurs as this RNA exits the nucleus before such RNA can be acquired by the translation machinery.

The major capsid protein has a molecular weight of 24 kDa (p24) in the mature virus particle (Schüpbach *et al.* 1984). The purified p24 will self-assemble into tube-like structures that have the approximate diameter of the central core particle. Mutations in the sequences which specify p24 prevent the formation of virus particles (Göttlinger and Haseltine, unpublished observations).

III. Replicative Proteins

In addition to structural proteins of the inner core virus particle contains enzymes required for virus replication. These enzymes include the virus-specified polymerase (sometimes called reverse transcriptase), virus ribonuclease H, and integrase activities. The location of these enzymes within the viral core particle is not known with certainty. The molar ratio of structural to nonstructural replication enzymes is about 20 to 1 (Wilson *et al.*, 1988). Noninfectious virus particles of normal appearance can be assembled in the absence of all of these replicative enzymes.

The membrane that surrounds the virus particle is derived from the outer membrane of the infected cells by budding of the core particles. The lipid bilayer that surrounds the virus particle contains at least two viral proteins. The viral protein, p17, is anchored to the inner surface of the membrane via a long chain fatty acid, myristilic acid, which is attached to the N-terminus of the protein (Veronese *et al.*, 1988). The other membrane protein is called the envelope glycoprotein. The envelope glycoprotein is a transmembrane protein in which almost three-fourths of the protein lies exterior to the cell and only one-fourth of the protein is located in the interior of the cell. The envelope protein is highly glycosylated by cellular enzymes during its biosynthesis (Allan et al., 1985a; Robey *et al.*, 1985).

The virion proteins are encoded in three long open reading frames. The 5' open reading frame (*gag*) specifies the inner membrane protein, the major capsid protein, and the nucleocapsid proteins. The *gag* gene is followed by the longest open reading frame of the virus which specifies the viral protease, polymerase, ribonuclease H, and integrase activities. The first two open reading frames overlap by about 200 nucleotides (Muesing *et al.*, 1985). The open reading frame from which the envelope glycoprotein is made is called *env* and is located about 1100 nucleotides from the 3' end of the *pol* gene.

The *gag/pol* open reading frame is translated from a different mRNA from that from which the *env* protein is made. Translation of *gag/pol* occurs from full-length viral RNA species. Initiation of protein synthesis occurs at the AUG codon about 330 nucleotides from the 5' end of the mRNA. Two major protein products are made by translation of this mRNA, the product of the translation of the complete *gag* open reading frame, a polynucleotide 512 amino acids long, and a Gap/Pol fusion protein 1447 amino acids long (Muesing *et al.*, 1985). The Gap/Pol fusion protein is made as the ribosomes undergo a minus one frame shift during the process of translation of the C-terminal region of *gag* (Jacks *et al.* 1988). The frame shift occurs at amino acid 434 of the Gag protein. The minus one frame shift occurs about once in every 20 times the ribosomes encounter the frame shift sequence. Minus one frame shift occurs at the same sequence in *in vitro* translation extracts programmed with viral RNA (Jacks *et al.*, 1988).

The Gag and Gag/Pol precursors are modified both during and after translation. Myristilic acid is added to the N-terminus of the nascent protein. A mutation which changes the penultimate N-terminal amino acid of the precursor of the Gag and Gag/Pol proteins from a glycine to alanine residue prevents addition of myristilic acid. This mutation inactivates the virus for replication (Göttlinger et al., 1989). In such cases the virus particle does not form on the inner surface of the membrane and viral core proteins are not exported from the cell.

The Gag and Gag/Pol precursors are cleaved in a post-translational event by the virtually specified protease enzyme (Kräusslich and Wimmer, 1988). The viral protease is encoded near the 5' end of the Pol precursor, amino acids 69–167 for BH10 and 57–155 for HXBc2. The protease is active only as a dimer (Nutt et al., 1988). The three-dimensional structure of the protein has been determined at the atomic resolution by X-ray crystallography (Navia et al., 1989; Wlodawer et al., 1989). The structure of the dimeric protease resembles other aspartyl proteases. However, the active site of the enzyme is distributed between the two subunits. This arrangement gives rise to the speculation that the initial events that produce the first protease enzyme occur upon the association of two Gag/Pol precursor proteins. Self-cleavage may result from juxtaposition of these two protease subunits. The initial cleavage event is thought to be a self-cleavage event whereby the protease frees itself of the Gag/Pol precursor polypeptide.

The viral protease recognizes the specific amino acid sequences within the Gag and Gag/Pol precursors. The cleavage site exists between the p17 and p24 proteins of *gag*. There are several sites of cleavage between p24 and the C-terminal Gag proteins (Henderson et al., 1988). Another protease cleavage site exists within the p15 protein giving rise to p9 and p6 proteins which are present in the mature virus particle (Henderson et al., 1988).

Cleavage sites flank both ends of the protease gene itself yielding a mature protease subunit 99 amino acids long. The site for protease cleavage also exists 288 amino acids from the C-terminus of the Gag/Pol precursor yielding the 288-amino acid long integrase protein (Lightfoote et al., 1986).

Virus is released from the cell by budding (Fig. 2). The ribonuclear core structure assembles at the cell membrane. Envelope proteins aggregate over the budding structure. The initial crescent form of the assembly virus particle closes into a spherical particle and separates from the cell, surrounded by a lipid membrane. After the particle leaves the cell, a nuclear protein complex condenses in the center of the particle. The immature budded particle contains the viral capsid precursor proteins complexed with the nucleic acids. The HIV-1 core condenses as a distinct blunt conical shape. The spherical form of the particle partially collapses as the condensation of the core occurs. The final phase of the viral maturation is dependent upon proteolytic cleavage. Mutants defective for protease function bud immature viral particles from the cell (Göttlinger et al., 1989; Kohl et al., 1988; Peng et al., 1989). The condensation process in the structure of the virus particle

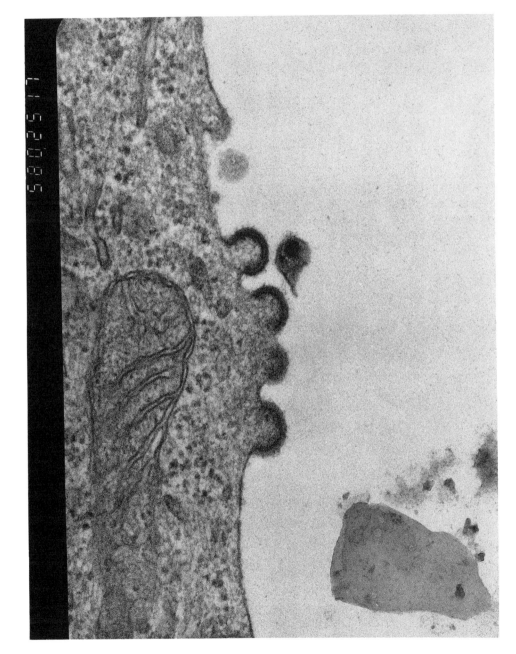

does not occur in protease-defective mutants. Such virus particles are noninfectious even though they contain viral RNA and active reverse transcriptase (Göttlinger *et al.* 1989; Kohl *et al.*, 1988; Peng *et al.*, 1989).

The budding process is not dependent upon envelope glycoproteins. Viruses defective for production of envelope protein can bud. Such envelope-defective mutants are not infectious from CD4$^+$ T-cells (Göttlinger, Sodroski, and Haseltine, unpublished observations).

IV. The Envelope Protein

The envelope protein of HIV is of interest as it provides a potential target for antiviral drug and vaccine development. Moreover, the envelope glycoprotein has been demonstrated to be toxic for CD4$^+$ lymphocytes (Lifson *et al.*, 1986a,b; Sodroski *et al.*, 1986b).

The envelope protein is made as a single polypeptide chain. It is heavily modified by the addition of complex sugar side chains via the cellular machinery. The protein is transported to the cell surface and cleaved. The heavily glycosylated N-terminal portion of the protein is located entirely external to the plasma membrane, whereas the more lightly glycosylated C-terminus is half in and half out of the membrane and is called the transmembrane protein (Allan *et al.*, 1985a,b; Robey *et al.*, 1985; Sodroski *et al.*, 1986b). The protein is anchored to the membrane by a short lipophilic segment in the center of the molecule (Kowalski *et al.*, 1987) (Fig. 3).

Dimers and possibly larger aggregates of the Env protein form via interaction of the exterior protein of the transmembrane protein (Pinter *et al.*, 1989).

The envelope protein specifies two activities, binding to the CD4 cellular surface protein and membrane-to-membrane fusion (Kowalski *et al.*, 1987; Larder *et al.*, 1989; Lifson *et al.*, 1986a,b; McDougal *et al.*, 1986; Sodroski *et al.*, 1986b). It is by this means that the virus gains entry to the cells (Bedinger *et al.*, 1988; Stein *et al.*, 1987). Cells that express envelope protein will also bind to and fuse with uninfected CD4$^+$ cells. The process of binding and fusion can be reiterated, leading to the formation of giant multinucleated syncytia (Lifson *et al.*, 1986a,b; Sodroski *et al.*, 1986b). There are some reports that viruses may gain entry into some cell types via a route that is independent of the envelope protein (Gartner *et al.*, 1986; Takeda and Tauzon, 1988; Homsy *et al.*, 1989). Little is known regarding such alternative routes of entry.

The cytotoxic effect of HIV-1 is dependent on the surface concentration of CD4 as well as on the surface concentration of envelope glycoprotein. Only a low concentration of surface CD4 is needed for infection. However, cells that contain only a small amount of CD4 in the surface are not killed by

Figure 2. Electron micrograph of HIV-1 budding from the surface of a cell. The envelope glycoprotein spikes are evident on the surface of the particle. Magnification × 103,000.

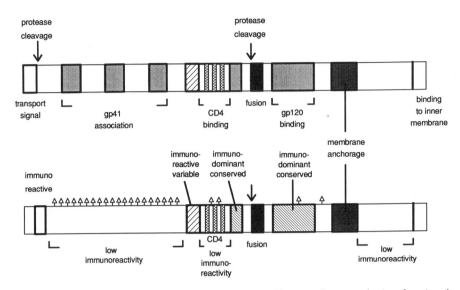

Figure 3. Schematic diagram of the Env protein. The top diagram depicts functional regions of the Env protein as determined by analysis of phenotypic effects of site-directed mutants within the *env* gene. The lower panel is a simplified diagram of some structural immunological features of the Env protein. The small triangles clustered at the N-terminus of the Env protein represent sites of complex carbohydrate modification. The diagram also summarizes the regions of the protein which are of high and low immunoreactivity.

HIV infection, even when large amounts of the virus are made by such cells (DeRossi *et al.*, 1986; Feinberg *et al.*, 1986; Gartner, *et al.*, 1986). An equation that describes killing of a single cell by HIV-1 is:

[virion-associated envelope protein] × [CD4] × [cell fusion factor] =
single cell death

where the term "virion-associated envelope protein" represents the concentration of budding virus that contains a functional envelope glycoprotein, the term "CD4" represents the surface concentration of a functional CD4 molecule, and "cell fusion factor" represents the concentration of a cellular factor required for efficient fusion. The latter factor is included in the equation because the effect of the envelope fusion reaction is not solely determined by the surface concentration of CD4. It depends on a property of cells that varies from species to species and even among T lymphoid cell lines of human origin.

Envelope-mediated giant-cell formation also accounts for some cell death. The surface concentration of the envelope protein on the infected cell and the concentration of CD4 on the uninfected fusion partners determine the ultimate size of syncytia as well as the rate of syncytium formation. Under some conditions, syncytia may include up to 500 nuclei. By this means, a single infected cell, under appropriate conditions, can eliminate several hundred uninfected cells. Death of uninfected cells by syncytium formation may account for the disappearance of a large number of CD4+ T cells in HIV-1-infected patients under conditions in which only 1 in 10,000 of the CD4 cells express viral antigen. This mechanism may be summarized by the equation

$$[1 \times 10^{-4} \text{ cells expressing viral antigen}] \times$$
$$[100 - 500 \text{ uninfected CD4}^+ \text{ T cells}] = 1-5\% \text{ cell death}$$

Large CD4+ virus-infected syncytia have been found in abundance in rhesus macaque monkeys infected with simian immunodeviciency virus (SIV), a virus that is a close relative of HIV-1 (Benveniste et al., 1988). Syncytia are observed in lymph nodes, liver, thymus, spleen and brain of the infected animals. CD4+ HIV-1 syncytia have also been found on autopsy in the spleens of some HIV-infected people (Byrnes et al., 1983). Abundant syncytia are frequently found in the brain on autopsy of patients that have serious neurological complications of HIV-1 infection (Nielson et al., 1984; Price et al., 1988). Abundant CD4+ syncytia are not expected to be found in routine autopsies of AIDS patients. Late in the disease the number of CD4+ lymphocytes is extremely low (Ewing et al., 1985). However, the similarity of pathogenesis of SIV and HIV infection indicates that syncytia formation is a significant pathogenic process in humans as well as in monkeys.

The regions of the envelope glycoprotein that specify specific functions have been identified by analysis of the functional consequences of mutations introduced in known positions in the sequences that specify the envelope glycoprotein (Kowalski et al. 1987; Lasky et al., 1987). The amino acid sequences that comprise the CD4 binding site are located in three contiguous regions of conserved amino acid sequence near the C-terminus of the exterior glycoprotein. These three regions are postulated to fold into a pocket that serves as a binding site for CD4. The region that specifies membrane-to-membrane fusion is a short hydrophobic amino acid sequence near the N-terminus of the transmembrane protein. The short lipophilic sequence near the center of the transmembrane protein anchors the entire envelope protein to the cell membrane. Two extensive regions near the N-terminus of the two proteins form the interface over which noncovalent attachment of the two subunits occurs.

It is possible to construct a dynamic model for the binding and fusion reactions from this information (Fig. 4). Juxtaposition of the two membranes is postulated to occur by high-affinity association between the exterior glycoprotein of one membrane of the CD4 molecule embedded in the second membrane. A binding-induced conformation change is then postulated to occur such that the hydrophobic N-terminus of the transmembrane protein is inserted into the apposed membrane. It is likely that the exterior envelope

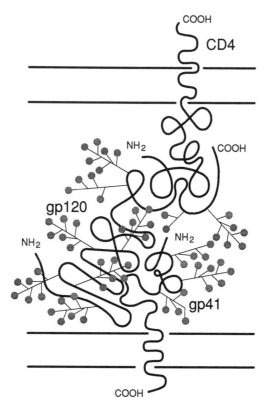

Figure 4. A schematic model of the complex formed between the viral membrane and cell membrane created by the HIV-1 envelope glycoprotein to CD4. The exterior glycoprotein, gp120, acts as a bridge between the membranes. It binds to CD4 anchored in the cell membrane and also to gp41 anchored in the virus membrane. Attachment of gp120 to CD4 and the consequent juxtaposition of the cell and virus membranes is the initial step of virus infection. The gp120 exterior glycoprotein is heavily modified by the addition of complex carbohydrates.

glycoprotein is displaced at this time. Disruption of the surface of the juxtaposed membranes by the transmembrane protein is postulated to initiate the fusion reaction.

The structure of the envelope protein and the life cycle of HIV helps us to understand how the virus evades the immune response. One of the earliest paradoxes regarding HIV infection was that the disease was observed to progress despite an immune response to viral antigens. Progressive disease was particularly puzzling because a high antibody titer to conserved regions on the envelope glycoprotein was reported in early experiments (Sarngadharan *et al.*, 1984 Barin *et al.*, 1985). This paradox was underscored when it was found that most human sera have a very poor ratio of virus-neutralizing activity as compared with the concentration of antibodies that bind the envelope glycoprotein (Robert-Guroff *et al.*, 1985; Weiss *et al.*, 1985). For example, human sera do not completely inhibit the envelope-mediated syncytium formation reaction (Lifson *et al.*, 1986b; Sodroski *et al.*, 1986b). Most sera of HIV-1-infected people have no effect on this reaction whereas the sera from others only slow the kinetics of syncytium formation. These observations indicate that the immunodominant epitopes of the envelope protein are not important for function or not accessible to antibodies when part of a functional envelope protein.

The immunodominant epitopes of HIV-1 have been located (Matthews *et al.*, 1986; Wang *et al.*, 1986; Gann *et al.*, 1987; Palker *et al.*, 1987). They include the C-terminus of the exterior glycoprotein, the N-terminal one-third of the transmembrane protein, and the sort-region-located N-terminal of the CD4-binding region of the exterior glycoprotein. Neither the CD4-binding region nor the fusion domain is immunodominant.

The C-terminus of the exterior glycoprotein is the site of proteolytic cleavage by a cellular enzyme (McCune *et al.*, 1988; Robey *et al.*, 1985; Sodroski *et al.*, 1986b). For this reason, it is very likely to be exposed to the surface. The sequence of the cleavage site is well conserved among HIV-1 isolates (McCune *et al.*, 1988). However, antibodies to this region do not inhibit envelope function, whether such antibodies are derived from patient antisera or have been raised to synthetic peptides corresponding to the predicted amino acid sequence of the region.

The N-terminal one-third of the transmembrane protein is located at the interface of the exterior and transmembrane envelope proteins. Although the sequence is highly conserved among HIV-1 isolates (Gallagher, 1987), it is unlikely to be exposed to the surface in the assembled envelope complex. The region will be exposed once the exterior glycoprotein is shed. However, in the assembled functional protein this sequence should be sequestered.

The third immunodominant region is hypervariable in sequence (Palker *et al*, 1987). Mutants in this region, located just N-terminal to the CD4-binding site, are reported to permit resistance to antibodies raised to peptides and correspond in sequence to a particular strain of HIV-1 (Looney *et al.*, 1988).

The absence of high-titer antibodies to the important functional domains, the CD4-binding region and the fusion region, is a possible consequence of their being sequestered by tertiary folding of the envelope glycoprotein.

Patient antisera do not recognize extensive regions of amino acid sequence of the exterior glycoprotein. These sequences are heavily covered with complex sugars. The density of sugar residues is predicted to be so great that a contiguous "sugar dome" is expected to be formed over the surface of the protein. The complex sugars are added to the envelope protein by cellular enzyme systems and for that reason, are not antigenic.

In summary, it appears that the envelope protein is constructed to conceal the essential functional domains either by complex sugar addition or by tertiary confirmation.

The ability of the virus to lie dormant as a provirus without expressing viral RNA or protein is another mechanism of evasion of the immune response. In this state, the viral genetic information is invisible to the immune system. Additionally, it is reported that in macrophages HIV-1 virus particles bud to the interior vesicles and are not present on the surface of the cell (Orenstein et al., 1988). In such cases, the infected cell cannot be recognized by the immune system.

These considerations do not mean that it will be impossible to create a vaccine to prevent HIV-1 infection. However, these features of the virus do help to explain the failure of current vaccine attempts, and indicate that the effective vaccines must produce antibody responses different from those characteristic of natural infections. One path toward effective vaccines is indicated by the success of blocking HIV infections by reagents that inhibit the envelope glycoprotein–CD4 interaction. Soluble forms of the CD4 surface molecule and monoclonal antibodies that bind to certain epitopes of CD4 completely prevent HIV-1 infection by inhibiting binding of the envelope protein to CD4 (Klatzmann et al., 1984; McDougal et al., 1986; Sattentau et al., 1986; Smith et al., 1987; Deen et al., 1988; Hussey et al., 1988; Traunecker et al., 1988). The sequences that specify the binding are conserved among HIV-1 strains (Kowalski et al., 1987; Lasky et al., 1987). Antibody responses in humans that produce similar reactivities should prevent HIV-1 infections. However, such antibody responses may interfere with normal immune function, and must be explored with care.

V. Establishment of Infection

After the virus has gained entry into the cell, the RNA genome is converted into a DNA form. The steps of this process, sometimes called reverse transcription, are outlined in Fig. 5. Initiation of DNA synthesis begins as the viral polymerase adds nucleotides to the 3' end of a tRNA lysine that is hydrogen-bonded near the 5' end of the genome. An initial short DNA/RNA hybrid is formed as a product of the initial elongation reaction. The 3' portion of the RNA of the initial DNA/RNA hybrid is degraded via the action of ribonuclease H. Such degradation reveals the unpaired 3' end of the newly synthesized DNA. This DNA is complementary to the 3' end of the viral genome, as genomic RNA of HIV-1 and other retroviruses contains a terminal redundancy. The terminal redundancy of HIV-1 is 97 nucleotides long

---

Figure 5. Schematic diagram of reverse transcription. The conversion of genomic RNA to DNA begins when the viral polymerase initiates DNA replication on a DNA primer annealed near the 5' end of the viral RNA. A short initial segment of DNA is made. The viral ribonuclease H degrades the viral RNA opposite the newly formed short DNA segment revealing a "sticky end." The "sticky end" of the viral DNA anneals to the 3' end of the second viral RNA

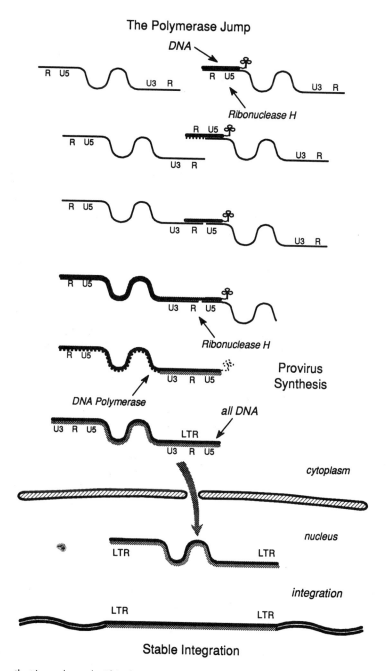

The Polymerase Jump

DNA

Ribonuclease H

Ribonuclease H

Provirus
Synthesis

DNA Polymerase

all DNA

cytoplasm

nucleus

integration

Stable Integration

genome that is packaged within the virion particle. This second RNA is copied into DNA. The viral RNA opposite the newly formed full-length single-stranded DNA is degraded by the viral ribonuclease. A portion of the viral RNA near the original 3′ end of the viral genome is used to initiate second strand RNA synthesis. A double-stranded DNA molecule results. The full-length proviral DNA contains a terminal redundancy. The formation of the linear pro-viral DNA occurs entirely in the cytoplasm. The proviral DNA migrates to the nucleus where integration may occur.

(Sanchez-Pescador *et al.*, 1985; Starcich *et al.*, 1985; Wain-Hobson *et al.*, 1985). The terminal redundant sequence does not include the 5' pol (A) which is located at the 3' end of the genome.

Elongation continues along the second genomic copy included in the viral particle until a complete complementary strand of the RNA genome is made. The details of the initiation of the second DNA strand for HIV-1 are not known with precision. Second strand synthesis begins near the 5' end of the initial cDNA copy, at a unique sequence rich in purines. This sequence is recognized by an endonuclease activity associated with ribonuclease H. The genomic RNA fragment produced by this reaction serves as a primer for second strand synthesis. Second strand synthesis proceeds through the tRNA primer binding site. The RNA complementary to this newly synthesized DNA is then degraded and a second jumping reaction occurs whereby formation of a long terminally redundant DNA molecule occurs. The RNA is then degraded as the polymerase continues to make a complete double-stranded copy of the viral nucleic acids. Study of DNA synthesis of the visna virus and animal lentivirus indicates that the second strand synthesis may initiate in more than one site (Blum *et al.* 1985). DNA synthesis occurs in the cytoplasm of the infected cell shortly after infection. The DNA form of viral genetic information is called the provirus. Proviral DNA can be detected within the cytoplasm of infected cells within several hours after virus binding (Kim *et al.*, 1989; Farnet and Haseltine, 1990).

The proviral DNA is present as a nucleoprotein complex in HIV (Farnet and Haseltine, unpublished observations). The HIV-1 proviral DNA migrates as a high-molecular-weight complex in sucrose gradients. In MoMuLV this complex includes the major capsid proteins (Bowerman *et al.*, 1989). Studies of other proviruses indicate that the major capsid protein remains associated with the proviral DNA throughout the process of DNA synthesis.

The newly formed linear molecules contain terminally redundant sequences which include, in addition to the redundant sequences at the end of the genome termed R, the genomic sequences which lie 3' to the site of the second strand initiation. These sequences are called U3. The entire terminal redundancy of HIV-1 is 634 base pairs long and is called the long terminal redundancy (LTR) (Sanchez-Pescador *et al.*, 1985; Starcich *et al.*, 1985; Wain-Hobson *et al.*, 1985).

The newly synthesized proviral DNA migrates to the nucleus as a nucleoprotein complex. Within the nucleus, circularization may occur either via blunt end joining of the full-length linear proviruses to yield a circular molecule with two adjacent LTR sequence, or by reciprocal recombination within LTR yielding a circular molecule that contains a single LTR.

Viruses defective in the polymerase and ribonuclease H activities do not complete proviral synthesis (Linial and Blair 1982, 1985; S. Goff, personal communication). Virus particles that contain mutations which affect particle maturation, including protease-defective mutants as well as mutations in the protease cleavage sites which prevent either cleavage events between p17 and p24 or between p24 and p15, are also inactive. The effect of such mutations may reflect, in part, the role of these proteins in proviral synthesis.

VI. Integration

Integration is an obligatory step in replication of HIV-1 in T cells. Mutations which prevent expression of the integrase protein fail to replicate (Terwilliger, Famet, and Haseltine, unpublished observations). A single round of cell replication postinfection is required for integration of some retroviruses (Varmus *et al.*, 1977; Humphries *et al.*, 1981). It may also be required for HIV replication as well. Recent studies indicate that the precursor of HIV-1 integration is a linear form of the provirus which remains as nuclear protein complex (Farnet and Haseltine, unpublished observations). The HIV-1 proviral DNA isolated from infected cells will integrate into heterologous double-stranded substrate DNA in an *in vitro* reaction (Farnet and Haseltine, unpublished observations). Details of integration for HIV are not known.

For murine and avian retroviruses, integration occurs in the events described in Fig. 6. Efficient retroviral integration requires specific *cis*- and *trans*-acting viral elements. The termini of all retroviruses contain inverted repeats of varying lengths that are highly conserved within retroviral species. Deletions that remove the terminal repeats greatly reduce or abolish

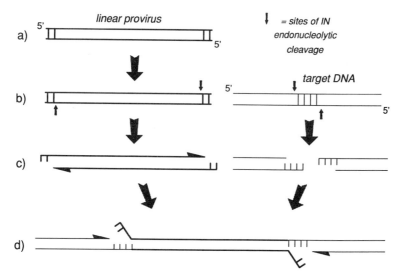

Figure 6. Schematic diagram of the intermediates in the integation reaction of retroviruses. A full-length linear provirus is made which contains staggered, recessed termini. The host DNA is cleaved. The recessed ends of the proviral DNA are covalently joined to the host DNA. The gaps and overlapping DNA are repaired. It is postulated that the viral integrase creates the staggered termini of the proviral DNA, cleaves the target DNA, and effects the covalent linkage of these two molecules. The reaction as depicted is derived from studies of murine and avian retroviruses, not from studies of HIV-1 itself.

integration and viral replication (Panganiban and Temin, 1983; Colicelli and Goff, 1985; Cobrinik *et al.*, 1987). One viral protein, the integrase protein, has also been shown to be essential for the establishment of proviral DNA. Viruses mutant in integrase synthesize all forms of unintegrated viral DNA (linear, circles containing one LTR, and circles containing two LTRs) at wild-type levels, but fail to integrate into the cellular genome (Donehower and Varmus, 1984; Schwartzberg *et al.*, 1984; Cobrinik *et al.*, 1987; Quinn and Grandgenett, 1988).

Recent experiments indicate that the linear viral DNA is the immediate precursor to the integrated provirus (Brown *et al.*, 1989; Fujiwara and Mizuuchi, 1988). The integrated provirus is precisely colinear with the unintegrated linear DNA molecule, except for the deletion of the terminal two base pairs from each end of the unintegrated precursor. Unintegrated linear viral DNA recovered from intracellular virion core particles consists of two types of molecules: blunt-ended duplex DNA molecules and molecules with termini having 3' strands recessed by two nucleotides (Brown *et al.*, 1989; Fujiwara and Mizuuchi, 1988). Time-course analysis of the terminal structure of the linear viral DNA of wild-type and integrase-defective viruses indicates that the blunt-ended molecules are precursors to the molecules with recessed ends, and that the integrase function is required for the generation of the recessed ends (Roth *et al.*, 1989). Integration can be envisioned to occur via the joining of the integrase-generated recessed 3' hydroxyl termini of the linear viral DNA to the target DNA, resulting in an integrated provirus that lacks the terminal two base pairs of the unintegrated molecule. Such a model is supported by in vitro integration reactions, where intermediates can be purified in which the recessed 3' ends of the LTR sequences are covalently joined to the target DNA while the 5' ends remain unjoined (Fujiwara and Mizuuchi, 1988). Like proviral structures formed *in vivo* linear viral DNA integrated *in vivo* is also bordered by short direct repeats of host DNA sequences originally present in single copy at the site of integration. The length of the direct repeat is retrovirus species-specific, and is presumed to result from staggered cleavage by the viral integrase protein at the integration target site.

Farnet and Haseltine have isolated a nucleoprotein complex from HIV-1-infected cells which integrates into heterologous target DNA (Farnet and Haseltine, 1990). The protein–DNA complex comigrates with ribosomes on sucrose gradients and contains linear duplex HIV-1 DNA. Cell-free integration proceeds to completion within 30 min. Exogenous energy sources such as ATP are not required for the integration reaction.

VII. Regulation of HIV Replication

Replication of HIV-1 is highly regulated, both *in vivo* and *in vitro*. The virus may establish three types of infection in infected people, active, controlled and silent (Fig. 7). During the active infection abundant viruses can be

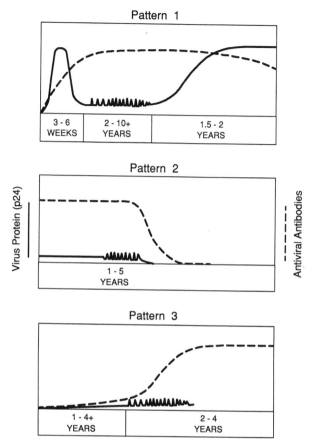

Figure 7. Schematic diagram of three patterns of HIV-1 infection. The concentration of plasma virus is depicted by the solid line. The concentration of anti-HIV-1 antibodies is depicted by the broken line. Pattern 1 infections are typical of most HIV-1-infected people. High levels of virus production follow initial infection, which in turn is followed by a prolonged period of low-level virus production. The concentration of plasma viruses rises progressively in the last several years of infection. Chronic high levels of anti-HIV-1 antibodies exist throughout most of the infection. Pattern 2 is rare. There are several reports that some people may revert to an HIV-1 antibody-negative status after being HIV-1 antibody-positive. The loss of virus production in such cases depicted here is hypothetical. Pattern 3 infections are currently under investigation. HIV-1 proviral DNA has been detected in some people who do not have detectable anti-HIV-1 antibodies. What portion of people exposed to HIV-1 exists in proviral positive HIV-1 antibody-negative state is a hotly debated topic. Some HIV-1 provirus-positive, HIV-1 antibody-negative people are later reported to produce HIV-1 antibodies.

detected in most body fluids. Prolonged active replication is correlated with progressive degenerative disease of the immune, central nervous, and other organ systems. Only very small amounts of virus are found during a state of controlled infection but very high levels of antibody to a broad array of viral proteins are present, indicative of low-level chronic virus production. No virus particles or anti-viral antibodies can be detected in silent infections but viral DNA can be detected using gene-amplification techniques. Live virus can be isolated from some people who harbor silent infections.

The complexity of virus replication in infected people is paralleled by the observation that multiple replication states exist in infected cells in culture. Virus information may exist within cells in a fully latent form in which neither viral RNA nor viral proteins are made. Similarly CD4$^+$ T cells and monocytes isolated from peripheral blood and other organs harbor viral DNA with either little or no RNA or protein expression. In other cell types, low levels of virus are produced continually. Under some circumstances virus replication can be prolific particularly if infected T cells are mitogenically stimulated. Under these cases replication of HIV-1 is markedly cytopathic in T cells provided that they express a high concentration of surface CD4. Cell lines capable of chronic high level HIV-1 production can be derived from the survivors of infection of CD4$^+$ T cell lines (Folks *et al.*, 1989). The virus produced by these cell lines is itself cytopathic for fresh CD4$^+$ T cell cultures. The concentration of CD4 is dramatically down-regulated in cells that abundantly produce high levels of HIV-1.

A complex and unprecedented set of *cis* and *trans* regulatory elements covers HIV replication. These regulatory elements serve to interact virus expression for the state of differentiation and activation of the infected cells. Some of the regulatory factors are specified by the infected cells and many others are encoded by the virus.

VIII. The Long Terminal Repeat (LTR)

The long terminal repeat (LTR) flanks the proviral DNA and contains signals that govern transcriptional initiation and termination. The LTR contains numerous sites that have been recognized as binding factors that facilitate RNA initiation including TATAA sequence, CAT sequence, and sequences which bind SP-1 (Corden *et al.*, 1980; Garcia *et al.*, 1980; Jones *et al.*, 1986) (Fig. 8). The LTR contains two sequences which recognize proteins that bind to DNA when T cells are activated, the NFκB and NFAT-1 sites (Shaw *et al.*, 1988; Nabel and Baltimore, 1987). These two sequences play an important role in the virus life cycle. In resting T cells very little, if any, RNA is made. When T cells are activated by specific or nonspecific stimuli, viral RNA is made and virus replication begins. The ability of the virus to lie dormant in circulating T cells can be attributed, at least in part, to the dependence on NFκB and NFAT-1 for viral transcription. Viruses defective

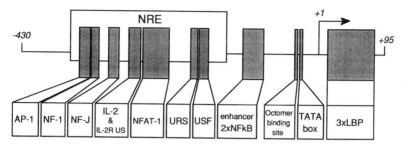

Figure 8. Schematic diagram of the HIV LTR. The diagram depicts some of the sequences which are homologous with other nucleic acid sequences which are known to bind to cellular proteins which regulate the rate of transcription initiation.

in NFAT-1 binding site replicate well. They remain responsive to T-cell mitogenic stimuli. Viruses defective in the NFκB site replicate poorly (Lu *et al.*, 1989). This sequence serves as the major transcription enhancer in T-cell lines. Despite the role of NFκB, the viruses defective in this site still replicate slowly and increase RNA synthesis in response to mitogenic stimuli (Lu and Haseltine, unpublished observations). Deletion of NFAT-1 and NFκB eliminates the ability of the viral LTR to respond to mitogenic stimuli (Lu and Haseltine, unpublished observations).

The LTR also contains sequences which suppress initiation called the negative regulatory element (NRE) (Rosen *et al.*, 1985; Skevila *et al.*, 1987). Deletion of this sequence yields virus which produces five times as much virus in T-cell lines and 30 times as much virus in monocyte cell lines (Lu *et al.* 1989). The NRE also contains sequences that recognize cellular proteins. The existence of the *cis*-acting negative sequence may help to explain the very low rate of virus replication, particularly in monocytes and macrophages. A second enhancer has recently been reported to lie between the end of the *pol* and the beginning of the *tat* genes (A. Burny, personal communication). This enhancer is also reported to be responsive to T-cell mitogenic stimuli.

IX. Positive Regulation

A. *Trans*-activator (*tat*)

There are two types of regulatory genes that accelerate virus replication. The *trans*-activator gene (*tat*) is a positive feedback regulator that increases the rate of its own synthesis and the synthesis of all viral proteins (Sodroski *et al.*, 1985b) (Fig. 9). It is an essential gene as *tat⁻* mutants do not replicate (Dayton *et al.*, 1986; Fisher *et al.*, 1986a,b).

There are two components to most genetic regulatory pathways, a diffusible effector molecule—often a protein—and a *cis*-acting responsive ele-

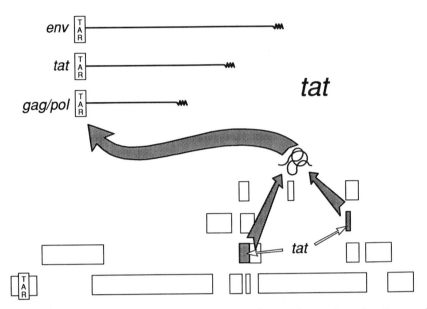

Figure 9. *Trans*-activation of viral gene expression by *tat*. The product of *tat* is a small protein which interacts either directly or indirectly with the TAR element, a stem-loop structure at the 5' end of all viral RNAs.

ment—often the nucleic acid sequence. In the case of the *tat* regulatory pathway, the effector is an 86-amino acid long protein (Arya *et al.*, 1985; Sodroski *et al*, 1985a). The proteins are made from a family of mRNAs in which HIV-1 leader sequences are joined by splicing to one or more internal exons. An 86-amino acid Tat protein is made by splicing an exon in the central region of the genome to a 3' splice acceptor. A smaller 72-amino acid Tat protein is made when the splice donor at the 3' end of the first *tat* coding exon is not used. The shorter *tat* product is active. Multiple noncoding exons may be present 5' to the first coding exon. These are derived from RNA sequences 5' to the first coding exon of *tat*.

The protein is located the nucleus and nucleolus of infected cells (Hauber *et al.*, 1987). The responsive element is a short segment of nucleic acid called the *trans*-acting responsive region (TAR) located between the +1 and +45 of the genome, +1 being the initial nucleotide of the viral RNA (Rosen *et al.*, 1985; Sodroski *et al.*, 1985a; Peterlin *et al.*, 1986; Hauber and Cullen, 1988). The TAR sequence can assume a stable double-stem loop configuration (Muesing *et al.*, 1987; Patarca and Haseltine, 1987). The TAR sequences serve to suppress viral RNA expression. It has been suggested that TAR sequences prevent RNA exit from the nucleus except in the presence of Tat protein (Haseltine, 1988). The double-stranded structure at the beginning of the 5' end of the RNA which serves as a mRNA also slows down the initiation process of protein synthesis in the absence of Tat (Parkin *et al.*, 1988). It has been reported that double-stranded TAR RNA can also induce interferons which act as inhibitors of HIV-1 replication (Edery *et al.*, 1989).

Current evidence indicates that the critical sequences of TAR are located at the tip of the first loop (Feng and Holland, 1989). Current evidence also indicates that the TAR sequence prevents efficient use of mRNA in the absence of the Tat protein. In the presence of the Tat protein, such inhibition is abrogated and RNA initiated by TAR sequence directs protein synthesis with extraordinary efficiency (Cullen, 1986. Rosen *et al.*, 1986. Wright *et al.*, 1986). There are several reports that Tat proteins bind directly to TAR RNA (Dingwall *et al.*, 1989).

The *tat* product may, under appropriate conditions, accelerate the production of viral proteins by several thousand fold. The magnitude of this effect reflects both low activity of TAR initiated in the absence of Tat and the efficient use of TAR RNA in the presence of Tat. The synthesis of all viral proteins, both those that comprise the virion and the regulatory proteins themselves, is affected by *tat* because all viral mRNAs contain the TAR sequence at the 5' end (Rosen *et al.*, 1985). In the absence of *tat* fragments of the 5' viral RNA, nucleotides 1–60 accumulate. Three possible explanations for this observation have been advanced: 1. Tat overcomes transcription attenuation; 2. Tat prevents nuclease degradation of viral RNA; 3. Tat permits rapid exit of viral RNA from the nucleus wherein nucleases degrade the RNA to a 60-nucleotide RNA-resistant fragment. Additional experiments are required to distinguish amongst these possibilities.

X. An Unusual Genetic Switch (*rev*)

The regulator of virion protein expression, *rev*, is a gene that positively regulates expression of virion proteins, but negatively regulates the expression of regulatory genes (Sodroski *et al.* 1986a; Feinberg *et al.*, 1986; Terwilliger *et al.*, 1988) (Fig. 10). The Rev protein is specified by two coding exons that overlap the *tat* coding exons. The initiation codons of *rev* lies 3' to that of *tat*. There are reported to be at least two splice acceptor sites between the *tat* and *rev* initiation codons that are probably used to make mRNAs to be used for the expression of *rev*. The same splice acceptor sites are used for both the *tat* and *rev* second coding exons.

There are two *cis*-acting sequences that are part of the *rev* regulatory pathways. One of these is a *cis*-acting repression sequence (CRS) (Rosen *et al.*, 1986). Multiple CRS elements exist in the viral coding sequences. There is at least one CRS element in the capsid coding sequences, one in the sequences that encode the replicative genes, and at least two or three in the

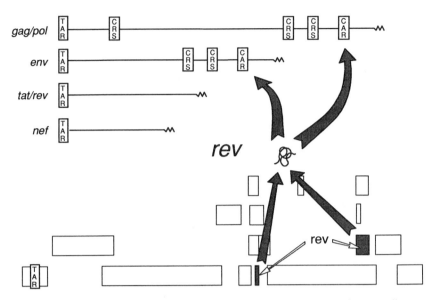

Figure 10. Schematic diagram of the action of *rev*. The *rev* gene encodes a small protein in *trans* which regulates the accumulation of unspliced or partially spliced messenger RNAs. The *cis*-acting sequence responsive to *rev* is called CAR or RRE. The Rev protein binds to CAR. The structural genes of HIV-1 contain elements which prevent the export of these introns from the nucleus unless *rev* is present. These sequences are called the CRS.

envelope glycoprotein. The genes that encode the regulatory proteins are devoid of CRS. When present in the RNA, the CRS prevent the message from being used as a substrate for protein synthesis. It is suspected that these sequences specify retention of the viral RNA in a nuclear compartment that contains splicing and powerful degradative activities as full-length RNA is not exported in the absence of *rev.*.

A sequence exists in the envelope gene that is required for Rev protein-dependent relief of CRS expression, the *cis*-acting antirepression sequences (CAR) also called *Rev* responsive element (RRE) (Rosen *et al.*, 1988; Malin *et al.*, 1988; Emerman *et al.*, 1989). The CAR element is responsive to the Rev protein. The interaction between the *rev* product and the CAR sequence overrides the repressive effects of CRS and permits use of mRNAs for protein synthesis. There is evidence that the interaction between the CAR and *rev* product permits transport of the RNA that contains CRS from the subnuclear compartment that contains splicing and degradative enzymes to a compartment in which splicing can no longer occur and from which the RNA may gain access to the translational machinery. RNA devoid of CAR which contains CRS is not translated and remains in the nucleus. The CAR and CRS affect expression of heterologous genes if they are located between the 3' coding exon and the site of poly(A) addition.

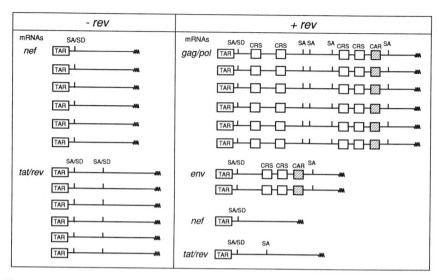

Figure 11. Schematic diagram of the phenotypic effect of *rev* mutations of the accumulation of viral RNA species. In the absence of *rev* multiply spliced RNA molecules accumulate in the cytoplasm and are translated. These highly spliced mRNA molecules specify regulatory proteins only. In the presence of Rev unspliced as well as partially spliced mRNAs accumulate and are translated. These mRNAs specify the core and envelope proteins of the virus particle.

In the absence of the Rev protein, RNA species that have had the viral structural genes removed by splicing accumulate (Feinberg *et al.*, 1986; Knight *et al.*, 1987) (Fig. 11). Removal of the virus structural genes by splicing also removes the CRS. These small RNAs are those that encode the regulatory proteins. In the presence of the Rev protein, full-length RNAs and RNA that encodes the Env proteins accumulate, and the concentration of mRNAs that specify regulatory proteins decreases.

The Rev protein binds to the RNA of the RRE (CAR) (Daly *et al.*, 1989; Dayton *et al.*, 1989; Zapp and Green, 1989; Heaphy *et al.*, 1990; Malim *et al.* 1990; Olsen *et al.*, 1990). Apparently, the Rev protein recognizes a duplex region of the RNA. Mutations of the RRE (CAR) which abolish Rev binding also abolish Rev-mediated effects on RNA.

XI. Tat–Rev–Env Fusion Protein (Tnv)

The 26-kDa protein is present in HIV-producing cells and is recognized by both anti-Tat and anti-Rev sera (Salfeld *et al.* 1989). This protein is formed by a splicing event in which the first coding exon of *tat* is joined by splicing to a segment derived from the RNA which in turn is spliced to the *rev* coding exon (Fig. 12). The resultant protein contains 71 N-terminal amino acids

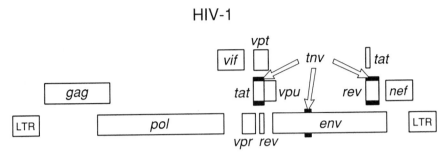

Figure 12. Schematic representation of the splicing events which form Tat, Tnv, and Rev proteins. The Tnv protein has an N-terminus derived from the first exon of *tat*, a middle region derived from an exon within the envelope gene which contains Env-derived amino acids, and a C-terminus identical to the C-terminus of the Rev protein.

derived from Tat, 40 amino acids derived from Env and 91 C-terminal amino acids derived from Rev. The protein has Tat but no Rev activity. The existence of this Tat–Rev–Env fusion protein called Tnv highlights the complexity of HIV-1 gene regulation.

XII. Negative Regulation

A. Negative Regulation Factor (Nef)

The negative regulation factor (Nef) is a protein that is located in the cytoplasm and retards virus replication (Allan *et al.*, 1985; Fisher *et al.*, 1986b; Terwilliger *et al.*, 1986; Luciw *et al.*, 1987). Nef is not required for replication as mutants defective in the gene replicate in T cells as well as monocytes. The 25–27-kDa protein is specified by a mRNA that has all of the sequences of the capsid, replicative, and envelope genes removed by splicing (Muesing *et al.*, 1985; Salfeld and Haseltine, unpublished observations). There are some reports that mutants defective in *nef* replicate more rapidly than *nef*[+] virus. However, this may not be the case for all cell HIV-1 strains or *nef*[+] alleles.

The Nef protein is reported to have interesting biochemical properties. It contains a fatty acid, myristic acid, at the N-terminus, suggesting that this form of the protein is embedded in the inner surface of the cell membrane (Fisher *et al.* 1987). The Nef protein is reported to bind GTP and have a GTPase activity (Guy *et al.*, 1987). The protein is also reported to be a protein kinase capable of autophosphorylation and is a substrate for protein kinase C (Guy *et al.*, 1987). The Nef protein may act as an analogue of signal-transducing proteins by modifying cellular proteins that regulate transcription inititation. It has been reported that Nef protein down-regulates LTR gene expression (Ahmad and Venkatesan, 1988; Ahmad *et al.*, 1989). LTR down-regulation by Nef may also vary amongst HIV-1 isolates and is not reported by all investigators. It has also been reported that Nef down-regulates CD4 expression (Guy *et al.*, 1987).

XIII. Virion Infectivity Factor (Vif)

HIV-1 also specifies a protein that increases the infectivity of the virus particle. Vif is not required for replication as mutants defective in the gene grow in T cells and monocytes. The gene that specifies this protein is called the virion infectivity factor gene (*vif*) (Lee *et al.*, 1986; Sodroski *et al.*, 1986c). The protein is made from a mRNA in which the *gag/pol* sequences have been removed by splicing. The 3′ sequences may also be removed by splicing in some mRNAs found in some HIV-1-infected cells. The protein is present in the cytoplasm of infected cells and can be found outside the cells as well (Sodroski *et al.*, 1986c). Viruses defective in this function have a normal appearance and bind with normal efficiency to the surface of susceptible cells. Current evidence indicates that the Vif protein increases the efficiency of the step that occurs after virus attachment but before the viral DNA is integrated into the host cell DNA (Fisher *et al.*, 1987; Strebel *et al.*, 1987). The titer of equivalent numbers of virus particles may differ by a factor of 1000 or more depending upon the presence or absence of an active *vif* product. The mechanism of Vif action is unknown.

It is likely that efficient cell-to-cell and possibly person-to-person spread of HIV-1 is partly a consequence of the increased infectivity of cell-free viruses attributable to the Vif protein.

XIV. Viral Protein U (Vpu)

Vpu encodes a protein 80–82 amino acids long (Cohen *et al.*, 1988; Matsuda *et al.*, 1988; Strebel *et al.*, 1988). The Vpu protein is not required for HIV-1 replication as mutants defective in the gene replicate in T cells and monocytes (Strebel *et al.*, 1989; Terwilliger *et al.*, 1989). The Vpu protein is specified by an open reading frame located 5′ to *env*. It is likely that Vpu and Env are made from the same polycistronic mRNA, as no splice acceptor sites have been detected in HIV-1-infected cells between the initiation codons of *vpu* and *env*.

The Vpu protein facilitates assembly and budding of viral particles. Five to ten times more virus is released from T cells infected with *vpu*+ as compared to isogenic *vpu*− virus (Cohen *et al.*, 1988; Strebel *et al.*, 1988; Terwilliger *et al.*, 1989) (Fig. 13). Evidently, cell-associated viral protein, except probably Env protein on the surface of CD4+ T cell, is required for cytopathic effect.

The combined action of the two viral genes *vif* and *vpu*, results in production of large amounts of highly infectious virus, facilitating cell-to-cell and person-to-person transmission.

-vpu   +vpu

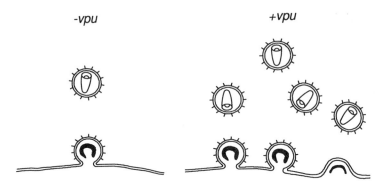

Figure 13. Schematic diagram of Vpu action. The action of the *vpu* products accelerates budding and release of viral proteins from the cell surface.

XV. Viral Protein R (Vpr)

HIV-1 specifies a second *trans*-activator, the product of *vpr*. This gene specifies a 15-kDa protein (Cohen *et al.*, 1990a). Vpr is not required for virus replication as mutants defective in the gene replicate in T cells and monocytes. Vpr is made from a multiply-spliced mRNA in which the HIV leader sequence is joined by splicing to the *vpu* coding exons. RNA 3' to *vpr* may also be removed from splicing in some *vpr* mRNAs. The Vpr protein acts in *trans* to accelerate the rate of viral protein production.

The product of *vpr* is capable of stimulating expression of homologous and heterologous genes in *trans* and may alter some cell functions (Cohen *et al.* 1990a). It is curious that, like *vpu* and *nef*, *vpr* is defective in many cultured HIV-1 isolates.

The Vpr protein is incorporated into virus particles (Cohen *et al.*, 1990b). Multiple copies of *vpr* are present within each particle. The Vpr protein produced by a separate plasmid can be incorporated into HIV-1 virion particles assembled from a provirus which is *vpr⁻* (Cohen and Haseltine, unpublished). Vpr is the only regulatory protein of HIV-1 to be incorporated within the virus particle, a finding that suggests that the protein may play a role in the establishment of infection.

XVI. Viral Protein T (Vpt)

Most HIV-1 strains contain an open reading frame which overlaps the first coding exons of *tat* and *rev* as well as the 5' part of the U open reading frame. When RNA derived from this region of the genome is used to program a reticulocyte *in vitro* protein synthesis extract, three proteins are made. Initiation codons yield protein products 15 and 17 kDa amino acids long. Initially an additional protein recognized by antisera raised to the N-terminus of Tat are also made. This protein contains N-terminal sequences

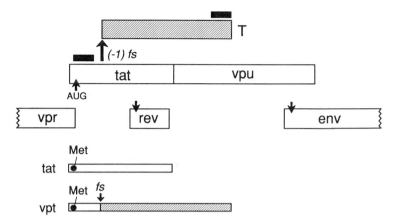

Figure 14. Schematic diagram of *vpt*. The *vpt* protein is initiated with the *tat* AUG. A minus one frame shift occurs with a frequency of about one in twenty at a defined site within the *tat* gene. After the frame shift event, the protein that is initiated with the *tat* AUG continues in a different open reading frame called T.

of Tat and C-terminal amino acids specified by the T open reading frame (Fig. 14). This fusion protein is called Vpt. The fusion event occurs via minus one frame shift at amino acid 29 of Tat. The frame shift occurs about 5% of the time that the Tat protein is translated. The fusion sequence at which the minus one frame shift occurs is similar to that of other minus one frame shifting events used by other retroviruses. The function of Vpt is unknown.

XVII. A Regulatory Network

The genes of HIV that affect the rate of viral replication also regulate one another. A diagram of some of the interactions among these regulatory proteins is depicted in Fig. 15.

Consider the interactions of separate components of this regulatory network. The *tat* product positively regulates *rev* and itself. The *rev* product negatively regulates *tat* and itself as it suppresses the accumulation of the spliced RNAs from which Tat and Rev are made. The result should be a steady-state level of both proteins. Similar positive regulation of *nef* by *tat* and negative regulation of *tat* by *nef* should result in steady-state levels of both proteins. Such regulation should permit controlled replication of the virus. The actual level of viral replication will be dependent on the intracellular conditions that determine the relative levels of each protein and the relative frequency of specific splicing events.

The consequences of interactions of the Rev and Nef pathways are different. The *rev* product suppresses the production of the *nef* gene and itself by preventing accumulation of spliced forms of RNA from which both

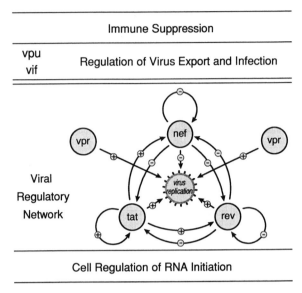

Figure 15. A schematic representation of potential interactions among factors which control HIV-1 replication. Cellular factors contribute to regulation of RNA transcription initiation. The viral regulatory genes *tat, rev, nef,* and *vpr* regulate virus replication and also regulate one another. Other viral genes, *vpu* and *vif,* regulate cellular export and infectivity of the virus particle respectively. The host immune system suppresses virus replication.

proteins are made. The *nef* product suppresses the synthesis of all viral RNAs by decreasing the rate of RNA initiation. The *rev–nef* interaction may provide an all-or-none switch for virus replication. If Nef protein accumulates first, then virus replication may be suppressed for prolonged periods. If Rev accumulates first, abundant or controlled replication may occur. Changes in intracellular conditions that affect relative *nef* and *rev* activities may either induce productive replication or permit productively infected cells to tum off replication and enter a latent state.

The *cis-* and *trans*-acting regulatory elements of RNA are sufficient to account for the complexity of virus–virus and virus–cell interactions that have been observed. Silent infections may be interpreted as establishment of latent infections via the action of the negative regulatory genes and infection of resting cell populations. Controlled replication can be interpreted in terms of steady-state levels of positive and negative regulatory proteins that, in turn, lead to controlled levels of viral synthesis. Prolific replication can be interpreted as either being due to perturbations of the regulatory network or to an increase in infections of populations of cells that can lead to high level productive infection.

The regulatory network may also help to account for the switch from silent to controlled infections and from controlled replication to silent infection.

XVIII. A Model for Progressive Disease

Progressive disease appears to be a consequence of the gradual rise in virus replication rate after a prolonged period of controlled replication. A dynamic model based on known properties of HIV-1 may account for the slow but inexorable switch from controlled to prolific replication (Fig. 16). One cell type in which prolific replication is known to occur is the activated $CD4^+$ T lymphocyte. Infection of such cells results in rapid production of prodigious quantities of virus (Barre-Sinoussi et al., 1983; Klatzmann et al., 1984; Popovic et al., 1984). However, infection of resting $CD4^+$ lymphocytes results in latent infection. At any time, the fraction of $CD4^+$ lymphocytes actually replicating in the blood is very small, approximately one in ten thousand (Harper et al., 1986). Consequently, most infections of this population result in latent infection. Over time, the low level of virus produced by infected monocytes and other cell types should eventually increase the fraction of $CD4^+$ cells that are latently infected. Over a prolonged period, the probability that a latently infected T cell will be activated at random will increase. Activation of the latently infected T cells should result in a large burst of virus release. This process should be autocatalytic. The result should eventually be a switch from a state of controlled replication to a state of continued high-level virus production.

XIX. Genomic Variability

Independent HIV isolates vary from one another by 20–25% of the nucleic acid sequence. Similar variation is noted in predicted amino acid sequences. The changes from one HIV strain to another occur by sequential mutation, changes in single bases, as well as small deletions and duplications. Such mutations are primarily introduced during replication. Replication of all retroviruses including HIV involves three separate conversions of nucleic acid. The virus polymerase converts the genomic viral RNA to DNA and the resultant RNA/DNA hybrid to a double-stranded DNA form. The nascent viral genome is made by copying the viral DNA by the cellular RNA polymerase. All three conversion steps introduce errors. The viral polymerase and cellular RNA polymerases do not have the ability to remove incorrectly base-paired nucleotides once incorporated. Moreover, viral DNA is made in the cytoplasm whereas most of the cellular enzymes that recognize mismatched nucleotides, small deletions, or duplications are present in the nucleus. No enzymatic means is known for correction of RNA that is incorrectly copied from DNA.

The error frequency during replication may be expressed as:

$$E = 2 \times [\text{error frequency of viral polymerase}] + [\text{error frequency of RNA polymerase}] + [\text{error frequency of cellular RNA replication}] \times N$$

The last term of this equation reflects the error frequency of cellular DNA replication, where N represents the number of divisions of a cell that

CD4+ T Cells

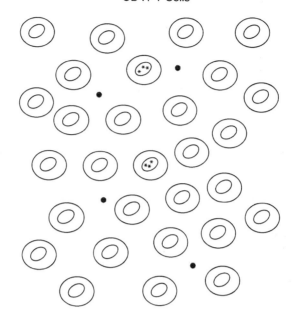

Progressive Infection

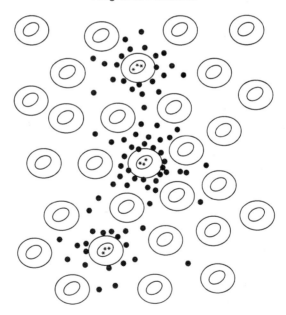

Figure 16. A model for progressive disease. (A) Early in infection very few CD4+ T cells harbor latent HIV-1 proviral DNA. Late in infection many CD4+ cells harbor HIV-1 proviral DNA. Early in infection there is a low probability that random activation of the T-cell population will trigger active replication in a latently infected CD4 T cell. (B) Late in infection random activation of the CD4+ T-cell population has a much greater probability of activating a latent provirus. Accelerated infection in CD4+ T cell deaths may result.

harbors a latent provirus. Under most conditions, this term will be very small because cellular DNA replication has an error frequency on the order of one mistake per billion nucleotides copied.

The error frequency for a single round of replication from a DNA to DNA form of an avian leukosis virus has been determined to be approximately one point mutation per 10,000 nucleotides and one deletion or duplication per 10,000 nucleotides. This mutation rate means that on average there is one point mutation and one deletion or duplication introduced per viral genome per round of replication. Variation of retroviruses is an inevitable consequence of replication—the more rounds of replication the more mutation (Meyerhans *et al.* 1989). The rapid rate of variation in HIV isolates therefore represents numerous rounds of replication that intervene between isolates.

It is likely that most mutants will result in nonviable virus. Attempts to clone infectious provirus usually result in a >10 to 1 yield of full-length defective to nondefective proviruses. The defective viruses are likely due to mutations that arise in regions of the virus genome essential for function.

Changes that occur in nonessential regions may be tolerated. Examination of variation within individual genes reveals that some regions of the genes are highly variable and others are not. It is clear that the regions of conservation of protein sequences represent functional domains of these proteins. For example, the envelope glycoprotein is, overall, one of the most variable proteins of the virus. However, subregions of this protein include the most conserved in the entire virus and others vary only slightly from strain to strain. Mutational and functional analysis of the envelope glycoprotein shows that the conserved regions encode functions such as CD4 recognition, membrane anchorage, and membrane-to-membrane fusion (Kowalski *et al.*, 1987; Lasky *et al.*, 1987; Jameson *et al.*, 1988).

Viewed from this perspective, natural variation of virus isolates appears to be adventitious. This view accords with the observation that despite marked sequence variation of HIV isolates, the transmission, pathogenesis and incubation period of the disease is similar regardless of the infecting strain.

Growth rate variants may be experimentally selected. Most HIV strains now studied are selected to grow in culture. Viruses that do not grow in culture are difficult to study. It is significant that approximately half of the HIV-1 strains sequenced today are obviously defective in one or both of the negative regulatory genes, *vpu* and *nef*. In addition to mutations that introduce termination codons or that remove initiation codons, it is possible that more subtle damage is present in the regulatory genes of viruses that have been selected to grow in culture. For example, mismatch mutations may be able to produce a full-length protein with reduced activity. It is also conceivable that mutations that accelerate virus growth in culture may occur in the positive as well as negative regulatory pathways. It is likely that the large pool of naturally arising variants that can be found in infected people, particularly during late stages of disease, provides a rich source of variants to be selected by the experimenter.

The rapid rate of variation also raises the possibility that variants of HIV-1 resistant to antiviral drugs may arise in treated patients. The reports of AZT-resistant strains in patients undergoing treatment may reflect selection of such variants (Larder *et al.*, 1989).

XX. Related Viruses

Genomic organization of other primate retroviruses related to HIV has been described (Meyers *et al.*, 1989). The general organization of the viral genomes is similar to that of HIV-1. All of these viruses contain, in addition to the *gag, pol,* and *env* virus genes, open reading frames that share substantial similarity with the *vif, tat, rev* and *nef* genes. *Tat* and *rev* functions have been described for several HIV-2 and SIV-1 strains.

Significant differences in genomic organization exist amongst the viruses. The organization of the *cis*-acting transcription sequences within the viral LTR differs substantially amongst these viruses. Similarly HIV-2 and SIV may contain at least two TAR-like sequences (Arya and Gallo, 1988; Feng and Holland, 1989). Two sets of mRNAs are made in HIV-2 and SIV-infected cells, one of which contains both TAR-like sequences and a second set in which 3' TAR-like sequences have been removed from the mRNAs by splicing (Colombini *et al.*, 1989; Salfeld and Haseltine, unpublished observations).

Several of the HIV-1-related primate viruses lack the *vpr* open reading frame and none contain the *vpu*, which is unique to HIV-1. Instead, in SIV and HIV-2, the *tat-* and *rev*-like open reading frames are juxtaposed to *env*. Some African green monkey SIV isolates do not specify *vpr*, whereas other SIV isolates from macaques and mangabeys do contain this open reading frame (Meyers *et al.*, 1989).

SIV and HIV-2 contain an open reading frame, an undesignated *vpx* not found in HIV-1, beginning immediately upstream of *vpr*. It is reported that *vpx* is not an essential gene for replication of SIV and HIV-2 in established T-lymphocytic lines, whereas mutations in *vpx* severely impair virus production in human peripheral blood lymphocytes (Guyader *et al.*, 1989; Yu *et al.*, 1988). *nef* of HIV-2 and SIV differs slightly from that of HIV-1 as it overlaps the *env* open reading frame substantially, whereas *nef* of HIV-1 begins after the termination of the *env* gene. An open reading frame similar to *vpt* is not found in HIV-2 or SIV strains.

The significance of the differences in genomic organization of HIV-1 and HIV-2 and SIV is not fully understood.

XXI. Summary

Some of the major features of the acquired immune deficiency syndrome may be understood in terms of the characteristics of the virus. Life-long infection is a consequence of the life cycle of retroviruses, the formation of stably integrated viral genetic information into host-cell DNA. The silent infection, controlled replication, and prolific replication may be understood in terms of the interactions of the positive and negative regulatory genes that control virus growth. Selective infectivity and selective cytotoxicity of HIV-1 are primarily the consequences of the properties of the envelope glycoprotein and its interactions with the surface CD4 molecule. The ability of HIV-1 to enter a state of prolific replication in the presence of an antiviral immune system and other regions appear to be covered by a dense cloud of sugar molecules. Concealment of the virus by regulated growth, by budding to the interior surfaces of macrophages as well as concealment by a sugar coating, may help to explain the failure to protect chimpanzees from infection by candidate vaccines.

References

Ahmad, N., and Venkatesan, S. (1988). *Science* **241,** 1481–1485.

Ahmad, N., Maltra, R. K. and Venkatesan, S. (1989). *Proc. Natl. Acad. Sci. U.S.A.* **86,** 6111–6115.

Allan, J. S., Coligan, J. E., Barin, P., McLane, M. P., Sodroski, J. G., Rosen, C. A., Haseltine, W. A., Lee, T. H., and Essex, M. (1985a). *Science* **228,** 1091–1094.

Allan, J. A., Coligan, J. E., Lee, T. H., McLane, M. P., Kanki, P. S., Groopman, J. E., and Essex, M. (1985b). *Science* **230,** 810–815.

Arya, S. K., and Gallo, R. C. (1988). *Proc. Natl. Acad.Sci. U.S.A.* **85,** 9753.

Arya, S. K., Guo, C., Joseph, S. F., and Wong-Staal, F. (1985). *Nature (London)* **340,** 367–371.

Barin, F., McLane, M. F., Allan, J. S., Lee, T. H., Groopman, J., and Essex, M. (1985). *Science* **228,** 1094–1096.

Barre-Sinoussi, F., Chermann, J. C., Rey, F., Tsai, C.-C., Ochs, H. D., Ward, J. M., Kuller, L., Knott, W., Hill, R.W., Gale, M. I., and Thouless, M.E. (1983) *Science* **220,** 868–871.

Bedinger, P., Moriarty, A., Von Borstel, Donovan, N. J., Steimer, K. S., and Littman, D. R. (1988). *Nature (London)* **334,** 162–165.

Benveniste, R. E., Morton, W. R., Clark, E. A., Fulevin-Wasserman, G., Arthos, J., Rosenberg, J., Maddon, P. J., Axel, R., and Sweet, R. W. (1988). *J. Virol.* **62,** 2091–2101.

Blum, H. E., Harris, J. D., Ventura, P., Walker, D., Staskus, K., Retzel, E., and Haase, A. T. (1985). *Virology* **142,** 270–277.

Bowerman, B., Brown, P. O., Bishop, J. M., and Varmus, H. E. (1987). *Genes & Development* **3,** 469–478.

Brown, P. O., Bowerman, B., Varmus, H. E., and Bishop, J. M. (1989). *Proc. Natl. Acad. Sci. U.S.A.* **86,** 2525–2529.

Byrnes, R. K., Chan, W. C., Spira, T. J., Ewing, E. P., and Chandler F. W. (1983). *JAMA* **250,** 1313–1317.

Cobrinik D., Katz R., Terry, R., Skalka A. M., and Leis J. (1987). *J. Virol.* **61,** 1999–2008.

Cohen, E. A., Terwilliger, E. F., Sodroski, J. G., and Haseltine, W. A. (1988). *Nature* (London) **334,** 532-534.

Cohen, E. A., Dehni, G., Sodroski, J. G., and Haseltine, W. A. (1990b). *J. Virol.* (in press).

Cohen E. A., Terwilliger, E. F. Sodroski, J. G., and Haseltine, W. A. (1988). *Nature (London)* **334,** 532–534.

Cohen, E. A., Terwilliger, E. F., Jalinoos, Y., Proulx, J., Sodroski, J. G., and Haseltine, W. A. (1990a) *J. AIDS* (in press).

Colicelli, J., and Goff, S. P. (1985). *Cell* **42,** 573–580.

Columbini, S., Arya, S. K., Restz, M. S., Jagodzinski, L., Leaver, B., and Wong-Staal, F. (1989). *Proc Natl. Acad. Sci* **86,** 4813

Corden, J., Wasylk, B., Buschwalder, A., Sassone-Corsi, P., Kedinger, C., and Chambon, P. (1980). *Science* **209,** 1406–1414.

Cullen, B. (1986). *Cell* **46,** 973.

Daly, T. J., Cook, K. S., Gray, G. S., Maione, T. E., and Rusche, J. R. (1989). *Nature (London)* **342,** 816–819.

Dayton, A. I., Sodroski, J. G., Rosen, C. A., Goh, W. C., and Haseltine, W. A. (1986). *Cell* **44,** 941–947.

Dayton, E. T., Powell, D. M., and Dayton, A. I. (1989). *Science* **246,** 1625–1629.

Deen, K. C., McDougal, J. S., Inacker, R. *et al.* (1988). *Nature (London)* **331,** 82–84.

DeRossi, A., Franchini, G., Aldovini, A., Del Mistro, A., Chieco-Bianchi, L., Gallo, R., and Wong-Staal F. (1986). *Proc. Natl. Acad. Sci. U.S.A.* **83,** 4297–4301.

Dingwall, C., Ernberg, I., Galt, M. J., Green, S. M., Heaphy, S., Kam, J., Lowe, A. D., Singh M., Skinner, M. A., and Valerio R. (1989). *Proc. Natl. Acad. Sci. U.S.A.* **86,** 6925–6929.

Donehower, L. A., and Varmus, H. E. (1984). *Proc. Natl. Acad. Sci U.S.A.* **81,** 6461–6465.

Edery, I., Petryshyn, R., and Sonenberg, N. (1989). *Cell* **56,** 303–312.

Emerman, M., Vazeux, R., and Peden, K. (1989). *Cell* **67** 1155–1165.

Ewing E. P., Chandler, F. W., Spira, T. J., Byrnes, R. K., and Chan W. C. (1985). *Arch Pathol. Lab. Med.* **109,** 977–981.

Farnet, C., and Haseltine W. A. (1990). *Proc. Natl. Acad. Sci. U.S.A.* **87,** 4164–4168.

Feinberg, M. B., Jarret, R. F., Aldovini, A., Gallo, R. C., and Wong-Staal, F. (1986). *Cell* **46,** 807–817.

Feng, S., and Holland, E. C. (1989). *Nature (London)* **344,** 165.

Fisher, A. G., Feinberg, M. B., Josephs, S. F., Harper, M. E., Marselle, L. M., Reyes, G., Gonda, M. E., Aldovini, A., Pebouk, C., Gallo, R. C., and Wong-Staal, F. (1986a) *Nature (London)* **320,** 367–371.

Fisher, A. G., Ratner, L., Mitsuya, H., Marselle, L. M., Harper, M. E., Broder, S., Gallo, R. C., and Wong-Staal, F. (1986b). *Science* **233,** 655–658.

Fisher, A. G., Ratner, L., Mitsuya, H., Marselle, L. M., Harper, M. E., Broder, S., Gallo, R. C., and Wong-Staal, F. (1987). *Science* **237,** 888.

Folks, T. M., Clouse, K. A., Justement, J., Rabson, A., Duh, E., Kehol, J. H., and Fauci. A. S. (1989). *Proc. Natl. Acad. Sci U.S.A.* **86,** 2365–2368.

Fujiwara, T., and Mizuuchi, K. (1988). *Cell* **54,** 497–504.

Gallagher, W. R. (1987). *Cell* **50,** 37.

Gann, J., Nelson, J., and Oldstone, M. (1987). *J. Virol.* 61, 2639–2641.

Garcia, J. A., Wa, F. K., Mitsuyasu, R., and Gaynor, R. B. (1980). *EMBO J.* **6,** 3761- 3770.

Gartner, S., Markovits, P., Markotvitz, D. M., Kaplan, M. H., Gallo, R. C., and Popovic, M. (1986). *Science* **233,** 215–219.

Gorelick, R. J., Henderson, L. E., Hanser, J. P., and Rein, A. (1988). *Proc. Natl. Acad Sci. U.S.A.* **85,** 8420–8424.

Göttlinger, H. G., Sodroski, J. G., and Haseltine, W. A. (1989). *Proc. Natl. Acad. Sci. U.S.A.* **86,** 5781–5785.

Green, L. M., and Berg, J. M. (1989). *Proc. Natl. Acad. Sci. U.S.A.* **86,** 4047–4051.

Guy, B., Kieny, M. P., Riviere, Y., LePeuch, C., Dott, K., Girard, M., Montagnier, L., and Lecoq, J. P. (1987). *Nature (London)* **330,** 266–269.

Guyader, M., Emerman, M., Montagnier, L., and Peden, K. (1989). *EMBOL J.* **8,** 1169–1175.

Harper, M. E., Marsell, L. M., Gallo, R. C., and Wong-Staal, F. (1986). *Proc. Natl. Acad. Sci. U.S.A.* **83**, 772–776.

Haseltine, W. A. (1988). *In* "The Control of Human Retrovirus Gene Expression" (B.R. Franza, B.R. Cullen, and F. Wong-Staal eds.), pp. 135–158. Cold Spring Harbor Laboratory, Cold Spring Harbor, New York.

Hauber, J., and Cullen, B. (1988). *J. Virol.* **62**, 673.

Hauber, J. Perkins, A., Heimer, E., and Cullen, B. (1987). *Proc. Natl. Acad. Sci U.S.A.* **84**, 6364.

Heaphy, S., Dingwall, C., Ernberg, I., Gait, M. J., Green, S. M., Karn, J., Lowe, A. D., Singh, M., and Skinner, M. A. (1990). *Cell* **60**, 685–693.

Henderson, L. E., Copeland, T. D., Sowder, R. C., Schultz, A. M., and Oroszlan, S. (1988). *In* "Human Retroviruses, Cancer and AIDS: Approaches to Prevention and Therapy" (D. Bolognesi, ed.), pp. 135–147. Liss, New York.

Homsy, J., Mwyer, M., Tateno, M., Clarkson, S., and Levy, J. A. (1989) *Science* **244** 1357–1360.

Humphries, E. H., Glover, C., and Reichmann, M. E. (1981). *Proc. Natl. Acad. Sci. U.S.A.* **78**, 2601–2605.

Hussey, R., Richardson, N., Kowalski, M., Brown, N. R., Chang, N-C., Siliciano, J., and Reinherz, E. C. (1988). *Nature (London)* **331**, 76–79.

Jacks, T., Power, M. D., Masierz, F. R., Luciw, P. A., Bar, P. J., and Varmus, H. E. (1987). *Nature (London)* **331**, 280–283.

Jameson, B. A., Rao, P. E., Kahl, L. I., Hahn, B. H., Shaw, G. M., Hood, L. E., and Kent, S. B. (1988) *Science* **240**, 1355–1379.

Jentoft, J. E., Smith, L. M., Fu, X., Johnson, M., and Leis, J. (1988). *Proc. Natl. Acad. Sci. U.S.A.* **85**, 7094–7098.

Jones, K., Kadonaga, J., Luciw, P.,. and Tijian, R. (1986). *Science* **232**, 755–759.

Kim, S., Byrn, R., Groopman, J., and Baltimore, D. (1989). *J. Virol.* **63**, 3708–3713.

Klatzmann, D., Champagne, E., Charmaret, S., Gruest, J., Guetard, D., Gluckman, J-C., and Montagnier, L. (1984). *Nature (London)* **312**, 767–768.

Knight, D. M., Floerfelt, F. A., and Ghrayeb, J. (1987). **236**, 837–840.

Kohl, N. E., Emini, E. A., Schleif, W. A., Davis, L. J., Heimback, J. C., Dixon, R. A. F., Scolnick, E. M., and Sigal, I. S. (1988). *Proc. Natl. Acad. Sci. U.S.A.* **85**, 4686–4690.

Kowalski, M., Potz, J., Basiripour, L., Goh, W. C., Terwilliger, E., Dayton, A., Rosen, C., Haseltine, W., and Sodroski, J. (1987). *Science* **237**, 1351–1355.

Kräusslich, H.-G., and Wimmer, E. (1988). *Annu.. Rev. Biochem.* **57**, 701–754.

Larder, B. A., Darby, G., and Richman, D. D. (1989). *Science* **243**, 1731–1734.

Lasky, L., Nakamura, G., Smith, D., Fennie, C., Shimasaki, C., Patzer, E., Berman, P., Gregory, T., and Capon, D. (1987). *Cell* **50**, 975–985.

Lee, T. H., Coligan, J. E., Allan, J. S., McLane, M. F., Groopman, J. E., and Essex, M. (1986). *Science* **231**, 1546–1549.

Lever, A., Göttlinger, H., Haseltine, W., and Sodroski, J. (1989). *J. Virol.* **63**, 4085–4087.

Lifson, J., Feinberg, M., Reyes, G., Rubin, L., Banapour, B., Charabarti, S., Moss, B., Wong-Staal, F., Steimer, K. S., and Engleman, E. C. (1986a). *Nature (London)* **323**, 725–728.

Lifson, J., Reyes, G., McGrath, M., Stein, B., and Engleman, E. (1986b). *Science* **232**, 1123–1127.

Lightfoote, M. M., Coligan, J. E., Folks, T. M., Fauci, A. S., Martin, M. A., and Venkatesan, S. (1986). *J. Virol.* **60**, 771–775.

Linial, M., and Blair, D. (1982). *In* "RNA Tumor Viruses" (R. Weiss, N. Teich, H. Varmus, and J. Coffin, eds.), Vol. 1, pp. 649–783. Cold Spring Harbor Laboratory, Cold Spring Harbor, New York.

Linial, M., and Blair, D. (1985). *In* "RNA Tumor Viruses" (R. Weiss, N. Teich, H. Varmus, and J. Coffin, eds.), Vol. 2, pp. 147–185. Cold Spring Harbor Laboratory, Cold Spring Harbor, New York.

Looney, D., Fisher, A., Putney, S., Rushes, J. R., Redfield, R., Burke, D., Gallo, R. C., and Wong-Staal, F. (1988). *Science* **241**, 357–359.

Lu, Y., Stenzel, M., Sodroski, J. G., and Haseltine, W. A. (1989). *J. Virol.* **63**, 4115–4119.

Luciw, P. A., Cheng-Mayer, C., and Levy, J. A. (1987). *Proc. Natl. Acad. Sci. U.S.A.* **84**, 1434–1438.

Malim, M. H., Hauber, J., Penrick, R., and Cullen, B. R. (1988). *Nature (London)* **335**, 181–183.

Malim, M. H., Tiley, L. S., McCarn, D. F., Rusche, J. R., Hauber, J., and Cullen, B. R. (1990). *Cell* **60**, 675–683.

Matsuda, Z., Chou, M., Matsuda, M., Huang, J., Cheng, Y., Redfield, R., Mayer, K., Essex, M., and Lee, T. H. (1988). *Proc. Natl Acad. Sci. U.S.A.* **85**, 6968–6972.

Matthews, T., Langlois, A., Robey, W., Chang, N., Gallo, R. C., Fischinger, P., and Bolognesi, D. (1986). *Proc. Natl. Acad. Sci. U.S.A.* **83**, 9709–9713.

McCune, J., Rabin, L., Feinberg, M., Lieberman, K., Kosek, J., Reyes, G., and Weissman, I. (1988). *Cell* **53**, 55–67.

McDougal, J., Kennedy, M., Sligh, J., Cort, S., Mawle, A., and Nicholson, J. (1986). *Science* **231**, 382–385.

Meyerhans, A., Cheynier, R., Albert, J., Seth, M., Kwok, S., Sninsky, J., Morfeld Manson, L., and Asjö, B. (1989). *Cell* **58**, 901–910.

Meyers, G., Josephs, S. F., Berzofsky, J. A., Rabson, A. B., Smith, T. P., and Wong-Staal, F. (eds.) (1989). "Human Retroviruses and AIDS," Los Alamos National Laboratory, Los Alamos, NM.

Muesing, M. A., Smith, D. H., Cabradilla, C. D., Benton, C. V., Lasky, L. A., and Capon, D. J. (1985). *Nature (London)* **313**, 480–488.

Muesing, M. A., Smith, D. H., and Capon, D. J. (1987). *Cell* **48**, 691.

Nabel, G., and Baltimore, D. (1987). *Nature (London)* **326**, 711–713.

Navia, M. A., Fitzgerald, P. M. D., McKeever, B. M., Leu, C.-T., Heimback, J. C., Herber, W. K., Sigal, I. S., Darke, P. L., and Springer J. P. (1989). *Nature (London)* **337**, 615.

Nielson, S. L., Petito, C. K., Urmacher, D. C., and Posner, J. B. (1984). *Am. J. Clin. Pathol.* **82**, 678.

Nutt, R. F., Brady, S. F., Darke, P. L., Ciccarone, T. M., Colton, C. D., Nutt, E. M., Rodkey, J. A., Bennett, C. D., Waxman, L. H., Sigal, I. S., Anderson, P. S., and Veber, D. F. (1988). *Proc. Natl. Acad. Sci. U.S.A.* **85**, 7129–7133.

Olsen, H. S., Nelbock, P., Cochrane, A. W., and Rosen, C. A. (1990). *Science* **247**, 845–848.

Orenstein, J., Meltzer, M., Phipps, T., and Gendelman, H. (1988). *J. Virol.* **62**, 2578–2586.

Palker, T., Matthews, T., Clark, M., Cianciolo, G., Randall, R., Langleis, A., White, G., Safai, B., Synderman, R., Bolognesi, D., and Haynes, B. (1987). *Proc. Natl. Acad. Sci. U.S.A.* **84**, 2479–2483.

Panganiban, A. T., and Temin, H. M. (1983). *Nature (London)* **306**, 155–160.

Parkin, N. T., Cohen, E. A., Darveau, A., Rosen, C., Haseltine, W., and Sonenberg, N. (1988). *EMBO J.* **7**, 2831–2837.

Patarca, R., and Haseltine, W. A. (1987). *AIDS Res. Hum. Retroviruses* **3**, 1.

Peng, C., Ho, B. K., Chang, T. W., and Chang, N. T. (1989). *J. Virol.* **63**, 2550–2556.

Peterlin, B. M., Luciw, P. A., Barr, P. J., and Walker, M. D. (1986). *Proc. Natl. Acad. Sci. U.S.A.* **83**, 9734–9738.

Pinter, A., Honnen, W. J., Tilley, S. A., Bona, C., Zaghouani, H., Gorny, M. K., and Zolla-Pazner, S. (1989). *J. Virol.* **63**, 2674–2679.

Popovic, M., Samgadharan, M. G., Read, E., and Gallo R. C. (1984). *Science* **224**, 497–500.

Price, R., Brew, B., Sidtis, J., Rosenblum, M., Scheck, A., and Cleary, P. (1988). *Science* **239**, 586–592.

Quinn, T. P., and Gradgenett, D. P. (1988). *J. Virol.* **62**, 2307–2312.

Robert-Guroff, M., Brown, M., and Gallo, R. C. (1985). *Nature (London)* **316**, 72–74.

Robey, W. G., Safai, B., Oroszlan, S., Arthur, L., Gonda, M., Gallo, R., and Fischinger, P. (1985). *Science* **228**, 593–595.

Rosen, C. A., Sodroski, J. G., and Haseltine, W. A. (1985b). *Cell* **41**, 813–823.

Rosen, C. A., Sodroski, J. G., Goh, W. C., Dayton, A. I., Lipke, J., and Haseltine, W. A. (1986). *Nature* (*London*) **319**, 555–559.

Rosen, C. A., Terwilliger, E., Dayton, A., Sodroski, J. G., and Haseltine, W. A. (1988). *Proc. Natl. Acad. Sci. U.S.A.* **85**, 2071–2075.

Roth, M. J., Schartzberg, P. L., and Goff, S. P. (1989). *Cell* **58**, 47–54.

Ruben, S., Perkins, A., Purcell, R., Joung, K., and Rosen, C. R. (1989). *J. Virol.* **63**, 1–8.

Salfeld, J., Gottlinger, H. G., Sia, R., Park, R., Sodroski, J. G., and Haseltine, W. A. (1990). *EMBO J.* **9**, 965–970.

Sanchez-Pescador, R., Power, M. D., Barr, P. J., Steimer, K. S., Stempien, M. M., Brown-Shimer, S. L., Gee, W. W., Renard, A., Randolf, A., Levy, J. A., Dina, D., and Luciw, P. A. (1985). *Science* **227**, 484–492.

Sarngadharan, M., Popovic, M., Bruch, L., Schupbach, J., and Gallo, R. C. (1984). *Science* **224**, 506–508.

Sattentau, Q., Dalgleish, A., Weiss, R., and Beverly, P. (1986). *Science* **234**, 1120–1123.

Schwartzberg, P., Colicelli, J., and Goff, S. P. (1984). *Cell* 1043–1052.

Schiff, L. A., Nibert, M. L., and Fields, B. N. (1988). *Proc. Natl. Acad. Sci. U.S.A.* **85**, 4195–4199.

Schüpbach, J., Popovic, M., Gilden, R. V., Gonda, M. A., Sarngadharan, M. G., and Gallo R. C. (1984). *Science* 503–505.

Skevila, M., Josephs, S. F., Dukovich, M., Peffer, N., Wong-Staal, F., and Greene, W. C. (1987). **238**, 1575–1578.

Smith, D., Byrn, R., Marsters, S., Gregory, T., Groopman, J., and Capon, D. (1987). *Science* **238**, 1704–1707.

Sodroski, J., Patarca, R., Rosen, C., and Haseltine, W. A. (1985a). *Science* **229**, 74–77.

Sodroski, J., Rosen, C., Wong-Staal, F. Popovic, M., Arya, S., Gallo, R. C., and Haseltine, W. A. (1985b). *Science* **227**, 171–173.

Sodroski, J. G., Goh, W. C., Rosen, C., Dayton, A., Terwilliger, E., and Haseltine, W. A. (1986a). *Nature* (*London*) **321**, 412–417.

Sodroski, J. G., Goh, W. C., Rosen, C., Campbell, K., and Haseltine, W. A. (1986b). *Nature* (*London*) **322**, 470–474.

Sodroski, J. G., Goh, W. C., Rosen, C., Tarter, A., Portetelle, D., Bumy, A., and Haseltine, W. A. (1986c). *Science* **231**, 1549–1553.

Starcich, B., Ratner, L., Josephs, S. F., Okamoto, T., Gallo, R. C., and Wong-Staal, F. (1985). *Science* **227**, 538–540.

Stein, B., Gowda, S., Lifson, J., Penhallow, R., Bensch, K., and Engelman, E. (1987). *Cell* **49**, 659–668.

Strebel, K., Daugherty, D., Clouse, K., Cohen, D., Folks, T., and Martin, M. A. (1987). *Nature* (*London*) **328**, 728.

Strebel, K., Klimkait, T., and Martin, M. A. (1988). *Science* **241**, 1221–1223.

Strebel, K., Klimkait, T., Maldarelli, F., and Martin, M. A., (1989). *J. Virol.* **63**, 3784–3791.

Takeda, A., and Tauzon, C. U. (1988). *Science* **242**, 580–583.

Terwilliger, E., Sodroski, J. G., Rosen, C. A., and Haseltine, W. A. (1986). *J. Virol.* **60**, 754–760.

Terwilliger, E. F., Sodroski, J. G., Haseltine, W. A., and Rosen, C. A. (1988). *J. Virol.* **62**, 655–668.

Terwilliger, E. F., Cohen, E. A., Lu, Y., Sodroski, J. G., and Haseltine, W. A. (1989). *Proc. Natl. Acad. Sci. U.S.A.* **86**, 5163–5167.

Traunecker, A., Lekew, and Karjalainen, K. (1988). *Nature* (*London*) **331**, 84–86.

Varmus, H. E., Padgett, T., Heasley, S., Simon, G., and Bishop, J. M. (1977). *Cell* **11**, 307–319.

Veronese, F. M., Copeland, T. D., Oroszlan, S., Gallo, R. C., and Sarngadharan, M. G. (1988). *J. Virol.* **62**, 795–801.

Wain-Hobson, S., Sonigo, P., Danos, O., Cole, S., and Alizon, M. (1985). *Cell* **40**, 9–17.

Wang, J., Steel, S., Wisniewolski, R., and Wang C. (1986). *Proc. Natl. Acad. Sci. U.S.A.* **83**, 6159–6163.

Weiss, R., Clapham, P., Cheingsong-Popov, R., Dalgleish, A., Carne, C., Weller, I., and Tedder, R. S. (1985). *Nature (London)* **316**, 69–72.

Wilson, W., Braddock, M., Adams, S. E., Rathjen, P. D., Kingsman, S. M., and Kingsman, A. J. (1988). *Cell* **55**, 1159–1169.

Wlodawyer, A., Miller, M., Jaskolski, M., Sathyanarayana, B. K., Baldwin, E., Weber, I. T., Selk, L. M., Clawson, L., Schneider, J., and Kent, S. B. H. (1989). *Science* **245**, 616.

Wright, C. M., Felber, B. K., Paskalis, H., and Paskalis, G. N. (1986). *Science* **234**, 988.

Yu, X.-F, Ito, S., Essex, M., and Lee. T. H. (1988). *Nature (London)* **335**, 262.

Zapp, M. L., and Green, M. R. (1989). *Nature (London)* **342**, 714–716.

Part II

Biology

Biology of Human T-cell Leukemia Virus (HTLV) Infection

Isao Miyoshi

I. Introduction

Since the first discovery of murine leukemia virus by Gross (1951), RNA tumor viruses (retroviruses) have been implicated in the etiology of many naturally occurring leukemias and lymphomas in various animal species. However, it was not until 1980–1981 when the first human retroviruses, then called human T-cell leukemia virus (HTLV) and adult T-cell leukemia virus (ATLV), were isolated in the United States (Poiesz *et al.*, 1980b, 1981) and Japan (Hinuma *et al.*, 1981; Miyoshi *et al.*, 1981b; Yoshida *et al.*, 1982), respectively, from patients with cutaneous T-cell lymphoma and adult T-cell leukemia (ATL) (Uchiyama *et al.*, 1977). These viruses were detected in T-cell lines established with the use of T-cell growth factor (IL-2) (Gazdar *et al.*, 1980; Poiesz *et al.*, 1980a) or by cocultivation with human cord blood leukocytes (Miyoshi *et al.*, 1979; 1981a). Both virus isolates were shown to be almost identical (Popovic *et al.*, 1982; Watanabe *et al.*, 1984) and collectively designated HTLV-I. Subsequently, HTLV-II was isolated from patients with a T-cell variant of hairy cell leukemia (Kalyanaraman *et al.*, 1982; Rosenblatt *et al.*, 1986). Extensive epidemiological, molecular, and biological studies have determined the etiological role of HTLV-I in ATL which is endemic in southwestern Japan, the Caribbean islands, and parts of Africa, as reviewed by Yoshida (1983), Wong-Staal and Gallo (1985) and Yamamoto and Hinuma (1985). HTLV-I has also been linked to the cause of tropical spastic paraparesis (TSP) (Gessain *et al.*, 1985) or HTLV-I-associated myelopathy (HAM) (Osame *et al.*, 1986).

II. Biological Activity of HTLV-I *in Vitro*

A. Lymphocyte Transformation by Cocultivation

The lymphocyte-transforming capacity of HTLV-I was first detected through our serendipitous discovery that feeder-layer leukocytes from newborn cord blood became immortalized when cocultivated with leukemic cells from ATL patients (Miyoshi *et al.*, 1981a,b, 1982d). The fact that the immortalized cells were T cells immediately suggested that a certain transforming agent other than Epstein-Barr virus (EBV) might have been transmitted from ATL cells *in vitro*, because spontaneous transformation of normal cord blood T cells was considered unlikely. The unexpected observation was confirmed by deliberate attempts to transform peripheral and cord blood T cells by cocultivation with lethally irradiated HTLV-I-producing cell lines (Miyoshi *et al.*, 1981c; Yamamoto *et al.*, 1982a; Popovic *et al.*, 1983a,b). In most cases, the transformed cells proliferated indefinitely as cell lines without dependence on exogenous IL-2 and exhibited morphological characteristics of primitive lymphoid cells with multinucleated giant cells. These cell lines were usually positive for T4 (helper T cell) antigens, HLA-DR determinants, and IL-2 receptors. This surface phenotype was very similar to that of patients' ATL cells (Hattori *et al.*, 1981). Infrequently, suppressor T cells were also transformed by HTLV-I (Markham *et al.*, 1984; Koeffler *et al.*, 1984). Most of the transformed cell lines were persistent producers of HTLV-I with a few nonproducer cell lines that failed to express viral antigens despite the presence of the HTLV-I genome (Salahuddin *et al.*, 1983). The second human retrovirus, HTLV-II, was likewise shown to have the same lymphocyte-transforming capacity (Chen *et al.*, 1983). These transformed T-cell lines constitutively liberated multiple lymphokines (Chen *et al.*, 1983; Salahuddin *et al.*, 1984; Koeffler *et al.*, 1984).

Furthermore, we have found that the cocultivation technique can be successfully used to isolate HTLV-I from healthy individuals who are seropositive for HTLV-I (Miyoshi *et al.*, 1983b). It is to be emphasized that although fresh leukemic cells from ATL patients and lymphocytes from healthy virus carriers completely lack viral expression, they can be induced to express viral antigens by short-term *in vitro* culture (Miyoshi *et al.*, 1982a,b; Hinuma *et al.*, 1982; Akagi *et al.*, 1982). A colony assay has been proposed for the quantitative analysis of *in vitro* T-cell transformation by HTLV-I and HTLV-II (Graziano *et al.*, 1987; Aboud *et al.*, 1987). Grassmann *et al.* (1989) succeeded in immortalizing human cord blood and thymic lymphocytes by recombinant *Herpesvirus saimiri* containing an insert of the pX region of HTLV-I. These IL-2-dependent transformed cell lines possessed the same phenotype as lymphocytes transformed by HTLV-I and expressed the known pX region-encoded proteins including *tax* and *rex*. This is the first demonstration of the transforming ability of the HTLV-I pX region and its functional role in determining the phenotype of the transformed cells.

B. Lymphocyte Transformation by Cell-Free Virus

In contrast to the consistency and rapidity with which lymphocytes can be transformed by cocultivation, attempts to transform lymphocytes by cell-free HTLV-I and HTLV-II were repeatedly unsuccessful (Miyoshi *et al.*, 1982d; Yamamoto *et al.*, 1982a; Chen *et al.*, 1983) or could be accomplished with much less efficiency (de Rossi *et al.*, 1985). The reason for this failure is not clear but may be due to the fragility of the virus envelope. Both viruses appear to be transmitted primarily by cell-to-cell infection and are distinct from EBV and human immunodeficiency virus (HIV) which can be transmitted by cell-free virus. It is postulated from these results that transmission of virus-infected lymphocytes, but not cell-free virus, would play an important role in the natural infection of HTLV-I and HTLV-II in humans.

C. Virus Transmission to Non-T Cells

Most of the early studies suggested that HTLV-I and HTLV-II are tropic for human T lymphocytes, particularly those with helper T-cell phenotype. However, some EBV-transformed B-cell lines established from patients with ATL or hairy T-cell leukemia were found to be infected with HTLV-I (Yamamoto *et al.*, 1982b; Mann *et al.*, 1984; Miyamoto *et al.*, 1984) or HTLV-II (Chen *et al.*, 1983). This dual infection probably represents an incidental *in vitro* infection of B cells by these retroviruses, since no disease of B cells carrying HTLV-I or HTLV-II has been reported.

Clapham *et al.* (1983) first showed that a human osteogenic sarcoma cell line can be productively infected by cocultivation and cell-free HTLV-I, indicating that the spectrum of this virus is not strictly limited to lymphocytes. Similarly, HTLV-I could be stably propagated to a human gastric cancer cell line (Akagi *et al.*, 1988). Furthermore, human endothelial cells were found to be susceptible to productive infection by HTLV-I (Ho *et al.*,

1984; Hoxie *et al.*, 1984). All these virus-infected nonlymphoid cells contained syncytial giant cells. It is possible, but remains to be proven, that normal endothelial cells are indeed infected *in vivo* by HTLV-I from circulating virus-bearing T cells and serve as a virus reservoir. Nevertheless, these findings indicate that HTLV-I has a wider host cell range than was initially assumed, but its *in vitro* transforming capacity seems to be restricted to T cells.

Nagy *et al.* (1983) studied HTLV-I-induced cell fusion by cocultivation with a variety of nonlymphoid indicator cells and suggested that HTLV-I envelope glycoproteins interact with the membranes of many cell types from diverse mammalian species. Serum antibodies from ATL patients inhibited the syncytium-forming activity of different HTLV-I isolates. Hoshino *et al.* (1983) also induced syncytia in a clonal cell line of cat kidney fibroblasts by cocultivation with fresh leukemic cells from ATL patients. However, no syncytia were induced by concentrated cell-free virus prepared from HTLV-I-producing cell lines.

Plaque assays of vesicular stomatitis virus (VSV) bearing envelope antigens of HTLV-I and HTLV-II also indicated the presence of retrovirus receptors on various mammalian nonlymphoid cells (Clapham *et al.*, 1984; Hoshino *et al.*, 1985). Following penetration and uncoating, the VSV genome contained in the pseudotype virions replicated to produce nonpseudotype progeny and cytopathic plaques could be used to titrate neutralizing antibodies specific to the retrovirus encoding the envelope glycoproteins. Neutralization of VSV (HTLV-I) pseudotypes was more specific and more sensitive than assays of syncytium inhibition. All HTLV-I isolates from Japan, the United States and the United Kingdom represented the same serotype. Human—mouse somatic hybrids were used to determine which human chromosome was required to confer susceptibility to VSV (HTLV-I) pseudotype infection and the receptor gene was localized to chromosome 17 (Sommerfelt *et al.*, 1988).

D. Transformation of Animal Lymphocytes

Shortly after the discovery of the human lymphocyte-transforming capacity of HTLV-I, we wondered if the virus would transform nonhuman primate lymphocytes *in vitro*. To our surprise, Japanese monkey (*Macaca fuscata*) lymphocytes were immortalized by cocultivation with virus-producing MT-2 cells (Miyoshi *et al.*, 1982c). Subsequently, lymphocytes from various species of animals such as the rabbit (Miyoshi *et al.*, 1983a), cat (Hoshino *et al.*, 1984), rat (Tateno *et al.*, 1984), and hamster (Akagi *et al.*, 1986) were similarly transformed by HTLV-I. These transformed cell lines were chromosomally normal, at least during early stages of transformation, and exhibited phenotypic characteristics similar to those of human HTLV-I-transformed cell lines (Table I). For unknown reasons, HTLV-I-transformed hamster lymphoid cell lines that carried integrated provirus were completely

Table I

Characteristics of Human and Animal Lymphoid Cell Lines
Transformed by HTLV-I

Cell line	T-cell antigen	Ia antigen	IL-2 receptor	HTLV-I antigen	HTLV-I virion
Human	+	+	+	+	+
Monkey	+	+	+	+	+
Cat	+	nt[a]	nt	+	+
Rabbit	+	+	nt	+	+
Rat	+	+	+	+	+
Hamster	+	nt	nt	−	−

[a] nt, Not tested.

negative for viral expression (Akagi *et al.*, 1986). This nonproducer state appeared to be comparable to the restricted expression of HTLV-I in a few human T-cell lines transformed by HTLV-I (de Rossi *et al.*, 1985). In our laboratory, mouse lymphocytes were repeatedly refractory to HTLV-I-induced transformation (unpublished), although they have been reported to be transiently infectable by HTLV-I (Sinangil *et al.*, 1985). Animal leukemia viruses do not usually transform their target leukocytes *in vitro* and such a wide host species range is unique to HTLV-I. This difference is probably accounted for by the extra genomic region called pX which is missing from most animal retroviruses.

E. Tumorigenicity of Transformed Lymphocytes

Human and animal lymphocytes immortalized by HTLV-I were tested for their ability to grow in heterologous, allogeneic, syngeneic, or autologous hosts. Human, rabbit, and hamster lymphoid cell lines transformed by HTLV-I produced lethal tumors when transplanted into the peritoneal cavity of newborn hamsters (Miyoshi *et al.*, 1982d, 1984a; Akagi *et al.*, 1986). The recipient animals had to be immunosuppressed by antilymphocyte serum to obtain progressive growth of the implanted human and rabbit cells. In contrast, a nonproducer hamster lymphoid cell line could be serially transplanted in allogeneic newborn hamsters without immunosuppression and caused leukemic disease in immunosuppressed adult hamsters (Eguchi *et al.*, 1988). These transplanted cells, except for the nonproducer hamster cells, continued to release HTLV-I particles *in vivo* and could be readily recultured as cell lines *in vitro*. *In vitro* HTLV-I-transformed rabbit lymphocytes failed to grow following autologous transplantation, suggesting that viral antigens expressed on the transformed cells rendered them immunogenic so as to be rejected by the hosts (Akagi *et al.*, 1985). Seto *et al.* (1988a) established an HTLV-I-transformed T cell line from a F1 hybrid rabbit between inbred strains of B/J and Chbb:HM. This cell line proved to be lethal when implanted into syngeneic newborn and adult rabbits. Similarly, rat lymphoid cell lines transformed by HTLV-I were transplantable not only into syngeneic rats but also into nude mice (Tateno *et al.*, 1984). These transplantation studies indicate that HTLV-I-transformed lymphocytes are potentially malignant and that their growth pattern *in vivo* is dependent on the host immune status. It is interesting to note that transgenic mice carrying the *tat* gene developed multiple soft tissue tumors between 13 and 17 weeks of age (Nerenberg *et al.*, 1987). Other founder mice exhibited extensive thymic depletion and growth retardation, but no lymphoid tumors were observed.

III. Transmission of HTLV-I to Monkeys

After we found that monkey and rabbit lymphocytes could be infected and immortalized by HTLV-I *in vitro*, we were interested to see if this virus could be transmitted to these animals. In our preliminary experiment, a Japanese monkey seronegative for HTLV-I was inoculated intravenously with Ra-1 cells (Miyoshi, 1984). Ra-1 is a male rabbit lymphoid cell line established by cocultivation with MT-2 cells, and is persistently infected with HTLV-I (Miyoshi *et al.*, 1983a). The monkey became HTLV-I antibody-positive after 2 weeks and remained so for the observation period of 6 months. After seroconversion, the peripheral lymphocytes were cultured in the presence of IL-2. The cultured cells, whose karyotypes were macaque, expressed HTLV-I antigens and virus particles. Yamamoto *et al.* (1984) also attempted to infect six cynomolgus monkeys (*Macaca fuscicularis*) with HTLV-I by inoculating MT-2 cells or cell-free MT-2 virus. All six animals developed antibodies to HTLV-I after 2–4 weeks and their cultured lymphocytes were reported to have expressed HTLV-I p19 and p24 for up to 24 weeks after infection. However, they did not mention if HTLV-I particles were detected from the infected monkeys. It is doubtful if cynomolgus monkeys were successfully infected by cell-free HTLV-I in view of the general failure of cell-free transmission of HTLV-I *in vitro*.

IV. Transmission of HTLV-I to Rabbits

A. Inoculation of HTLV-I-Producing Cells and Cell-Free Virus

Noninbred Japanese white rabbits were inoculated intravenously with human and rabbit HTLV-I-producing lymphoid cell lines, MT-2 and Ra-1, with or without mitomycin-C treatment (Miyoshi *et al.*, 1985; Akagi *et al.*, 1985). Autologous rabbit lymphocytes transformed by HTLV-I *in vitro* were also used for inoculation. Most of these rabbits developed antibodies to

HTLV-I within 10 days and their antibody titers often rose to 1:1280, as determined by indirect immunofluorescence. These high-titered sera, when reacted with [^{3}H] leucine-labeled lysates of MT-2 cells, specifically immuno-precipitated HTLV-I gp68, p28, p24, and p19 (Miyoshi *et al.*, 1984b). Im-munofluorescence analysis of frozen sections revealed the presence of MT-2 cells in the blood vessels of lungs taken 2 and 6 h after inoculation of MT-2 cells. Peripheral lymphocytes were obtained at various time intervals and cultured with or without IL-2. MT-2 cells could be regrown when peripheral lymphocytes were cultured 2 h after inoculation of MT-2 cells. Long-term cultures of rabbit lymphocytes contained cells positive for HTLV-I antigens and harbored virus particles. These cultures were shown to be different from the inoculated allogeneic or autologous cells with respect to the sex chromo-somes and proviral integration pattern. Thus, none of the HTLV-I-produc-ing cell lines proliferated progressively when inoculated into rabbits; how-ever, HTLV-I was transmitted to the host lymphocytes prior to being rejected by the host immune system.

Seto *et al.* (1988b) suggested that neonatal inoculation of HTLV-I-pro-ducing lymphoid cell lines derived from an inbred strain of rabbits may result in immunological tolerance and seronegative virus carriers. It remains to be determined if such a seronegative, genome-positive state exists in the natu-ral infection of HTLV-I in humans. Rabbits inoculated intravenously with concentrated HTLV-I virions prepared from the MT-2 cell line likewise seroconverted for HTLV-I but their antibody titers gradually diminished over several months (unpublished). Peripheral lymphocytes from these rab-bits ceased to grow within 2 months in the presence of IL-2. Transfusion of 20 ml of blood from these rabbits did not cause seroconversion in normal rabbit recipients. These results suggest that cell-free HTLV-I is not infec-tious to rabbits and contradict the claim of Yamamoto *et al.* (1984) that cynomolgus monkeys can be infected with cell-free virus. It is noteworthy that rabbits can also be infected by inoculation of cell-free HIV and HIV-infected cells (Filice *et al.*, 1988; Kulaga *et al.*, 1989).

B. Transmission by Blood Transfusion

Detection of HTLV-I particles in lymphocyte cultures from seropositive healthy individuals including blood donors suggested a risk of HTLV-I infec-tion by blood transfusion (Miyoshi *et al.*, 1982a,b; Gotoh *et al.*, 1982). A retrospective study by Okochi *et al.* (1984) indicated a high seroconversion rate (63.4%) in the recipients of blood and blood cell components, but no seroconversion in recipients of plasma from seropositive donors. We per-formed a transfusion experiment using rabbits to substantiate the risk of blood-borne transmission of HTLV-I (Kotani *et al.*, 1986). As shown in Fig. 1, HTLV-I could be serially transmitted to rabbits of the opposite sex by transfusion of 20 ml of whole blood or washed blood cell suspension (12/12) but not by infusion of cell-free plasma (0/2) from virus-infected rabbits. We

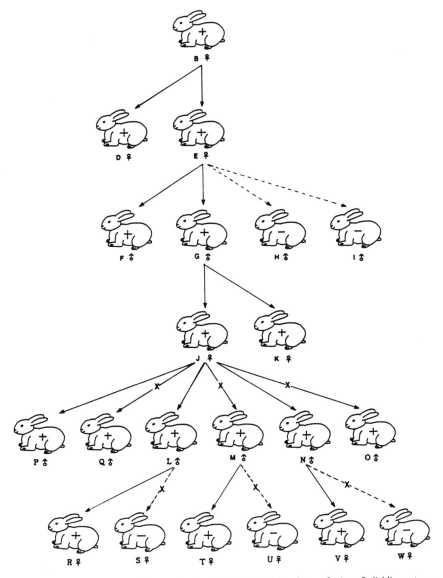

Figure 1. Scheme of serial transmission of HTLV-I by blood transfusion. Solid lines, transfusion of whole blood or washed blood cell suspension (fresh or stored for 1–2 weeks at 4°C); dotted lines, infusion of cell-free plasma; crossed solid lines, transfusion of X-irradiated fresh blood; crossed dotted lines, transfusion of X-irradiated stored blood. Seroconversion is indicated by +. HTLV-I-producing lymphoid cell lines were established from rabbits E, F, J, and L. (From Kotani *et al.*, 1986.)

then investigated whether X-irradiation (6000 rad) would prevent the blood-borne transmission of HTLV-I. Virus transmission did occur (3/3) when the blood was transfused immediately after irradiation but not (0/3) when the irradiated blood was transfused after storage for 1–2 weeks at 4°C. Seroconversion occurred 2–4 weeks after blood transfusion and HTLV-I-carrying lymphoid cell lines were grown from representative seroconverted rabbits at different passages. These results imply that irradiation *per se* does not inhibit virus induction from the genome-carrying lymphocytes and that the irradiated lymphocytes probably die within 1 week and can no longer transmit the virus.

C. Oral Transmission

Results of seroepidemiologic surveys suggested natural transmission of HTLV-I from mother to child and between spouses (Tajima *et al.*, 1982; Hino *et al.*, 1985; Kajiyama *et al.*, 1986). Detection of viral antigens in cultured cells from semen and mothers' milk indicated that the virus would be transmitted by sexual intercourse and breast feeding (Kinoshita *et al.*, 1984; Nakano *et al.*, 1984). Hino's group succeeded in transmitting HTLV-I to common marmosets (*Callithrix jacchus*) by oral inoculation of short-term cultured lymphocytes from ATL patients or concentrated milk from postpartum seropositive mothers (Yamanouchi *et al.*, 1985; Kinoshita *et al.*, 1985). One marmoset in each experiment seroconverted for HTLV-I after 2.5 months, and short-term cultures of lymphocytes from both marmosets expressed viral antigens. In our experiment, one of four female rabbits given twice-weekly oral inoculation of HTLV-I-producing Ra-1 cells of male rabbit origin became seropositive for HTLV-I after 8 weeks (Uemura *et al.*, 1986). Peripheral lymphocyte cultures yielded an HTLV-I-carrying lymphoid cell line from the seroconverted rabbit, but not from the seronegative ones. It is likely that some of the virus-infected cells penetrated the mucosal barrier during their passage from the oral cavity to the gastrointestinal tract. These results support the concept of milk-borne transmission of HTLV-I in humans.

D. Mother-to-Offspring Transmission

Mother-to-offspring transmission of HTLV-I was studied in our rabbit model (Uemura *et al.*, 1987). Four virus-infected rabbits (two males and two females) were individually mated with four noninfected rabbits. Two virus-infected females mated with noninfected males gave birth to seven offspring while two noninfected females mated with infected males delivered five offspring (Fig. 2). Four of the seven offspring born to the virus-infected mothers seroconverted for HTLV-I when they were 6–13 weeks old. None of the five offspring born to the noninfected mothers became seropositive during the observation period of 6 months. Peripheral lymphocytes were

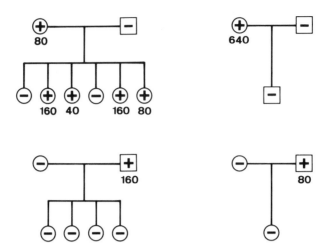

Figure 2. Parent-to-offspring transmission of HTLV-I in rabbits. Squares, males; circles, females. Figures below + denote immunofluorescence antibody titers to HTLV-I. All four seroconverted offspring yielded HTLV-I-carrying lymphoid cell lines. (From Uemura *et al.*, 1987.)

cultured in the presence of IL-2, and HTLV-I-producing lymphoid cell lines were established from all four seroconverted offspring. These data indicate that HTLV-I can be transmitted to offspring from virus-infected female rabbits but not from virus-infected male rabbits, which is in agreement with the pattern of transmission of HTLV-I among families (Tajima *et al.*, 1982; Hino *et al.*, 1985; Kajiyama *et al.*, 1986).

We then examined whether newborn rabbits born to a virus-infected rabbit were already infected at birth with HTLV-I. The virus could not be detected or isolated in cultures of their splenic lymphocytes (Uemura *et al.*, 1987). This is consistent with the absence of HTLV-I antigens in cultured cord blood lymphocytes from neonates born to seropositive women (Nakano *et al.*, 1984; Hino *et al.*, 1985).

Furthermore, we investigated the effect of foster-nursing on vertical transmission of HTLV-I (Hirose *et al.*, 1988). Virus-infected and noninfected female rabbits were mated with noninfected male rabbits. Rabbits born to virus-infected mothers were foster-nursed by noninfected females within 24 hr of age and vice versa. Many newborn rabbits died after transfer to foster mothers and only four litters survived the foster-nursing procedure (Fig. 3). Two of five rabbits born to noninfected mothers and fostered by virus-infected females seroconverted for HTLV-I after 7 weeks. In contrast, all seven rabbits born to virus-infected mothers and fostered by noninfected females remained seronegative for the observation period of 6 months. HTLV-I-carrying T-cell lines were derived from the two seroconverted rabbits.

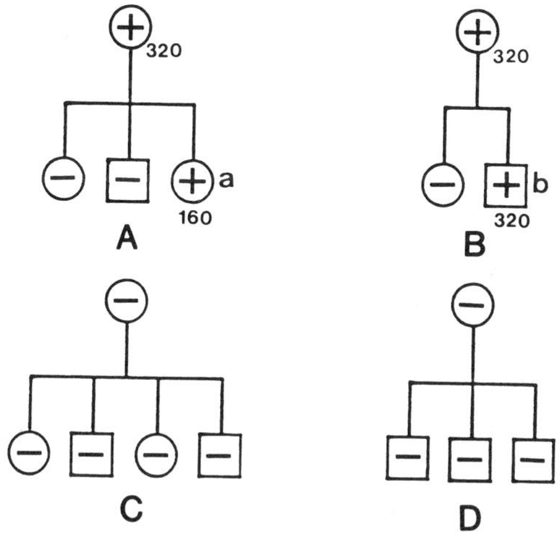

Figure 3. Effect of foster-nursing on HTLV-I transmission in rabbits. Litters A and B were born to noninfected mothers and foster-nursed by virus-infected females, and vice versa for litters C and D. HTLV-I-producing lymphoid cell lines were established from seroconverted offspring a and b. (From Hirose *et al.*, 1988.)

Although the number of animals used is small, these findings suggest that HTLV-I is transmitted by way of milk from mother to offspring and make the transplacental transmission of HTLV-I less likely. Pilot studies in some HTLV-I-endemic areas of Japan have shown that mother-to-child transmission of HTLV-I can be prevented by feeding babies powdered milk or freeze-thawed mothers' milk (Ando *et al.*, 1987, 1989; Hino *et al.*, 1987). Alternatively, mothers' milk heated at 56°C for 30 min may be given to babies, since this heat treatment is shown to inactivate HTLV-I and virus-carrying cells (Yamato *et al.*, 1986).

V. Immunization against HTLV-I in Rabbits

A. Active Immunization

Since rabbits were found to be highly susceptible to HTLV-I infection, we have made immunization attempts against challenge infection by blood transfusion in this animal system (Takehara *et al.*, 1989). Two groups of three rabbits were each immunized with heat-inactivated HTLV-I or a synthetic *env* peptide, designated env175–196, which contained the putative epitope for a monoclonal antibody to HTLV-I gp46. All six rabbits developed antibodies to HTLV-I with immunofluorescence titers of 1:20 to 1:640 and the presence of antibodies to viral proteins including gp68 and gp46 detected by radioimmunoprecipitation assay. Immunized rabbits were then challenged with transfusion of 20 ml of blood from HTLV-I-infected rabbits of the opposite sex. After transfusion challenge, the antibody titers further rose in both groups and antibodies to HTLV-I p24 and p19 newly appeared in the env175–196 group. To see whether the rabbits were protected by active immunization, peripheral lymphocytes harvested 1–6 months after challenge were cultured with added IL-2. HTLV-I-carrying lymphoid cell lines of recipient origin were established from all six immunized and challenged rabbits. Six normal rabbits transfused from these immunized and challenged rabbits seroconverted for HTLV-I after 2–4 weeks. When pre-challenge sera were titrated for neutralizing antibody against VSV pseudo-type bearing envelope glycoproteins of HTLV-I, none of the sera neutralized the pseudotype virions at a dilution of 1:10.

Nakamura *et al.* (1987) reported that cynomolgus monkeys immunized with HTLV-I *env* gene peptides produced in *Escherichia coli* were protected from virus challenge. Shida *et al.* (1987) also reported that inoculation of rabbits with recombinant vaccinia viruses containing the HTLV-I *env* gene protected them against HTLV-I challenge. In both experiments, a virus-producer cell line, MT-2, was used for challenge. Protection from challenge was based solely on the absence of detectable viral antigens in short-term peripheral lymphocyte cultures and was not supported by other means, such as transfusion assay from immunized and challenged animals to normal ones. It is, therefore, difficult to exclude a low level of HTLV-I infection in these studies.

B. Passive Immunization

Three rabbits were infused twice, one week apart, with hyperimmune anti-HTLV-I IgG prepared from rabbits that had developed high antibody titers to HTLV-I as a result of inoculation of Ra-1 cells (Takehara *et al.*, 1989). They were challenged by transfusion of 20 ml of blood from HTLV-I-infected rabbits within 5 min of the first immunization. Serum antibody titers were 1:640 immediately after the first and second immunizations and then gradually declined to a level of 1:10 or lower over a period of 4 months. The prechallenge sera were shown to have antibodies to gp68, gp46, p28, p24, and p19 as well as VSV (HTLV-I) pseudotype neutralizing antibody titers of 1:250 to 1:1250. One to four months after transfusion challenge, lympho-cytes were obtained on two occasions and cultured in the presence of IL-2. None of the cultures expressed HTLV-I antigens and no lymphoid cell lines were established. Transfusion assay from these immunized and challenged rabbits failed to seroconvert three normal recipient rabbits. In contrast, three control rabbits similarly transfused from the same HTLV-I-infected rabbits as used for transfusion challenge seroconverted for HTLV-I after 2–4 weeks, and HTLV-I was detected in lymphoid cell lines established from these rabbits.

In the active immunization experiment described above, rabbits immu-nized with heat-inactivated HTLV-I or a synthetic *env* peptide were not protected from HTLV-I infection, although both immunogens provoked the production of antibodies to envelope glycoproteins. Results of VSV (HTLV-I) pseudotype neutralization assays indicated that the active immunization group had no detectable neutralizing antibody, whereas the passive immunization group had high neutralizing antibody titers. The contrasting results following challenge of these two immunization groups, therefore, appear to be explained at least in part by the difference in neutralizing antibody titers.

There is no evidence for the presence of extracellular HTLV-I virions in the blood of seropositive individuals or rabbits. HTLV-I genome-carrying lymphocytes transmit the virus apparently by cell-to-cell contract by a mechanism which is not well understood. Antibodies to HTLV-I are known to inhibit virion- induced syncytium formation (Nagy *et al.*, 1983; Hoshino *et al.*, 1983) and neutralize VSV (HTLV) pseudotypes (Clapham *et al.*, 1984; Hoshino *et al.*, 1985). An attempt to immortalize normal lymphocytes by cocultivation with MT-2 cells was unsuccessful in the continued presence of 20% serum from seropositive persons (unpublished). Our passive immunization experiment suggests that cell-to-cell infection of HTLV-I can be blocked *in vivo* by neutralizing antibody and provide hope not only for clinical application but also for effective vaccine development. On the other hand, the situation seems to be more complex in HIV infection, since it has been reported that neither active nor passive immunization protected chimpanzees against HIV challenge (Hu *et al.*, 1987; Berman *et al.*, 1988; Prince *et al.*, 1988).

VI. Concluding Remarks

HTLV belongs to a family of T-cell tropic human type-C retroviruses, of which HTLV-I is the prototype. HTLV-I, the first human leukemia virus to be discovered, has a unique property of infecting and immortalizing not only human T cells but also T cells from various species of animals *in vitro*. Moreover, the virus infects monkeys and rabbits, rendering them seropositive virus carriers as in humans. The rabbit model has proved particularly useful for demonstrating the transmission routes of HTLV-I, although no virus-induced disease, hematological or neurological, has been so far observed. For full control of ATL and TSP/HAM, more progress must be made in three aspects of virus infection, pathogenesis, and treatment. First, it is important to interrupt the routes of virus transmission, especially from one generation to the next. Since 1986, all blood donors have been screened in Japanese blood banks, and transfusion-associated transmission of HTLV-I is considered almost nonexistent in Japan. Screening and counseling are also advisable in prenatal clinics to prevent milk-borne transmission of HTLV-I. Our passive immunization experiment in rabbits suggests that neutralizing antibody is effective in preventing cell-to-cell infection of HTLV-I. If this is the case, we have to clarify the level and duration of protection provided by maternally transmitted antibodies in babies. In the long run, it may be possible to administer HTLV-I immune globulin or a vaccine capable of inducing neutralizing antibody under certain circumstances. Second, it is important to understand better the mechanism of viral leukemogenesis and myelopathy development so that we can predict and circumvent disease among asymptomatic virus carriers. Third, a new approach of therapy should be introduced to improve the patient's survival and prognosis.

Acknowledgments

This work was supported by grants from the Japanese Ministry of Education, Science, and Culture, Japan Medical Association, and Uehara Memorial Foundation. The author thanks Miss Haruko Kawamura for preparing this manuscript.

References

Aboud, M., Golde, D. W., Bersch, N., Rosenblatt, J. D., and Chen, I. S. Y. (1987). *Blood* **70,** 432–436.

Akagi, T., Ohtsuki, Y., Takahashi, K., Kubonishi, I., and Miyoshi, I. (1982). *Lab. Invest.* **47,** 406–408.

Akagi, T., Takeda, I., Oka, T., Ohtsuki, Y., Yano, S., and Miyoshi, I. (1985). *Jpn. J. Cancer Res. (Gann)* **76,** 86–94.

Akagi, T., Takata, H., Ohtsuki, Y., Takahashi, K., Oka, T., Yano, S., and Miyoshi, I. (1986). *Int. J. Cancer* **37,** 775–779.

Akagi, T., Yoshino, T., Motoi, M., Takata, H., Yano, S., Miyoshi, I., Oka, T., and Ohtsuki, Y. (1988). *Jpn. J. Cancer Res. (Gann)* **79,** 836–842.

Ando, Y., Nakano, S., Saito, K., Shimamoto, I., Ichijo, M., Toyama, T., and Hinuma, Y. (1987). *Jpn. J. Cancer Res. (Gann)* **78,** 322–324.

Ando, Y., Kakimoto, K., Tanigawa, T., Furuki, K., Saito, K., Nakano, S., Hashimoto, H., Moriyama, I., Ichijo, M., and Toyama, T. (1989). *Jpn. J. Cancer Res.* **80,** 405–407.

Berman, P. W., Groopman, J. E., Gregory, T., Clapham, P. R., Weiss, R. A., Ferriani, R., Riddle, L., Shimasaki, C., Lucas, C., Lasky, L. A., and Eichberg, J. W. (1988). *Proc. Natl. Acad. Sci. U.S.A.* **85,** 5200–5204.

Chen, I. S. Y., Quan, S. G., and Golde, D. W. (1983). *Proc. Natl. Acad. Sci. U.S.A.* **80,** 7006–7009.

Clapham, P., Nagy, K., Cheingsong-Popov, R., Exley, M., and Weiss, R. A. (1983). *Science* **222,** 1125–1127.

Clapham, P., Nagy, K., and Weiss, R. A. (1984). *Proc. Natl. Acad. Sci. U.S.A.* **81,** 2886–2889.

de Rossi, A., Aldovini, A., Franchini, G., Mann, D., Gallo, R. C., and Wong-Staal, F. (1985). *Virology* **143,** 640–645.

Eguchi, T., Kubonishi, I., Daibata, M., Yano, S., Imamura, J., Ohtsuki, Y., and Miyoshi, I. (1988). *Int. J. Cancer* **41,** 868–872.

Filice, G., Cereda, P. M., and Varnier, O. E. (1988). *Nature (London)* **335,** 366–369.

Gazdar, A. F., Carney, D. N., Bunn, P. A., Russell, E. K., Jaffe, E. S., Schecter, G. P., and Guccion, J. G. (1980). *Blood* **55,** 409–417.

Gessain, A., Barin, F., Vernant, J. C., Gout, O., Maurs, L., Calender, A., de Thé, G. (1985). *Lancet* **ii,** 407–410.

Gotoh, Y., Sugamura, K., and Hinuma, Y. (1982). *Proc. Natl. Acad. Sci. U.S.A.* **79**, 4780–4782.

Grassmann, R., Dengler, C., Müller-Fleckenstein, I., Fleckenstein, B., McGuire, K., Dokhelar, M.-C., Sodroski, J. G., and Haseltine, W. A. (1989). *Proc. Natl. Acad. Sci. U.S.A.* **86**, 3351–3355.

Graziano, S. L., Lehr, B. M., Merl, S. A., Ehrlich, G. D., Moore, J. L., Hallinan, E. J., Hubbell, C., Davy, F. R., Vournakis, J., and Poiesz, B. J. (1987). *Cancer Res.* **47**, 2468–2473.

Gross, L. (1951). *Proc. Soc. Exp. Biol. Med.* **76**, 27–32.

Hattori, T., Uchiyama, T., Toibana, T., Takatsuki, K., and Uchino, H. (1981). *Blood* **58**, 645–647.

Hino, S., Yamaguchi, K., Katamine, S., Sugiyama, H., Amagasaki, T., Kinoshita, K., Yoshida, Y., Doi, H., Tsuji, Y., and Miyamoto, T. (1985). *Jpn. J. Cancer Res. (Gann)* **76**, 474–480.

Hino, S., Sugiyama, H., Doi, H., Ishimaru, T., Yamabe, T., Tsuji, Y., and Miyamoto, T. (1987). *Lancet* **ii**, 158–159.

Hinuma, Y., Nagata, K., Hanaoka, M., Nakai, M., Matsumoto, T., Kinoshita, K., Shirakawa, S., and Miyoshi, I. (1981). *Proc. Natl. Acad. Sci. U.S.A.* **78**, 6476–6480.

Hinuma, Y., Gotoh, Y., Sugamura, K., Nagata, K., Goto, T., Nakai, M., Kamada, N., Matsumoto, T., and Kinoshita, K. (1982). *Gann* **73**, 341–344.

Hirose, S., Kotani, S., Uemura, Y., Fujishita, M., Taguchi, H., Ohtsuki, Y., and Miyoshi, I. (1988). *Virology* **162**, 487–489.

Ho, D. D., Rota, T. R., and Hirsch, M. S. (1984). *Proc. Natl. Acad. Sci. U.S.A.* **81**, 7588–7590.

Hoshino, H., Shimoyama, M., Miwa, M., and Sugimura, T. (1983). *Proc. Natl. Acad. Sci. U.S.A.* **80**, 7337–7341.

Hoshino, H., Tanaka, H., Shimotohno, K., Miwa, M., Nagai, M., Shimoyama, M., and Sugimura, T. (1984). *Int. J. Cancer* **34**, 513–517.

Hoshino, H., Clapham, P. R., Weiss, R. A., Miyoshi, I., Yoshida, M., and Miwa, M. (1985). *Int. J. Cancer* **36**, 671–675.

Hoxie, J. A., Matthews, D. M., and Cines, D. B. (1984). *Proc. Natl. Acad. Sci U.S.A.* **81**, 7591–7595.

Hu, S.-L., Fultz, P. N., McClure, H. M., Eichberg, J. W., Thomas, E. K., Zarling, J., Singhal, M. C., Kosowski, S. G., Swenson, R. B., Anderson, D. C., and Todaro, G. (1987). *Nature (London)* **328**, 721–723.

Kajiyama, W., Kashiwagi, S., Ikematsu, H., Hayashi, J., Nomura, H., and Okochi, K. (1986). *J. Infect. Dis.* **154**, 851–857.

Kalyanaraman, V. S., Sarngadharan, M. G., Robert-Guroff, M., Miyoshi, I., Blayney, D., Golde, D., and Gallo, R. C. (1982). *Science* **218**, 571–573.

Kinoshita, K., Hino, S., Amagasaki, T., Ikeda, S., Yamada, Y., Suzuyama, J., Momita, S., Toriya, K., Kamihira, S., and Ichimaru, M. (1984). *Gann* **75**, 103–105.

Kinoshita, K., Yamanouchi, K., Ikeda, S., Momita, S., Amagasaki, T., Soda, H., Ichimaru, M., Moriuchi, R., Katamine, S., Miyamoto, T., and Hino, S. (1985). *Jpn. J. Cancer Res. (Gann)* **76**, 1147–1153.

Koeffler, H. P., Chen, I. S. Y., and Golde, D. W. (1984). *Blood* **64**, 482–490.

Kotani, S., Yoshimoto, S., Yamato, K., Fujishita, M., Yamashita, M., Ohtsuki, Y., Taguchi, H., and Miyoshi, I. (1986). *Int. J. Cancer* **37**, 843–847.

Kulaga, H., Folks, T., Rutledge, R., Truckenmiller, M. E., Gugel, E., and Kindt, T. J. (1989). *J. Exp. Med.* **169**, 321–326.

Mann, D. L., Clark, J., Clarke, M., Reitz, M., Popovic, M., Franchini, G., Trainor, C. D., Strong, D. M., Blattner, W. A., and Gallo, R. C. (1984). *J. Clin. Invest.* **74**, 56–62.

Markham, P. D., Salahuddin, S. Z., Macchi, B. Robert-Guroff, M., and Gallo, R. C. (1984). *Int. J. Cancer* **33**, 13–17.

Miyamoto, K., Tomita, N., Ishii, A., Nishizaki, T., and Togawa, A. (1984). *Gann* **75**, 655–659.

Miyoshi, I. (1984). *Acta Haematol. Jpn.* **47**, 1535–1541.

Miyoshi, I., Kubonishi, I., Sumida, M., Yoshimoto, S., Hiraki, S., Tsubota, T., Kobashi, H., Lai, M., Tanaka, T., Kimura, I., Miyamoto, K., and Sato, J. (1979). *Jpn. J. Clin. Oncol.* **9** (suppl.), 485–494.

Miyoshi, I., Kubonishi, I., Yoshimto, S., and Shiraishi, Y. (1981a). *Gann* **72**, 978–981.

Miyoshi, I., Kubonishi, I., Yoshimoto, S., Akagi, T., Ohtsuki, Y., Shiraishi, Y., Nagata, K., and Hinuma, Y. (1981b). *Nature (London)* **294**, 770–771.

Miyoshi, I., Yoshimoto, S., Kubonishi, I., Taguchi, H., Shiraishi, Y., Ohtsuki, Y., and Akagi, T. (1981c). *Gann* **72**, 997–998.

Miyoshi, I., Fujishita, M., Taguchi, H., Ohtsuki, Y., Akagi, T., Morimoto, Y. M., and Nagasaki, A. (1982a). *Lancet* **i**, 683–684.

Miyoshi, I., Taguchi, H., Fujishita, M., Niiya, K., Kitagawa, T., Ohtsuki, Y., and Akagi, T. (1982b). *Gann* **73**, 339–340.

Miyoshi, I., Taguchi, H., Fujishita, M., Yoshimoto, S., Kubonishi, I., Ohtsuki, Y., Shiraishi, Y., and Akagi, T. (1982c). *Lancet* **i**, 1016.

Miyoshi, I., Taguchi, H., Kubonishi, I., Yoshimoto, S., Ohtsuki, Y., Shiraishi Y., and Akagi, T. (1982d). *Gann Monogr.* **28**, 219–228.

Miyoshi, I., Yoshimoto, S., Taguchi, H., Kubonishi, I., Fujishita, M., Ohtsuki, Y., Shiraishi, Y., and Akagi, T. (1983a). *Gann* **74**, 1–4.

Miyoshi, I., Taguchi, H., Ohtsuki, Y., and Shiraishi, Y. (1983b). *In* "Oncogenes and Retroviruses" (F. Rauscher, and T. E. O'Connor, eds.), pp. 243–249. Alan R. Liss, New York.

Miyoshi, I., Kubonishi, I., Yoshimoto, S., Ohtsuki, Y., and Akagi, T. (1984a). *Gann* **75**, 482–484.

Miyoshi, I., Yoshimoto, S., Fujishita, M., Kubonishi, I., Yamashita, M., Yamato, K., Taguchi, H., Ohtsuki, Y., and Akagi, T. (1984b). *In* "Manipulation of Host Defense Mechanisms" (T. Aoki, E. Tsubura, and I. Urushizaki, eds.), pp. 144–151. Excerpta Medica, Amsterdam.

Miyoshi, I., Yoshimoto, S., Kubonishi, I., Fujishita, M., Ohtsuki, Y., Yamashita, M., Yamato, K., Hirose, S., Taguchi, H., Niiya, K., and Kobayashi, M. (1985). *Int. J. Cancer* **35**, 81–85.

Nagy, K., Clapham, P., Cheingsong-Popov, R., and Weiss, R. A. (1983). *Int. J. Cancer* **32**, 321–328.

Nakamura, H., Hayami, M., Ohta, Y., Ishikawa, K., Tsujimoto, H., Kiyokawa, T., Yoshida, M., Sasagawa, A., and Honjo, S. (1987). *Int. J. Cancer* **40**, 403–407.

Nakano, S., Ando, Y., Ichijo, M., Moriyama, I., Saito, S., Sugamura, K., and Hinuma, Y. (1984). *Gann* **75**, 1044–1045.

Nerenberg, M., Hinrichs, S. H., Reynolds, R. K., Khoury, G., and Jay, G. (1987). *Science* **237**, 1324–1329.

Okochi, K., Sato, H., and Hinuma, Y. (1984). *Vox Sang.* **46**, 245–253.

Osame, M., Usuku, K., Izumo, S., Ijichi, N., Amitani, H., Igata, A., Matsumoto, M., and Tara, M. (1986). *Lancet* **i**, 1031–1032.

Poiesz, B. J., Ruscetti, F. W., Mier, J. W., Woods, A. M., and Gallo, R. C. (1980a). *Proc. Natl. Acad. Sci. U.S.A.* **77**, 6815–6819.

Poiesz, B. J., Ruscetti, F. W., Gazdar, A. F., Bunn, P. A., Minna, J. D., and Gallo, R. C. (1980b). *Proc. Natl. Acad. Sci. U.S.A.* **77**, 7415–7419.

Poiesz, B. J., Ruscetti, F. W., Reitz, M. S., Kalyanaraman, V. S., and Gallo, R. C. (1981). *Nature (London)* **294**, 268–271.

Popovic, M., Reits, M. S. Jr., Sarngadharan, M. G., Robert- Guroff, M., Kalyanaraman, V. S., Nakao, Y., Miyoshi, I., Minowada, J., Yoshida, M., Ito, Y., and Gallo, R. C. (1982). *Nature (London)* **300**, 63–66.

Popovic, M., Sarin, P. S., Robert-Guroff, M., Kalyanaraman, V. S., Mann, D., Minowada, J., and Gallo, R. C. (1983a). *Science* **219**, 856–859.

Popovic, M., Lange-Wantzin, G., Sarin, P. S., Mann, D., and Gallo, R. C. (1983b). *Proc. Natl. Acad. Sci. U.S.A.* **80**, 5402–5406.

Prince, A. M., Horowitz, B., Baker, L., Shulman, R. W., Ralph, H., Valinsky, J., Cundell, A., Brotman, B., Boehle, W., Rey, F., Piet, M., Reesink, H., Lelie, N., Tersmette, M., Miedema, F., Barbosa, L., Nemo, G., Nastala, C. L., Allan, J. S., Lee, D. R., and Eichberg, J. W. (1988). *Proc. Natl. Acad. Sci. U.S.A.* **85**, 6944–6948.

Rosenblatt, J. D., Golde, D. W., Wachsman, W., Giorgi, J. V., Jacobs, A., Schmidt, G. M., Quan, S., Gasson, J. C., and Chen, I. S. Y. (1986). *N. Engl. J. Med.* **315**, 372–377.

Salahuddin, S. Z., Markham, P. D., Wong-Staal, F., Franchini, G., Kalyanaraman, V. S., and Gallo, R. C. (1983). *Virology* **129**, 51–64.

Salahuddin, S. Z., Markham, P. D., Lindner, S. G., Gootenberg, J., Popovic, M., Hemmi, H., Sarin, P. S., and Gallo, R. C. (1984). *Science* **223**, 703–707.

Seto, A., Kawanishi, M., Matsuda, S., Ogawa, K., and Miyoshi, I. (1988a). *Jpn. J. Cancer Res. (Gann)* **79**, 335–341.

Seto, A., Kawanishi, M., Matsuda, S., and Ogawa, K. (1988b). *J. Exp. Med.* **168**, 2409–2414.

Shida, H., Tochikura, T., Sato, T., Konno, T., Hirayoshi, K., Seki, M., Ito, Y., Hatanaka, M., Hinuma, Y., Sugimoto, M., Takahashi-Nishimaki, F., Maruyama, T., Miki, K., Suzuki, K., Morita, M., Sashiyama, H., and Hayami, M. (1987). *EMBO J.* **6**, 3379–3384.

Sinangil, F., Harada, S., Purtilo, D. T., and Volsky, D. J. (1985). *Int. J. Cancer* **36**, 191–198.

Sommerfelt, M. A., Williams, B. P., Clapham, P. R., Solomon, E., Goodfellow, P. N., and Weiss, R. A. (1988). *Science* **242**, 1557–1559.

Tajima, K., Tominaga, S., Suchi, T., Kawagoe, T., Komoda, H., Hinuma, Y., Oda, T., and Fujita, K. (1982). *Gann* **73**, 893–901.

Takehara, N., Iwahara, Y., Uemura, Y., Sawada, T., Ohtsuki, Y., Iwai, H., Hoshino, H., and Miyoshi, I. (1989). *Int. J. Cancer* **44**, 332–336.

Tateno, M., Kondo, N., Itoh, T., Chubachi, T., Togashi, T., and Yoshiki, T. (1984). *J. Exp. Med.* **159**, 1105–1116.

Uchiyama, T., Yodoi, J., Sagawa, K., Takatsuki, K., and Uchino, H. (1977). *Blood* **50**, 481–492.

Uemura, Y., Kotani, S., Yoshimoto, S., Fujishita, M., Yano, S., Ohtsuki, Y., and Miyoshi, I. (1986). *Jpn. J. Cancer Res. (Gann)* **77**, 970–973.

Uemura, Y., Kotani, S., Yoshimoto, S., Fujishita, M., Yamashita, M., Ohtsuki, Y., Taguchi, H., and Miyoshi, I. (1987). *Blood* **69**, 1255–1258.

Watanabe, T., Seiki, M., and Yoshida, M. (1984). *Virology* **133**, 238–241.

Wong-Staal, F., and Gallo, R. C. (1985). *Nature (London)* **317**, 395–403.

Yamamoto, N., and Hinuma, Y. (1985). *J. Gen. Virol.* **66**, 1641–1660.

Yamamoto, N., Okada, M., Koyanagi, Y., Kannagi, M., and Hinuma, (1982a). *Science* **217**, 737–739.

Yamamoto, N., Matsumoto, T., Koyanagi, Y., Tanaka, Y., and Hinuma, Y. (1982b). *Nature (London)* **299**, 367–369.

Yamamoto, N., Hayami, M., Komuro, A., Schneider, J., Hunsmann, G., Okada, M., and Hinuma, Y. (1984). *Med. Microbiol. Immunol.* **172**, 57–64.

Yamanouchi, K., Kinoshita, K., Moriuchi, R., Katamine, S., Amagasaki, T., Ikeda, S., Ichimaru, M., Miyamoto, T., and Hino, S. (1985). *Jpn. J. Cancer Res. (Gann)* **76**, 481–487.

Yamato, K., Taguchi, H., Yoshimoto, S., Fujishita, M., Yamashita, M., Ohtsuki, Y., Hoshino, H., and Miyoshi, I. (1986). *Jpn. J. Cancer Res. (Gann)* **77**, 13–15.

Yoshida, M. (1983). *Gann* **74**, 777–789.

Yoshida, M., Miyoshi, I., and Hinuma, Y. (1982). *Proc. Natl. Acad. Sci. U.S.A.* **79**, 2031–2035.

Receptors for Human Retroviruses

Robin A. Weiss

I. Introduction

There has been a resurgence of interest in cell surface receptors for human viruses, since the CD4 cell differentiation antigen was shown to be the receptor for human immunodeficiency virus type 1 (HIV-1) (Dalgleish *et al.*, 1984; Klatzmann *et al.*, 1984). Other viruses besides HIV have adopted for their own use cell surface antigens performing normal physiological or differentiation functions. Influenza virus utilizes sialic acid residues on several glycoproteins or glycolipids (Weis *et al.*, 1988). Epstein Barr virus binds to the CR2 glycoprotein which is the receptor for the complement component C3d (Fingeroth *et al.*, 1984); poliovirus binds to a new member of the immunoglobulin superfamily of membrane proteins discovered through gene transfection (Mendelsohn *et al.*, 1989); the major group of rhinoviruses use a related protein, the ICAM-1 cell adhesion molecule (Greve *et al.*, 1989; Staunton *et al.*, 1989). Cytomegalovirus (CMV) may bind to Class I major histocompatibility (MHC) antigens by using β_2-microglobulin as a bridging molecule (Grundy *et al.*, 1987).

Retroviruses show a great variety of specificities of host range and cell tropism (Weiss, 1985). This is determined both at the cell surface level and in restriction of replication following viral penetration, e.g., the *Fv-1* host restriction of murine leukemia virus (MLV) in mice (Jolicoer and Baltimore, 1976). The murine receptor gene for ecotropic MLV (MLV-E) has been cloned using a strategy of gene transfer to receptor-negative cells followed by infection of a retrovirus vector containing a dominant selectable marker

(Albritton *et al.*, 1989). A similar method was used to identify the human gene encoding the receptor for gibbon ape leukemia virus (O'Hara et al., 1990).

Many animal retroviruses can infect human cells in culture, although they are not known to occur as natural infections *in vivo*. For example, xenotropic MLV and feline leukemia virus subgroup B (FeLV-B) readily replicate in human cells. Despite man's proximity to cats and mice, however, these viruses are not known to infect humans. By analyzing interference patterns between twenty different retroviruses plating on human cells, we recently delineated at least eight distinct receptors (Sommerfelt and Weiss, 1990).

Among the human retroviruses, little is known about the receptor for the putative spumavirus or foamy virus, except that it will replicate in many cell types *in vitro* (Loh *et al.*, 1977). The oncoviruses, human T-cell leukemia virus types I and II (HTLV-I, HTLV-II), utilize a common receptor not yet identified biochemically. The lentiviruses, HIV-1 and HIV-2, as well as simian immunodeficiency viruses (SIV) all use CD4 as the major, high-affinity receptor on lymphocytes and monocytes (Sattentau *et al.*, 1988), although other modes of virus entry are known and will be described.

Retroviruses bind to cell surface receptors via their outer envelope glycoproteins, known as surface (SU) proteins. Following binding, membrane fusion is probably effected by a hydrophobic region of the transmembrane (TM) envelope protein. Fusion of many retroviruses takes place at neutral pH (McClure *et al.*, 1990) and is reflected by the capacity of the retrovirus to induce syncytial, multinucleated cells (Sommerfelt and Weiss, 1990). However, MLV-E (Andersen and Nexo, 1983) and murine mammary tumor virus (Redmond *et al.*, 1984) normally require the acid environment of endosomes to effect penetration. The mode of entry of HIV and HTLV into cells following binding to the primary receptor is as yet poorly understood.

II. Detection of Receptors

Cell types that are susceptible to human retrovirus infection presumably express cell surface receptors. However, these may be difficult to detect or enumerate. Direct binding assays of virions and of purified or recombinant envelope glycoproteins can be employed to detect receptors. Fluorescence techniques have been used to detect virion binding (Sinangil *et al.*, 1984; McDougal *et al.*, 1985; Krichbaum-Stenger *et al.*, 1987). Coprecipitation of iodinated cell surface proteins with viral glycoproteins was successful in discerning CD4 with HIV gp120 (McDougal *et al.*, 1986).

Functional steps in binding and internalization of human retroviruses have employed cell fusion assays whereby the retrovirus induces the formation of multinucleated syncytia. This was observed with both HTLV-I

(Hoshino *et al.*, 1983; Nagy *et al.*, 1983) and HIV (Dalgleish *et al.*, 1984; Montagnier *et al.*, 1984). Another useful method is to construct viral pseudotypes bearing the envelope specificity of the retrovirus under study. Vesicular stomatitis virus (VSV) pseudotypes bearing the envelope glycoproteins of HTLV-I or HTLV-II (Clapham *et al.*, 1984) and of HIV-1 (Dalgleish *et al.*, 1984). Maddon *et al.*, 1986) provide a simple plaque assay for the presence or absence of receptors. This has been particularly useful for HTLV where cell-free virions of the retrovirus itself are difficult to obtain.

Retroviral pseudotypes that package a dominant selectable marker can also be exploited for receptor studies. This is how the genetic basis of chicken cell receptors for avian leukemia viruses were first elucidated (Rubin, 1965; Vogt and Ishizaki, 1965) using the *src* gene of defective Bryan strain Rous sarcoma virus. The modern equivalent is to use recombinant retroviral vectors containing drug resistance markers, as exploited by Albritton *et al.* (1989) and O'Hara *et al.* (1990) for receptor studies.

III. The HTLV Receptor

The molecular identity of the cell surface receptor for HTLV-I and HTLV-II is still not known. Syncytial and VSV pseudotype assays indicate that the receptor is widely expressed on human cell lines in culture, and also several other mammalian species including cat, rat, and hamster cell lines (Hoshino *et al.*, 1983; Nagy *et al.*, 1983, 1984). Human osteosarcoma cells (Clapham *et al.*, 1983) and endothelial cells (Ho *et al.*, 1984; Hoxie *et al.*, 1984) can be productively infected with HTLV-I, as can B cells (Longo *et al.*, 1984), T cells (Miyoshi *et al.*, 1981; Popovic *et al.*, 1983), and myeloid leukemia cells (D'Onofrio *et al.*, 1988). Rabbit and rat lymphocytes have also been immortalized by HTLV-I (Miyoshi *et al.*, 1981; Tateno *et al.*, 1989). Although the malignant cells of adult T-cell leukemia lymphoma are derived from CD4+ T-helper lymphocytes as are the cells immortalized *in vitro* by HTLV-I, it is clear that many other proliferating cell types express receptors in culture.

The cell tropism exhibited by HTLV-I in cell transformation does not appear, therefore, to be chiefly determined at the receptor level. This was also ascertained by fluorescent virion binding (Sinangil *et al.*, 1985; Krichbaum-Stenger *et al.*, 1987). Virion attachment to separated fresh blood mononuclear cells indicated a preferential binding to activated T lymphocytes rather than resting cells and even more strongly to proliferating cells in culture. Among human T cells there was a good correlation between virion binding and short-term infection (Krichbaum-Stenger *et al.*, 1987), but this was less apparent in a wider range of cell types (Sinangil *et al.*, 1985). It also appears that CD4+ cells are the principal infected population in peripheral blood cells of asymptomatic subjects or patients with tropical spastic paraparesis (Richardson *et al.*, 1990).

Although Sinangil *et al.* (1985) observed binding of HTLV-I to Balb/c mouse splenocytes, our studies indicated that the sensitivity of mouse cells to the plating of VSV (HTLV-I) pseudotypes is approximately 1% that of human cells (Nagy *et al.*, 1984). This observation allowed us to use human–mouse somatic cell hybrids to map the human gene determining high sensitivity to HTLV-I. We found that human chromosome 17 carried the gene determining the HTLV-I receptor (Sommerfelt *et al.*, 1988). The gene is probably located on cen-q1.2 region of the chromosome.

The HTLV-II receptor gene cosegregates with that for HTLV-I (Sommerfelt *et al.*, 1988). Cross-interference studies indicate that HTLV-I and HTLV-II share the same receptor. This has been adduced from competition to virion binding (Krichbaum-Stenger *et al.*, 1987), and interference to syncytium formation and to VSV pseudotypes (Sommerfelt and Weiss, 1990). Related retroviruses of chimpanzees and macaques (STLV) also share the HTLV receptor, whereas bovine leukosis virus, a related *tax*- and *rex*-containing retrovirus, recognizes a different receptor on human cells (Sommerfelt and Weiss, 1990).

The nature of the HTLV receptor remains to be elucidated. Various approaches to expression cloning and screening antibodies to chromosome 17-specific determinants are being employed to identify this receptor.

IV. The Role of CD4 in HIV Infection and Pathogenesis

The evidence that the CD4 differentiation antigen is the high-affinity binding receptor for HIV-1 on T lymphocytes is well known (Sattentau and Weiss, 1988). Monoclonal antibodies (mAbs) specific for CD4 block HIV-1 infection (Klatzmann *et al.*, 1984), syncytium formation and pseudotype plating (Dalgleish *et al.*, 1984), binding of HIV virions (McDougal *et al.*, 1985) and recombinant gp120 (Lasky *et al.*, 1987). CD4 can be coprecipitated with gp120 after virion binding to the cell surface (McDougal *et al.*, 1986). Finally, expression of the cloned CD4 cDNA in otherwise CD4$^-$ cells such as HeLa renders them sensitive to HIV-1 binding, cell fusion, and infection (Maddon *et al.*, 1986).

The CD4 antigen belongs to the immunoglobulin superfamily (Maddon *et al.*, 1986, 1988), with four immunoglobulin-like domains (V1–V4). Mapping the recognition site on CD4 for HIV gp120 has been accomplished with monoclonal antibodies and site-specific mutagenesis (Sattentau *et al.*, 1986, 1989; McClure *et al.*, 1987; Berger *et al.*, 1988; Peterson and Seed, 1988; Landau *et al.*, 1988; Clayton *et al.*, 1988; Mizukami *et al.*, 1988; Arthos *et al.*, 1989). From these studies, it is clear that the gp120 binding site is contained in the N-terminal V1 domain. Conservative changes switching murine CD4 codons in place of human ones (murine CD4 does not bind gp120 but has a similar overall domain organization) indicate that an epitope between amino acid residues 40 and 60 is required. (Peterson and Seed, 1988; Arthos *et al.*, 1989). Another domain at residues 80–100 may also be important (Lifson *et al.*, 1988; Jameson *et al.*, 1988).

The CD4 antigen is important in immune recognition. $CD4^+$ T lymphocytes, whether with helper or cytotoxic functions, interact specifically with class II major histocompatibility antigens (MHC II) (Doyle and Strominger, 1987; Gay *et al.*, 1987). CD4 may act as an accessory adhesion molecule that recognizes a nonpolymorphic region of the MHC-II–T-cell receptor complex (Swain, 1983). However, CD4 may also play a signaling role through the association of the lymphocyte-specific p56 tyrosine kinase with its cytoplasmic domain (Shaw *et al.*, 1989). The binding of gp120 to CD4 blocks immune reactions (Habeshaw and Dalgleish, 1989) and may also uncouple p56 (A. Pelchen-Matthews and M. Marsh, personal communication).

Relatively few peripheral blood lymphocytes are infected with and express HIV in infected subjects, so it has remained a puzzle as to how HIV infection eventually induces such a severe depletion of $CD4^+$ cells and the ensuing immune deficiency (see Rosenberg and Fauci, Chapter 10). The infection of precursor cells, the short survival of virus-expressing cells, and the induction by HIV infection of growth suppressors such as tumor necrosis factor each help to explain the total depletion of T cells when only a small proportion appear to be infected at any one time. Receptor interaction may also play a role (Habeshaw and Dalgleish, 1989). In culture, a single HIV producing cell may fuse up to 100 "bystander" $CD4^+$ uninfected cells into giant syncytia, which subsequently die. Giant cells are rarely seen *in vivo*, but would be rapidly cleared from the circulation. It is not known whether cell fusion is important in HIV pathogenesis. Soluble gp120 shed from HIV virions and from infected cells would also be expected to exhibit immunosuppressive effects *in vivo*. Soluble gp120 has not been detected in human plasma but would presumably be rapidly and specifically adsorbed to $CD4^+$ cells in the circulation and fixed tissues.

V. Exploiting CD4–gp120 Interactions for Vaccines and Therapy

Our knowledge of CD4–gp120 coupling has led to many studies aimed at blocking the interaction in order to prevent HIV infection or its subsequent spread within the body.

The site on gp120 thought to be crucial for CD4 recognition (Lasky *et al.*, 1987; Cordonnier *et al.*, 1989) does not appear to be highly immunogenic in infected humans. Probably this epitope is folded within the gp120 molecule. However, mAbs raised to this epitope have a neutralizing effect (Sun *et al.*, 1989). We have expressed this epitope on the surface of polio virus Sabin type 1 virions, following the constructs described for the HIV gp41 neutralization site (Evants *et al.*, 1989). Sera raised against these polio virus chimeras bearing gp120 peptides of the CD4-recognition site block CD4–gp120 interaction and both HIV-1 and HIV-2 infection. This epitope may therefore be relevant to vaccine development because of its functional and conserved nature (McKeating and Willey, 1989).

An alternative approach to blocking CD4–gp120 interaction for vaccine development has been to raise anti-idiotypic antibodies to anti-CD4 mAbs to the site recognized by gp120. The Leu3a anti-CD4 mAb binds to the same site that appears to be necessary for gp120 binding (Peterson and Seed, 1988). Mirror-image, anti-idiotypic sera to the antigen-combining site of Leu3a, therefore, might recognize gp120 and neutralize HIV by preventing binding to CD4. Weak neutralization of many HIV strains has been reported (Chanh *et al.*, 1987; Dalgleish *et al.*, 1987). Potent neutralizing monoclonal anti-idiotypic antibodies have not, however, been obtained, and the value of the receptor anti-idiotype approach remains in doubt (Beverley *et al.*, 1989).

More progress has been made in exploiting recombinant, soluble forms of CD4 as potential therapeutic agents. By synthesizing truncated forms of CD4 secreted by cells, a soluble CD4 protein (sCD4) that powerfully neutralizes HIV infectivity can be made (Smith *et al.*, 1987; Fisher *et al.*, 1988; Hussey *et al.*, 1988; Deen *et al.*, 1988; Traunecker *et al.*, 1988). Because the CD4-recognition site is conserved on HIV-1, HIV-2, and SIV, sCD4 molecules neutralize diverse strains of virus (Clapham *et al.*, 1989). A recent report curiously records an enhancement of infection by sCD4 for SIV_{agm} (Allan *et al.*, 1990). The V1 domain containing the gp120 binding site is as potent in blocking HIV as sCD4 representing the whole of the extramembranous domains (Arthos *et al.*, 1989). A synthetic peptide may also be inhibitory (Lifson *et al.*, 1988).

Various sCD4 products are currently undergoing phase I clinical trial in HIV-infected subjects and AIDS patients. Because sCD4 is rapidly cleared from plasma, hybrid constructs with the Fc portion of immunoglobulin molecules are being employed to maintain an effective therapeutic dose (Capon *et al.*, 1989; Traunecker *et al.*, 1989). Soluble CD4 does not appear to be immunotoxic itself (Liu and Liu, 1988; Watanabe *et al.*, 1989) but it is too early to tell whether it will form a useful part of the HIV therapeutic armamentarium. Primary strains of HIV-1 are less sensitive to inactivation by sCD4 (Daar *et al.*, 1990).

A further exploitation of CD4-gp120 interaction is to employ sCD4 linked to or fused with bacterial or plant toxins to destroy HIV-infected cells. sCD4–toxins will bind to cells expressing surface gp120 with high affinity, thereby killing sources of HIV production (see Chapter 18; Chaudhary *et al.*, 1988; Till *et al.*, 1988). This type of molecule works potently on HIV-infected cells in culture but its pharmaceutical use is, to my mind, more questionable.

VI. HIV Cell Tropism and Alternative Receptors

CD4 antigen is the principal, high-affinity receptor for HIV-1 on T lymphocytes. HIV-2 and various SIV isolates also recognize CD4 and their infection can be blocked by anti-CD4 antibodies (Sattentau *et al.*, 1988). Primate CD4 antigens are highly conserved around the gp120 binding site; even New-World monkey T cells are susceptible to HIV1 and HIV-2 infection *in vitro* via CD4 (McClure *et al.*, 1988). Rhesus monkeys are being used to test sCD4 in SIV therapy (Watanabe *et al.*, 1989). The affinity of HIV-2 for CD4 appears to be lower than that of HIV-1 (Clapham *et al.*, 1989; Moore, 1990). We have also been able to select variants of HIV-1 *in vitro* that are relatively resistant to sCD4 (J. A. McKeating *et al.*, unpublished results) which may not augur well for sCD4 therapeutics.

The CD4 antigen is also the principal HIV receptor expressed on monocytic cell lines (Åsjö *et al.*, 1987; Clapham *et al.*, 1987) and on monocyte and macrophage cultures (Collman *et al.*, 1989; Meltzer *et al.*, 1990. Soluble CD4 blocks monocyte infection as well as T-lymphocyte infection (Clapham *et al.*, 1989; Meltzer *et al.*, 1990). Monocytes and macrophages are thought to be important reservoirs of HIV infection *in vivo*, and their infection by HIV contributes significantly to pathogenesis (Gartner *et al.*, 1986; Pauza *et al.*, 1988; Gendelman *et al.*, 1989). Macrophages and microglial cells (of monocytic origin) may be the main route of entry and reservoir of HIV infection in the brain (Koenig *et al.*, 1986).

Dendritic cells may also be important in HIV infection (Patterson and Knight, 1988). These cells and their derivatives such as Langerhans cells in the skin and mucous membranes express CD4 (Wood *et al.*, 1983), and anti-CD4 antibodies block dendritic cell infection by HIV *in vitro*. Thus phagocytic and antigen-presenting cells of the immune system become infected by HIV via CD4. Langerhans cells appear to be a target for infection *in vivo* (Tschaler *et al.*, 1987). Certain strains of HIV in culture are more T-lymphotropic or more monocyte-tropic (Gartner *et al.*, 1986, Cheng-Mayer *et al.*, 1987; Collman *et al.*, 1989). It is not yet known whether such tropism involves differences in CD4-receptor recognition. Some lymphoblastic B-cell lines transformed by Epstein-Barr virus can be infected in culture by HIV (Montagnier *et al.*, 1984). These cells also depend on CD4 for HIV infection (Dalgleish *et al.*, 1984; Monroe *et al.*, 1988); the significance of B-cell infection is not known.

Human glial, neural, fibroblastic, and epithelial cell lines can be infected in culture at low efficiency by HIV (Chiodi *et al.*, 1987; Christofinis *et al.*, 1987; Cheng-Mayer *et al.*, 1987; Dewhurst *et al.*, 1987; Weber *et al.*, 1989). In several of these cases, HIV infection appears to be CD4 independent, because neither anti-CD4 antibodies nor sCD4 block infection (Clapham *et al.*, 1989; Harouse *et al.*, 1989; Tateno *et al.*, 1989; Weber *et al.*, 1989).

These cells may therefore express an alternative receptor for HIV, though it would not appear to be an efficient or high-affinity interaction as observed for CD4.

Several groups have demonstrated enhanced infection of monocytes and other cell types by HIV–antibody complexes. Infection by opsonized HIV takes place via Fc gamma receptors (Takeda et al., 1988; Joualt et al., 1989) or complement receptors (Robinson et al., 1988). The enhancement is less than 10-fold and is therefore much less than observed for flaviviruses. However, dilution of human antibodies beyond the end point for HIV neutralization does yield significant enhancement. Fc receptor-mediated infection of CD4$^+$ monocyte can proceed in the presence of CD4 inhibitors (Homsy et al., 1989). Infection via Fc or complement receptors has implications for vaccines as the generation of enhancing antibodies should be avoided.

Recently, we have shown that Fc receptors induced by CMV can render otherwise HIV-resistant human fibroblasts sensitive to infection by opsonized HIV (McKeating et al., 1990). Thus Fc receptors may expand HIV cell tropism as well as enhance infectious titers.

The host range and cell tropism of HIV can be extended by phenotypic mixing of the envelope with glycoproteins of other viruses. HIV pseudotypes with coats of amphotropic murine leukemia virus or of herpes simplex virus will infect and replicate in CD4$^-$ cells (Lusso et al., 1990; Zhu et al., 1990). While there is no direct evidence, it is plausible that this type of complementation could occur in vivo in mixed infection with other enveloped viruses. This can best be envisaged with viruses such as herpes simplex and cytomegalovirus where trans-activation of viral genome expression (Ho et al., 1990) and infection of common cell types via herpes-induced Fc receptors would allow dual expression of HIV and the complementing envelope. In this way, infection and spread in vivo might be exacerbated.

VII. Mechanism of HIV Entry into Cells

Although the binding of HIV to CD4 represents the first event of infection, subsequent steps in membrane fusion and internalization of the virus into cells are poorly understood. It would appear that binding to CD4 is not sufficient to trigger internalization. Maddon et al., (1986) showed that mouse cells expressing human CD4 bound HIV as avidly as human cells, but no cell fusion or HIV infection was apparent. Furthermore, vesicular stomatitis (VSV) pseudotypes bearing the HIV envelope glycoproteins were unable to plate on CD4$^+$ murine cells, although once internalized these cells are highly permissive for VSV. It is unlikely that mouse cells dominantly suppress HIV internalization as other nonprimate cells expressing human CD4 are also nonpermissive for HIV infection and cell fusion (Clapham et al., 1991), as is the human U87 cell line (Chesebro et al., 1990). It is therefore reasonable to assume that a second human component is necessary to trigger membrane fusion and effect HIV entry. This "second receptor" molecule may not be lineage-specific as diverse human cells of hemopoietic, epithelial, and fibroblastic origin become HIV permissive when CD4 is expressed. However, an attempt to detect HIV-permissive cells among CD4$^+$ human–murine T-cell hybrids was not successful (Tersmette *et al.*, 1989).

There is some evidence that HIV enters cells via receptor-mediated endocytosis (Bauer *et al.*, 1987; Pauza and Price, 1988). Electron microscopy also shows HTLV-I particles in coated pits and endosomes (Timar *et al.*, 1987). But cell constructs with truncated CD4 cytoplasmic domains, which are insensitive to endocytosis triggered by phorbol esters, remain equally sensitive to HIV infection and to HIV-mediated cell fusion (Bedinger *et al.*, 1988; Maddon *et al.*, 1988). Such cells, however, still exhibit constitutive CD4 endocytosis (Pelchen-Matthews *et al.*, 1989).

The observation that CD4$^+$ murine cells are insensitive to gp120-mediated cell fusion (Maddon *et al.*, 1986, 1988) indicates that a cell-surface trigger is absent from these cells following HIV adsorption to CD4. HIV-1 entry is not inhibited by raising endosomal pH (Stein *et al.*, 1987; McClure *et al.*, 1988). This is also true of HIV2 and HTLV-I and -II (McClure *et al.*, 1990). Therefore the trigger is probably not the acid environment of the endosome as is common with many enveloped viruses (Marsh and Helenius, 1989). In the case of ecotopic murine leukemia virus, proteolytic cleavage of the SU (gp70) protein is required during virus entry (Andersen and Skov, 1989). Cleavage of HIV gp120 (in the V3 neutralization loop) might be the trigger to membrane fusion and internalization with HIV (Stephens *et al.*, 1990). In that case one would postulate that the second receptor may be a cell surface proteinase (McClure *et al.*, 1990; Stephens *et al.*, 1990). Clapham *et al.* (1991) have noted different secondary cell surface factors for the infection of CD4$^+$ cells by HIV-1, HIV-2, and SIV.

A second cell surface component, necessary for membrane fusion but not for HIV binding, would be an attractive target for antiviral therapy. It is possible that sulfated sugars act by blocking this receptor. Further work is needed to elucidate the early postbinding events in human retrovirus infection.

References

Albritton, L. M., Tseng, L., Scadden, D., and Cunningham, J. M. (1989). *Cell* **57**, 659–666.
Allan, J. S., Strauss, J., and Buck, D. W. (1990). *Science* **247**, 1084–1088.
Andersen, K. B., and Nexo, B. A. (1983). *Virology* **125**, 85–98.
Andersen, K. B., and Skov, H. (1989). *J. Gen. Virol.* **70**, 1921–1927.
Arthos, J., Deen, C. K., Chaikin, M. A., Fornwald, J. A., Sattentau, Q. J., Clapham, P. R., Weiss, R. A., McDougal, J. S., Pietropaolo, C., Maddon, P. J., Truneh, A., Axel, R., and Sweet, R. W. (1989). *Cell* **57**, 469–481.
Åsjö, B., Ivhed, I., Gidlung, M., Fuerstenberg, S., Fenyo, E. M., Nilsson, K., and Wigzell, H. (1987). *Virology* **157**, 359–365.

Bauer, P. G., Barth, O. M., and Pereira, M. S. (1987). *Mem. Inst. Oswaldo Cruz* **82**, 449–450.

Bedinger, P., Moriarty, A., von-Borstel, R. C., Donovan, N. J., Steimer, K. S., and Littman, D. R. (1988). *Nature (London)*. **334**, 162–165.

Berger, E. A., Fuerst, T. R., and Moss, B. (1988). *Proc. Natl. Acad. Sci. U.S.A.* **85**, 2357–2361.

Beverley, P. C. L., Healey, D., Broadhurst, K., and Sattentau, Q. J. (1989). *J. Autoimmun.* **2** (Suppl.) 243–249.

Capon, D. J., Chamow, S. M., Mordenti, J., Marsters, S. A., Gregory, T., Mitsuya, H. J., Byrn, R. A., Lucas, C., Wurm, F. M., Groopman, J. E., Broder, S., and Smith, D. H. (1989). *Nature (London)* **337**, 525–527.

Chanh, T. C., Dreesman, G. R., and Kennedy, R. C. (1987). *Proc. Natl. Acad. Sci. U.S.A.* **84**, 389–385.

Chaudary, V. K., Mizukami, T., Fuerst, T. R., FitzGerald, D. J., Moss, B., Pastan, I., and Berger, E. A. (1988). *Nature (London)* **335**, 369–372.

Chesebro, B., Buller, R., Portis, J., and Wehrly, K. (1990). *J. Virol.* **64**, 215–221.

Cheng-Mayer, C., Rutka, J. T., Rosenblum, M. L., McHugh, T., Stites, D., and Levy, J. A. (1987). *Proc. Natl. Acad. Sci. U.S.A.* **84**, 3526–3530.

Chiodi, F., Fuerstenberg, G., Gidlund, M., Åsjö, B., and Fenyo, E. M. (1987). *J. Virol.* **61**, 1244–1247.

Christofinis, G., Papadaki, L., Sattentau, Q. J., Ferns, R. B., and Tedder, R. S. (1987). *AIDS* **1**, 229–234.

Clapham, P. R., Nagy, K., Cheingsong-Popov, R., and Weiss, R. A. (1983). *Science* **222**, 1125–1127.

Clapham, P. R., Nagy, K., and Weiss, R. A. (1984). *Proc. Natl. Acad. Sci. U.S.A.* **81**, 3083–3086.

Clapham, P. R., Weiss, R. A., Dalgleish, A. G., Exley, M., Whitby, D., and Hogg, N. (1987). *Virology* **158**, 44–51.

Clapham, P. R., Weber, J. N., Whitby, D., McIntosh, K., Dalgleish, A. C., Maddon, P. J., Deen, K. C., Sweet, R. W., and Weiss, R. A. (1989). *Nature (London)* **337**, 368–370.

Clapham, P. R., Blanc, D., and Weiss, R. A. (1991). *Virology* (in press).

Clayton, L. K., Hussey, R. E., Steinbrick, R., Ramachandran, H., Hussain, Y., and Reinherz, E. L. (1988). *Nature (London)* **335**, 363–366.

Collman, R., Hassan, N. F., Walker, R., Godfrey, B., Cutilli, J., Hastings, J. C., Friedman, H., Douglas, S. D., and Nathanson, N. (1989). *J. Exp. Med.* **170**, 1149–1156.

Cordonnier, A., Montagnier, L., and Emerman, M. (1989). *Nature (London)* **340**, 571–574.

Daar, E. S., Li, X. L., Moudgil, T., and Ho, D. D. (1990). *Proc. Natl. Acad. Sci. U.S.A.* **87**, 6574–6578.

Dalgleish. Q. G., Beverley, P. C. L., Clapham, P. R., Crawford, D. H., Creaves, M. F., and Weiss, R. A. (1984). *Nature (London)* **312**, 763–767.

Dalgleish, A. G., Thomson, B. J., Chanh, T. C., Malkovsky, M., and Kennedy, R. C. (1987). *Lancet* **ii**, 1047–1050.

Deen, K. C., McDougal, J. S., Inacker, R., Folena-Wasserman, G., Arthos, J., Rosenberg, J., Maddon, P. J., Axel, R., and Sweet, R. W. (1988). *Nature (London)* **331**, 82–84.

Dewhurst, S., Bresser, J., Stevenson, M., Sakai, R. K., Evinger-Hodges, M. J., and Volsky, D. (1987). *FEBS Lett.* **213**, 138–143.

D'Onofrio, C., Perno, C. F., Mazzetti, P., Graziano, G., Calio, R., and Bonmassar, E. (1988). *Brit. J. Cancer* **57**, 481–488.

Doyle, C., and Strominger, J. L. (1987). *Nature (London)* **330**, 256–259.

Evans, D. J., McKeating, J., Meredith, J. M., Burke, K. L., Katrak, K., John, A., Ferguson, M., Minor, P. D., Weiss, R. A., and Almond, J. W. (1989). *Nature (London)* **339**, 385–388.

Fingeroth, J. D., Weiss, J. J., Tedder, T. F., Strominger, J. L., Biro, P. A., and Fearon, D. T. (1984). *Proc. Natl. Acad. Sci. U.S.A.* **81**, 4510–4514.

Fisher, R. A., Bertonis, J. M., Meier, W., Johnson, V. A., Costopoulos, D. S., Liu, T., Tizard,

R., Walke, B. D., Hirsch, M. S., Schooley, R. T., and Flavell, R. A. (1988). *Nature* (*London*) **331**, 76–78.

Gartner, S., Markovitz, P., Markovitz, D. M., Kaplan, M. H., Gallo, R. C., and Popovic, M. (1986). *Science* **233**, 215–218.

Gay, D., Maddon, P., Sekaly, R., Talle, M. A., Godfrey, M., Long, E., Goldstein, G., Chess, L., Axel, R., Kappler, J., and Marrack, P. (1987). *Nature* (*London*) **328**, 626–628.

Gendelman, H. E., Orenstein, J. M., Baca, L. M., Weiser, B., Burger, H., Kalter, D. C., and Meltzer, M. S. (1989). *AIDS* **3**, 475–482.

Greve, J. M., Davis, G., Meyer, A. M., Forte, C. P., Connolly Yost, S. C., Marlor, C. W., Karmarck, M. E., and McClelland, A. (1989). *Cell* **56**, 839–847.

Grundy, J. A., McKeating, J. A., Ward, P. J., Sanderson, A. R., and Griffiths, P. D. (1987). *J. Gen. Virol.* **68**, 793–803.

Habeshaw, J. A., and Dalgleish, A. G. (1989). *J. AIDS* **2**, 457–468.

Harouse, J. M., Kunsch, C., Hartle, H. T., Laughlin, M. A., Hoxie, J. A., Wigdahl, B., and Gonzalez-Scarano, F. (1989). *J. Virol.* **63**, 2527–2533.

Ho, D. D., Rota, T. R., and Hirsch, M. S. (1984). *Proc. Natl. Acad. Sci. U.S.A.* **81**, 7588–7590.

Ho, W.-Z., Harouse, J. M., Rando, R. F., Gönczöl, E., Srinivasan, A., and Plotkin, S. A. (1990). *J. Gen. Virol.* **71**, 97–103.

Homsy, J., Meyer, M., Tateno, M., Clarkson, S., and Levy, J. A. (1989). *Science* **244**, 1357–1360.

Hoshino, H., Shimoyama, M., Miwa, M., and Sugimura, T. (1983). *Proc. Natl. Acad. Sci. U.S.A.* **80**, 7337–7341.

Hoxie, J. A., Matthews, D. M., and Cines, D. B. (1954). *Proc. Natl. Acad. Sci. U.S.A.* **81**, 7591–7594.

Hussey, R. E., Richardson, N. E., Kowalski, M., Brown, N. R., Chang, H-C., Siliciano, R. F., Dorfman, T., Walker, B., Sodroski, K., and Reinherz, E. L. (1988). *Nature* (*London*) **331**, 78–81.

Jameson, B. A., Rao, P. E., Kong, L. I., Hahn, B. H., Shaw, G. M., Hood, L. E., and Kent, S. B. H. (1988). *Science* **240**, 1335–1339.

Jolicoeur, P., and Baltimore, D. (1976). *Proc. Natl. Acad. Sci. U.S.A.* **81**, 4510–4514.

Joualt, T., Chapuis, F., Oliver, R., Parravicini, C., Babraouhi, E., and Gluckmann, J. C. (1989). *AIDS* **3**, 125–131.

Klatzmann, D., Champagne, E., Chamaret, S., Gruest, J., Guetard D., Hercend, T., Gluckman, J.-C., and Montagnier, L. (1984). *Nature* (*London*) **312**, 767–768.

Koenig, S., Gendelman, H. E., Orenstein, J. M., Dal Canto, M. C., Pezeshkpour, G. H., Yungbluth, M., Janotta, F., Aksamit, A., Martin, M. A., and Fauci A. S. (1986). *Science* **233**, 1089–1092.

Krichbaum-Stenger, K., Poiesz, B. J., Keller, P., Ehrlich, G., Gavalchin, J., Davis, B. H., and Moore, J. L. (1987). *Blood* **70**, 1303–1311.

Landau, N., Warton, M., and Littman, D. R. (1988). *Nature* (*London*) **334**, 159–162.

Lasky, L. A., Nakamura, G., Smith, D., Fennie, C., Shimasaki, C., Patzer, E., Berman, P., Gregory, T., and Capon, D. J. (1987). *Cell* **50**, 975–985.

Lifson, J. D., Hwang, K. M., Nara, P. L., Fraser, B., Padgett, M., Dunlop, N. M., and Eiden, L. E. (1988). *Science* **241**, 712–716.

Liu, M. A., and Liu, T. (1988). *J. Clin. Invest.* **82**, 2176–2180.

Loh, P. C., Achong, B. C., and Epstein, M. A. (1977). *Intervirology* **8**, 204–217.

Longo, D. L., Gelmann, E. P., Cossman, J., Young, R. A., Gallo, R. C., O'Brien, S. J., and Matis, L. A. (1984). *Nature* (*London*) **310**, 505–506.

Lusso, P., Veronese, F. di M., Ensoli, B., Franchini, G., Jemma, C., DeRocco, S. E., Kalyanaraman, V. S., and Gallo, R. C. (1990). *Science* **247**, 848–852.

McClure, M. O., Sattentau, Q. J., Beverley, P. C. L., Hearn, J. P., Fitzgerald, A. K., Zuckerman, A. J., and Weiss, R. A. (1987). *Nature* (*London*) **330**, 487–489.

McClure, M. O., Marsh, M., and Weiss, R. A. (1988). EMBO J. **7**, 513–518.

McClure, M. O., Sommerfelt, M. A., Marsh, M., and Weiss, R. A. (1990). J. Gen. Virol. **71**, 767–773.

McDougal, J. S., Mawle, A., Cort, S. P., Nicholson, J. K. A., Cross, G. D., Sheppler-Campbell, J. A., Hicks, D., and Sligh, J. (1985). J. Immunol. **135**, 3151–3162.

McDougal, J. S., Kennedy, M. S., Sligh, J. M., Cort, S. P., Mawle, A., and Nicholson, J. K. A. (1986). Science **231**, 382–385.

McKeating, J. A., and Willey, R. L. (1989). AIDS **3** (suppl. 1), S35–S41.

McKeating, J. A., Griffiths, P. D., and Weiss, R. A. (1990). Nature (London) **343**, 659–661.

Maddon, P. J., Dalgleish, A. G., McDougal, J. S., Clapham, P. R., Weiss, R. A., and Axel, R. (1986). Cell **47**, 333–348.

Maddon, P. J., McDougal, J. S., Clapham, P. R., Dalgleish, A. G., Sumayah, J., Weiss, R. A., and Axel, R. (1988). Cell **54**, 865–874.

Marsh, M., and Helenius, A. (1989). Adv. Virus Res. **36**, 107–151.

Meltzer, M. S., Skillman, D. R., Hoover, D. L., Hanson, B. D., Turpin, J. A., Kalter D. C., and Gendelman, H.-E. (1990). Immunol. Today **11**, 217–223.

Mendelsohn, C. L., Wimmer, E., and Racaniello, V. R. (1989). Cell **56**, 855–865.

Miyoshi, I., Kubonishi, I., Yoshimoto, S., Akagi, T., Outsuki, Y., Shimaishi, Y., Nagata, K., Nagata, K., and Hinuma, Y. (1981). Nature (London) **294**, 770–771.

Mizukami, T., Fuerst, T. R., Berger, E. A., and Moss, B. (1988). Proc. Natl. Acad. Sci. U.S.A. **85**, 9273–9277.

Monroe, J. E., Calender, A., and Mulder, C. (1988). J. Virol. **62**, 3497–3500.

Montagnier, L., Gruest, J., Chamaret, S., Dauguet, C., Axler, C., Guétard, D., Nugeyre, M. T., Barré Sinoussi, F., Chermann, J.-C., Brunet, J. B., Klatzmann, D., and Gluckmann, J. C. (1984). Science **225**, 63–66.

Moore, J. (1990). AIDS **4**, 297–305.

Nagy, K., Clapham, P. R., Cheingsong-Popov, R., and Weiss, R. A. (1983). Int. J. Cancer **32**, 321–328.

Nagy, K., Weiss, R. A., Clapham, P. R., and Cheingsong-Popov, R. (1984). In "Human T-cell Leukemia/Lymphoma Virus" (R.C. Gallo, ed.), pp. 121–131. Cold Spring Harbor Laboratory Press, Cold Spring Harbor, New York.

O'Hara, B., Johann, S. V., Klinger, H. P., Blair, D. G., Rubinson, H., Dunne, K. J., Sass, P., Vitek, S. M., and Robins, T. (1990). Cell Growth Diff. **1**, 119–127.

Patterson, S., and Knight, S. C. (1988). J. Gen. Virol. **68**, 1177–1181.

Pauza, C. D., and Price, T. M. (1988). J. Cell Biol. **107**, 959–968.

Pelchen-Matthews, A., Armes, J. E., and Marsh, M. (1989). EMBO J. **8**, 3641–3649.

Peterson, A., and Seed, B. (1988). Cell **54**, 65–71.

Popovic, M., Lange-Wantzing G., Sarin, P. S., Mann, D., and Gallo, R. C. (1983). Proc. Natl. Acad. Sci. U.S.A. **80**, 5402–5406.

Redmond, S., Peters, G., and Dickson, C. (1984). Virology **133**, 393–402.

Richardson, J., Edwards, A. J., Cruickshank, J. K., Rudge, P., and Dalgleisch, A. G. (1990). J. Virol. **64**, 5682–5687.

Robinson, W. E., Montefiore, D. C., and Mitchell, W. M. (1988). Lancet **i**, 790.

Rubin, H. (1965). Virology **26**, 270–276.

Sattentau, Q. J., and Weiss, R. A. (1988). Cell **52**, 631–633.

Sattentau, Q. J., Dalgleish, A. G., Weiss, R. A., and Beverley, P. C. L. (1986). Science **234**, 1120–1123.

Sattentau, Q. J., Clapham, P. R., Weiss R. A., Beverley, P. C. L., Montagnier, L., Alhalabi, M. F., Gluckmann, J.-C., and Klatzmann, D. (1988). AIDS **2**, 101–105.

Sattentau, Q. J., Arthos, J., Deen, K., Hanna, N., Healey, D., Beverley, P. C. L., Sweet, R., and Truneh, A. (1989). J. Exp. Med. **170**, 1319–1334.

Shaw, A. S., Amrein, K. E., Hammond, C., Stern, D. F., Sefton, B. M., and Rose, J.K. (1989). Cell **59**, 627–636.

Sinangil, F., Harada, S., Purtilo, D. T., and Volsky, D. J. (1985). *Int. J. Cancer* **36**, 191–198.

Smith, D. H., Byrn, R. A., Masters, S. A., Gregory, T., Groopman, J. E., and Capon, D. J. (1987). *Science* **238**, 1704–1707.

Sommerfelt, M. A., and Weiss, R. A. (1990). *Virology* **176**, 58–69.

Sommerfelt, M. A., Williams, B. P., Clapham, P. R., Solomon, E., Goodfellow, P. N., and Weiss, R. A. (1988). *Science* **242**, 1557–1559.

Staunton, D. E., Merluzzi, V. J., Rothlein, R., Barton, R., Marlin, S. D., and Springer, T. A. (1989). *Cell* **56**, 849–853.

Stein, B. S., Gowda, S. D., Lifson, J. D., Penhallon, R. C., Bensch, K. G., and Engelman, E. G. (1987). *Cell* **490**, 659–668.

Stephens, P. E., Clements, G., Yarranton, G. T., and Moore, J. (1990). *Nature (London)* **343**, 219.

Sun, N. C., Ho, D. D., Sun, C. R., Liou, R. S., Gordon, W., Fung, M. S., Li-X. L., Ting, R. C., Lee, T. H., Chang, N. T. *et al.* (1989). *J. Virol.* **63**, 3579–3585.

Swain, S. L. (1983). *Immunol. Rev.* **74**, 129–142.

Takeda, A., Tuazon, C. U., and Ennis, F. A. (1988). *Science* **242**, 580–583.

Tateno, M., Gonzalez-Scarano, F., and Levy, J. A. (1989). *Proc. Natl. Acad. Sci. U.S.A.* **86**, 4287–4291.

Tersmette, M., Van Dongen J. J. M., Clapham, P. R., DeGoede, R. E. Y., Wolvers-Tettero, I. L. M., Geurts Van Kessl, A., Huisman, J. G., Weiss, R. A., and Miedema, F. (1989). *Virology* **168**, 367–373.

Till, M. A., Ghetie, V., Gregory, T., Patzer, E. J., Porter, J. P., Uhr, J., Capon, D. J., and Vitetta, E. S. (1988). *Science* **242**, 1166–1168.

Timar, J., Nagy, K., Robertson, D., and Weiss, R. A. (1987). *J. Gen. Virol.* **68**, 1011–1020.

Traunecker, A., Läke, W., and Karjalainen, K. (1988). *Nature (London)* **331**, 84–86.

Traunecker, A., Schneider, J., Kiefer, H., and Karjalainen, K. (1989). *Nature (London)* **339**, 68–70.

Tschaler, E., Groh, V., Popovic, M., Mann, D. L., Konrad, K., Safai, B., Eron, F., Veronese, F. D., Wolff, K., and Stingl, G. (1987). *J. Invest. Dermatol.* **88**, 233–239.

Vogt, P. K., and Ishizaki, R. (1965). *Virology* **26**, 664–672.

Watanabe, M., Reiman, K. A., Delong, P. A., Lin, T., Fisher, R. A., and Letvin, N. L. (1989). *Nature (London)* **337**, 267–270.

Weber, J. N., Clapham, P. R., McKeating, J. A., Stratton, M., Robey, E., and Weiss, R. A. (1989). *J. Gen. Virol.* **70**, 2653–2660.

Weis, W., Brown, J. H., Cusack, S., Paulson, J. C., Skehel, J. J., and Wiley, D. C. (1988). *Nature (London)* **333**, 426–431.

Weiss, R. A. (1985). *In* "RNA Tumor Viruses" (R. A. Weiss, N. M. Teich, H. E. Varmus, and J. Coffin, eds.) Vol. 1, pp. 209–260. Cold Spring Harbor Laboratory Press, Cold Spring Harbor, New York.

Wood, G. S., Warner, N. L., and Warnke, R. A. (1983). *J. Immunol.* **131**, 212–218.

Zhu, Z., Chen, S. S. L., and Huang, A. S. (1990). *J. AIDS* **3**, 215–219.

Immunopathologic Mechanisms of Human Immunodeficiency Virus (HIV) Infection

Zeda F. Rosenberg and Anthony S. Fauci

I. Introduction

The most striking feature of infection with HIV *in vivo* is a dramatic reduction in the number of T4 cells, a specific subset of helper/inducer T lymphocytes that expresses the CD4 antigen (Lane and Fauci, 1985). An initial decline in circulating T4 cells is thought to occur at the time of acute infection with HIV, coincident with early viral replication. During this first stage of infection, although T4 cell numbers decrease somewhat, they may remain within the normal range. Weeks to months after initial infection with HIV, an immune response is generated, virus replication appears to be suppressed, and the infected individual enters into a clinically asymptomatic state. The average length of time between initial infection with HIV and the development of symptoms is approximately 10 years (Bacchetti and Moss,

1989). During this second stage, virus replication may occur at low levels with numbers of T4 cells remaining stable or slowly declining. The third stage of HIV infection is characterized by increased virus replication and a substantial and sometimes precipitous decline in T4 cell numbers.

Since the T4 lymphocyte plays a central role in the majority of human immune responses, the ultimate destruction of the T4 cell population results in a critical immune deficiency that allows for the development of a wide range of opportunistic infections and tumors. This chapter describes how HIV selectively infects particular immune system cells, including T4 cells and monocyte/macrophages, and examines the potential mechanisms by which HIV interferes with normal immune cell function and ultimately kills T4 cells. In addition, this chapter presents what is currently known about the effects of HIV on neurologic function, and also addresses the factors that are involved in the induction of HIV expression that occurs when HIV-infected individuals progress from the asymptomatic phase of infection to full-blown disease.

II. Mechanisms of T4 Cell Depletion

A. The Role of the CD4 Receptor in HIV Infection

It has been well established that HIV preferentially infects T4 lymphocytes (Klatzmann et al., 1984; McDougal et al., 1985). Indeed, it is the T4 (CD4) surface molecule that acts as the principal receptor for HIV and provides the initial site of contact between HIV and the T4 cell surface. The importance of CD4 in the life cycle of HIV derives from several lines of experimental evidence. Monoclonal antibodies to CD4 have been shown to block both HIV envelope-induced T4 cell fusion and T4 cell infection (Dalgleish et al., 1984; Klatzmann et al., 1984). In addition, antibodies to both CD4 and HIV were able to coprecipitate the viral envelope glycoprotein (gp120) and CD4 from infected cell lysates (McDougal et al., 1986). By transfecting the CD4 gene into CD4⁻ nonmurine cells previously incapable of supporting HIV infection, Maddon et al. (1986) have shown that the CD4 molecule is sufficient to render these cells infectible by HIV. Subsequent experiments by these researchers and others have demonstrated that soluble CD4 can inhibit HIV infection and syncytium formation in vitro (Smith et al., 1987; Deen et al., 1988; Hussey et al., 1988; Traunecker et al., 1988; Fisher et al., 1988).

B. Direct Virus-Induced T4 Cell Cytopathicity

It was observed early on that infection of T4 cells in vitro results in rapid and extensive cell killing (Popovic et al., 1984). In those instances in which cells survived initial infection, stimulation of the surviving cells with phyto-hemagglutinin (PHA) resulted in both increased virus expression and cell death (Zagury *et al.*, 1986). In both acute and chronic HIV infection of T cells, it appears that cell killing is associated with the production of large amounts of virus. Several researchers have demonstrated that HIV-induced cell killing may be the result of injury to the cell membrane (Leonard *et al.*, 1988; Lynn *et al.*, 1988). The insertion of large amounts of HIV envelope protein into the lipid bilayer of the cell membrane and subsequent virus budding can cause the formation of microholes in the cell membrane. These researchers have shown that the weakening of the membrane can result in severe ionic imbalance and, ultimately, cell death. Premature cell death has been observed in both HIV-infected single or multinucleated cells.

It has also been observed that high levels of heterodisperse RNAs amass in HIV-infected T cells. It has been hypothesized that the accumulation of these abnormal high molecular weight RMA species may result in a reduction in the amount of routine cellular messenger RNA species and lead to cell death (Koga *et al.*, 1988). Another factor that may interfere with normal cellular protein synthesis is the production of high levels of viral core proteins which can consume as much as 40% of total protein synthesis (Somasundaran and Robinson, 1988). HIV-associated cytopathicity has also been associated with an accumulation of unintegrated viral DNA (Shaw *et al.*, 1984).

The high affinity of gp120 for CD4 may play a role in cytopathicity as a result of intracellular binding of envelope proteins to CD4 molecules (Hoxie *et al.*, 1986). A role for CD4 in cell killing is supported by the experiments of De Rossi *et al.* (1986) which demonstrated that only HTLV-I-infected T4 cell clones and not T8 cell clones were killed upon superinfection with HIV. The virus that was released from the superinfected T8 cell clone was cytopathic for T4 cells, suggesting that it is the interaction of HIV with CD4 that results in a cytopathic effect. A role for the HIV envelope glycoprotein in cytopathicity is supported by experiments which indicated a correlation of expression of gp120 and T4 cell death (Sodroski *et al.*, 1986; Fisher *et al.*, 1986).

C. Indirect Methods of T4 Cell Cytopathicity

Infection with HIV may result in the destruction of T4 cells through a variety of indirect mechanisms. For example, uninfected T4 cells can be killed by fusing with HIV-infected T4 cells to form syncytia (Lifson *et al.*, 1986a,b; Sodroski *et al.*, 1986; Yoffe *et al.*, 1987). These multinucleated giant cells are formed as a result of the binding of several uninfected T4 cells via their surface CD4 molecules to the gp120 molecule that is expressed on the surface of the infected cell. The syncytium, which may be comprised of multiple uninfected cells surrounding a single infected cell, usually dies within 48 hr after formation. It has recently been reported that syncytia formation may involve a leukocyte adhesion receptor (LFA-1) in addition to CD4 and gp120 (Hildreth and Orentas, 1989). The relevance of syncytia formation to T4 cell death remains unclear at present since several investigators have found a discordance between cytotoxicity and syncytia formation *in vitro* (Somasundaran and Robinson, 1987; Leonard *et al.*, 1988).

Another process that may result in the elimination of uninfected T4 cells involves autoimmune events. In this regard, it has been shown that antibodies to gp120 can direct an antibody-dependent cellular cytotoxic (ADCC) response against uninfected $CD4^+$ T cells that have adsorbed gp120 to their surface membrane CD4 molecules (Lyerly *et al.*, 1987). In a similar vein, it has been shown that T4 cells can capture, process, and present soluble gp120 as a processed peptide in association with Class II major histocompatibility (MHC) molecules. Thus, although gp120-specific, $CD4^+$ cytotoxic T lymphocytes (CTL) may function in a protective manner against MHC Class II-restricted HIV-infected cells, these same CTL may also kill uninfected T4 cells that have bound free gp120 molecules to their CD4 receptors, internalized and processed it, and have presented the processed peptide to the CTL (Lanzavecchia *et al.*, 1988; Siliciano *et al.*, 1988) (Fig. 1).

A possible role for autoimmune phenomena in the pathogenesis of HIV infection has been suggested by the presence of anti-lymphocyte antibodies in patients with AIDS (Dorsett *et al.*, 1985). Others have reported that the sera of AIDS patients contain a cytotoxic autoantibody that reacts with an antigen on stimulated T4 cells and, in combination with complement, induces cytotoxicity of these cells *in vitro* (Stricker *et al.*, 1987). The generation of anti-lymphocytic antibodies may be due to the antigenic similarity that exists between the CD4 binding sites on the HIV envelope and Class II MHC molecules. The Class II MHC molecule has recently been shown to be the natural ligand for CD4 (Gay *et al.*, 1987; Doyle and Strominger, 1987). Specific anti-HIV antibodies may cross-react with Class II MHC molecules present on the surface of T lymphocytes (Ziegler and Stites, 1986). In this regard, it has been shown that the HIV transmembrane glycoprotein (gp41) shares a region of homology with the human Class II MHC molecule, and that patient anti-HIV sera react with native Class II antigens. It has been postulated that these antibodies may be involved in ADCC or complement mediated killing (Golding *et al.*, 1988).

Other mechanisms of indirect depletion of T4 cells include: infection by HIV of a T4 precursor cell; selective depletion of a subset of T4 lymphocytes or $CD4^+$ nonlymphoid cells which is trophic for T4 cells and thus critical for the propagation of the T4 cell pool; or induction of secretion of soluble substances toxic to T4 cells by HIV-infected T4 cells and/or monocytes (Fauci, 1987, 1988). Table I presents a summary of the potential mechanisms of direct and indirect HIV-induced cytopathicity.

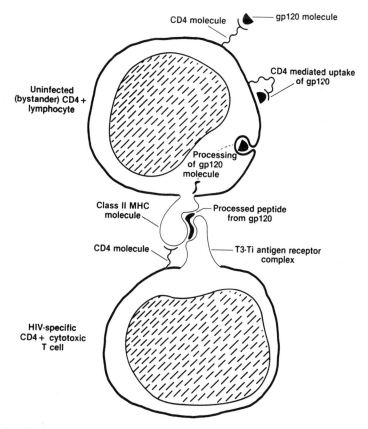

Figure 1. Elimination of CD4+ lymphocytes by CD4+ cytotoxic T cells. (Reprinted with permission from Rosenberg and Fauci, 1989.)

Table I

Potential Mechanisms of Depletion of T4 Cells in HIV Infection

1. Direct cytopathic effect of HIV
2. Infection by HIV of a T4 cell precursor
3. Selective depletion of a subset of T4 lymphocytes or CD4+ nonlymphoid cells which is trophic for T4 cells and thus critical for the propagation of the entire T4 cell pool
4. Induction of secretion of soluble substances toxic to T4 cells by HIV-infected T4 cells and/or monocytes
5. Syncytia formation between uninfected and infected T4 cells
6. Autoimmune phenomena

III. Impairment of Immune System Function

A. Functional Abnormalities of T4 cells

A large body of evidence has been amassed that convincingly shows that HIV can interfere with normal T4 cell function in the absence of cytopathicity. Lane *et al.* (1985) initially reported that AIDS patients had a selective defect in T-cell proliferation to soluble antigens such as tetanus toxoid. While decreased T-cell proliferative responses to mitogens could be restored by normalizing the proportion of T4 cells in culture, providing additional T4 cells had no effect on the proliferative response to soluble antigen. To rule out the possibility that a defect in antigen-presenting cells was responsible for the diminished proliferative response, experiments were conducted with sets of identical twins where one sibling had AIDS and the other was HIV-negative. The monocytes from the HIV-negative sibling were unable to promote a proliferative response to tetanus toxoid when cocultured with T cells from his diseased brother (Fauci, 1987).

Other researchers have also reported diminished T-cell proliferative responses to diverse soluble antigens, such as influenza, *Candida albicans,* and *Cryptococcus neoformans,* and have noted, in certain cases, that these functional defects could be detected early in the course of HIV infection, prior to the development of symptoms (Shearer *et al.*, 1985; Giorgi *et al.*, 1987; Hoy *et al.*, 1988; Miedema *et al.*, 1988). Inhibition of T-cell proliferation has been shown to occur when the sera of HIV-infected individuals were added to the T-cell cultures (Cunningham-Rundles *et al.*, 1983; Hennig and Tomar, 1984; Israel-Biet *et al.*, 1988). It has been reported that one consequence of exposure to the serum inhibitory factor is a decrease in the level of interleukin (IL)-2 receptors (Donnelly *et al.*, 1987).

There are several potential mechanisms to explain how HIV infection can cause a functional impairment of otherwise healthy T4 cells. A noncytopathic infection of T4 cells with HIV *in vivo* may result in the diversion of normal cellular processes and interference with normal cell functions. In this regard, it has been demonstrated that HIV-infected T cells progressively lose the ability to mount an antigen-specific response prior to cell death (Lyerly *et al.*, 1986).

It has become increasingly clear that direct infection of a T4 cell is not necessary to render it functionally impaired. It has been demonstrated that a noncytopathic exposure of peripheral blood cells *in vitro* resulted in a reduction in antigen-specific, and to a lesser degree, mitogen-specific responses (Margolick *et al.*, 1987). Furthermore, exposure of T cells to killed virus-infected cells, noninfectious HIV, or purified envelope glycoproteins was shown to result in the abrogation of proliferative responses to mitogens, antigens, and anti-CD3 (Sandstrom *et al.*, 1986; Shalaby *et al.*, 1987; Mann *et al.*, 1987; Amadori *et al.*, 1988; Krowka *et al.*, 1988; Chanh *et al.*, 1988;

Chirmule *et al.*, 1988). Since antigen-specific responses to CD4⁺ T lymphocytes depend upon the interaction of the CD4 molecule on the T cell and the Class II MHC molecule on the antigen-presenting cell, the binding of gp120 to CD4 may interfere with this interaction and suppress the antigen-induced proliferative response. In support of this hypothesis, Diamond *et al.*, (1988) have shown that gp120 can inhibit a CD4-dependent response to antigen in a murine T-cell hybridoma which expresses the human CD4 molecule. In addition, Clayton *et al.* (1989) have demonstrated that the interaction of CD4-transfected cells with MHC Class II-bearing lymphocytes is eliminated upon exposure to soluble gp120. It has also been reported that exposure of CD4⁺ T cells to anti-CD4 monoclonal antibody (mAb) results in a secondary reduction in CD3/T cell receptor (TCR) cell surface expression (Cole *et al.*, 1989). Thus, gp120-related cross-linking of CD4 could result in a decrease in antigen-specific responses by down-modulation of the T-cell receptor.

The presence of antibodies to Class II MHC molecules may also result in inhibition of T-cell responses. As mentioned above, the homology between the CD4-binding site on gp120 and Class II MHC molecules may result in the generation of anti-gp120 antibodies that can bind to Class II MHC (Ziegler and Stites, 1986). In addition, antibodies to a region of the HIV transmembrane glycoprotein, gp41, that cross-react with Class II MHC may interfere with CD4–Class II MHC interactions that occur during normal immune responses (Golding *et al.*, 1988). In this regard, it has been shown that proliferative responses of normal CD4⁺ T cells to tetanus toxoid and alloantigens were significantly impaired in the presence of sera containing gp41/Class II cross-reactive antibodies (Golding *et al.*, 1989). Bost *et al.* (1988) have demonstrated the presence of anti-IL-2 antibodies in the sera of HIV-infected individuals. They suggest that these antibodies are generated against a region of gp120 that is homologous to an epitope on the IL-2 molecule and represent an immunosuppressive mechanism for HIV-induced pathogenesis.

B. Effects of HIV on Other Immune Cells

Abnormalities in immunological function that are independent of T helper cell effects have been documented in HIV infection. For example, HIV-infected individuals possess an increased number of peripheral blood B lymphocytes that spontaneously secrete immunoglobulin (Ig) (Lane *et al.*, 1983; Pahwa *et al.*, 1984). It has been hypothesized that these polyclonally activated B cells exhibit abnormal responses to activation signals due to their chronically excited state (Katz *et al.*, 1986). It has been shown recently that B cells from asymptomatic HIV-infected men failed to produce Ig in response to pokeweed mitogen (Miedema *et al.*, 1988). Although there is no evidence that B cells are infected by HIV *in vivo*, several investigators

(Pahwa *et al.*, 1985; Schnittman *et al.*, 1986) have shown that HIV can directly induce B-cell proliferation and Ig secretion, suggesting that exposure of B cells to HIV results in polyclonal activation.

Certain functions of monocyte/macrophages have also been reported to be impaired in HIV infection. These include decreased chemotaxis, monocyte-dependent T-cell proliferation, Fc receptor function, and C3-receptor-mediated particle clearance (reviewed in Rosenberg and Fauci, 1989b). Monocytes from individuals in the early asymptomatic stage of HIV infection have been shown to exhibit decreased accessory function in the presence of normal T-helper activity (Miedema *et al.*, 1988). Although HIV can infect monocyte/macrophages both *in vitro* and *in vivo* (see Section IV), direct infection of these cells with HIV is not thought to be the cause of their functional impairment. Too few HIV-infected monocyte/macrophages are found in the peripheral blood to explain the universal decline in cell function (Schnittman *et al.*, 1989).

Like B cells and monocyte/macrophages, neutrophils also exhibit impaired function in HIV infection including reductions in chemotaxis, bacterial killing, and phagocytosis (Ellis *et al.*, 1988). Similarly, natural killer cells show aberrant function in HIV-infected individuals. These cells have impaired *in vitro* cytotoxic capabilities which can be complemented by the addition of exogenous IL-2, concanavalin A, or phorbol ester and calcium ionophore to cell cultures (Reddy *et al.*, 1984; Rook *et al.*, 1985; Bonavida *et al.*, 1986). In addition, natural killer and ADCC functions have been shown to be suppressed in HIV infection both in advanced disease as well as in asymptomatic carriers (Rook *et al.*, 1985; Ljunggren *et al.*, 1989). Functional abnormalities of peripheral blood dendritic cells have also been reported in symptomatic HIV-infected patients (Eales *et al.*, 1988). These cells are CD4[+] and have been shown to be infectible by HIV *in vitro* (Patterson and Knight, 1987).

Epidermal Langerhans cells are another class of immune cells that appear to be affected during HIV infection. Initially, it was noted that the number of epidermal Langerhans cells in symptomatic HIV-infected patients was lower than that of healthy controls (Belsito *et al.*, 1984). An inverse correlation between the level of Langerhans cells in the epidermis of HIV-infected individuals and the stage of disease has been reported (Dreno *et al.*, 1988). HIV has been observed in Langerhans cells in the skin of infected patients (Tschachler *et al.*, 1987) and a recent report has described the isolation of HIV from such cells, suggesting that Langerhans cells may serve as a reservoir for HIV in the body (Rappersberger *et al.*, 1988). However, since other investigators have failed to detect HIV infection of these cells *in vivo*, it remains unclear at present what role Langerhans cells play in the pathogenesis of HIV infection (Kanitakis *et al.*, 1989).

C. HIV Infection of Other Cell Types

HIV infection of a variety of different cell types has been reported both *in vivo* and *in vitro*. The common denominator of infection of many of these human cells is the presence of CD4 protein or RNA message which has been demonstrated in cells such as B cells and glial cells (Maddon *et al.*, 1986; Funke *et al.*, 1987). These HIV-infectible cells include EBV^+ and EBV^- B-cell lines, colorectal cells, glial cells, glioma cells, neuroretinal cells, cervical cells, capillary endothelial cells, oligodendroglial cells, and astrocytes (reviewed in Rosenberg and Fauci, 1989b). HIV infection of cultured human thymic epithelium has recently been described and shown to be blocked by mAbs to CD4 (Numazaki *et al.*, 1989). Other cells that have been reported to be infected by HIV include the glomerular and tubular epithelium from patients with HIV-associated nephropathy, and enterochromaffin cells from the rectal mucosa of an AIDS patient. These cells were shown to contain HIV genetic material by nucleic acid hybridization (Cohen *et al.*, 1989; Levy *et al.*, 1989). *In vitro* HIV infection of cells from the human choroid plexus has also been reported (Harouse *et al.*, 1989b). It is thought that human eosinophils may be susceptible to HIV infection since these cells express CD4 protein and bind HIV gp120 (Lucey *et al.*, 1989).

Although the CD4 surface protein is thought to be the principal, if not sole, receptor for HIV, there have been several reports of HIV infection of $CD4^-$ cells. Weiss *et al.* (1988) have described low-level HIV infection of tumor cells from both glial and muscle origin that could not be blocked by anti-CD4 mAbs. It has also been reported that HIV can infect human fibroblastoid cell lines which do not express CD4 protein or mRNA (Tateno *et al.*, 1989). Similarly, $CD4^-$ glioblastoma and medulloblastoma cell lines, and primary fetal neural cells can be infected with HIV, even in the presence of anti-CD4 antibodies (Wigdahl *et al.*, 1987; Harouse *et al.*, 1989a), suggesting that another receptor may be involved in HIV infection of neural cells.

The presence of certain types of hematological abnormalities in AIDS patients raises the likelihood that bone marrow cells may be infected with HIV *in vivo*. Evidence from a variety of observations supports this possibility including: inhibition of hematopoietic colony growth both in the presence and absence of anti-HIV sera (Donahue *et al.*, 1987; Stella *et al.*, 1987; Leiderman *et al.*, 1987); the presence of HIV mRNA in myeloid precursor cells from AIDS patients (Busch *et al.*, 1986); and *in vitro* HIV infection of purified bone marrow precursor cells (Folks *et al.*, 1988). During *in vitro* infection of bone marrow progenitor cells, the cells remain viable and produce HIV predominantly intracellularly. Other investigators have shown that differentiation of HIV-infected immature bone marrow cells toward $CD4^+$ cells is accompanied by increased viral replication (Lunardi-Iskandar *et al.*, 1989).

Clearly, HIV can infect a diversity of cells both *in vivo* and *in vitro*. Although the list of potential cellular targets is long, the preponderance of evidence suggests that the CD4$^+$ T lymphocyte is the principal target cell for HIV *in vivo* and infection of these cells with HIV is the common denominator of the pathogenesis of HIV-induced disease. Another cell that is thought to be significantly involved in HIV infection is the monocyte/macrophage.

IV. Role of the Monocyte/Macrophage in HIV Infection

Like other members of the lentivirus family, HIV can readily infect the CD4$^+$ monocyte/macrophage both *in vitro* and *in vivo* (reviewed in Rosenberg and Fauci, 1989b). HIV infection of monocyte/macrophages exhibits different characteristics from HIV infection of T4 cells. While HIV infection of T4 cells *in vitro* usually results in widespread syncytia formation and death of cells in culture, HIV infection of monocyte/macrophages appears to be both noncytopathic and persistent. In addition, HIV infection of monocyte/macrophages can result in the intracellular production of mature virions, virions that bud from the membranes of intracytoplasmic vesicles rather than from the cell surface. Whether or not a monocyte/macrophage produces intracellular or extracellular virus depends upon the state of differentiation of the cell (G. Poli and A. S. Fauci, unpublished data). In this regard, it has recently been reported that monocyte differentiation is associated with increased HIV replication (Griffin *et al.*, 1989).

The resistance of the monocyte/macrophage to the lytic effects of HIV and the phenomenon of intracellular virion production combine to render the monocyte/macrophage an ideal potential reservoir of HIV in the body. Intracellular virion production would allow for the cell to escape immune surveillance since no virion proteins would be expressed that would target the cell for immune destruction. In addition, the infected monocyte/macrophage may be instrumental in the spread of HIV to other organ systems, particularly the lungs and the brain. As the infected cell migrates to different areas of the body, differentiation may occur and HIV replication may be triggered, thus spreading HIV to surrounding tissues.

Recent experiments by Schnittman *et al.* (1989) have shown that the CD4$^+$ T cell is the main reservoir of HIV infection in the peripheral blood. Using gene amplification, only a minority of HIV-infected individuals were found to contain HIV sequences in circulating monocytes. Since there is abundant evidence that tissue macrophages harbor HIV, it is likely that these cells rather than circulating monocytes serve as reservoirs for the virus.

V. Mechanisms of HIV-Induced Neurological Disease

HIV-induced neurological disease, otherwise known as AIDS encephalopathy or AIDS dementia complex, is a common finding in HIV infection. However, at present, the mechanisms of neuropathogenesis remain obscure. Although HIV can be routinely isolated from the brain and cerebrospinal fluid of HIV-infected individuals (Levy *et al.*, 1985; Ho *et al.*, 1985; Gartner *et al.*, 1986), evidence of direct infection of neural cells *in vivo* is lacking. There have been isolated reports of HIV infection of various neural tissues (see section IIIB), but the vast majority of studies have shown that the predominant cell type in the brain that is infected by HIV is the macrophage (Koenig *et al.*, 1986; Gabuzda *et al.*, 1986; Gartner *et al.*, 1986; Wiley *et al.*, 1986; Vazeux *et al.*, 1987; Meyenhofer *et al.*, 1987).

The question remains as to how the presence of HIV-infected macrophages results in neuropathology. One mechanism may involve the release by infected macrophages of monokines or other factors that impair neurons or cause local inflammation (Fauci, 1988; Price *et al.*, 1988). In this regard, elevated serum levels of tumor necrosis factor (TNF) have recently been found in pediatric patients with progressive encephalopathy (Mintz *et al.*, 1989). The authors postulate that circulating TNF may be responsible for the myelin damage observed in AIDS. TNF may also be responsible for inducing HIV expression in the brain (see Section VI). Other cytotoxic factors may be produced as a result of HIV stimulation of astrocytes (Robbins *et al.*, 1987).

In addition to the release of HIV-induced factors that are neurotoxic, HIV may actually interfere with the action of factors that are normally neurotropic. It has been reported that HIV can interfere with the growth-promoting activity of neuroleukin when added to cultures of sensory neurons (Lee *et al.*, 1987). Since HIV gp120 and neuroleukin share regions of sequence homology, it is thought that HIV may compete for the neuroleukin-binding sites on the sensory neurons. In a similar manner, HIV is thought to interfere with vasoactive intestinal peptide which functions as a neurotransmitter (Brenneman *et al.*, 1988).

Coinfection of cells by HIV and other heterologous viruses may represent yet another potential route for HIV-induced neuropathology. It has recently been reported that cytomegalovirus (CMV) infection of astrocytes that were transfected with the promoter region of HIV linked to the chloramphenicol acetyltransferase (CAT) gene resulted in activation of CAT expression (Duclos *et al.*, 1989). The authors postulate that HIV causes a nonproductive infection of astrocytic cells *in vivo* and that coinfection with CMV can enhance HIV expression in these cells. Since HIV and CMV have been observed to infect the same cells in the brain in AIDS patients (Nelson *et al.*, 1988), it is possible that these viruses interact *in vivo* to enhance HIV infection and, potentially, CMV disease.

VI. Activation of Latent HIV Infection

One component of the pathogenesis of HIV infection that has yet to be determined is the precise mechanism of activation of HIV *in vivo*. The long latency period *in vivo* between initial infection and the development of disease suggests that HIV replication is restricted until some factor or factors allow for unchecked viral expression. While the immune response may play a role initially in suppressing the spread of the virus, *in vitro* experiments suggest that conditions which result in the activation of HIV-infected cells may also stimulate HIV replication. It has been shown that exposure of HIV-infected T cells from AIDS patients to mitogens, such as PHA, results in augmentation of HIV expression and can lead to cell death (reviewed in Rosenberg and Fauci, 1989a). The mechanism of PHA induction of HIV expression is thought to involve the production of cellular activation factors that bind to the HIV promoter region and initiate viral transcription. One specific area of the HIV promoter that appears to be an important target for mitogen activation is the tandemly repeated NFκB-binding sequence (Nabel and Baltimore, 1987; Kawakami *et al.*, 1988).

The importance of cellular activators in HIV infection is supported by several different experiments. Exposure of cells to mitogens during the course of *de novo* HIV infection of normal lymphocytes *in vitro* has been shown to result in an enhancement of HIV replication (McDougal *et al.*, 1985; Folks *et al.*, 1986). It has also been reported that mitogen or anti-CD3 mAb activation of T4 cells is required for initial HIV infection (Gowda *et al.*, 1989). Exposure of normal peripheral blood cells to soluble antigens can also result in an enhancement of HIV replication (Margolick *et al.*, 1987; Zack *et al.*, 1988).

Heterologous viruses represent another class of HIV activators. Using cotransfection experiments, it has been shown by a number of investigators that the products of the immediate early genes of several DNA viruses can activate the HIV promoter (reviewed in Rosenberg and Fauci, 1989a). These heterologous viruses include herpes simplex virus type 1 (HBV-1), cytomegalovirus, Epstein-Barr virus, and hepatitis B virus (HBV). A variety of studies designed to test the effects of infectious heterologous viruses (HSV-1 and human herpes virus type 6) on the HIV promoter, immediate early heterologous virus (HSV-1) genes on intact HIV, and simultaneous infection of heterologous viruses (HSV-1) and HIV showed similar activation of HIV expression.

The mechanism of action of heterologous virus-induced enhancement of HIV expression appears to be similar to that observed for mitogen stimulation of HIV-infected cells. Specifically, certain heterologous viral genes induce cellular proteins that bind to the HIV promoter at varying sites and activate HIV transcription (reviewed in Rosenberg and Fauci, 1989a). It has been found that the region of the HIV long terminal repeat (LTR) that is responsive to the HBV X gene contains the NFκB enhancer sequences (Twu *et al.*, 1989a,b). In addition, the HTLV-I *tax* gene product has been shown to activate the HIV LTR by inducing a cellular protein, HIVEN86A, that binds to the NFκB-binding site (Bohnlein *et al.*, 1989).

Coinfection with heterologous viruses may also cause a stress response in HIV-infected cells. It has been shown that exposure of HIV-infected cells to substances such as mitomycin-C or ultraviolet (UV) light that induce stress responses in cells results in the upregulation of HIV expression through activation of the HIV promoter in HIV-CAT-transfected cells (Valerie *et al.*, 1988). In addition, UV light has been shown to accelerate HIV replication during *de novo* infection of T-cell lines and to enhance HIV replication in chronically infected T cells (Valerie *et al.*, 1988; Stanley *et al.*, 1989).

A substantial body of data has recently been accrued that suggests that certain cytokines which are produced by immunologically activated cells may play an important role in the upregulation of HIV expression. Early experiments showed that cytokine-containing supernatants from mitogen-activated T cells or monocytes were able to enhance HIV replication in chronically HIV-infected T-cell and promonocytic cell lines (Folks *et al.*, 1987; Clouse *et al.*, 1989a). Further experiments have identified TNF-α as the active component of the cytokine-rich supernatants that induces HIV from both chronically infected T cells and monocytes (Clouse *et al.*, 1989a; Matsuyama *et al.*, 1988; Folks *et al.*, 1989). It has been shown that TNF causes a cytocidal effect on T cells chronically infected with HIV and enhances viral replication and syncytia formation with uninfected T4 cells (Matsuyama *et al.*, 1989a,b). Similarly, experiments with *de novo* infected cells suggest that TNF increases HIV replication and cytopathicity of acute HIV infection (Ito *et al.*, 1989). It has also been reported that TNF enhances HIV replication in peripheral blood mononuclear cells from HIV-infected individuals (Michihiko *et al.*, 1989).

Other cytokines that may be involved in the activation of HIV include granulocyte-macrophage colony-stimulating factor (GM-CSF) and IL-6. GM-CSF has been shown to augment HIV replication in acute infection of peripheral blood monocytes and in chronic infection of promonocytic cells (Folks *et al.*, 1987; Koyanagi *et al.*, 1988). In addition, it has been demonstrated that IL-6 can activate HIV and, like GM-CSF, acts with TNF to synergistically induce HIV expression in chronically infected promonocytic cells (G. Poli and A. S. Fauci, personal communication) (Fig. 2). The mechanism of action of TNF-α induction of HIV expression appears to involve a pathway common to mitogen and heterologous virus induction of HIV, namely the enlistment of cellular proteins that bind to the NFκB enhancer sequences on the HIV LTR (Osborn *et al.*, 1989; Okamoto *et al.*, 1989; Duh *et al.*, 1989).

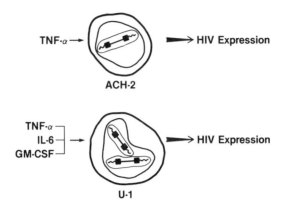

Figure 2. Effects of various cytokines on HIV expression in chronically infected T-cell (ACH-2) and promonocytic (U-1) cell lines.

VII. Subversion of the Immune System by HIV

It is becoming increasingly clear that initial infection with HIV and activation of HIV from a latent to a productive infection is intrinsically involved with the functioning of the human immune system. As was discussed earlier, activation of CD4$^+$ lymphocytes with either PHA or anti-CD3 mAb was shown to be required for initial HIV infection (Gowda *et al.*, 1989). In addition, immunological activation of HIV-infected cells with PHA can lead to HIV expression and cell death (Zagury *et al.*, 1986). Since PHA stimulation of HIV-infected T cells leads to IL-2 and IL-2 receptor expression, it has been suggested that repeated antigenic activation of HIV-infected T cells by microorganisms may result in a triggering of HIV replication and gradual loss of T4 cells (Zagury *et al.*, 1986). Antigenic stimulation of neighboring uninfected T4 cells may render them more susceptible to HIV infection. It has also been shown that certain viral antigens can stimulate monokine production from normal human monocytes. These monokines, in turn, can augment HIV expression in latently or chronically infected cells (Clouse *et al.*, 1989b).

Activation of T cells has been correlated with the production of a cellular protein, NFκB, that binds to specific sites in the HIV promoter and is involved in the initiation of viral RNA transcription (Nabel and Baltimore, 1987). NFκB has also been shown to be important in HIV expression in monocytes/macrophages as well since monocyte differentiation is associated both with the appearance of NFκB-binding activity, which is constitutively produced in mature monocyte/macrophages, and with HIV replication (Griffin *et al.*, 1989). As mentioned in Section VI, TNF-α has been shown to upregulate HIV expression through the induction of cellular proteins that bind to the NFκB region of the HIV LTR. It also appears that TNF-α can induce IL-2 receptor gene expression in some T cells (Hackett *et al.*, 1988) and that activation of the IL-2 receptor gene by TNF-α involves the induction of cellular proteins that bind specifically to the NFκB region of the IL-2R promoter in primary human T cells (Lowenthal *et al.*, 1989).

TNF-α has been shown to be present in significant levels in the sera of AIDS patients (Lahdevirta *et al.*, 1988). A recent study has shown that the binding of HIV to the CD4 receptor on monocyte/macrophages results in the induction of TNF-α (Merrill *et al.*, 1989). It has also been shown that TNF-α can enhance its own production by PMA (G. Poli and A. S. Fauci, unpublished observations). Moreover, it has been observed that higher numbers of TNF-α receptors are present on the surface of HIV-infected cells when compared to their uninfected counterparts (Folks *et al.*, 1989). Thus, TNF-α may alone, or in synergy with other cytokines, function in an autocrine/ paracrine loop to enhance replication of HIV and facilitate spread of the virus to uninfected cells (Fig. 3). If this theory proves correct *in vivo*, HIV

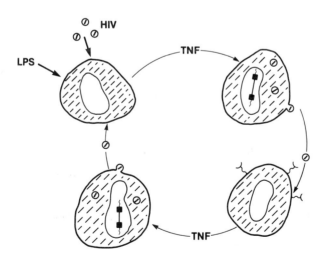

Figure 3. The role of tumor necrosis factor (TNF) in the propagation of HIV infection. LPS, lipopolysaccharide.

will have been able to subvert the usual mechanisms for cytokine-induced immunological activation to its own advantage.

References

Amadori, A., Faulkner-Valle, G. P., De Rossi, A., Zanovello, P., Collavo, D., and Chieco-Bianchi, L. (1988). *Clin. Immunol. Immunopathol.* **46**, 37–54.

Bacchetti, P., and Moss, A. R. (1989). *Nature (London)* **338**, 251–253.

Belsito, D. V., Sanchez, M. R., Baer, R. L., Valentine, F., and Thorbecke, G. J. (1984). *N. Engl. J. Med.* **310**, 1279–1282.

Bohnlein, E., Siekevitz, M., Ballard, D. W., Lowenthal, J. W., Rimsky, L., Bogerd, H., Hoffman, J., Wano, Y., Franza, B. R., and Greene, W. C. (1989). *J. Virol.* **63**, 1578–1586.

Bonavida, B., Katz, J., and Gottlieb, M. (1986). *J. Immunol.* **137**, 1157–1163.

Bost, K. L., Hahn, B. H., Saag, M. S., Shaw, G. M., Weigent, D. A., and Blalock, J. E. (1988). *Immunology* **65**, 611–615.

Brenneman, D. E., Westbrook, G. L., Fitzgerald, S. P., Ennist, D. L., Elkins, K. L., Ruff, M. R., and Pert, C. B. (1988). *Nature (London)* **335**, 639–642.

Busch, M., Beckstead, J., Gantz, D., and Vyas, G. (1986). *Blood* **68** (Suppl. 1), 122a.

Chanh, T. C., Kennedy, R. C., and Kanda, P. (1988). *Cell Immunol.* **111**, 77–86.

Chirmule, N., Kalyanaraman, V., Oyaizu, N., and Pahwa, S. (1988). *J. AIDS* **1**, 425–430.

Clayton, L. K., Sieh, M., Pious, D. A., and Reinherz, E. L. (1989). *Nature (London)* **339**, 548–551.

Clouse, K. A., Powell, D., Washington, I., Poli, G., Strebel, K., Farrar, W., Barstad, P., Kovacs, J., Fauci, A. S., and Folks, T. M. (1989a). *J. Immunol.* **142**, 431–438.

Clouse, K. A., Robbins, P. B., Fernie, B., Ostrove, J. M., and Fauci, A. S. (1989b). *J. Immunol.* **143**, 470–475.

Cohen, A. H., Sun, N. C., Shapshak, P., and Imagawa, D. T. (1989). *Mod. Pathol.* **2**, 125–128.

Cole, J. A., McCarthy, S. A., Rees, M. A., Sharrow, S. O., and Singer, A. (1989). *J. Immunol.* **143**, 397–402.

Cunningham-Rundles, S., Michelis, M. A., and Masur, H. (1983). *J. Clin. Immunol.* **3**, 156–165.

Dalgleish, A. G., Beverley, P. C., Clapham, P. R., Crawford, D. H., Greaves, M. F., and Weiss, R. A. (1984). *Nature (London)* **312**, 763–767.

De Rossi, A., Franchini, G., Aldovini, A., Del Mistro, A., Chieco-Bianchi, L., Gallo, R. C., and Wong-Staal, F. (1986). *Proc. Natl. Acad. Sci. U.S.A.* **83**, 4297–4301.

Deen, K. C., McDougal, J. S., Inacker, R., Folena-Wasserman, G., Arthos, J., Rosenberg, J., Maddon, P. J., Axel, R., and Sweet, R. W. (1988). *Nature (London)* **331**, 82–84.

Diamond, D. C., Sleckman, B. P., Gregory, T., Lasky, L. A., Greenstein, J. L., and Burakoff, S. J. (1988). *J. Immunol.* **241**, 3715–3717.

Donahue, R. E., Johnson, M. M., Zon, L. I., Clark, S. C., and Groopman, J. E. (1987). *Nature (London)* **26**, 200–203.

Donnelly, R. P., La Via, M. F., and Tsang, K. Y. (1987). *Clin. Exp. Immunol.* **68**, 488–499.

Dorsett, B., Cronin, W., Chuma, V., and Ioachim, H. L. (1985). *Am. J. Med.* **78**, 621–626.

Doyle, C., and Strominger, J. L. (1987). *Nature (London)* **330**, 256–259.

Dreno, B., Milpied, B., Bignon, J. D., Stalder, J. F., and Litoux, P. (1988). *Brit. J. Dermatol.* **118**, 481–486.

Duclos, H., Elfassi, E., Michelson, S., Arenzana-Seisdedos, F., Munier, A., and Virelizier, J. L. (1989). *AIDS Res. Hum. Retroviruses* **5**, 217–224.

Duh, E. J., Maury, W. J., Folks, T. M., Fauci, A. S., and Rabson, A. B. (1989). *Proc. Natl. Acad. Sci. U.S.A.* **86**, 5974–5978.

Eales, L. J., Farrant, J., Helbert, M., and Pinching, A. J. (1988). *Clin. Exp. Immunol.* **71**, 423–427.

Ellis, M., Gupta, S., Galant, S., Hakim, S., VandeVen, C., Toy, C., and Cairo, M. S. (1988). *J. Infect. Dis.* **158**, 1268–1276.

Fauci, A. S. (1987). *Clin. Res.* **35**, 503–510.

Fauci, A. S. (1988). *Science* **239**, 617–622.

Fisher, A. G., Ratner, L., Mitsuya, H., Marselle, L. M., Harper, M. E., Broder, S., Gallo, R. C., and Wong-Staal, F. (1986). *Science* **233**, 655–659.

Fisher, R. A., Bertonis, J. M., Meier, W., Johnson, V. A., Costopoulos, D. S., Liu, T., Tizard, R., Walker, B. D., Hirsch, M. S., Schooley, R. T., and Flavell, R. A. (1988). *Nature (London)* **331**, 76–78.

Folks, T., Kelly, J., Benn, S., Kinter, A., Justement, J., Gold, J., Redfield, R., Sell, K. W., and Fauci, A. S. (1986). *J. Immunol.* **136**, 4049–4053.

Folks, T. M., Justement, J., Kinter, A., Dinarello, C. A., and Fauci, A. S. (1987). *Science* **238**, 800–802.

Folks, T. M., Kessler, S. W., Orenstein, J. M., Justement, J. S., Jaffe, E. S., and Fauci, A. S. (1988). *Science* **242**, 919–922.

Folks, T. M., Clouse, K. A., Justement, J., Rabson, A., Duh, E., Kehrl, J. H., and Fauci, A. S. (1989). *Proc. Natl. Acad. Sci. U.S.A.* **86**, 2365–2368.

Funke, I., Hahn, A., Rieber, E. P., Weiss, E., and Reithmuller, G. (1987). *J. Exp. Med.* **165**, 1230–1235.

Gabuzda, D. H., Ho, D. D., de la Monte, S. M., Hirsch, M. S., Rota, T. R., and Sobel, R. A. (1986). *Ann. Neurol.* **20**, 289–295.

Gartner, S., Markovits, P., Markovitz, D. M., Betts, R. F., and Popovic, M. (1986). *JAMA* **256**, 2365–2371.

Gay, D., Maddon, P., Sekaly, R., Talle, M. A., Godfrey, M., Long, E., Goldstein, G., Chess, L., Axel, R., Kappler, J., and Marrack, P. (1987). *Nature, (London)* **328**, 626–629.

Giorgi, J. V., Fahey, J. L., Smith, D. C., Hultin, L. E., Cheng, H. L., Mitsuyasu, R. T., and Detels, R. (1987). *J. Immunol.* **138**, 3725–3730.

Golding, H., Robey, F. A., Gates, F. T., 3rd., Linder, W., Beining, P. R., Hoffman, T., and Golding, B. (1988). *J. Exp. Med.* **167**, 914–923.

Golding, H., Shearer, G. M., Hillman, K., Lucas, P., Manischewitz, J., Zajac, R. A., Clerici, M., Gress, R. E., Boswell, R. N., and Golding, B. (1989). *J. Clin. Invest.* **83**, 1430–1435.

Gowda, S. D., Stein, B. S., Mohagheghpour, N., Benike, C. J., and Engleman, E. G. (1989). *J. Immunol.* **142**, 773–780.

Griffin, G. E., Leung, K., Folks, T. M., Kunkel, S., and Nabel, G. J. (1989). *Nature (London)* **339**, 70–73.

Hackett, R. J., Davis, L. S., and Lipsky, P. E. (1988). *J. Immunol.* **140**, 2639–2644.

Harouse, J. M., Kunsch, C., Hartle, H. T., Laughlin, M. A., Hoxie, J. A., Wigdahl, B., and Gonzalez-Scarano, F. (1989a). *J. Virol.* **63**, 2527–2533.

Harouse, J. M., Wroblewska, Z., Laughlin, M. A., Hickey, W. F. Schonwetter, B. S., and Gonzalez-Scarano, F. (1989b). *Ann. Neurol.* **25**, 406–411.

Hennig, A. K., and Tomar, R. H. (1984). *Clin. Immunol. Immunopathol.* **33**, 258–267.

Hildreth, J. E., and Orentas, R. J. (1989). *Science* **244**, 1075–1078.

Ho, D. D., Rota, T. R., Schooley, R. T., Kaplan, J. C., Allan, J. D., Groopman, J. E., Resnick, L., Felsenstein, D., Andrews, C. A., and Hirsch, M. S. (1985). *N. Engl. J. Med.* **313**, 1493–1497.

Hoxie, J. A., Alpers, J. D., Rackowski, J. L., Huebner, K., Haggarty, B. S., Cedarbaum, A. J., and Reed, J. C. (1986). *Science* **234**, 1123–1127.

Hoy, J. F., Lewis, D. E., and Miller, G. G. (1988). *J. Infect. Dis.* **158**, 1071–1078.

Hussey, R. E., Richardson, N. E., Kowalski, M., Brown, N. R., Chang, H. C., Siliciano, R. F., Dorfman, T., Walker, B., Sodroski, J., and Reinherz, E. L. (1988). *Nature (London)* **331**, 78–81.

Israel-Biet, D., Ekwalanga, M., Venet, A., Even, P., and Andrieu, J.-M. (1988). *Clin. Exp. Immunol.* **74**, 185–189.

Ito, M., Baba, M., Sato, A., Hirabayashi, K., Tanabe, F., Shigeta, S., and De Clercq, E. (1989). *Biochem. Biophys. Res. Commun.* **158**, 307–312.

Kanitakis, J., Marchand, C., Su, H., Thivolet, J., Zambruno, G., Schmitt, D., and Gazzolo, L. (1989). *AIDS Res. Hum. Retroviruses* **5**, 293–302.

Katz, I. R., Krown, S. E., Safai, B., Oettgen, H. F., and Hoffmann, M. K. (1986). *Clin. Immunol. Immunopathol.* **39**, 359–367.

Kawakami, K., Scheidereit, C., and Roeder, R. G. (1988). *Proc. Natl. Acad. Sci. U.S.A.* **85**, 4700–4704.

Klatzmann, D., Champagne, E., Chamaret, S., Gruest, J., Guetard, D., Hercend, T., Gluckman, J. C., and Montagnier, L. (1984). *Nature (London)* **312**, 767–768.

Koenig, S., Gendelman, H. E., Orenstein, J. M., Dal Canto, M. C., Pezeshkpour, G. H., Yungbluth, M., Janotta, F., Aksamit, A., Martin, M. A., and Fauci, A. S. (1986). *Science* **233**, 1089–1093.

Koga, Y., Lindstrom, E., Fenyo, E. M., Wigzell, H., and Mak, T. W. (1988). *Proc. Natl. Acad. Sci. U.S.A.* **85**, 4521–4525.

Koyanagi, Y., O'Brien, W. A., Zhao, J. Q., Golde, D. W., Gasson, J. C., and Chen, I. S. (1988). *Science* **241**, 1673–1675.

Krowka, J., Stitres, D., Mills, J., Hollander, H., McHugh, T., Busch, M., Wilhelm, L., and Blackwood, L. (1988). *Clin. Exp. Immunol.* **72**, 179–185.

Lahdevirta, J., Maury, C. P., Teppo, A. M., and Repo, H. (1988). *Am. J. Med.* **85**, 289–291.

Lane, H. C., and Fauci, A. S. (1985). *Annu. Rev. Immunol.* **3**, 477–500.

Lane, H. C., Masur, H., Edgar, L. C., Whalen, G., Rook, A. H., and Fauci, A. S. (1983). *N. Engl. J. Med.* **309**, 453–458.

Lane, H. C., Depper, J. M., Greene, W. C., Whalen, G., Waldmann, T. A., and Fauci, A. S. (1985). *N. Engl. J. Med.* **313**, 79–84.

Lanzavecchia, A., Roosnek, E., Gregory, T., Berman, P., and Abrignani, S. (1988). *Nature (London)* **334**, 530–532.

Lee, M. R., Ho, D. D., and Gurney, M. E. (1987). *Science* **237**, 1047–1051.

Leiderman, I. Z., Greenberg, M. L., Adelsberg, B. R., and Siegal, F. P. (1987). *Blood* **70**, 1267–1272.

Leonard, R., Zagury, D., Desportes, I., Bernard, J., Zagury, J. F., and Gallo, R. C. (1988). *Proc. Natl. Acad. Sci. U.S.A.* **85**, 3570–3574.

Levy, J. A., Shimabukuro, J., Hollander, H., Mills, J., and Kaminsky, L. (1985). *Lancet* **ii**, 586–588.

Levy, J. A., Margaretten, W., and Nelson, J. (1989). *Am. J. Gastroenterol.* **84**, 787–789.

Lifson, J. D., Feinberg, M. B., Reyes, G. R., Rabin, L., Banapour, B., Chakrabarti, S., Moss, B., Wong-Staal, F., Steimer, K. S., and Engleman, E. G. (1986a). *Nature (London)* **323**, 725–728.

Lifson, J. D., Reyes, G. R., McGrath, M. S., Stein, B. S., and Engleman, E. G. (1986b). *Science* **232**, 1123–1127.

Ljunggren, K., Karlson, A., Fenyo, E. M., and Jondal, M. (1989). *Clin. Exp. Immunol.* **75**, 184–189.

Lowenthal, J. W., Ballard, D. W., Bohnlein, E., and Greene, W. C. (1989). *Proc. Natl. Acad. Sci. U.S.A.* **86**, 2331–2335.

Lucey, D. R., Dorsky, D. I., Nicholson-Weller, A., and Weller, P. F. (1989). *J. Exp. Med.* **169**, 327–332.

Lunardi-Iskandar, Y., Nugeyre, M. T., Georgoulias, V., Barré-Sinoussi, F., Jasmin, C., and Chermann, J. C. (1989). *J. Clin. Invest.* **83**, 610–615.

Lyerly, H. K., Weinhold, K. J., Cohen, O. J., and Bolognesi, D. P. (1986). *Surg. Oncol.* **37**, 404–407.

Lyerly, H. K., Matthews, T. J., Langlois, A. J., Bolognesi, D. P., and Weinhold, K. J. (1987). *Proc. Natl. Acad. Sci. U.S.A.* **84**, 4601–4605.

Lynn, W. S., Tweedale, A., and Cloyd, M. W. (1988). *Virology* **163**, 43–51.

Maddon, P. J., Dalgleish, A. G., McDougal, J. S., Clapham, P. R., Weiss, R. A., and Axel, R. (1986). *Cell* **47**, 333–348.

Mann, D. L., Lasane, F., Popovic, M., Arthur, L. O., Robey, W. G., Blattner, W. A., and Newman, M. J. (1987). *J. Immunol.* **138**, 2640–2644.

Margolick, J. B., Volkman, D. J., Folks, T. M., and Fauci, A. S. (1987). *J. Immunol.* **138**, 1719–1723.

Matsuyama, T., Hamamoto, Y., Kobayashi, S., Kurimoto, M., Minowada, J., Kobayashi, N., and Yamamoto, N. (1988). *Med. Microbiol. Immunol. (Berl.)* **177**, 181–187.

Matsuyama, T., Yoshiyama, H., Hamamoto, Y., Yamamoto, N., Soma, G., Mizuno, D., and Kobayashi, N. (1989a). *AIDS Res. Hum. Retroviruses* **5**, 139–146.

Matsuyama, T., Hamamoto, Y., Soma, G., Mizuno, D., Yamamoto, N., and Kobayashi, N. (1989b). *J. Virol.* **63**, 2504–2509.

McDougal, J. S., Mawle, A., Cort, S. P., Nicholson, J. K., Cross, G. D., Scheppler-Campbell, J. A., Hicks, D., and Sligh, J. (1985). *J. Immunol.* **135**, 3151–3162.

McDougal, J. S., Kennedy, M. S., Sligh, J. M., Cort, S. P., Mawle, A., and Nicholson, J. K. (1986). *Science* **231**, 382–385.

Merrill, J. E., Koyanagi, Y., and Chen, I. S. Y. (1989). *J. Virol.* **63**, 4404–4408.

Meyenhofer, M. F., Epstein, L. G., Cho, E. S., and Sharer, L. R. (1987). *J. Neuropathol. Exp. Neurol.* **46**, 474–484.

Michihiko, S., Yamamoto, N., Shinozaki, F., Shimada, K., Soma, G., and Kobayashi, N. (1989). *Lancet* **i**, 1206–1207.

Miedema, F., Petit, A. J., Terpstra, F. G., Schattenkerk, J. K., de Wolf, F., Al, B. J., Roos, M., Lange, J. M., Danner, S. A., Goudsmit, J., and Schellekens, P. Th. A. (1988). *J. Clin. Invest.* **82**, 1908–1914.

Mintz, M., Rapaport, R., Oleske, J. M., Connor, E. M., Koenigsberger, M. R., Denny, T., and Epstein, L. G. (1989). *Am. J. Dis. Child.* **143**, 771–774.

Nabel, G., and Baltimore, D. (1987). *Nature (London)* **326**, 711–713.

Nelson, J. A., Reynolds-Kohler, C., Oldstone, M. B., and Wiley, C. A. (1988). *Virology* **165**, 286–290.

Numazaki, K., Bai, X. Q., Goldman, H., Wong, I., Spira, B., and Wainberg, M. A. (1989). *Clin. Immunol. Immunopathol.* **51**, 185–195.

Okamoto, T., Matsuyama, T., Mori, S., Hamamoto, Y., Kobayashi, N., Yamamoto, N., Josephs, S. F., Wong-Staal, F., and Shimotohno, K. (1989). *AIDS Res. Hum. Retroviruses* **5**, 131–138.

Osborn, L., Kunkel, S., and Nabel, G. J. (1989). *Proc. Natl. Acad. Sci. U.S.A.* **86**, 2336–2340.

Pahwa, S. G., Quilop, M. T., Lange, M., Pahwa, R. N., and Grieco, M. H. (1984). *Ann. Intern. Med.* **101**, 757–763.

Pahwa, S., Pahwa, R., Saxinger, C., Gallo, R. C., and Good, R. A. (1985). *Proc. Natl. Acad. Sci. U.S.A.* **82**, 8198–8202.

Patterson, S., and Knight, S. C. (1987). *J. Gen. Virol.* **68**, 1177–1181.

Popovic, M., Sarngadharan, M. G., Read, E., and Gallo, R. C. (1984). *Science* **224**, 497–500.

Price, R. W., Brew, B., Sidtis, J., Rosenblum, M., Scheck, A. C., and Cleary, P. (1988). *Science* **239**, 586–592.

Rappersberger, K., Gartner, S., Schenk, P., Stingl, G., Groh, V., Tschachler, E., Mann, D. L., Wolff, K., Konrad, K., and Popovic, M. (1988). *Intervirology* **29**, 185–194.

Reddy, M. M., Chinoy, P., and Grieco, M. H. (1984). *J. Biol. Response. Mod.* **3**, 379–386.

Robbins, D. S., Shirazi, Y., Drysdale, B. E., Lieberman, A., Shin, H. S., and Shin, M. L. (1987). *J. Immunol.* **139**, 2593–2597.

Rook, A. H., Hooks, J. J., Quinnan, G. V., Lane, H. C., Manischewitz, J. F., Macher, A. M., Masur, H., Fauci, A. S., and Djeu, J. Y. (1985). *J. Immunol.* **134**, 1503–1507.

Rosenberg, Z. F., and Fauci, A. S. (1989a). *AIDS Res. Hum. Retroviruses* **5**, 1–4.

Rosenberg, Z. F., and Fauci, A. S. (1989). *Adv. Immunol.* **47**, 377–431.

Sandstrom, E. G., Andrews, C., Schooley, R. T., Byington, R., and Hirsch, M. S. (1986). *Clin. Immunol. Immunopathol.* **40**, 253–258.

Schnittman, S. M., Lane, H. C., Higgins, S. E., Folks, T., and Fauci, A. S. (1986). *Science* **233**, 1084–1086.

Schnittman, S. M., Psallidopoulos, M. C., Lane, H. C., Thompson, L., Baseler, M., Massari, F., Fox, C. H., Salzman, N. P., and Fauci, A. S. (1989). *Science* **245**, 305–308.

Shalaby, M. R., Krowka, J. F., Gregory, T. J., Hirabayashi, S. E., McCabe, S. M., Kaufman, D. S., Stites, D. P., and Ammann, A. J. (1987). *Cell Immunol.* **110**, 140–148.

Shaw, G. M., Hahn, B. H., Arya, S. K., Groopman, J. E., Gallo, R. C., and Wong-Staal, F. (1984). *Science* **226**, 1165–1171.

Shearer, G. M., Salahuddin, S. Z., Markham, P. D., Joseph, L. J., Payne, S. M., Kriebel, P., Bernstein, D. C., Biddison, W. E., Sarngadharan, M. G., and Gallo, R. C. (1985). *J. Clin. Invest.* **76**, 1699–1704.

Siliciano, R. F., Lawton, T., Knall, C., Karr, R. W., Berman, P., Gregory, T., and Reinherz, E. L. (1988). *Cell* **54**, 561–575.

Smith, D. H., Byrn, R. A., Marsters, S. A., Gregory, T., Groopman, J. E., and Capon, D. J. (1987). *Science* **238**, 1704–1707.

Sodroski, J., Goh, W. C., Rosen, C., Campbell, K., and Haseltine, W. A. (1986). *Nature (London)* **322**, 470–474.

Somasundaran, M., and Robinson, H. L. (1987). *J. Virol.* **61**, 3114–3119.

Somasundaran, M., and Robinson, H. L. (1988). *Science* **242**, 1554–1557.

Stanley, S. K., Folks, T. M., and Fauci, A. S. (1989). *AIDS Res. Hum. Retroviruses* **5**, 375–384.

Stella, C. C., Ganser, A., and Hoelzer, D. (1987). *J. Clin. Invest.* **80**, 286–293.

Stricker, R. B., McHugh, T. M., Moody, D. J., Morrow, W. J., Stites, D. P., Shuman, M. A., and Levy, J. A. (1987). *Nature (London)* **327**, 710–713.

Tateno, M., Gonzalez-Scarano, F., and Levy, J. A. (1989). *Proc. Natl. Acad. Sci. U.S.A.* **86**, 4287–4290.

Traunecker, A., Luke, W., and Karjalainen, K. (1988). *Nature (London)* **331**, 84–86.

Tschachler, E., Groh, V., Popovic, M., Mann, D. L., Konrad, K., Safai, B., Eron, L., diMarzo Veronese, F., Wolff, K., and Stingl, G. (1987). *J. Invest. Dermatol.* **88**, 233–237.

Twu, J. S., Rosen, C. A., Haseltine, W. A., and Robinson, W. S. (1989a). *J. Virol.* **63**, 2857–2860.

Twu, J. S., Chu, K., and Robinson, W. S. (1989b). *Proc. Natl. Acad. Sci. U.S.A.* **86**, 5168–5172.

Valerie, K., Delers, A., Bruck, C., Thiriart, C., Rosenberg, H., Debouck, C., and Rosenberg, M. (1988). *Nature (London)* **333**, 78–81.

Vazeux, R., Brousse, N., Jarry, A., Henin, D., Marche, C., Vedrenne, C., Mikol, J., Wolff, M., Michon, C., Rozenbaum, W., Bureau, J-F., Montagnier, L., and Brahic, M. (1987). *Am. J. Pathol.* **126**, 403–410.

Weiss, R. A., Clapham, P. R., McClure, M. O., McKeating, J. A., McKnight, A., Dalgleish, A. G., Sattentau, Q. J., and Weber, J. N. (1988). *J. AIDS* **1**, 536–541.

Wigdahl, B., Guyton, R. A., and Sarin, P. S. (1987). *Virology* **159**, 440–445.

Wiley, C. A., Schrier, R. D., Nelson, J. A., Lampert, P. W., and Oldstone, M. B. (1986). *Proc. Natl. Acad. Sci. U.S.A.* **83**, 7089–7093.

Yoffe, B., Lewis, D. E., Petrie, B. L., Noonan, C. A., Melnick, J. L., and Hollinger, F. B. (1987). *Proc. Natl. Acad. Sci. U.S.A.* **84,** 1429–1433.

Zack, J. A., Cann, A. J., Lugo, J. P., and Chen, I. S. (1988). *Science* **240,** 1026–1029.

Zagury, D., Bernard, J., Leonard, R., Cheynier, R., Feldman, M., Sarin, P. S., and Gallo, R. C. (1986). *Science* **231,** 850–853.

Ziegler, J. L., and Stites, D. P. (1986). *Clin. Immunol. Immunopathol.* **41,** 305–313.

Human T-cell Leukemia Virus (HTLV) Infection: A Clinical Perspective

Kiyoshi Takatsuki

I. Discovery of Adult T-cell Leukemia/Lymphoma

Adult T-cell leukemia/lymphoma (ATL) was the first human cancer found to be caused by a retrovirus. It was around the year 1973 that we came to recognize the existence of ATL, previously an unknown disease. In this introductory chapter, I would like to give a retrospective view concerning the background of our studies on ATL.

Many clinicians in Japan probably feel that the descriptions of diseases in U.S. and European textbooks frequently differ from the features of the diseases observed in their practices in Japan. In addition, a large portion of the Japanese textbooks of medicine are nothing more than translations of U.S. or European textbooks. These Japanese textbooks do not quite meet the requirements for treating Japanese patients. We can mention many examples in the field of hematology. One of these is that in Japan the incidence of chronic lymphocytic leukemia is quite low, being only 1.0% of all hematologic malignancies according to recent statistics. However, what differs between Japan and Europe/the United States is not only the incidence but also the detailed symptoms and signs of diseases. That is, many different points can be found between Japan and Europe/the United States when the symptomatology of the diseases is studied in depth. Some diseases have a higher incidence in Japan than in other countries. Hence, knowledge of them is indispensable to Japanese clinicians. However, some of these diseases are not described at all in Japanese textbooks. I have felt for a long time that this situation is strange and unreasonable. This vague feeling that an autochthonous pathology probably occurred in Japan must be considered the first factor in the background of our study of ATL.

The second major factor in our ATL study is the recent progress made in immunology. Having an interest in immunoglobulin abnormalities, I studied multiple myeloma and related diseases in the days when even the word "immunoglobulins" was not yet in use. In those days, the central theme of immunology was to clarify the structure of antibodies. Meanwhile, immunology advanced rapidly, and myeloma was defined as a B-cell malignancy. Therefore, it was quite natural to enlarge the objectives of our clinical immunology studies to all lymphoproliferative diseases.

At this stage of our research, joint studies with young researchers were initiated. This was the third and the most important element of the background in our ATL study. I became acquainted with Dr. Junji Yodoi (Institute for Virus Research, Kyoto University) and Dr. Takashi Uchiyama (Department of Internal Medicine, Kyoto University School of Medicine) during their postgraduate training. Dr. Yodoi prepared an antiserum against human thymocytes and invited us to direct our attentions to T cells. In the course of examining patients with various lymphoproliferative disorders, we arrived at the conclusion that ATL was a disease which had not been described anywhere before. It may not be pertinent to describe here the detailed history of the discovery of this disease and characterization of its features in our laboratory. In essence, we came to recognize that T-cell leukemia had a high incidence among Japanese adults and that most of the patients with this disease were from Kyushu. This discovery was made from our bedside observations rather than from laboratory work. Thereafter, Dr. Uchiyama went to the National Cancer Institute, Bethesda, U.S.A., to work with Dr. Thomas A. Waldmann and raised a monoclonal antibody (named "Tac" antibody), which later played an important role in our ATL studies. At the end of 1981, I moved from Kyoto to Kumamoto, which is located in the middle of Kyushu. Our studies advanced remarkably in Kumamoto, due mainly to the efforts of excellent young co-workers including Dr. Kazunari Yamaguchi and Dr. Toshio Hattori.

The discovery of ATL ushered in some dramatic developments in oncology and, unexpectedly, neurology. Chronologically speaking, this disease was internationally recognized in 1977 (Takatsuki et al., 1977; Uchiyama et al., 1977). We stated in our first report that attempts to elucidate leukemogenesis in this disease should be directed towards exploring the genetic background and a possible viral involvement. HTLV (human T-cell leukemia virus), the pathogen of ATL, was first reported by Dr. R. C. Gallo and his co-workers (National Cancer Institute, Bethesda, U.S.A.) in 1980 and 1981. They isolated HTLV from cultured cells taken from one patient with an aggressive variant of mycosis fungoides and from another with Sézary syndrome (Poiesz *et al.*, 1980, 1981). Although both patients were said to have cutaneous T-cell lymphoma (mycosis fungoides/Sézary syndrome), they had some unusual features which, in retrospect, link them to the clinical entity now called ATL. In Japan, coculturing of ATL cells with umbilical cord blood lymphocytes was first done successfully by Dr. I. Miyoshi's group (Kochi Medical School), who obtained the cell line MT-1 (Miyoshi *et al.*, 1980). Dr. Y. Himuma and his co-workers (Kyoto University Institute for Virus Research) demonstrated that ATL patients have antibodies against presumed viral antigens ATLA (ATL-associated antigens) on MT-1 cells (Hinuma *et al.*, 1981). Subsequently, a retrovirus was isolated, characterized, and named ATLV (adult T-cell leukemia virus) by Dr. M. Yoshida and his co-workers (Cancer Institute) (Yoshida *et al.*, 1982). Since it was shown that HTLV and ATLV are, in fact, identical (Seiki *et al.*, 1983), the term HTLV-I (human T-lymphotropic virus type-I) has been commonly used. Immediately after the discovery of ATL and HTLV-I, the acquired immunodeficiency syndrome (AIDS) began to be recognized, and a retrovirus, the human immunodeficiency virus (HIV), was also found to be a causative agent in this disease. The advances made in the worldwide study of AIDS owe much to the knowledge gained about the relation between ATL and HTLV-I. For example, almost identical methods for isolation of HTLV-I were used for isolation of HIV. It is also noteworthy that HIV is tropic not only to helper T cells but also to neural tissues.

II. Prototypic ATL

ATL shares some features with Sézary syndrome but is distinct from it. We studied 35 patients in Kyoto (Takatsuki *et al.*, 1985), 18 males and 17 females. Age at onset ranged from 27 to 73 years, with a median of 52 years. The predominant physical findings were peripheral lymph node enlargement (86%), hepatomegaly (72%), splenomegaly (51%), and skin lesions (42%). The white blood cell count (WBC) ranged from 10,000 to 500,000/mm³. Leukemic cells resembled Sézary cells, having indented or lobulated nuclei. Hypercalcemia was frequently associated with this condition. Skin lesions were one of the characteristic manifestations of this disease. Histological examination of the skin revealed common dermal, subcutaneous, and, frequently, epidermal infiltrations similar to that of Pautrier's microabscesses. The survival time ranged from 1 month to more than 6 years. The most striking aspect in our study was the clustering of the patients' birth places; 22 of 35 were born in Kyushu, 11 of them in the Kagoshima Prefecture, the southernmost part of Kyushu island. Most of them had grown up in their places of birth and moved later to their present locations. This peculiar geographical distribution led us to consider this leukemia to be a new disease. Since 1982, we have studied 187 patients with ATL in Kumamoto, 113 males and 74 females (1.5 : 1), whose age at onset ranged from 27 to 82 years, with a median of 55 years. They had peripheral lymph node enlargement (72%), hepatomegaly (47%), splenomegaly (26%), skin lesions (53%), and hypercalcemia (28%). Other symptoms and findings at onset were abdominal pain, diarrhea, pleural effusion, ascites, cough, sputum, and abnormal chest X-rays. Survival time ranged from 2 weeks to more than 1 year. The causes of death were pulmonary complications including *Pneumocystis carinii* pneumonia (Yoshioka *et al.*, 1985), hypercalcemia (Kiyokawa *et al.*, 1987), cryptococcal meningitis, disseminated herpes zoster, disseminated intravascular coagulopathy, and others.

III. Classification of ATL

In the light of new virological evidence, it became necessary to reexamine the concept of ATL. Variations in the clinical features of atypical ATL suggested the need to classify the spectrum of ATL into five types: acute, chronic, smoldering, crisis, and lymphoma (Kawano *et al.*, 1985; Yamaguchi *et al.*, 1983, 1986a).

Acute ATL corresponds to the prototypic ATL as described above with acute or subacute progression. Most patients in this group are resistant to chemotherapy and die rapidly. Combination chemotherapy, for example, vincristine, cyclophosphamide, prednisolone, adriamycin, and, sometimes, methotrexate, had been used. In general, a poor prognosis is indicated by elevations of serum lactic dehydrogenase (LDH), calcium, and bilirubin as well as by a high WBC.

Chronic ATL is considered to be a form of T-cell type chronic lymphocytic leukemia (T-CLL) in which the leukemic cells contain the proviral DNA of HTLV-I. In a few patients, slight lymphadenopathy and hepatosplenomegaly are observed. An elevation in serum LDH is also noted in a few patients, but this form is not associated with hypercalcemia or hyperbilirubinemia.

Smoldering ATL is characterized by the presence of a few ATL cells (0.5–3%) in the peripheral blood over a long period of time (Yamaguchi *et al.*, 1983). Patients with smoldering ATL frequently have skin lesions and/or other diseases which lead physicians to scrutinize the peripheral blood smears. The serum values of LDH are within normal range and there is no associated hypercalcemia. Lymphadenopathy, hepatosplenomegaly, and bone marrow infiltration are usually not present.

Crisis in chronic or smoldering ATL indicates the progression of the disease to acute ATL.

Lymphoma type of ATL is characterized by the absence of leukemic cells in the peripheral blood; this is considered to be a form of T-cell-type malignant lymphoma related to HTLV-I (Yamaguchi *et al.*, 1986). Biopsy of lymph nodes shows the histology of a diffuse nonHodgkin lymphoma. Serum LDH and calcium levels are frequently high. Most patients have a poor prognosis.

Practically speaking, there are many patients with a transitional disease state which cannot be definitely allocated to any of these five types. It is particularly difficult to draw a clear distinction between patients with smoldering ATL and healthy HTLV-I carriers, because Southern blot analysis is not sensitive enough to detect the monoclonal integration of HTLV-I proviral DNA in the cells. Individuals with polyclonal integration of HTLV-I proviral DNA seem to represent an intermediate state between smoldering ATL (monoclonal integration) and healthy HTLV-I carriers (with anti-HTLV-I antibodies but no detectable HTLV-I proviral DNA). Patients with this intermediate state of HTLV-I infection may be at risk for progression to ATL (Yamaguchi *et al.*, 1988). In addition, there are some unusual cases in our series. For example, we reported a patient with ATL who had five episodes of exacerbation and spontaneous remission during a period of 6 years (Kawano *et al.*, 1984).

IV. Diagnosis of ATL

Diagnosis of ATL is usually suggested by clinical and hematological characteristics and readily confirmed by showing positive anti-HTLV-I antibodies in the serum and typical phenotypic markers of leukemic cells. In questionable cases it should be, however, confirmed by demonstrating the presence of HTLV-I provirus genome in leukemic cells. Serum specimens from all patients with ATL have anti-HTLV-I antibodies. In Japan, anti-HTLV-I antibodies are examined by enzyme-linked immunosorbent assay or particle agglutination assay as a screening test, and positive results are confirmed by indirect immunofluorescence assay and/or strip radioimmunoassay based on the Western blot techinque. There is no difference between the antibody pattern of ATL patients and that of HTLV-I carriers.

Phenotypic markers of ATL cells are analyzed by laser flow cytometry using a variety of monoclonal antibodies. Almost all ATL cells are positive for CD3 and CD4 antigens and negative for CD8 and CD1 antigens, suggesting that they are derived from mature helper T cells. In addition, ATL cells express activated surface T-cell markers such as interleukin 2 (IL-2) receptor and HLA-DR antigens (Hattori *et al.*, 1981). There are a few reports of CD8$^+$ cells (CD4$^+$CD8$^+$ and CD4$^-$CD8$^+$). The presence of HTLV-I proviral DNA in peripheral blood mononuclear cells and/or lymph node cells from all patients with ATL can be determined by Southern blot (Yamaguchi *et al.*,

1984). Without exception, tumor cells are monoclonal with respect to the site of provirus integration (Yoshida *et al.*, 1984).

There are a few patients with "ATL not associated with HTLV-I" (Shimoyama *et al.*, 1986). These patients have a CD4+ T-cell malignancy not caused by HTLV-I infection and develop a disease state very similar to ATL. It is, however, evident that the clinical manifestations and the hematologic features of ATL are not caused by HTLV-I infection *per se* but by malignant proliferation of CD4+ peripheral T cells. It is natural that an indistinguishable disease occurs when CD4+ peripheral T cells undergo malignant transformation by any mechanism(s) other than HTLV-I infection. In our experience patients with "ATL not associated with HTLV-I" certainly occur but are rare instances and have no geographic predilection; thus, these do not seem to represent a disease entity for the moment.

V. Other Diseases Accompanying HTLV-I Infection

HTLV-I infection is a direct cause of ATL. However, infection with this virus has been found to be an indirect cause of or a contributing factor to many other diseases, such as chronic lung diseases, opportunistic lung infections, cancer of other organs (Asou *et al.*, 1986), monoclonal gammopathy (Matsuzaki *et al.*, 1987), chronic renal failure (Lee *et al.*, 1987), strongyloidiasis (Nakada *et al.*, 1987), nonspecific intractable dermatomycosis, and nonspecific lymphadenopathy. The association of HTLV-I infection with these diseases is considered to be due, to some extent, to the immunodeficiency induced by HTLV-I infection. The following is an example of our studies.

In the early stage of our ATL study in Kumamoto, we unexpectedly became aware of the frequent association of other malignancies with smoldering ATL (5 of 18 cases in our series). This prompted us to screen the seroprevalence of HTLV-I in patients with cancer, in cooperation with the 11 major general hospitals in the Kumamoto Prefecture (Asou *et al.*, 1986). Among 394 cancer patients who had not had blood transfusion, 61 (15.5%) were found to be positive for HTLV-I antibodies. The prevalence was significantly higher in males older than 40 and females of all ages compared to age- and sex-matched healthy seronegative individuals. There was no significant correlation between the site of malignancy and antibody prevalence. These results suggest that HTLV-I infection may contribute to increase the risk of developing other malignancies.

The recently highlighted disease HAM (HTLV-I-associated myelopathy) is thought to be synonymous with TSP (tropical spastic paraparesis). Clarification of the mechanism responsible for its onset is now an important theme.

VI. ATL Cells

In the early days of our studies, the E-rosette formation technique using sheep erythrocytes was employed for identification of T cells. This has been largely replaced at present by laser flow cytometry with use of monoclonal antibodies. Usually, ATL cells are positive for the surface markers CD2(T11), CD3(T3), CD4(T4), CD25(Tac), and HLA-DR, and negative to the surface marker CD8(T8) (Hattori et al., 1981). This finding suggests that ATL cells originate from the CD4$^+$ (helper/inducer) subset of mature T cells. The reduction in the CD3 level and expression of CD24 and HLA-DR indicate that ATL cells are in an activated state (Matsuoka et al., 1986). CD3 is known to form a complex with the T-cell receptor (TCR) and to play an important role in the transmission of antigen recognition information by T cells. The reduction in CD3 means a reduction in TCR. TCR consist of α and β chains, while CD3 is made of γ, δ, and ε chains. ATL cells are also being studied at the gene level. It was demonstrated that the genes of the β and γ chains were rearranged in almost all cases of ATL, and that there was no rearrangement pattern specific to ATL (Matsuoka et al., 1988).

ATL cells have various chromosome aberrations, but no ATL-specific abnormality is known (Ueshima et al., 1981; Sanada et al., 1985). Generally, chromosome aberrations are more frequently seen in acute cases than in chronic ones. A comparative cytogenetic study of lymphoma-type ATLA and T-cell lymphoma not associated with HTLV-I has revealed that there is no definite difference in the numerical and structural chromosomal abnormalities observed in these two groups, indicating that HTLV-I does not play a specific role in chromosome aberrations (Sanada et al., 1987).

It is quite possible that ATL cells release various active substances (cytokines) when activated, and that these released substances modify the pathologic features of ATL patients. For example, ATL cells release biologically active IL-1, which is not generally recognized as a product of T cells (Wano et al., 1987). We are studying the validity of the working hypotheses that postulate that a cytokine that stimulates osteoclasts is released from ATL cells in ATL patients showing hypercalcemia (Kiyokawa et al., 1987), and that a cytokine that promotes the differentiation and growth of granulocytes is released from ATL cells in ATL patients showing neutrophilia (Yamamoto et al., 1986a,b). In studying ATL, we should remember that many of the cell lines called "cultured cells" do not originate from patients' ATL cells. These "cultured cells" originate from normal T cells which were infected with HTLV-I and grew because the incubation conditions were favorable for their growth. These cultured cells are monoclonal, but they cannot be regarded as tumor cells. It is technically difficult to obtain cultured strains from the patients' fresh ATL cells, but this is sometimes successful. The cell line SKT-1B, established in our laboratory, is an example of an ATL cell line thus obtained (Koito et al., 1987).

VII. Prevention of HTLV-I Infection and Treatment of ATL

To prevent infection with HTLV-I in Japan all samples of donated blood have been subjected to HTLV-I antibody testng by gelatin particle agglutination assay since November 1986 on a nationwide scale. In order to prevent vertical transmission, HTLV-I seropositive mothers are now instructed to refrain from breast feeding in some local communities within the framework of a pilot study. This measure will also be implemented on a nationwide scale. Heating and freezing of milk are both known to be effective in eliminating the infectivity of HTLV-I, but these measures do not seem to be practical. Transmission between spouses is transmission between persons of the same generation; hence, there is hardly any risk of ATL onset even when this form of infection occurs. Therefore, what is required for the prevention of HTLV-I infection in the next generation is to have HTLV-I carrying mothers refrain from breast feeding. Experiments have disclosed that vaccination against HTLV-I is possible. However, there seems to be no necessity for vaccination.

The results of ATL treatment in the past were rather dismal (Shimoyama et al., 1988). Usually, acute and lymphoma types of ATL are treated with chemotherapy, but conventional chemotherapy using multiple drugs is effective only transiently. It has been reported that DCF (2'-deoxycoformycin), a potent inhibitor of adenosine deaminase, was effective in patients with ATL (Daenen et al., 1984). In our experience, five ATL patients refractory to conventional chemotherapy were treated with DCF and two of them showed a good response (Yamaguchi et al., 1986b). A cooperative trial of DCF therapy of ATL is currently in progress.

References

Asou, N., Kumagai, T., Ue ihara, S., et al. (1986). Cancer **58**, 903–907.

Daenen, S., Rojer, R. A., Smith, J. W., et al. (1984). Br. J. Haematol. **58**, 273–277.

Hattori, T., Uchiyama, T., Toibana, T., et al. (1981). Blood **58**, 645–647.

Hinuma, Y., Nagata, K., Hanaoka, M., et al. (1981). Proc. Natl. Acad. Sci. U.S.A. **78**, 6476–6480.

Kawano, F., Tsuda, H., Yamaguchi, K., et al., (1984). Cancer **54**, 131–134.

Kawano, F., Yamaguchi, K., Nishimura, H., et al. (1985). Cancer **55**, 851–856.

Kiyokawa, T., Yamaguchi, K., Takeya, M., et al. (1987). Cancer **59**, 1187–1191.

Koito, A., Shirono, K., Suto, H., et al. (1987). Jpn. J. Cancer Res. **78**, 365–371.

Lee, S. Y., Matsushita, K., Machida, J., et al. (1987). Cancer **60**, 1474–1478.

Matsuoka, M., Hattori, T., Chosa, T., et al. (1986). Blood **67**, 1070–1076.

Matsuoka, M., Hagiya, M., Hattori, T., et al. (1988). Leukemia **2**, 84–90.

Matsuzaki, H., Yamaguchi, K., Kagimoto, T., et al. (1985). Cancer **56**, 1380–1383.

Miyoshi, I., Kumonishi, I., Sumida, M., et al. (1980). Gann **71**, 155–156.

Nakada, K., Yamaguchi, K., Furugen, S., et al. (1987). Int. J. Cancer **40**, 145–148.

Poiesz, B. J., Ruscetti, F. W., Gazdar, A. F., et al. (1980). Proc. Natl. Acad. Sci. U.S.A. **77**, 7415–7419.

Poiesz, B. J., Ruscetti, F. W., Reitz, M. S., *et al.* (1981). *Nature (London)* **294**, 268–271.
Sanada, I., Tanaka, R., Kumagai, E., *et al.* (1985). *Blood* **65**, 649–654.
Sanada, I., Ishii, T., Matsuoka, M., *et al.* (1987). *Hemat. Oncol.* **5**, 157–166.
Seiki, M., Hattori, S., Hirayama, Y., *et al.* (1983). *Proc. Natl. Acad. Sci. U.S.A.* **80**, 3618–3622.
Shimoyama, M., Kagami, Y., Shimamoto, K., *et al.* (1986). *Proc. Natl. Acad. Sci. U.S.A.* **83**, 4254–4528.
Shimoyama, M., Lymphoma Study Group (1988). *J. Clin. Oncol.* **6**, 128–141.
Takatsuki, K., Uchiyama, T., Sagawa, K., *et al.* (1977). *In* "Topics in Hematology" (S. Seno, F. Takaku, and S. Irino, eds.), pp. 73–77. Excerpta Medica, Amsterdam.
Takatsuki, K., Yamaguchi, K., Kawano, F., *et al.* (1985). *Cancer Res.* **45** (Suppl), 4644–4645.
Uchiyama, T., Yodoi, J., Sagawa, K., *et al.* (1977). *Blood* **50**, 481–491.
Ueshima, Y., Fukuhara, S., Hattori, T., *et al.* (1981). *Blood* **58**, 420–425.
Wano, Y., Hattori T., Matsuoka, M., *et al.* (1987). *J. Clin. Invest.* **80**, 911–916.
Yamaguchi, K., Nishimura, H., Kohrogi, H., *et al.* (1983). *Blood* **62**, 758–766.
Yamaguchi, K., Seiki, M., Yoshida, M., *et al.* (1984). *Blood* **63**, 1235–1240.
Yamaguchi, K., Yoshioka, R., Kiyokawa, T., *et al.* (1986a). *Hematol. Oncol.* **4**, 59–65.
Yamaguchi, K., Lee, S. Y., Oda, T., *et al.* (1986b). *Leukemia Res.* **10**, 989–993.
Yamaguchi, K., Kiyokawa, T., Nakada, K., *et al.* (1988). *Brit. J. Haematol.* **68**, 169–174.
Yamamoto, S., Hattori, T., Asou, N., *et al.* (1986a). *Jpn. J. Cancer Res.* **77**, 858–861.
Yamamoto, S., Hattori, T., Matsuoka, M., *et al.* (1986b). *Blood* **67**, 1714–1720.
Yoshioka, R., Yamaguchi, K., Yoshinaga, T., and Takatsuki, K. (1985). *Cancer* **55**, 2491–2494.
Yoshida, M., Miyoshi, I., and Hinuma, Y. (1982). *Proc. Natl. Acad. Sci. U.S.A.* **79**, 2031–2035.
Yoshida, M., Seiki, M., Yamaguchi, K., and Takatsuki, K. (1984). *Proc. Natl. Acad. Sci. U.S.A.* **81**, 2534–2537.

Part III

Epidemiology

Epidemiology of Human T-cell Leukemia Virus Type I (HTLV-I)

Stefan Z. Wiktor and William A. Blattner

I. Introduction

Several decades of intensive search for a human retrovirus culminated in the isolation of HTLV-I in 1980 (Poiesz *et al.*, 1980). Since the discovery of HTLV-I, much has been learned about its epidemiology. First and foremost, studies have linked the virus with two distinct clinical entities: adult T-cell leukemia (ATL) and HTLV-I-associated myelopathy/tropical spastic paraparesis (HAM/TSP), a chronic degenerative neurologic syndrome. Secondly, research efforts have characterized the worldwide epidemiological distribution, and established the modes of transmission.

THE HUMAN RETROVIRUSES

II. Diagnosis of Infection

Theoretically, viral isolation of HTLV-I should be the "gold standard" for defining infection. However, this is technologically complex and often fails to detect virus in seropositive individuals. Consequently, infection is defined by the presence of specific anti-HTLV-I antibodies, and a number of techniques have been developed to accurately detect these antibodies. These techniques include screening assays, such as enzyme-linked immunosorbent assay (ELISA) and particle agglutination, and confirmatory assays, such as radioimmunologic precipitation assay (RIPA), immunofluorescence, and immunoblotting. Some of the early confirmatory tests used in epidemiologic studies lacked adequate specificity and provided data that have not been reproducible. But in recent years, improvements in HTLV-I serologic assays have resulted in the standardization of serologic results, which, in turn, has led to the effective screening of blood donors in a number of countries, including the United States. Current screening for HTLV-I follows a similar algorithm to that developed for HIV testing (Anderson et al., 1989). Samples are first tested by a whole-virus ELISA. Samples that are repeatedly screen-positive are confirmed by immunoblotting using a whole-virus extract as antigen. Criteria for immunoblot positivity vary, but most standards require the demonstration of viral specific bands to two gene products, usually to the core and envelope proteins (Centers for Disease Control, 1988). RIPA analysis is often required to demonstrate the presence of antibodies to envelope, as these are frequently not detected on immunoblots.

Researchers are using recombinant proteins to develop new assays that promise to have improved capability of detecting viral antibodies. One approach is the addition of a recombinant envelope protein to a standard immunoblot to improve the ability to detect envelope (Lillehoj et al., 1990). Another approach is the use of viral type-specific epitopes to develop assays that can distinguish between HTLV-I and HTLV-II (Lipka et al., 1990).

Despite improvements, serologic assays may underestimate the true prevalence of HTLV-I infection since they require the presence of an immune response to the virus. The small number of circulating HTLV-I-infected lymphocytes in healthy carriers is often below the detection threshold of techniques such as DNA hybridization. The recent development of DNA amplification by polymerase chain reaction (PCR) promises to overcome these obstacles (Saiki et al., 1988). By greatly amplifying the number of copies of viral DNA, this technique is theoretically able to detect one infected cell in a million, and researchers using this technique have already identified a few individuals who are infected by HTLV-I yet lack detectable antibody to the virus (Ehrlich et al., 1989). The use of PCR testing in large-scale studies should address the frequency and importance of this viral latency phenomenon. In addition, PCR testing has also made it possible to distinguish between HTLV-I and HTLV-II, two viruses that are closely related and which have been difficult to distinguish serologically.

III. Geographic Distribution

The global distribution of HTLV-I is marked by two sharply demarcated endemic areas, Japan and the Caribbean, where the epidemiology has been well described. In addition, evidence of infection is found in a number of other regions but prevalence in these regions is generally lower and the epidemiology is not as well characterized.

A. Asia

A number of prevalence surveys established the presence of HTLV-I in the southern islands of Japan particularly in the islands of Kyushu, Shikoku, Okinawa, and surrounding smaller islands (Fig. 1) (Hinuma *et al.*, 1982; Clark *et al.*, 1985; Tajima *et al.*, 1987). A few pockets of seropositivity are found in more northern islands of Japan, but these are probably the result of migration of individuals from the endemic islands. Hinuma (1986) also reported high rates of seropositivity in the Ainu, an aboriginal population inhabiting Japan's northernmost island. In primary endemic islands, considerable clustering of infection occurs within small geographic areas with high

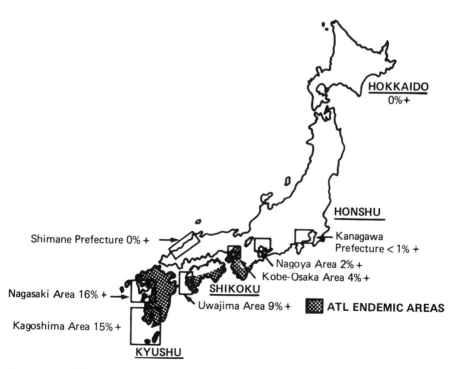

Figure 1. HTLV-1 seroprevalence and distribution of adult T-cell leukemia (ATL) cases in Japan. (Reproduced from the *Journal of Experimental Medicine,* 1983, vol. 157, pp. 248–258, by copyright permission of the Rockefeller University Press.)

rates and low rates in adjacent villages. In some surveys higher rates are associated with more remote isolated regions (Hinuma *et al.*, 1982), and in the Myazaki region of Kyushu, coastal communities tended to have a higher prevalence (Mueller *et al.*, 1990). The reasons for this microclustering remain unclear.

Few studies have focused on the presence of HTLV-I in Asia outside of Japan. Surveys from Taiwan and the Peoples Republic of China failed to identify regions of endemicity. Surveys in Vietnam, the Philippines, and Malaysia have identified only a handful of seropositives, these usually having historic ties to Japan. Conflicting reports from Papua New Guinea exemplify the difficulties in defining HTLV-I endemicity. Early surveys demonstrated a prevalence of 26% among healthy villagers (Kazura *et al.*, 1987), but retesting of the same sera by newer serologic tests failed to confirm any seropositivity (Weber *et al.*, 1989). Since then, newer studies from Papua New Guinea have identified rates ranging from 3 to 13% and a possible case of HAM/TSP (Currie *et al.*, 1989). These conflicting reports may be the result of differences in the groups tested and in the serologic assays used.

B. Caribbean

The Caribbean basin is the other primary endemic region for HTLV-I. An island-wide seroprevalence survey undertaken in Jamaica found a relatively uniform prevalence rate of 6% throughout the island (Murphy, 1990). While clustering does not occur as it does in Japan, individuals residing at higher altitudes tended to have a lower prevalence (Maloney *et al.*, 1990). Data from other surveys suggest that seropositivity is associated with lower socioeconomic classes, so it is unclear whether the geographic variations seen in Jamaica represent different environmental exposure at different altitudes or simply class differences. Varying rates of HTLV-I prevalence are found in a number of other islands in the Caribbean, and Martinique and Haiti have similar rates to Jamaica. In Trinidad where the overall prevalence is 3.7%, seropositivity is limited almost exclusively to individuals of African descent (Miller *et al.*, 1986), despite the fact that individuals of African and Indian descent have shared a common environment for over 100 years.

C. Africa

After Japan and the Caribbean, parts of Africa also appear to have large reservoirs of infection. Population-based studies have identified possible endemic areas within Gabon, Chad, Cameroon, and Guinea, with higher seroprevalence in more remote regions of these countries (Delaporte *et al.*, 1989a). A study from the Ivory Coast found highest rates among female prostitutes, suggesting that frequent sexual contact may increase the risk of acquiring infection (Verdier *et al.*, 1989). This higher rate may also result from the in-migration of prostitutes from areas of higher endemicity, since a study of female prostitutes from Kinshasa, Zaire failed to identify any sexual practices that were associated with seropositivity. Prostitutes who originated from Equateur, a remote northwestern region of Zaire, had a higher seroprevalence than did prostitutes from all other regions of Zaire (Wiktor *et al.*, 1990a). Kayembe and colleagues (1990) reported a cluster of 20 HTLV-I-associated cases of HAM/TSP in this same region of Zaire adding further evidence that this area is endemic for HTLV-I. Despite this finding and the identification of a large number of seropositives in Africa, there are few other reports of African cases of HAM/TSP or ATL (Williams *et al.*, 1984). Shorter life span and poor access to health care may explain this apparent deficit, particularly for HAM/TSP and for ATL, the former of which is marked by a gradual onset of symptoms and the latter of which is rapidly fatal. When cases of HAM/TSP were actively sought in a village of Gabon, 2 of 84 adults had clinical syndromes consistent with HAM/TSP (Delaprote *et al.*, 1989b).

D. The Americas and Europe

Several Latin American countries have reported evidence of HTLV-I infection. In Colombia, seropositive individuals are clustered along the Pacific coast, with higher rates among blacks (Maloney *et al.*, 1989). This area also has reported a high prevalence of HAM/TSP (Zaninovic *et al.*, 1988). The highest rate of seropositivity in Panama surfaced among the Guaymi Indians, an isolated tribe in the northwestern region adjacent to Costa Rica (Reeves *et al.*, 1990). Recent data indicate that these individuals are infected with HTLV-II rather than HTLV-I (Lairmore *et al.*, 1990), suggesting that HTLV-II may have its origins among Indian groups rather than in Africa, as had been hypothesized for HTLV-I (Gallo *et al.*, 1986). HTLV-I prevalence reported in Brazil ranges between 1 and 13% with the highest rates among male prostitutes and individuals infected with HIV (Cortes *et al.*, 1989).

The overall prevalence of HTLV-I in the United States is very low. Four of 10,000 (0.025%) blood donors tested positive in a 1988 survey (Williams *et al.*, 1988). In spite of this low rate, intravenous drug abusers (IVDAs) represent a group at extraordinarily high risk of acquiring HTLV-I/II. Some 49% of black IVDAs attending a methadone clinic in New Orleans were seropositive in 1985 (Weiss *et al.*, 1987). Unusual serologic reactivity among IVDAs raised the suspicion that this group may be infected with HTLV-II rather than HTLV-I (Agius *et al.*, 1988). Recently, PCR testing confirmed these suspicions and revealed that approximately 80% of seropositive New Orleans IVDAs were actually infected with HTLV-II (Lee *et al.*, 1989). This is the first group to show a prevalence for the virus and this finding may shed light on the natural history of HTLV-II. It is unclear how HTLV-II entered into the drug-using community since no naturally endemic area for this virus has been confirmed.

The epidemiology of HTLV-I in Europe is similar to that in the U.S. The overall rate is low, with seropositivity limited to IVDAs and migrants from endemic areas (Greaves *et al.*, 1984b; Gradilone *et al.*, 1986). A cluster of ATL and seropositivity among Caucasians in southern Italy suggests that this area may be endemic for HTLV-I (Manzari *et al.*, 1985). Further virologic analyses of some of these seropositive individuals demonstrated a possible new retrovirus which has been named HTLV-V (Manzari *et al.*, 1987).

IV. Demographic Factors

Most of the existing information concerning the epidemiology of HTLV-I comes from studies undertaken in the endemic areas of southern Japan and the Caribbean basin. These studies reveal that the distribution of HTLV-I is strongly influenced by factors of age, race, sex, and geography. In endemic areas, the overall prevalence is between 5 and 10%. Prevalence in children is low, at approximately 1%, but starts to increase during the teenage years. This age-related increase is more marked for females than males, so much so that by the age of 40 women are twice as likely to be infected as men (Fig. 2).

Several hypotheses attempt to explain this age-related increase. One posits that individuals acquire infection throughout adult life, primarily through sexual intercourse. Under this hypothesis, the higher rate of female infection results from the more efficient transmission of virus from males to females (Kajiyama *et al.*, 1986). Another hypothesis proposes that this age-related increase represents a cohort effect: older individuals had a higher probability of being exposed to the virus when they were young than do young people today. As most HTLV-I surveys are cross-sectional and recent, this hypothesis is difficult to address. But it is supported by recent longitudinal data from Japan, where sera were collected over a 20-year

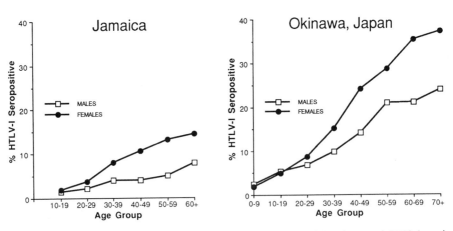

Figure 2. HTLV-I prevalence by age and sex in Jamaica (Murphy *et al.* 1989c) and Okinawa, Japan (Kajiyama *et al.*, 1986).

period among young women (Ueda *et al.*, 1989). HTLV-I prevalence among women of a similar age was lower in 1988 than in 1978 or 1968. In Barbados, however, HTLV-I prevalence in sera collected in 1972 showed identical prevalence rates to those of more recent studies (Riedel *et al.*, 1989). A final hypothesis explains this age-related curve by suggesting that individuals can be infected early in life even though they only develop antibodies to HTLV-I many years after infection. The recent development of more sensitive assays such as PCR to detect presence of viral DNA rather than presence of antibody may answer this question.

V. Modes of Transmission

Epidemiologic studies of family members of ATL cases provided the first clues as to the routes of transmission of HTLV-I. Rates among healthy relatives of ATL cases were fourfold higher than among nonfamily members (Robert-Guroff *et al.*, 1983; Sarin *et al.*, 1983). Studies documented the transmission from parent to child and among spouses, especially from husband to wife. These early findings have been repeated and amplified (Table I).

A. Sexual Transmission

Several studies suggest that sexual transmission of HTLV-I occurs, and that this transmission is more efficient from males to females. One large study of spouse pairs estimated that over 60% of female spouses of HTLV-I-positive husbands would become infected over a 10-year period, while less

Table I

Modes of Transmission of HTLV-I

I. Mother-to-child
   A. 20% of offspring of seropositive mothers are seropositive
   B. Breast-feeding most likely route of transmission
   C. Transmission associated with higher maternal antibody titer
II. Sexual transmission
   A. Primarily male-to-female and male-to-male transmission
   B. Cofactors
      (1) Older age and higher antibody titer of seropositive person
      (2) Females—large number of male partners, positive syphilis serology
      (3) Males—ulcerative genital lesions
III. Parenteral transmission
   A. Transfusion-associated
      (1) Associated with transmission of cellular products
      (2) Seroconversion occurs in 40–60% of transfusion recipients
   B. Intravenous drug abuse and needle sharing
      (1) HTLV-II is predominant virus in some U.S. drug abuser groups

than 1% of husbands would seroconvert over a similar time period (Kaji-yama *et al.*, 1986). A study of sexual transmission among patients attending sexually transmitted disease clinics in Jamaica elucidated sex-specific risk factors for transmission (Murphy *et al.*, 1989b). For women, a history of syphilis and many sexual partners was associated with seropositivity, while for men, only a history of syphilis and a history of genital ulcers were risk factors. This finding suggests that for men, disruption of normal genital mucosa is an important cofactor for acquiring infection. The risk of acquiring HTLV-I per sexual act is probably low and requires repeated exposure to an infected partner. Longitudinal data from a study in Myazaki, Japan, further demonstrated that given an infected partner, male to female transmission is five times more likely than transmission in the opposite direction (Mueller *et al.*, 1990). Such transmission was associated with older age of the infected partner and with higher antibody titer, suggesting that viral load and transmissibility increase with duration of infection. Recent data from Okinawa document the rare occurrence of female-to-male-transmission of HTLV-I from infected Okinawan women married to Caucasian U.S. Marines (Brodine *et al.*, 1990).

Homosexual men in endemic areas are at increased risk of acquiring HTLV-I, presumably through anal intercourse. In Trinidad, 15% of homosexual men were seropositive for HTLV-I (and 40% for HIV) as opposed to 2.4% of the general population (Bartholomew *et al.*, 1987). The relatively higher rate of HIV than HTLV-I among homosexuals is an indication of HIV's much higher efficiency of transmission.

B. Mother-to-Child Transmission

Several lines of evidence emphasize the importance of mother-to-child transmission of HTLV-I as the principal mode of childhood infection. Cross-sectional seroprevalence surveys indicate that children seroconvert in the first 2 years of life, and, thereafter, rates remain stable until the early teenage years, when presumably sexual transmission begins to take place. Cross-sectional family studies from Japan have shown that more than 90% of infected school-age children have mothers who were seropositive, and that 20–25% of children born of infected mothers themselves became infected (Sugiyama *et al.*, 1986; Hino *et al.*, 1985).

Breast milk, in particular, has been implicated in transmission of HTLV-I from mother to child. Short-term culture of breast milk lymphocytes obtained from HTLV-I-infected mothers showed the presence of HTLV-I antigen (Kinoshita *et al.*, 1984). Using an animal model, marmoset monkeys were successfully infected through oral administration of infected human breast milk lymphocytes (Kinoshita *et al.*, 1985). The close cell association of this virus with T lymphocytes presumably prevents it from crossing the placental barrier, as only one case has been reported of successful culturing of HTLV-I from cord blood (Komuro *et al.*, 1983).

Breast-fed babies from infected mothers are far more likely to become infected than are bottle-fed babies (Ando *et al.*, 1987). However, approximately 3% of bottle-fed babies will seroconvert indicating that *in utero* and peripartum transmission probably occur as well, but at a lower efficiency than breast milk transmission (Hino *et al.*, 1990). As described with heterosexual transmission, markers of higher viral load such as higher antibody titer and antigen expression are associated with more frequent mother-to-infant transmission (Sugiyama *et al.*, 1986; Kinoshita *et al.*, 1987; Wiktor *et al.*, 1990b).

C. Parenteral Transmission

Parenteral transmission by transfusion and by intravenous drug use is also well documented, and, in an attempt to prevent this mode of transmission, blood donors are being screened in endemic areas such as Japan as well as in the United States. Okochi and colleagues (1984) described seroconversion in 63% of recipients of cellular components, and in 0% of recipients of plasma. Seroconversion occurred at a mean of 40 days after transfusion with a range of 20–90 days. In a similar study, none of the 565 recipients of HTLV-I-negative units seroconverted, indicating that current serologic assays efficiently screen for the presence of HTLV-I (Nishimura *et al.*, 1989).

The high rates of HTLV-I/II among U.S. drug users is strong evidence of the efficiency of needle sharing as a route of transmission. The recent finding of HTLV-II among drug users raises the possibility that the parenteral transmission of HTLV-II is even more efficient than that of HTLV-I. Apparently, HTLV-I/II was present in drug users well before the appearance of HIV in the early 1980s since HTLV-I/II prevalence rates in IVDAs from several cities in 1972 were similar to rates seen today. None of these individuals were seropositive for HIV (Biggar *et al.*, 1991).

D. Environmental Cofactors

Descriptive studies have raised the possibility that ecologic factors influence the rate of seropositivity in populations suggesting a role for insect transmission of HTLV-I by some, and environmental factors associated with immune activation as a cofactor by others. In Japan, individuals who had antibodies to filariasis, which is transmitted by mosquitoes, were more likely to be HTLV-I seropositive (Tajima *et al.*, 1983). But, a similar association was also found with strongyloides, a parasite that is not transmitted by insects (Nakada *et al.*, 1984). These associations may be the result of an uncharacterized immunologic perturbation caused by one pathogen that predisposes an individual to acquire the other. This theory may also explain the association of HTLV-I with markers of lower socioeconomic status such as poor housing ascribed by Miller *et al.* (1986) as suggesting a role for insects. However, as reported by Blattner (1990), these markers may reflect immunologic perturbations that predispose to infection with a number of pathogens. Migrant studies also demonstrate the importance of environmental factors in HTLV-I infection. Kruickshank and colleagues (1990) were unable to document infection in any British-born offspring of seropositive Jamaican mothers, while 22% of offspring born to seropositive mothers in Jamaica were seropositive.

Biologically it is unlikely that HTLV-I is transmitted by insects since it is so closely cell-associated and is present in very low titers. Furthermore, HTLV-I seropositivity is not associated with presence of antibodies to a number of arthropod-borne viruses (Murphy *et al.*, 1989a).

VI. Disease Associations

A. Adult T-cell Leukemia

ATL was the first clinical outcome to be linked with HTLV-I. This aggressive form of leukemia was first reported as a clinical entity by Takatsuki *et al.* (1976) prior to the discovery of the virus. He described a T-cell lymphocytic leukemia characterized by malignant skin involvement, lymphadenopathy, hypercalcemia, bone marrow involvement, and a uniformly fatal outcome (Table II). An infectious etiology was suggested by the fact that many of the patients with ATL were born in Kyushu. Following the discovery of HTLV-I, serologic testing identified a high rate of HTLV-I seropositivity among patients with ATL (Robert-Guroff *et al.*, 1982; Blattner *et al.*, 1982).

ATL is found in all areas where HTLV-I is endemic. For example, in Japan cases of ATL occur predominantly in the endemic southern islands

Table II
Characteristics of HTLV-I-Associated Leukemia/Lymphoma[a]

Demographic features	
Male/female ratio	1.07 : 1
Mean age	40–45 years
Age range	15–84 years
Place of birth	Southern Japan
	Caribbean basin
	South America (some regions)
	Southern United States (blacks)
	Equatorial Africa
	Seychelles
Clinical Features	
Leukemia	50–85%
Hypercalcemia	50–60%
Bone marrow involvement	50%
Generalized lymphadeno-	
pathy	80–90%
Skin involvement	20–45%
Hepatosplenomegaly	50%
Lytic bone lesions	15%

[a]Reproduced with permission from Murphy and Blattner, 1988.

(Fig. 1). Owing to the short clinical course of this disease, its prevalence approximates the incidence. The annual incidence of ATL is estimated at 3.5/100,000 population in Japan (Tajima and Kuroishi, 1985) and 2/100,000 in Jamaica, and the lifetime risk for a seropositive individual is approximately 4% (Murphy et al., 1989c). The male to female ratio is equal, and the mean age of onset is in the fifth and sixth decades of life after which the incidence drops sharply (Tajima et al., 1990).

The incidence in the United States is not known as cancer registries do not distinguish ATL from other nonHodgkin's lymphomas and leukemias. In a series of case reports, U.S. cases clustered mainly among blacks in the southeastern U.S., among immigrants from the Caribbean or Japan, and in individuals with a history of travel to endemic areas (Levine et al., 1988; Ratner and Poiesz, 1989).

The incubation period for this disease is probably in the order of decades. Migrants from endemic areas, whose only exposure to HTLV-I was through their mothers, have developed the disease 20–30 years after leaving the endemic area (Greaves et al., 1984b). Modeling HTLV-I seroprevalence and ATL incidence has led to the suggestion that the major risk obtains in individuals exposed at a young age (Murphy et al., 1989c).

The epidemiologic link between HTLV-I and ATL is well established. Between 50 and 70% of all nonHodgkin's lymphoma cases in Jamaica are HTLV-I associated (Blattner et al., 1983; Manns, unpublished), and approximately 80–90% of clinically defined ATL cases are seropositive compared to a background rate of 5–15%. The few cases of antibody-negative ATL are undergoing molecular characterization to search for the presence of viral genome, but these cases may also represent a nonviral etiology of ATL.

The original syndrome as described by Takatsuki is now referred to as acute ATL. Clinical studies have since identified several subtypes of ATL (Fig. 3). Smoldering ATL, often indistinguishable from mycosis fungoides, is characterized by an indolent clinical course, cutaneous involvement, mild lymphadenopathy, and/or splenomegaly. Chronic ATL is another variant type with an indolent clinical course and a high percentage of circulating malignant cells. These cases overlap with some cases of T-cell chronic lymphocytic leukemia. Another variant of ATL is the lymphomatous type. This syndrome overlaps completely with T-cell nonHodgkin's lymphoma except for the presence of monoclonally integrated HTLV-I virus in the malignant cells. The distinction between these subtypes of ATL and the other clinical syndromes can only be made by detection of HTLV-I antibodies in serum of the patients, and by the identification of monoclonally integrated viral genome in the malignant cells.

B. Tropical Spastic Paraparesis

Gessain and colleagues (1985) made the link between HTLV-I and the chronic neurological syndrome, tropical spastic paraparesis during a survey of neurological diseases in Martinique. The syndrome was first described in

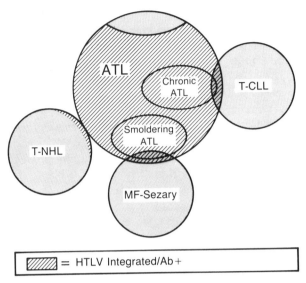

Figure 3. The clinical spectrum of HTLV-I-associated hematologic malignancies. Most patients present with acute adult T-cell leukemia (ATL). Chronic ATL overlaps with T-cell chronic lymphocytic leukemia (T-CLL), and smoldering ATL overlaps with mycosis fungoides–Sézary syndrome. (Reprinted with permission from Blattner and Gallo, 1986.)

the 1960s as Jamaican neuropathy (Montgomery *et al.*, 1964), and is characterized by a chronic, slowly progressive dysfunction of the pyramidal tracts (Table III). Osame *et al.* (1986b) described a similar syndrome in Japan and named it HTLV-I-associated myelopathy (HAM). The current recommended nomenclature is HAM/TSP (WHO, 1989).

Patients suffer spastic paraparesis of the limbs, particularly of the legs, bladder and bowel dysfunction, and hyperreflexia. Pathologically, the disease is marked by demyelinating plaques of the long tracts of the spinal cord. Although there is considerable overlap in symptoms with multiple sclerosis, the two diseases differ by the absence of cognitive symptoms in TSP and the lack of a waxing and waning course since HAM/TSP is characteristically a slowly progressive disease process (Roman, 1988).

Since the disease is one of long duration and is often missed by clinicians, there are no good estimates of incidence and prevalence. Japan reports the largest number of cases of HAM/TSP, with case occurrence estimated to be 1 in every 1000–2000 HTLV-I-positive carriers (Osame *et al.*, 1989). Other studies from the Caribbean estimate the lifetime risk of acquiring HAM/TSP for a seropositive individual to be between 1% and 5%. There is an excess of cases among women with the female to male ratio being approximately 2:1, and the mean age of disease onset is in the fourth to fifth decade of life. In Japan, 25% of HAM/TSP cases gave a history of blood transfusion (Osame *et al.*, 1986a). The minimum time between transfusion

Table III
Clinical Diagnosis of HAM/TSP[a]

Main neurologic manifestations

Chronic spastic paraparesis which usually progresses slowly, sometimes remains static after initial progression.

Weakness of the lower limbs, more marked proximally.

Bladder disturbance is usually an early feature, and constipation usually occurs later; impotence or decreased libido is common.

Sensory symptoms such as tingling, pins and needles, burning, etc. are more prominent than objective physical signs.

Low lumbar pain with radiation to the legs is common.

Vibration sense is frequently impaired, proprioception less often affected.

Hyperreflexia of the lower limbs, often with clonus and Babinski's sign.

Exaggerated jaw jerk in some patients.

Less frequent neurologic findings

Cerebellar signs; optic atrophy; deafness; nystagmus; other cranial nerve deficits; hand tremor; absent or depressed ankle jerk.

Convulsions, cognitive impairment, dementia, or impaired consciousness are rare.

Laboratory diagnosis

Presence of HTLV-I antibodies or antigens in blood and cerebrospinal fluid (CSF).

CSF may show mild lymphocyte pleocytosis.

Lobulated lymphocytes may be present in blood and/or CSF.

Mild to moderate increase of protein may be present in CSF.

Viral isolation when possible from blood and/or CSF.

---

[a]Recommendations of the Scientific Group on HTLV-I Infections and its associated diseases, WHO Regional Office for the Western Pacific (WHO, 1989).

and development of disease was 18 weeks in a well-characterized case report from France (Gout *et al.*, 1990). The absence of a similar association with ATL suggests that unlike ATL the incubation period for this disease can be short.

Patients with HAM/TSP have a higher HTLV-I antibody titer when compared to ATL cases. When lymphocytes from HAM/TSP patients are placed into culture, they show a higher level of spontaneous proliferation than do those of asymptomatic seropositives and ATL patients (Kramer *et al.*, 1989). These findings have led to the hypothesis that patients with HAM/TSP demonstrate a heightened immune response to HTLV-I, and that the pathologic changes are the result of an autoimmune process against the central nervous system (Kramer and Blattner, 1989).

C. Other Diseases

Researchers have looked for an association between HTLV-I and a number of other illnesses, particularly with autoimmune disorders. No association has surfaced for systemic lupus erythematosus (Murphy *et al.*, 1988), or with sarcoidosis (Blayney *et al.*, 1983). A recent report suggests that HTLV-I may play a role in the pathogenesis of polymyositis as 11/13 Jamaican patients with this disease were HTLV-I positive (Morgan *et al.*, 1989). Another possible association is with certain types of polyarthropathy, where atypical lymphocytes compatible with ATL-like cells were found in the synovial fluid of patients (Nishioka *et al.*, 1989). The report of two cases of B-cell chronic lymphocytic leukemia by Mann and colleagues (1987) also suggests a possible role for HTLV-I in this syndrome where the B-cell tumor represents HTLV-I-antigen-committed lymphoid cells. Other as yet unconfirmed associations include a single case of small cell cancer of the lung with monoclonally integrated HTLV-I in the tumor and a pulmonary infiltrative disease recognized in some Japanese HAM/TSP cases.

D. Immunodeficiency

Several lines of experimental data raise the possibility that HTLV-I may cause subtle immune dysfunction even among asymptomatic carriers. After infection with HTLV-I, certain T helper cell lines stimulate B cells to proliferate, resulting in polyclonal immunoglobulin production (Popovic *et al.*, 1984). Lymphocytes from asymptomatic carriers tend to proliferate spontaneously when placed in culture medium (Kramer *et al.*, 1989).

Epidemiologic data support the role of HTLV-I as an immune suppressor. Even prior to developing leukemia, ATL patients occasionally acquire infections, such as pneumocystis pneumonia, that are usually associated with immunosuppression. Asymptomatic HTLV-I carriers demonstrate skin test anergy more often than do seronegative individuals (Tachibana *et al.*, 1988). This phenomenon is more marked in older individuals, suggesting that the immune suppression becomes more marked with longer duration of infection with HTLV-I. Several small studies indicate that HTLV-I carriers are more likely to acquire a number of infections, including tuberculosis, leprosy, and strongyloidiosis. The small numbers of individuals studied and the lack of prospective data make it difficult to determine the true role HTLV-I plays in the pathogenic process of these infections. One indication implicating HTLV-I is that patients with strongyloides who were HTLV-I positive responded less well to therapy than did those who were negative (Terry, unpublished). Well-designed epidemiologic studies with close laboratory links will be necessary in order to elucidate some of the more subtle immunological effects of HTLV-I.

VII. Conclusion

In the ten years since the discovery of HTLV-I there have been rapid strides in understanding the role of HTLV-I as a human pathogen. In addition, its major modes of transmission have been defined and prospective studies are starting to elucidate the long-term natural history of HTLV-I. For example,

HTLV-I-associated lymphoproliferative diseases probably result from early life infection, with many decades between exposure and disease outcome. By contrast, the latency for HAM/TSP appears shorter, raising the possibility of a different natural history and pathogenesis, possibly involving indirect immunologic mechanisms.

Future studies of this virus promise to help elucidate new understanding of the pathogenetic process for disease occurrence as well as providing new insights that may lead to the recognition of other types of HTLV-related viruses of long latency associated with a number of chronic diseases of unknown etiology.

References

Agius, G., Biggar, R. J., Alexander, S. S., Waters, D. J., Drummond, J. E., Murphy, E. L., Weiss, S. H., Levine, P. H., and Blattner, W. A. (1988). *J. Infect. Dis.* **158**, 1235–1244.

Anderson, D. W., Epstein, J. S., Lee, T. H., Lairmore, M. D., Saxinger, C., Kalyanaraman, V. S., Slamon, D., Parks, W., Poiesz, B. J., Pierik, L. T., Lee, H., Montagna, R., Roche, P. A., Williams, A., and Blattner, W. (1989). *Blood* **74**, 2585–2591.

Ando, Y., Nakano, S., Saito, K., Shimamoto, I., Ichijo, M., Tayama, T., and Hinuma, Y. (1987). *Jpn. J. Cancer Res.* **78**, 322–324.

Bartholomew, C., Saxinger, W. C., Clark, J. W., Gail, M., Dudgeon, A., Mahabir, B., Hull-drysdale, B., Cleghorn, F., Gallo, R. C., and Blattner, W. A. (1987). *JAMA* **257**, 2604–2608.

Biggar, R. J., Buskell-Bales, Z., Yakshe, P. N., Caussy, D., Gridley, G., and Seeff, L. (1991). *J. Infect. Dis.* **163**, 57–63.

Blattner, W. A. (1990). *In* ''Human Retrovirology: HTLV'' (W. A. Blattner, ed.), pp. 251–266. Raven Press, New York.

Blattner, W. A., and Gallo, R. C. (1986). *In* '' Proceedings of the XIIth Symposium of the International Association for Comparative Research on Leukemia and Related Diseases'' (F. Deinhardt, ed.), pp. 361–382. Hamburg.

Blattner, W. A., Kalyanaraman, V. S., Robert-Guroff, M., Lister, T. A., Galton, D. A., Sarin, P. S., Crawford, M. H., Catovsky, D., Greaves, M., and Gallo, R. C. (1982). *Int. J. Cancer* **30**, 257–264.

Blattner, W. A., Gibbs, W. N., Saxinger, C., Robert-Guroff, M., Clark, J., Lofters, W., Hanchard, B., Campbell, M., and Gallo, R. C. (1983). *Lancet* **ii,** 61–64.

Blayney, D., Rohatgi, P. K., Hines, W., Robert-Guroff, M., Saxinger, W. C., Blattner, W. A., and Gallo, R. C. (1983). *Ann. Intern. Med.* **99**, 409.

Brodine, S. K., Blattner, W. A., Holmberg, J., Molgaard, C. A., Oldfield, E. C., and Corwin, A. L. (1990). Paper presented at the Third Annual Retrovirology Conference, Maui, Hawaii.

Centers for Disease Control (1988). *Morbid. Mortal.* **37**, 736–747.

Clark, J. W., Robert-Guroff, M., Ikehara, O., Henzan, E., and Blattner, W. A. (1985). *Cancer Res.* **45**, 2849–2852.

Cortes, E., Detels, R., Aboulafia, D., Xi Ling, L., Moudgil, T., Alam, M., Bonecker, C., Gonzaga, A., Oyafuso, L., Tondo, M., Boite, C., Hammershlak, N., Capitani, C., Slamon, D., and Ho, D, (1989). *N. Engl. J. Med.* **320**, 953–958.

Currie, B., Hinuma, Y., Imai, J., Cumming, S., and Doherty, R. (1989). *Lancet* **ii,** 1137.

Delaporte, E., Peeters, M., Durand, J. P., Dupont, A., Schrijvers, D., Bedjabaga, L., Honre, C., Ossari, S., Trebucq, A., Josse, R., and Merlin, M. (1989a). *J. AIDS* **2**, 410–413.

Delaporte, E., Peeters, M., Simoni, M., and Piot, P. (1989b). *Lancet* **ii,** 1226.

Ehrlich, G. D., Glaser, J. B., Abbott, M. A., Slamon, D. J., Keith, D., Sliwkowski, M., Brandis, J., Keitelman, E., Teramoto, Y., Papsidero, L., Simpkins, H., Sninsky, J. J., and Poiesz, B. J. (1989). *Blood* **74**, 1066–1072.

Gallo, R. C., Sliski, A. H., de Noronha, C. M., and de Noronha, F. (1986). *Nature (London)* **320**, 219.

Gessain, A., Barin, F., Vernant, J. C., Gout, O., Maurs, L., Calender, A., and De The, G. (1985). *Lancet* **ii**, 407–410.

Gout, O., Baulac, M., Gessain, A., Semah, F., Saal, F., Peries, J., Cabrol, C., Foucault-fretz, C., Laplane, D., Sigaux, F., and De The, G. (1990). *N. Engl. J. Med.* **322**, 383–388.

Gradilone, A., Zani, M., Garillari, G., Modesti, M., Agliano, A. M., Maiorano, G., Ortona, L., Frati, L., and Manzari, V. (1986). *Lancet* **ii**, 753–754.

Greaves, M. F., Verbi, W., Tilley, R., Lister, T. A., Habeshaw, J., Guo, H. G., Trainor, D. C., Robert-Guroff, M., Blattner, W. A., Reitz, M., and Gallo, R. C. (1984a). *Int. J. Cancer* **33**, 795–806.

Greaves, M. F., Verbi, W., Tilley, R., Lister, T. A., Robert-Guroff, M., Blattner, W. A., Reitz, M. and Gallo, R. C. (1984b). *In* "Human T-cell Leukemia/lymphoma Viruses" (R. C. Gallo, M. E. Essex, and L. Gross, eds), pp. 297–306. Cold Spring Harbor Laboratory, Cold Spring Harbor, New York.

Hino, S., Yamguchi, K., Katamine, S., Sugiyama, H., Amagasaki, T., Kinoshita, K., Yoshida, Y., Doi, H., Tsuji, Y., and Miyamoto, T. (1985). *Jpn. J. Cancer Res.* **76**, 474–480.

Hino, S., Kubota, K., Doi, H., and Miyamoto, T. (1990). Paper presented at the Third Annual Retrovirology Conference, Maui, Hawaii.

Hinuma, Y. (1986). *Aids Res.* **2** (Suppl. 1), s17–s22.

Hinuma, Y., Komoda, H., Chosa, T., Kondo, T., Kohakura, M., Takenaka, T., Kikuchi, M., Ichimaru, M., Yunoki, K., Sato, I., Matsuo, R., Takiuchi, Y., Uchino, H., and Hanaoka, M. (1982). *Int. J. Cancer* **29**, 631–635.

Kajiyama, W., Kashiwagi, S., Ikematsu, H., Hayashi, J., Nomura, H., and Okochi, K. (1986). *J. Infect. Dis.* **154**, 851–857.

Kayembe, K., Goubau, P., Desmyter, J., Vlietinck, E., and Carton, H. (1990). *J. Neurol. Neurosurg. Psychiatry* **53**, 4–10.

Kazura, J. W., Saxinger, W. C., Wenger, J., Forsyth, K., Lederman, M. M., Gillespie, J. A., Carpenter, C. C., and Alpers, M. A. (1987). *J. Infect. Dis.* **155**, 1100–1107.

Kinoshita, K., Hino, S., Amagasaki, T., Ikeda, S., Yamada, Y., Suzuyama, J., Myomata, S., Toriya, K., Kamihira, S., and Ichimaru, M. (1984) *Jpn. J. Cancer Res.* **75**, 103–105.

Kinoshita, K., Yamanouchi, K., Ikeda, S., Momita, S., Amagasaki, T., Soda, H., Ichimaru, M., Moriughi, R., Katamine, S., Miyamoto, T., and Hino, S. (1985). *Jpn. J. Cancer Res.* **76**, 1147–1153.

Kinoshita, K., Amagasaki, T., Hino, S., Doi, H., Yamagouchi, K., Ban, N., Momita, S., Ikeda, S., Kamihira, S., and Ichimaru, M. (1987). *Jpn. J. Cancer Res.* **78**, 674–680.

Komuro, A., Hayami, M., Fuji, H., Miyahara, S., and Hirayama, M. (1983). *Lancet* **i**, 240.

Kramer, A., and Blattner, W. A. (1989). *In* "Concepts in Viral Pathogenesis" (A. L. Notkins, and M. B. A. Oldstone, eds.), pp. 204–214. Springer-Verlag, New York.

Kramer, A., Jacobson, S., Murphy, E. L., Wiktor, S. Z., Cranston, B., Figueroa, J. P., Hanchard, B., Mcfarlin, D., and Blattner, W. A. (1989). *Lancet* **ii**, 923–924.

Kruickshank, J. K., Richardson, J. H., Morgan, O. St. C., Porter, J., Klenerman, P., Knight, J., Newell, A. L., Rudge, P., and Dalgleish, A. G. (1990). *Brit. Med. J. (Clin. Res.)* **300**, 300–304.

Lairmore, M. D., Jacobson, S., Kaplan, J. E., Roberts, B. D., De, B. K., Levine, P., and Reeves, W. C. (1990). Paper presented at the Third Annual Retrovirology Conference, Maui, Hawaii.

Lee, H., Swanson, P., Shorty, V. J., Zack, J. A., Rosenblatt, J. D., and Chen, I. S. Y. (1989). *Science* **244**, 471–475.

Levine, P. H., Jaffe, E., Manns, A., Murphy, E., Clark, J., and Blattner, W. A. (1988). *Yale J. Biol. Med.* **61**, 215–222.

Lillehoj, E. P., Alexander, S. A., Chang-Chih, T., Dubrule, C. J., Adams, R., Manns, A., Wiktor, S. Z., Blattner, W. A., Cyrus, S., Decker, A., and Swenson, S. (1990). *J. Clin. Micro.* **28,** 2653–2658.

Lipka, J. J., Bui, K., Reyes, G. R., Moeckli, R., Wiktor, S. Z., Blattner, W. A., Murphy, E. L., Shaw, G. M., Hanson, C. V., Sninsky, J. J., and Foung, S. K. H. (1990). *J. Infect. Dis.* **162,** 353–357.

Maloney, E. M., Ramirez, H. C., Levin, A., and Blattner, W. A. (1989). *Int. J. Cancer* **440,** 419–423.

Maloney, E. M., Murphy, E. L., Figueroa, J. P., Gibbs, W. N., Cranston, B., Hanchard, B., and Blattner, W. A. (1990). Paper presented at the Third Annual Retrovirology Conference, Maui, Hawaii.

Mann, D. L., Desantis, P., Mark, G., Pfeifer, A., Newman, M., Gibbs, N., Popovic, M., Sarngadharan, M. G., Gallo, R., Clark, J., and Blattner, W. A. (1987). *Science* **236,** 1103–1106.

Manzari, V., Gradilone, A., Barillari, G., Zani, M., Collati, E., Pandolfi, F., DeRossi, G., Liso, V., Babbo, P., Robert-Guroff, M., and Frati, L. (1985). *Int. J. Cancer* **36,** 557–559.

Manzari, V., Gismondi, A., Barillari, G., Morrone, S., Modesti, A., Albonici, L., Demarchis, L., Fazio, V., Gradilone, A., Zani, M., Frati, L., and Santoni, A. (1987). *Science* **238,** 1581–1583.

Miller, G. J., Pegram, S. M., Kirkwood, B. R., Beckles, G. L., Byam, N. T., Clayden, S. A., Kinlen, L. J., Chan, L. C., Carson, D. C., and Greaves, M. F. (1986). *Int. J. Cancer* **38,** 801–808.

Montgomery, R. D., Cruickshank, E. K., Robertson, W. B., and McMenemy, W. H. (1964). *Brain* **87,** 425–462.

Morgan, O. St. C., Rodgers-Johnson, P., Mora, C., and Char, G. (1989). *Lancet* **ii,** 1184–1186.

Mueller, N., Tachibana, N., Stuver, S. O., Okayama, O., Ishizaki, J., Shishime, E., Murai, K., Shioiri, S., and Tsuda, K. (1990). *In* "Human Retrovirology: HTLV" (W. A. Blattner, ed.), pp. 281–294. Raven Press, New York.

Murphy, E. L. (1990). *In* "Human Retrovirology: HTLV" (W. A. Blattner, ed.), pp. 295–306. Raven Press, New York.

Murphy, E. L., and Blattner, W. A. (1988). *Ann. Neurol.* **23** (Suppl.), 5174–5180.

Murphy, E. L., Deceulaer, K. D., Williams, W., Clark, J. W., Saxinger, C., Gibbs, W. N., and Blattner, W. A. (1988). *J. AIDS.* **1,** 18–22.

Murphy, E. L., Calisher, C. H., Figueroa, J. P., Gibbs, W. N., and Blattner, W. A. (1989a). *N. Engl J. Med.* **320,** 1146.

Murphy, E. L., Figueroa, J. P., Gibbs, W. N., Barthwaite, A., Holding-Cobham, M., Waters, D., Cranston, B., Hanchard, B., and Blattner, W. A. (1989b). *Ann. Intern. Med.* **111,** 555–560.

Murphy, E. L., Hanchard, B., Figueroa, J. P., Gibbs, W. N., Lofters, W. S., Campbell, M., Goedert, J. J., and Blattner, W. A. (1989c). *Int. J. Cancer* **43,** 250–252.

Nakada, K., Kohakura, M., Komoda, H., and Hinuma, Y. (1984). *Lancet* **i,** 633.

Nishimura, Y., Yamaguchi, K., Kiyokawa, T., Takatsuki, K., Imamura, Y., and Fujiwara, H. (1989). *Transfusion* **29,** 372.

Nishioka, K., Maruyama, I., Sato, K., Kitjima, I., Najkima, Y., and Osame, M. (1989). *Lancet* **ii,** 441.

Okochi, K., Sato, H., and Hinuma, Y. (1984). *Vox Sang.* **46,** 245–253.

Osame, M., Isumo, S., Igata, A., Matsumoto, M., Matsumoto, T., Sonoda, S., Tara, M., and Shibata, Y. (1986a). *Lancet* **ii,** 104–105.

Osame, M., Usuku, K., Izumo, S., Ijichi, N., Amitani, H., Igata, A., Matsumoto, M., and Tara, M. (1986b). *Lancet* **i,** 1031–1032.

Osame, M., Kaplan, J., Igata, A., Kubota, H., Nishitani, H., and Janssen, R. (1989). *Proceedings of the V International Conference on AIDS, Montreal, 1989,* Abstr. Th.A.P.17.

Poiesz, B. J., Ruscetti, F. W., Gazdar, A. F., Bunn, P. A., Minna, J. D., and Gallo, R. C. (1980). *Proc. Natl. Acad. Sci. U.S.A.* **77,** 7415–7419.

Popovic, M., Flomenberg, N., Volkman, D. J., Mann, D., Fauci, S. A., Dupont, B., and Gallo, R. C. (1984). *Science* **226**, 459–462.

Ratner, L., and Poiesz, B. J. (1989). *Medicine* **7**, 401–422.

Reeves, W. C., Levine, P. H., Cuevas, M., Quiroz, E., Maloney, E., and Saxinger, W. C. (1990). *Am. J. Trop. Med. Hyg.* **42**, 374–379.

Riedel, D. A., Evans, A. S., and Blattner, W. A. (1989). *J. Infect. Dis.* **159**, 603–609.

Robert-Guroff, M., Nakao, Y., Notake, K., Ito, Y., Sliski, A., and Gallo, R. C. (1982). *Science* **215**, 975–978.

Robert-Guroff, M., Kalyanaraman, V. S., Blattner, W. A., Popovic, M., Sarngadharan, M. G., Maeda, M., Blayney, D., Catovsky, D., Bunn, P. A., Shibata, A., Nakao, Y., Ito, Y., Aoki, T., and Gallo, R. C. (1983). *J. Exp. Med.* **157**, 248–258.

Roman, G. C. (1988). *Ann. Neurol.* **23** (Suppl.), s113–s120.

Ruscetti, F. W., Robert-Guroff, M., Ceccherini-nelli, L., Minowada, J., Popovic, M., and Gallo, R. C. (1983). *Int. J. Cancer* **31**, 171–180.

Saiki, R. K., Gelfand, D. H., Stoffel, S., Scharf, S. J., Higuchi, R., Horn, G. T., Mullis, K. B., and Ehrlich, H. (1988). *J. Virol. Methods* **23**, 21–31.

Sarin, P. S., Aoki, T., Shibata, A., Ohnishi, Y., Aoyagi, Y., Miyakoshi, H., Emura, I., Kalyanaraman, V. S., Robert-Guroff, M., Popovic, M., Sarngadharan, M., Nowell, P. C., and Gallo, R. C. (1983). *Proc. Natl. Acad. Sci. U.S.A.* **80**, 2370–2374.

Sugiyama, H., Doi, H., Ymaguchi, K., Tusi, Y., Miyamoto, T., and Hino, S. (1986). *J. Med. Virol.* **20**, 253–260.

Tachibana, N., Okayama, A., Ishizaki, J., Yokota, T., Shshime, E., Murai, K., Shioiri, S., Tsuda, K., Essex, M., and Mueller, N. (1988). *Int. J. Cancer* **42**, 829–831.

Tajima, K., and Kuroishi, T. (1985). *Jpn. J. Clin. Oncol.* **15**, 423–430.

Tajima, K., Fujita, K., Tsukidate, S., Oda, T., Tominaga, S., Suchi, T., and Hinuma, Y. (1983). *Gann* **74**, 188–191.

Tajima, K., Kamura, S., Shin-ichiro, I., Nagatomo, M., Kinoshita, K., and Ikeda, S. (1987). *Int. J. Cancer* **40**, 741–746.

Tajima, K., The T- and B-Cell Malignancy Study Group, and co-authors (1990). *Int. J. Cancer* **45**, 237–243.

Takatsuki, K., Uchiyama, T., Sagawa, K., Yodoi, J. (1976). *Proceedings of the 16th International Congress of Hematology Kyoto, Japan,* 73–77.

Ueda, K., Kushuhara, K., Tokugawa, K., Miyazaki, C., Yoshida, C., Tokumura, K., Sonoda, S., and Takahashi, K. (1989). *Lancet* **ii**, 979.

Verdier, M., Denis, F., Sangare, A., Barin, F., Gershy-damet, G., Rey, J. L., Soro, B., Leonard, G., Mounier, M., and Hugon, J. (1989). *J. Infect. Dis.* **160**, 363–370.

Weber, J., Banatvala, N., Clayden, S., McAdam, K., Palmer, S., Moulsdale, H., Tosswill, J., Dilger, P., Thorpe, R., and Amann, S. (1989). *J. Infect. Dis.* **159**, 1025–1028.

Weiss, S. H., Saxinger, W. C., Ginzburg, H. M., Mundon, F. K., and Blattner, W. A. (1987). *Proc. ASCO* **6**, 4.

WHO (1989). *Wkly. Epidem. Rec.* **49**, 382–383.

Wiktor, S. Z., Mann, J. M., Nzilambi, N., Francis, H., Piot, P., Blattner, W. A., and Quinn, T. C. (1990a). *J. Infect. Dis.* **161**, 1073–1077.

Wiktor, S. Z., Pate, E., Murphy, E. L., Champagnie, E., Ramlal, A., and Blattner, W. A. (1990b). *AIDS* **6**, 136.

Williams, A. E., Fang, C. T., Slamon, D. J., Poiesz, B. J., Sandler, S. G., Darr, W. F., Shulman, G., Mcgowan, E. I., Douglas, D. K., Bowman, R. J., Peetoom, F., Kleinman, S. H., Lenes, B., and Dodd, R. (1988). *Science* **240**, 643–646.

Williams, C. K., Saxinger, W. C., Alabi, G. O., Junaid, T. A., Blayney, D. W., Greaves, M. F., Gallo, R. C., and Blattner, W. A. (1984). *Brit. Med. J.* **288**, 1495–1496.

Zaninovic, V., Arango, C., Biojo, R., Mora, C., Rodgers-Johnson, P., Concha, M., Corral, R., Barreto, P., Borrero, I., Garruto, R. M., Gibbs, C. J., and Gajdusek, D. C. (1988). *Ann. Neurol.* **23** (Suppl.), s127–s132.

Human Immunodeficiency Virus (HIV) Infections in the United States

Allison L. Greenspan, Ruth L. Berkelman,
Timothy J. Dondero, Jr., and James W. Curran

I. Introduction

Human immunodeficiency virus (HIV) infection has emerged over the past decade as a formidable public health challenge with staggering medical, ethical, legal, and economic implications. The current epidemic was already under way in the United States by 1981 when acquired immunodeficiency syndrome (AIDS), the end-stage clinical manifestation of infection with HIV, was first recognized. Between 1981 and 1989, more than 100,000 AIDS cases were reported to the Centers for Disease Control (CDC)—35,238 in 1989 alone. According to Public Health Service estimates, 52,000–57,000 AIDS cases will be diagnosed in 1990, and the number of newly diagnosed AIDS cases will continue to increase (Centers for Disease Control, 1990). By 1993, the cumulative case count is expected to reach 390,000–480,000, with the epidemic claiming more than 285,000 lives.

Since 1981, the natural history and modes of transmission of HIV have been illuminated by the availability of results from epidemiologic studies carried out in this country and others. HIV infection is a progressive disease process, with severe immunosuppression and AIDS as its end point. The time from HIV infection to clinical illness can be months to years. Once infected, symptomatic and asymptomatic persons can transmit the virus to others through three main modes of spread: sexual contact, blood-to-blood contact, and perinatal transmission from mother to infant.

Ongoing analysis of reported U.S. AIDS cases since 1981 has yielded valuable information on the distribution of severe HIV-related illness and has demonstrated an enormous diversity in the HIV/AIDS epidemic by geographic area, age and sex, race and ethnicity, and behavioral patterns. The epidemic that America faces today is notably different from that of 9 years ago: the geographic focus has widened, the profile of persons at greatest risk is changing, and the relative importance of different routes of transmission is shifting.

II. AIDS Case Surveillance

AIDS is a reportable disease in all states and U.S. territories. As of December 31, 1989, state and local health departments had reported 117,781 men, women, and children with AIDS. The number of AIDS cases reported each year has steadily, and inexorably, increased (Fig. 1). The rate of increase has slowed, however, from 215% between 1982 and 1983, to 117% in 1984, 84%

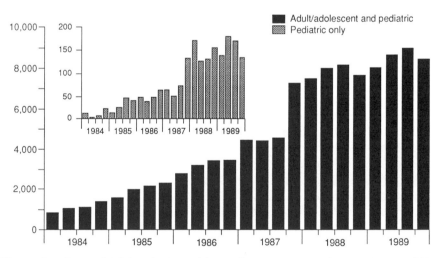

Figure 1. Cases of AIDS in the United States, by quarter year of report, January 1984 through December 1989. The September 1987 revision of the case definition for AIDS surveillance resulted in an abrupt increase in reported cases in the fourth quarter of 1987.

in 1985, 60% in 1986 and 1987, and 34% in 1988. The 60% increase in 1987, which remained unchanged from the year before, coincided with the revision of the surveillance definition for AIDS. This revision expanded the range of reportable illnesses, reflecting changes in diagnostic practices and the wider availability and use of HIV antibody testing.

AIDS cases have been reported from all 50 states, the District of Columbia, and four U.S. territories, but about two-thirds of cases have come from five states: New York, New Jersey, Florida, Texas, and California. Annual incidence rates by state varied in 1989 from 0.6 cases per 100,000 persons in South Dakota to 33.5 per 100,000 in New York; the rate in Puerto Rico was 44.9 per 100,000 persons (Fig. 2). Half the states have reported fewer than 600 cases each, with a range of from fewer than 25 in North Dakota and South Dakota to more than 20,000 in California and New York. AIDS case counts are highest in the most populous metropolitan areas; those with more than a million people make up 41% of the nation's population but account for 75% of all AIDS cases (Centers for Disease Control, 1989c). New York City continues to report the largest number of cases. At the end of 1988, the number of cases reported from New York City (18,022) exceeded the combined total from San Francisco (6602), Los Angeles (6010), Houston (2578), and Newark (2460) (Greenberg et al., 1989).

Although the geographic distribution of AIDS cases is expected to be uneven for some time, areas that are less densely populated and outside the current high-incidence states are beginning to share the AIDS case burden. The proportion of cases in urban areas under 500,000 population rose from 12% in pre-1986 years to 19% in 1988, most noticeably in the Midwest and South Atlantic regions. In 1988 only 32% of the nation's AIDS cases were reported from the Middle Atlantic states of New York, New Jersey, and Pennsylvania, compared to 54% four years earlier (Centers for Disease Control, 1989c).

Since its discovery as the cause of AIDS, HIV infection has been associated with an increasing number of disease manifestations. The AIDS case definition used for national surveillance has been modified twice to reflect the growing understanding of the disease. In 1985 the AIDS surveillance definition was expanded to include disseminated histoplasmosis, non-Hodgkin's lymphoma, and chronic isosporiasis in persons with evidence of HIV infection (Centers for Disease Control, 1985). In 1987 the definition was revised again to include a broader spectrum of HIV-associated diseases (including HIV wasting syndrome, HIV encephalopathy, and disseminated tuberculosis) and to allow for the presumptive diagnoses of certain other conditions (Centers for Disease Control, 1987a). Of cases reported between September 1987 and December 1988, 29% would not have been classified as AIDS under the old definition at the time of report (Centers for Disease Control, 1989c). In particular, the 1987 revision resulted in an increased proportion of women, blacks and Hispanics, and intravenous drug users

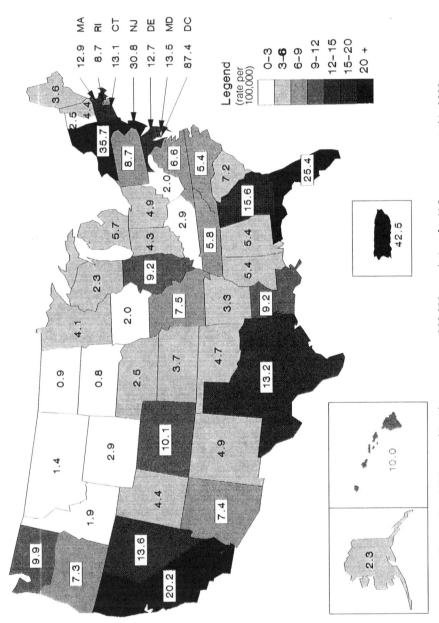

Figure 2. AIDS annual incidence rates per 100,000 population, for U.S. cases reported in 1989.

(IVDUs) among all reported AIDS patients. Since many persons reported as AIDS cases based only on the revised definition will subsequently develop illnesses that would have met the criteria of the old definition, the effect of the revision is, in part, to identify some AIDS patients earlier in the course of their illness.

III. Surveillance of HIV Infection

The AIDS epidemic is the most visible and severe manifestation of a much larger epidemic of HIV infection. For every person with AIDS, many more suffer from HIV-related conditions and an even greater number are infected but have no symptoms at all. As the total AIDS case count exceeds 100,000, the number of HIV-infected persons nationwide is estimated at 1.0 million (Centers for Disease Control, 1990). Since HIV infection often precedes AIDS by many years, trends in reported AIDS cases may not completely reflect current patterns of HIV infection. The epidemic of HIV infection in the United States is, in fact, an evolving composite of many partially over-lapping epidemics, each with its own dynamics and rate of spread (Curran *et al.*, 1988). To supplement AIDS case surveillance and more effectively as-certain and monitor the levels and trends of HIV infection within population groups and geographic areas, CDC has undertaken a "family" of surveys and studies, each of which highlights a different element of the HIV/AIDS epidemic (Dondero *et al.*, 1988). Together with information from HIV infec-tion reporting, routine serologic testing programs, and other ongoing epide-miologic studies, these focused serosurveys will help state and local health officials to determine the extent of HIV infection in selected groups and areas and to target and evaluate prevention activities.

IV. Exposure Categories

As has been the case since the epidemic's earliest days, most AIDS patients are male—90% of the reported cases in adults and adolescents (Table I). However, the proportion of adult cases in women has been steadily increas-ing, from 7% before 1984 to 10% in 1989. Cases occur over the full range of ages, but 80% of AIDS patients are diagnosed between 20 and 45 years of age (Table I). About 2% of AIDS cases are in children under age 13. AIDS cases in adolescents are increasing at a rate similar to that of other age groups (Manoff *et al.*, 1989).

The vast majority of adult AIDS patients are homosexual or bisexual men (61%), IVDUs (21%), or both (7%) (Table II). An additional 2% of AIDS patients are recipients of blood transfusions from HIV-infected persons, and 1% are patients with hemophilia or related disorders. Another 5% of cases are attributed to heterosexual contact with an HIV-infected partner. The

Table I

AIDS Cases by Age and Sex through December 1989
in United States

Age at diagnosis (years)	Male	Female	Total	(%)
Under 5	850	793	1,643	(1.4)
5–12	232	120	352	(0.3)
13–19	367	94	461	(0.4)
20–24	4,346	744	5,090	(4.3)
25–29	16,744	2,222	18,966	(16.1)
30–34	25,927	2,944	28,871	(24.5)
35–39	23,422	2,041	25,463	(21.6)
40–44	14,957	1,003	15,960	(13.6)
45–49	8,531	460	8,991	(7.6)
50–54	4,874	288	5,162	(4.4)
55–59	3,085	233	3,318	(2.8)
60–64	1,596	190	1,786	(1.5)
65 or older	1,326	392	1,718	(1.5)
Total	106,257	11,524	117,781	(100.0)

Table II

Adult AIDS Cases by Exposure Category and Sex through December 1989
in United States

Exposure category	Male	Female	Total	(%)
Male homosexual/bisexual contact	70,093	—	70,093	(61)
IV drug use (female and heterosexual male)	18,721	5,491	24,212	(21)
Male homosexual/bisexual contact and IV drug use	8,117		8,117	(7)
Hemophilia/coagulation disorder	1,034	28	1,062	(1)
Heterosexual contact:	2,308	3,322	5,630	(5)
Sex with IV drug user	816	2,055	2,871	
Sex with bisexual male	—	353	353	
Sex with person with hemophilia	4	48	52	
Born in Pattern II country[a]	1,193	439	1,632	
Sex with person born in Pattern II country	49	33	82	
Sex with transfusion recipient with HIV infection	22	59	81	
Sex with person with HIV infection, risk not specified	224	335	559	
Receipt of transfusion	1,768	1,062	2,830	(2)
Other/undetermined	3,134	708	3,842	(3)
Total	105,175	10,611	115,786	(100)

[a] Pattern II countries are countries in which most of the reported cases occur in heterosexuals.

means by which HIV infection was acquired is undetermined in about 3% of cases, generally because risk information is incomplete or unobtainable.

A. Homosexual/Bisexual Men

Homosexual and bisexual men were the first group in which AIDS was recognized in the United States. Data from studies of transmission among homosexual and bisexual men suggest that the risk of infection through sexual contact increases with the number of sex partners and the frequency of participation in certain sex practices, especially receptive anal intercourse (Darrow et al., 1987; Winkelstein et al., 1987). Although homosexual and bisexual men still account for the majority of AIDS patients, the proportion of cases from this group has decreased in recent years, from 63% of all AIDS cases before 1985 to 57% in 1989. The geographic distribution of AIDS cases in this group is also shifting. Before 1985, 56% of cases in homosexual men were reported from New York City, Los Angeles, and San Francisco, compared with 30% in 1988.

Surveys of homosexual men conducted since 1984 in 22 cities in 15 states show HIV infection rates ranging from 10% to 70%, with the highest rates in areas with the highest incidence of AIDS (Centers for Disease Control, 1987c). For two cohorts of homosexual men in San Francisco who are being followed over time, the incidence of new infections dropped sharply between 1982 and 1988, from more than 10% to less than 2% (Hessol et al., 1989; Winkelstein et al., 1988). These trends are consistent with reports of changes in sexual behavior in this group and of decreases in reported cases of syphilis and gonorrhea (Centers for Disease Control, 1988d; Winkelstein et al., 1988).

B. Intravenous Drug Users

HIV infection has frequently been transmitted by the sharing of contaminated needles or syringes among persons who inject intravenous drugs, primarily heroin and cocaine (Allen et al., 1989; Centers for Disease Control, 1989b). The IVDU-associated transmission of HIV has also spread from the drug users to their sex partners and their newborns. IVDU-associated AIDS cases made up one-third of all AIDS cases reported in 1989 and included 6139 (53%) heterosexual male IVDUs, 1831 (16%) female IVDUs, 2138 (19%) homosexual or bisexual male IVDUs, 388 (3%) men whose heterosexual partners were IVDUs, 733 (6%) women whose heterosexual partners were IVDUs, 249 (2%) children whose mothers were IVDUs, and 124 (1%) children whose mothers were sex partners of IVDUs. The 1121 persons who were heterosexual partners of IVDUs accounted for 57% of the total reported cases associated with heterosexual transmission of HIV; the 373 children whose mothers were IVDUs or sex partners of IVDUs accounted for 70% of the cases associated with perinatal HIV transmission reported in 1989.

Table III

Number and Rate per 100,000 Population of AIDS Cases Associated with IV Drug Use by Census Region and Race/Ethnicity (United States, 1988)[a]

Race/ethnicity	No. cases (rate)				
	Northeast	Midwest	South	West	Total[b]
White	1203 (2.9)	217 (0.4)	687 (1.2)	719 (2.2)	2826 (1.6)
Black	2929 (62.0)	294 (5.5)	1318 (9.5)	277 (12.5)	4818 (18.4)
Hispanic	1699 (65.2)	69 (5.4)	135 (3.0)	159 (2.5)	2062 (14.1)
Asian/Pacific Islander	6 (1.1)	0.(0.0)	0 (0.0)	6. (0.3)	12 (0.3)
Amerian Indian/ Alaskan Native	1 (1.2)	2 (0.8)	0 (0.0)	6. (0.8)	9 (0.6)
Unspecified	23	0	0	2	25
Total	5861 (11.9)	582 (1.0)	2140 (2.8)	1169 (2.7)	9762 (4.3)

[a] Source: Centers for Disease Control (1989b).

[b] Total cases and total rates exclude territories. Rates are based on the 1980 U.S. census.

The rate of IVDU-associated AIDS is higher for blacks and Hispanics than for whites. Except for the West, where 1988 rates for whites and Hispanics were similar, the divergence by race/ethnicity is evident in all regions of the country but is greatest in the Northeast (Table III). Compared with the incidence in whites, the higher incidence of IVDU-associated AIDS in blacks and Hispanics is the most significant contributing factor to their overall higher incidence of AIDS (Selik et al., 1988). This phenomenon is most striking in the Northeast, where 1988 case rates for IVDU-associated AIDS were dramatically higher in blacks and Hispanics than in whites and where AIDS cases in IVDUs, their sex partners, and their children exceeded other AIDS cases.

Data on HIV infection among IVDUs, derived primarily from studies conducted in drug treatment centers in 1986 and 1987, show great geographic variation in seroprevalence. Infection rates were highest in New York City, Puerto Rico, and the metropolitan areas of the Northeast, ranging from 10% in Providence, Rhode Island, to 65% in Brooklyn, New York. Lower rates (7–29%) were found in the South Atlantic states and in the metropolitan areas of Atlanta (10%), Detroit (8–13%), and San Francisco (7–13%). Rates were lowest (5% or less) in other areas of the West and in the Midwest and the South (Hahn et al., 1989).

C. Persons Infected through Heterosexual Contact

The annual number of cases attributed to heterosexual contact has steadily increased—from 55 reported in 1981–1982 and 107 reported in 1983 to 1954 reported in 1989—and the composition of the group has also changed. In 1983, U.S. residents born in countries where heterosexual transmission predominates accounted for nearly 80% of the heterosexually acquired cases reported. In contrast, only 20% of the heterosexual transmission cases reported in 1989 were in persons born in these countries.

Intravenous drug use is now the major path through which HIV is infecting heterosexual men and women in the United States (Centers for Disease Control, 1989d; Chamberland et al., 1989; Onorato et al., 1989). As of December 31, 1989, 5630 AIDS cases had been reported in which infection with HIV was reportedly acquired through heterosexual contact. Of these heterosexual transmission cases, 1632 (29%) were born in countries where heterosexual transmission is a major route of HIV infection. The other 71% of heterosexual transmission cases (3998) reported heterosexual contact with a partner with or at increased risk for HIV infection. Most of these heterosexual contacts (72%) were with IVDUs (Table II).

Data on heterosexuals attending sexually transmitted disease (STD) clinics corroborate the role of intravenous drug use in HIV transmission to heterosexual men and women. In an ongoing study begun in January 1988 in New York City, 47% of exclusively heterosexual STD clinic clients with a history of intravenous drug use and 13% of clients with a sex partner who used IV drugs were HIV-seropositive (Chiasson et al., 1989). A 1987 STD clinic survey in Baltimore detected HIV antibody in 15% of men and 22% of women with a history of IV drug use and in 11% of women who reported sexual contact with men who used IV drugs or were bisexual (Quinn et al., 1988). Among clients who did not report specific risks for HIV infection, seroprevalence was 4% in men and 5% in women in New York City and 3% in men and 2% in women in Baltimore. HIV infection in prostitutes is also strongly associated with IV drug use (Darrow et al., 1988); 75% of HIV-infected female prostitutes in six cities gave a history of intravenous drug use. The risk of HIV transmission from infected persons to their steady heterosexual partners without other risks varied in 26 studies that included at least 20 couples each; HIV seroprevalence in heterosexual partners ranged from 0 to 58%, with a median of 24% (Centers for Disease Control, 1987c, 1989f).

Heterosexual transmission, from men to women or women to men, occurs mainly through vaginal intercourse; some studies have suggested that anal intercourse has a higher risk of transmission (Padian et al., 1987), but others find no increased risk (Fischl et al., 1988; Simonsen et al., 1988; Laga et al., 1988). The risk of transmission through heterosexual contact is difficult to estimate. Transmission of HIV has been reported after only one sexual contact with an infected partner, but studies indicate that many persons remain uninfected despite hundreds of contacts with an infected partner. This suggests that unexplained biologic or behavioral factors may affect the efficiency of transmission (Padian et al., 1988; Peterman et al., 1988).

D. Recipients of Blood and Blood Products

By December 31, 1989, there were 4210 AIDS cases in transfusion recipients (3042) and persons with hemophilia (1168) (Tables II and IV). Transfusion of HIV-infected blood is a highly efficient mode of transmission; 89–100% of recipients of blood from an HIV-infected person are reported to become infected (Mosley *et al.*, 1987; Ward *et al.*, 1987). But with the routine screening of blood and plasma donors and the heat treatment of blood products since 1985, as well as the deferral of donors at high risk, the risk of new infections from transfusion or receipt of blood products has been greatly reduced.

A small risk of HIV infection remains for recipients of blood screened as negative for antibody to HIV. Thirteen transfusion recipients are reported to have become infected after receiving screened blood. In each case, the donors were tested soon after they were infected and before developing detectable antibodies. The risk of infection from a transfusion of screened blood was estimated at 1 in 40,000 transfused units by Ward *et al.*, (1988) and at 1 in 153,000 units by Cumming *et al.*, (1989) of the American Red Cross. Reports of seroconversions in six hemophilic persons have also been published, and CDC has evaluated more than 75 reports worldwide of HIV seroconversions possibly associated with heat-treated blood products (Cen-

Table IV

Pediatric AIDS Cases by Exposure Category through December 1989 in United States

Exposure category		Total	(%)
Hemophilia/coagulation disorder		106	(5)
Mother with/at risk for AIDS/HIV infection:		1614	(81)
IV drug use	826		
Sex with IV drug user	330		
Sex with bisexual male	37		
Sex with person with hemophilia	7		
Born in Pattern-II country[a]	172		
Sex with person born in Pattern-II country	7		
Sex with transfusion recipient with HIV infection	10		
Sex with person with HIV infection, risk not specified	64		
Receipt of transfusion	34		
Has HIV infection, risk not specified	127		
Receipt of transfusion		212	(11)
Undetermined		63	(3)
Total		1995	(100)

[a] Pattern II countries are countries in which most of the reported cases occur in heterosexuals.

ters for Disease Control, 1988c). The 18 infections meeting CDC's operational criteria for a probable association with heat-treated blood products were attributed to blood products produced before the initiation of donor screening for HIV or to products heated in the dry state at 60°C for 24–30 hr. These products are no longer available in the United States but have been replaced with products heated in the dry state at 68°C for 72 hr.

Transmission of HIV infection through artificial insemination and through transplantation of tissues, including kidney, liver, heart, pancreas, and bone, has also been reported (Centers for Disease Control, 1988e). The risk of HIV infection associated with these procedures has been greatly reduced, however, by HIV antibody screening of donors of tissue, organ allografts, and semen.

E. Health-Care Workers

The possibility of blood-borne transmission of HIV is a small but present occupational risk for health-care workers. Some 5% of all reported AIDS cases are in health-care workers; more than 94% of these cases were associated with risks incurred outside the work setting, particularly with sexual activity or intravenous drug use (Centers for Disease Control, 1988a). Documented occupational HIV infections in health-care workers have resulted from needlestick injury or from contact of mucous membranes or nonintact skin with blood from an HIV-infected person. Three studies in which health-care workers have been followed serologically after percutaneous exposures document seroconversion rates of approximately 0.4% after a needlestick injury (Fahey et al., 1989; Gerberding et al., 1988; Marcus, 1988). Although a few HIV infections have been reported following exposure of mucous membranes or skin to HIV-infected blood, the risk of transmission is likely much lower than the 0.4% associated with percutaneous exposure. The risk of occupational exposure can be reduced through strict adherence to universal precautions and recommended infection control guidelines and through avoidance of accidental injuries with needles and other sharp instruments (Centers for Disease Control, 1987b, 1988b).

F. Children

The AIDS epidemic has grave consequences for an increasing number of infants and children. The first cases of pediatric AIDS (cases in children under age 13) were reported to CDC in 1982, about 1 year after the initial adult cases were recognized. By December 31, 1989, 1995 cases in children had been reported, representing about 2% of the total number of AIDS cases (Table IV). The Public Health Service projects that more than 3000 cases in children will have been reported by 1991, with more than 1000 reported in 1991 alone (Morgan and Curran, 1986). An additional, but unknown, number of children will develop HIV-related illnesses that do not meet the currently defined criteria for AIDS but will still require intensive medical attention. It is estimated that by 1991 there will be at least 10,000–20,000 HIV-infected children in the United States (Novello *et al.*, 1989).

In 16% of all children with AIDS reported as of December 31, 1989, the source of HIV infection was either a blood transfusion or treatment for hemophilia. More than 80% of children who develop AIDS were infected perinatally (before, during, or soon after birth) (Table IV) (Rogers, 1988). Approximately half of the perinatally acquired cases have been traced to intravenous drug use by the child's mother, with most of the remainder associated with heterosexual HIV transmission to the mother. AIDS cases in children have been reported from 45 states, the District of Columbia, Puerto Rico, and the Virgin Islands; 56% of cases come from New York, New Jersey, and Florida, primarily from large urban areas. Three out of four children with AIDS are black or Hispanic, a racial pattern that reflects the increasing number of AIDS cases in black and Hispanic women, especially those who are IVDUs or heterosexual contacts of infected persons.

The timing of perinatal transmission is unknown, but a large proportion is likely to occur *in utero*—through passage of the virus across the placenta (Rogers, 1988). Transmission during labor and delivery—through exposure to maternal blood and vaginal secretions—probably also occurs but has been difficult to document. Several case reports have documented transmission through breastfeeding (Oxtoby, 1988). Studies in the United States, Europe, and Africa estimate the rate of transmission from HIV-infected mothers to their infants at about 30% (Elanche *et al.*, 1989; Rogers, 1989; Ryder *et al.*, 1989). However, studies of perinatal HIV transmission have been complicated by the lack of a reliable diagnostic test to detect or rule out HIV infection in newborns, since both infected and uninfected infants passively acquire HIV antibody from their mothers. Maternally acquired antibody may persist for as long as 15–18 months (Rogers, 1989). Recently, the polymerase chain reaction technique has facilitated early diagnosis of HIV infection in a subset of infants most likely to develop AIDS (Rogers *et al.*, 1989).

HIV antibody screening of blood routinely collected from newborns accurately measures seroprevalence among childbearing women. The variations in HIV prevalence in this group parallel those in female AIDS case rates, from less than 0.1% in California, Colorado, Michigan, New Mexico, and Texas to 0.2% in Massachusetts, 0.5% in Florida and New Jersey, and 0.7% in New York (Centers for Disease Control, 1989f).

G. Adolescents

Of the 117,781 AIDS cases reported as of December 31, 1989, 461 (0.4%) were in adolescents 13–19 years of age and 5090 (4.3%) were in young adults ages 20–24 years (Table I). Because of the long period between infection with HIV and diagnosis of AIDS, many of the young adults with AIDS were likely infected with HIV as adolescents.

Adolescent AIDS patients are concentrated in five states (New York, New Jersey, Florida, Texas, California) and Puerto Rico, with the majority (74%) in densely populated urban areas (Manoff *et al.*, 1989). Although males represent 81% of adolescent AIDS cases, adolescents with AIDS are more likely to be female (19%) and either black or Hispanic (53%) than are older AIDS patients (approximately 9% and 42%, respectively). Routes of exposure to HIV vary with age among adolescents. Exposure through therapeutically administered blood or blood products accounted for 67% of all cases reported through 1988 in 13–14-year-old AIDS patients, 57% in 15–16-year-olds, and 20% in 17–19-year-olds. Conversely, sexual contact or intravenous drug use accounted for 10% of cases in 13–14-year-olds, 26% of cases in 15–16-year-olds, and 56% in 17–19-year-olds. Sexual contact or intravenous drug use accounted for more than 90% of all cases in young adults 20–24 years old (Manoff *et al.*, 1989).

Data from the HIV antibody screening of civilian applicants for military service between October 1985 and December 1988, while not representative of the general population of adolescents, show an HIV prevalence among 17–19-year-old men and women of 0.04% and 0.03%, respectively (unpublished data, Department of Defense, 1988). The prevalence of infection increases progressively from the 17- to 19-year age group, reaching levels of 0.18% in 20–24-year-olds and 0.4% in 25–29-year-olds. Rates are higher for men than for women in all three age groups: 0.04%, 0.2%, and 0.45% for males and 0.03%, 0.09%, and 0.13% for females, respectively. Data from HIV antibody screening of entrants to the Job Corps program (16–21 years of age) show HIV infection rates approximately 10 times higher (0.4%) in disadvantaged youths (St. Louis *et al.*, 1989).

H. Investigations of Other Potential Modes of Transmission

The observed patterns of AIDS cases and HIV infection rates are consistent with the three recognized routes of HIV transmission: sexual contact with an infected person, parenteral exposure to infected blood or blood products (mainly through needle-sharing among IVDUs), and perinatal transmission from an infected woman to her fetus or infant. There is no epidemiologic or laboratory evidence of transmission in casual social situations, through air, food, or water, or by insects (Lifson, 1988; Office of Technology Assessment, 1987). No known cases of AIDS or HIV infection have resulted from sharing kitchens, bathrooms, laundries, eating utensils, or living space with an HIV-infected person.

V. Natural History

The rate of progressive immunosuppression, AIDS, and other clinical manifestations of HIV infection increases with the duration of infection. The

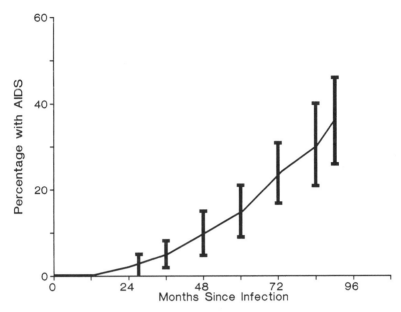

Figure 3. Kaplan–Meier survival curve showing the proportion of men developing AIDS by estimated duration of HIV infection, San Francisco City Clinic Cohort Study. Bars represent 95% confidence intervals.

interval between infection with HIV and onset of AIDS is variable and often long. Progression to AIDS in adults is rare in the first 3 years following infection (Fig. 3). In a cohort of HIV-infected homosexual men who initially participated in hepatitis B virus studies in San Francisco in 1978–80, the median period from infection with HIV to diagnosis of AIDS is estimated at 9.8 years. Only 15% of the men in this group have remained completely asymptomatic after a median of 100 months of follow up (Lifson *et al.*, 1989). Other studies have shown that the incubation period varies by age and is shorter in infants and older adults than in young adults and adolescents (Eyster *et al.*, 1987; Peterman *et al.*, 1985). Estimates derived from mathematical models suggest average incubation periods of 8–11 years for adults (Lifson *et al.*, 1988).

In children, the median interval between infection with HIV and diagnosis of AIDS was estimated in 1988 to be 8 months for children with high-risk mothers and 19 months for children who received transfusions. The time of onset of AIDS has ranged from 1 month to 9 years, with most children developing clinical symptoms within 3 years of infection (Rogers, 1988). Because of the relatively recent period of perinatal exposure, however, there has not yet been time to observe pediatric AIDS cases with long latency periods.

Pneumocystis carinii pneumonia and Kaposi's sarcoma are the two most frequently reported opportunistic diseases in adults with AIDS. *P. carinii* pneumonia is the most common life-threatening opportunistic infection in every transmission category, accounting for 61% of opportunistic diseases reported in adults. The percentage of AIDS patients initially seeking treatment for *P. carinii* pneumonia has increased from 37% of persons diagnosed in 1982 to 68% of those diagnosed in 1988. The percentage of AIDS patients with Kaposi's sarcoma varies by patient group; 22% of homosexual men have been reported with Kaposi's sarcoma compared to 2.8% of heterosexual male IVDUs, 2.3% of female IVDUs, and 1% of men with hemophilia (Beral *et al.*, 1989). The percentage of AIDS patients initially presenting with Kaposi's sarcoma has decreased from 28% of those diagnosed in 1982 to 10% of those diagnosed in 1988.

In the absence of therapy, the average survival time after developing AIDS has been approximately 12 months. Persons with Kaposi's sarcoma generally survive longer (2 years) than persons with *P. carinii* pneumonia (11 months) or other opportunistic diseases, except for candidiasis (6 months) (Payne *et al.*, 1988; Rothenberg *et al.*, 1987). Preliminary data suggest that therapeutic interventions, including zidovudine (AZT) and aerosolized pentamidine prophylaxis, have increased survival times in San Francisco, but the ultimate impact of these therapies is not yet clear (Lemp *et al.*, 1989).

VI. Mortality

More than half of all reported AIDS patients—and 85% of those diagnosed before 1986—are reported to have died. The actual case-fatality rate is higher, however, because of incomplete reporting of deaths. Because AIDS disproportionately affects persons in the 20–44-year-old age group, deaths due to AIDS lead to substantial decreases in life expectancy. In 1988, HIV infection/AIDS was ranked 15th among causes of death and was the 7th leading cause of premature mortality, as measured by years of potential life lost (YPLL) before age 65 (Centers for Disease Control, 1989a). AIDS deaths in 1988 represented 12% of deaths in men and 5% of deaths in women 25–34 years of age and 11% of deaths in men and 2% of deaths in women 35–44 years of age. The impact of the epidemic on patterns of premature death is particularly striking in areas of high AIDS incidence. In New York City, AIDS is now the leading cause of death for males aged 30–59 and for females aged 1–9 and 30–39 (New York State Department of Health, 1989).

In 1987 AIDS was the 9th leading cause of death in children 1–4 years of age (Kilbourne *et al.*, 1989). Fatality rates in children vary by age. Of children diagnosed under the age of 1 year, half died within 6 months of diagnosis. Of children over age 1, half died within approximately 20 months after diagnosis (Rogers, 1988).

VII. Impact on Racial and Ethnic Minorities

The disproportionate percentage of AIDS cases in minority men, women, and children is becoming increasingly clear. Although blacks and Hispanics make up less than 20% of the U.S. population, they accounted for 40% of all AIDS cases reported to CDC by the end of 1988. Roughly three-fourths of women and children with AIDS are black or Hispanic, 72% and 77%, respectively. Blacks and Hispanics had the highest AIDS incidence rates per 100,000 population in 1988 (34.9 and 28.9, respectively), followed by whites (9.6); Asians/Pacific Islanders (5.4); and American Indians/Alaskan Natives (2.2). The national incidence rate in 1988 was 13.7 AIDS cases per 100,000 population (Centers for Disease Control, 1989c).

VIII. HIV-2 Infection

A second human immunodeficiency retrovirus, HIV-2, was reported in 1986. HIV-2 is closely related to HIV-1 and causes illness that cannot at this time be distinguished from AIDS caused by HIV-1 (Clavel et al., 1986, 1987). Most cases of AIDS resulting from HIV-2 infection have been reported from countries in West Africa and have been associated with the same modes of transmission as HIV-1 (Clavel et al., 1987; Centers for Disease Control, 1989e; Horsburgh and Holmberg, 1988). Of the seven cases reported in the United States, six are known to be West African; the nationality of the seventh is unknown. All evidence suggests that these persons became infected through heterosexual contact with other infected West Africans. Of six additional cases that are under investigation, four are West Africans and two are of unknown nationality (Centers for Disease Control, 1989e).

IX. Conclusion

HIV infection is a serious public health problem that is widespread in certain populations in the United States, notably in homosexual and bisexual men, intravenous drug users, heterosexual partners of persons at risk, and infants born to infected women. The risk of HIV transmission through blood and blood products has been largely eliminated. Transmission among homosexual men has been greatly reduced in some areas but remains a vital concern because of the high prevalence rates in this group and the difficulty in reaching adolescent and minority men with successful education and prevention programs. The linkage of intravenous drug use and HIV infection has the most alarming implications for the course of the epidemic; increased heterosexual and perinatal transmission associated with intravenous drug use means that many more women and children in the nation's poor, drug-ravaged neighborhoods are being infected with HIV.

The extent of the problem calls for immediate attention to a comprehensive program of therapeutic, vaccine, and behavioral research and the application of therapeutic advances and counseling to all HIV-infected persons. Prevention of HIV infection requires widespread education of the public, especially persons at highest risk. Management of HIV-infected persons will often include medical care and social services, as well as treatment for drug abuse as appropriate. Continued counseling, family planning, and partner notification services are also needed. Nonetheless, given the large number of Americans currently infected and the difficulty in interrupting transmission among intravenous drug users, their sex partners, and their children, it is clear that morbidity and mortality associated with HIV infection will not be easily or completely checked. The continuing epidemic will remain an urgent, daunting, and increasingly pervasive problem for many years to come.

References

Allen, D. M., Onorato, I., Sweeney, P., and Jones, T. S. (1989). *Abstracts of the V International Conference on AIDS, Montreal 1989,* 56.

Beral, V., Peterman, T. A., Berkelman, R. L., Holmberg, S. D., and Jaffe, H. W. (1989). *Abstracts of the V International Conference on AIDS, Montreal 1989,* 50.

Blanche, S., Rouzioux, C., Guihard Moscato, M.-L., Veber, F., Mayaux, M.-J., Jacomet, C., Tricoire, J., Deville, A., Vial, M., Firtion, G., de Crepy, A., Douard, D., Robin, M., Courpotin, C., Ciraru-Vigneron, N., le Deist, F., Griscelli, C., and the HIV Infection in Newborns French Collaborative Study Group (1989). *N. Engl. J. Med.* **320,** 1643–1648.

Centers for Disease Control (1985). *Morbid. Mortal. Weekly Rep.* **34,** 373–375.

Centers for Disease Control (1987a). *Morbid. Mortal. Weekly Rep.* **36** (Suppl 1S), 1S-15S.

Centers for Disease Control (1987b). *Morbid. Mortal. Weekly Rep.* **36** (Suppl. 2S), 1S-18S.

Centers for Disease Control (1987c). *Morbid. Mortal. Weekly Rep.* **36** (Suppl. S-6), 1–48.

Centers for Disease Control (1988a). *Morbid. Mortal. Weekly Rep.* **37,** 229–234, 239.

Centers for Disease Control (1988b). *Morbid. Mortal. Weekly Rep.* **37,** 377–382, 387–388.

Centers for Disease Control (1988c). *Morbid. Mortal. Weekly Rep.* **37,** 441–450.

Centers for Disease Control (1988d). *Morbid. Mortal. Weekly Rep.* **37,** 486–489.

Centers for Disease Control (1988e). *Morbid. Mortal. Weekly Rep.* **37,** 597–599.

Centers for Disease Control (1989a). *Morbid. Mortal. Weekly Rep.* **38,** 27–9.

Centers for Disease Control (1989b). *Morbid. Mortal. Weekly Rep.* **38,** 165–170.

Centers for Disease Control (1989c). *Morbid. Mortal. Weekly Rep.* **38,** 229–236.

Centers for Disease Control (1989d). *Morbid. Mortal. Weekly Rep.* **38,** 423–424, 429–434.

Centers for Disease Control (1989e). *Morbid. Mortal. Weekly Rep.* **38,** 572–574, 579–580.

Centers for Disease Control (1989f). *Morbid. Mortal. Weekly Rep.* **38** (Suppl S-4), 1–38.

Centers for Disease Control (1990). *Morbid. Mortal. Weekly Rep.* **39,** 110–112, 117–119.

Chamberland, M., Conley, L., and Buehler, J. (1989). *Abstracts of the V International Conference on AIDS, Montreal 1989,* 66.

Chiasson, M. A., Stoneburner, R. L., Telzak, E., Hildebrandt, D., Schultz, S., and Jaffe, H. W. (1989). *Abstracts of the V International Conference on AIDS, Montreal 1989,* 117.

Clavel, F., Guetard, D., Brun-Vezinet, F., Chamaret, S., Rey, M.-A., Santos-Ferreira M.-O., Laurent, A. G., Dauguet, C., Katlama, C., Rouzioux, C., Klatzmann, D., Champalimaud, J.-L., and Montagnier, L. (1986). *Science* **233,** 343–346.

Clavel, F., Mansinho, K., Chamaret, S., Guetard, D., Favier, V., Nina, J., Santos-Ferreira, M.-O., Champalimaud, J.-L., and Montagnier, L. (1987). *N. Engl. J. Med.* **316,** 1180–1185.

Cumming, P. D., Wallace, E. L., Schorr, J. B., and Dodd, R. Y. (1989). *N. Engl. J. Med.* **321,** 941–946.

Curran, J. W., Jaffe, H. W., Hardy, A. M., Morgan, W. M., Selik, R. M., and Dondero, T. J. (1988). *Science* **239,** 610–616.

Darrow, W. W., Echenberg, D. F., Jaffe, H. W., O'Malley, P. M., Byers, R. H., Getchell, J. P., and Curran, J. W. (1987). *Am. J. Publ. Hlth.* **77,** 479–483.

Darrow, W. W., Bigler, W., Deppe, D., French, J., Gill, P., Potterat, J., Ravenholt, O., Schable, C., Sikes, R. K., and Wofsy, C. (1988). *Abstracts of the IV International Conference on AIDS, Stockholm, 1988,* 273.

Dondero, T. J., Jr., Pappaioanou, M., and Curran, J. W. (1988). *Publ. Hlth. Rep.* **103,** 213–220.

Eyster, M. E., Gail, M. H., Ballard, J. O., Al-Mondhiry, H., and Goedert, J. J. (1987). *Ann. Intern. Med.* **107,** 1–6.

Fahey, B. J., Schmitt, J. M., Saah, A. J., Lane, H. C., and Henderson, D. K. (1989). *Abstracts of the V International Conference on AIDS, Montreal 1989,* 725.

Fischl, M., Fayne, T., Flanagan, S., Ledan, M., Stevens, R., Fletcher, M., LaVoie, L., and Trapido, E. (1988). *Abstracts of the IV International Conference on AIDS, Stockholm, 1988,* 274.

Gerberding, J. L., Littell, C. G., Chambers, H. F., *et al.* (1988). *Program and Abstracts of the Twenty-eighth Interscience Conference on Antimicrobial Agents and Chemotherapy, Los Angeles 1988,* 169.

Greenberg, A. E., Thomas, P. A., Hindin, R. H., Greene, A. W., Rautenberg, E. L., and Schultz, S. J. (1989). *Abstracts of the V International Conference on AIDS, Montreal 1989,* 121.

Hahn, R. A., Onorato, I. M., Jones, T. S., and Dougherty, J. (1989). *JAMA* **261,** 2677–2684.

Hessol, N. A., O'Malley, P., Lifson, A., Cannon, L., Kohn, R., Barnhart, L., Harrison, J., Bolan, G., Doll, L., and Rutherford, G. (1989). *Abstracts of the V International Conference on AIDS, Montreal 1989,* 50.

Horsburgh C. R., Jr., and Holmberg, S. D. (1988). *Transfusion* **28,** 192–195.

Kilbourne, B. W., Rogers, M. F., and Bush, T. J. (1989). *Abstracts of the V International Conference on AIDS, Montreal 1989,* 66.

Laga, M., Taelman, H., Bonneux, L., Cornet, P., Vercauteren, G., and Piot, P. (1988). *Abstracts of the IV International Conference on AIDS, Stockholm, 1988,* 260.

Lemp, G. F., Payne, S. F., Neal, D. P., and Rutherford, G. W. (1989). *Abstracts of the V International Conference on AIDS, Montreal 1989,* 410.

Lifson, A. R. (1988). *JAMA* **259,** 1353–1356.

Lifson, A. R., Rutherford, G. W., and Jaffe, H. W. (1988). *J. Infect. Dis.* **158,** 1360–1367.

Lifson, A. R., Hessol, N., Rutherford, G. W., Buchbinder, S., O'Malley P., Cannon, L., Barnhart, L. Harrison, J., Doll, L., Holmberg, S., Jaffe, H. (1989). *Abstracts of the V International Conference on AIDS, Montreal, 1989,* 60.

Manoff, S. B., Gayle, H. D., Mays, M. A., and Rogers, M. F. (1989). *Pediatr. Infect. Dis. J.* **8,** 309–314.

Marcus, R. (1988). *N. Engl. J. Med.* **319,** 1118–1123.

Morgan, W. M., and Curran, J. W. (1986). *Publ. Hlth. Rep.* **101,** 459–465.

Mosley, J. W., and The Transfusion Safety Study Group (1987). *Abstracts of the III International Conference on AIDS, Washington, D.C. 1987,* TH 10.2.

New York State Department of Health (1989). *AIDS in New York State.* Albany, New York.

Novello, A. C., Wise, P. H., Willoughby, A., and Pizzo, P. A. (1989). *Pediatrics* **84,** 547–555.

Office of Technology Assessment, Congress of the United States. (1987). *Do Insects Transmit AIDS?* Staff Paper 1. Washington, D.C.

Onorato, I., Peterson, L., Pappaioanou, M., and Dondero, T. (1989). *Abstracts of the V International Conference on AIDS, Montreal, 1989,* 78.

Oxtoby, M. J. (1988). *Pediatr. Infect. Dis. J.* **7,** 825–835.

Padian, N., Marquis, L., Francis, D. P., Anderson, R. E., Rutherford, G. W., O'Malley, P. M., and Winkelstein, W. Jr. (1987). *JAMA* **258,** 788–790.

Padian, N., Glass, S., Marquis, L., Wiley, J., and Winkelstein, W. (1988). *Abstracts of the IV InternationaL Conference on AIDS, Stockholm, 1988,* 264.

Payne, S. F., Lemp, G. F., and Rutherford, G. W. (1988). *Program and Abstracts of the Twenty-eighth Interscience Conference on Antimicrobial Agents and Chemotherapy, Los Angeles, 1988,* 332.

Peterman, T. A., Jaffe, H. W., Feorino, P. M. *et al.* (1985). *JAMA* **254,** 2913–2917.

Peterman, T. A., Stoneburner, R. L., Allen, J. R., Jaffe, H. W., and Curran, J. W. (1988). *JAMA* **259,** 55–58.

Quinn, T. C., Glasser, D., Cannon R. O., Matuszak, D. L., Dunning, R. W., Kline, R. L., Campbell, C. H., Israel, E., Fauci, A. S., and Hook, E. W. III. (1988). *N. Engl. J. Med.* **318,** 197–203.

Report of the Second Public Health Service AIDS Prevention and Control Conference. (1988). *Publ. Hlth Rep.* **103** (Suppl), 10–18.

Rogers, M. F. (1988). *Pediatr. Ann.* **17,** 324–331.

Rogers, M. (1989). *Abstracts of the V International Conference on AIDS, Montreal, 1989,* 199.

Rogers, M. F., Ou, C.-Y., Rayfield, M., Thomas, P. A., Schoenbaum, E. E., Abrams, E., Krasinski, K., Selwyn, P. A., Moore, J., Kaul, A., Grimm, K. T., Bamji, M., Schochetman, G., and the New York City Collaborative Study of Maternal HIV Transmission and Montefiore Medical Center HIV Perinatal Transmission Study Group (1989). *N. Engl. J Med.* **320,** 1649–1654.

Rothenberg, R., Woelfel, M., Stoneburner, R., Milberg, J., Parker, R., and Truman, B. (1987). *N. Engl. J. Med.* **317,** 1297–1302.

Ryder, R. W., Nsa, W., Hassig, S. E., Behets, F., Rayfield, M., Ekungola, B., Nelson, A. M., Mulenda, U., Francis, H., Mwandagalirwa, K., Davachi, F., Rogers, M., Nzilambi, N., Greenberg, A., Mann, J., Quinn, T. C., Piot, P., and Curran, J. W. (1989). *N. Engl. J. Med.* **320,** 1637–1642.

Selik, R. M., Castro, K. G., and Pappaioanou, M. (1988). *Am. J. Publ. Hlth.* **78,** 1539–1545.

Simonsen, J. N., Cameron, D. W., Gakinya, M. N., Ndinya-Achola, J. O., D'Costa, L. J., Karasira, P., Cheang, M., Ronald, A. R., Piot, P., and Plummer, F. A. (1988). *N. Engl. J. Med.* **319,** 274–278.

St. Louis, M. E., Hayman, C. R., Miller, C., Anderson, J. E., Petersen, L. R., and Dondero, T. J. (1989). *Abstracts of the V International Conference on AIDS, Montreal, 1989,* 711.

Ward, J. W., Deppe, D. A., Samson, S., Perkins, H., Holland, P., Fernando, L., Feorino, P. M., Thompson, P., Kleinman, S., and Allen, J. R. (1987). *Ann. Intern. Med.* **106,** 61–62.

Ward, J. W., Holmberg, S. D., Allen, J. R., Cohn, D. L., Critchley, S. E., Kleinman, S. H., Lenes, B. A., Ravenholt, O., Davis, J. R., Quinn, M. G., and Jaffe, H. W. (1988). *N. Engl. J. Med.* **318,** 473–478.

Winkelstein, W. Jr., Lyman, D. M., Padian, N., Grant, R., Samuel, M., Wiley, J. A., Anderson, R. E., Lang, W., Riggs, J., and Levy, J. A. (1987). *JAMA* **257,** 321–325.

Winkelstein, W. Jr., Wiley, J. A., Padian, N. S., Samuel, M., Shiboski, S., Ascher, M. S., and Levy, J. A. (1988). *Am. J. Publ. Hlth.* **78,** 1472–1474.

Acquired Immunodeficiency Syndrome (AIDS): An International Perspective

Paul A. Sato, James Chin, and Jonathan M. Mann

I. Introduction

The worldwide scope of human immunodeficiency virus (HIV) infection and AIDS did not become fully apparent until the middle of the 1980s, by which time HIV had spread silently and extensively across several continents. The HIV/AIDS pandemic is now recognized as an unprecedented challenge to global health, and has generated an extraordinary global mobilization for the prevention and control of the pandemic. This paper will review the global epidemiology of HIV/AIDS at the end of the 1980s, describe the potential evolution of the pandemic in the 1990s, and consider the global response that has been mobilized to prevent and control HIV/AIDS.

II. Global Epidemiologic Patterns of HIV Infection and AIDS

Four broad yet distinct epidemiological patterns of HIV/AIDS are now recognized world wide (Fig. 1). These global epidemiologic patterns result from the same basic modes of transmission (sexual, parenteral, perinatal) but express variations related to when extensive spread of HIV began, the predominant mode of viral entry into each population, and the varying epidemiology of risk behaviors and practices.

In *Pattern I* (Australia, Canada, New Zealand, USA, and Western Europe), extensive spread of HIV began in the late 1970s to the early 1980s. Thus far, most HIV infections and AIDS cases have been among homosexual or bisexual males and urban intravenous drug users. Heterosexual transmission accounts for only small percentage of infections, but is increasing. Transmission through blood and blood products occurred up to the mid-1980s, but is now almost entirely prevented. Except among intravenous drug users (IVDU), transmission through unsterilized or inadequately sterilized skin-piercing instruments such as needles and syringes has not been a significant mode of spread for HIV. The male/female sex ratio has usually been close to 10:1; thus perinatal transmission (from an infected mother to her fetus or infant) is relatively uncommon in Pattern I areas. The prevalence rate of HIV infection in populations as a whole is estimated to be <1% but has been reported to exceed 50% among groups of individuals whose behavior place them at high risk of infection. As of mid-1990, WHO estimates that in Pattern I areas approximately, 1.5–1.75 million people were infected with HIV, and that an estimated cumulative total of close to 175,000 adult AIDS cases had occurred (Table I).

In *Pattern II* (sub-Saharan Africa), extensive spread of HIV is also thought to have started in the late 1970s to the early 1980s. HIV has been predominantly heterosexually transmitted in these populations. Intravenous drug use and homosexual or bisexual transmission account for at most a

Table I

Estimated Cumulative HIV Infections and Cumulative AIDS, Mid-1990

	Cumulative HIV infections	Cumulative AIDS
Pattern I	1.5–1.75 million	Close to 175,000
Pattern I/II	0.8–1.2 million	Close to 75,000
Pattern II	Over 5.0 million	Over 500,000
Pattern III	Close to 500,000	Over 1,000
Global total	8.0–10.0 million	Around 800,000

Figure 1. Four global epidemiologic patterns of HIV infection and AIDS can, in 1989, be distinguished. Pattern I (■) is found in North America, western Europe, Australia, and New Zealand; in these areas about 80–90% of the cases are homosexual/bisexual men, or intravenous drug users. Pattern II (◐) is found in sub-Saharan Africa; the primary mode of transmission in this area is heterosexual, and the ratio of infected males to infected females is close to unity. The epidemiologic pattern in Latin America is in evolution and is currently classified as Pattern I/II (◪). Pattern III (□) consists of those areas where few cases or infections have occurred to date. Boundaries on this map do not imply the expression of any opinion whatsoever on the part of WHO concerning the legal status of any country, territory, city, or area, or of its authorities, or concerning the delimitation of its frontiers or boundaries.

small proportion of infections. Transmission through contaminated blood and blood products remains a significant problem in many of the countries. Transmission via inadequately sterilized needles and syringes, as well as other skin-piercing instruments remains of concern, although it is considered to account for only a small proportion of infections. The prevalence rate of HIV infection on a national basis in a number of countries in Pattern II areas exceeds 1% and in some urban areas up to 25% of persons aged 15–49 years are infected. As of Mid-1990, WHO estimates that in Pattern II areas over 5.0 million people were HIV-infected, and that over 500,000 cumulative AIDS cases in adults had occurred (Table I).

Pattern I/II is now considered typical of most of Latin America, previously classified as Pattern I. As in Patterns I and II, extensive spread of HIV in Pattern I/II areas commenced during the late 1970s or early 1980s. Initially, the burden of HIV infection was essentially confined to homosexual or bisexual men with multiple male sexual partners, IVDU sharing injection equipment, and recipients of HIV-contaminated blood or blood products. However, since the middle 1980s, increasing transmission among heterosexuals has been noted, so that heterosexual transmission has increasingly become the predominant mode of HIV propagation in these populations. The male/female ratio of HIV infections acquired in the last several years in some Pattern I/II areas is approaching unity. As heterosexual transmission of HIV increases, the number of pediatric HIV infections resulting from perinatal transmission will inevitably increase. As of mid-1990, WHO estimates that in Pattern I/II areas approximately 0.8–1.2 million-people were HIV-infected, and that close to 75,000 cumulative adult AIDS cases had occurred (Table I).

Pattern III is those areas where HIV was not introduced until at least the early to mid 1980s. Only a relatively small number of AIDS cases (approximately 1% of all cases reported to WHO) have been reported from these areas (Asia, eastern Europe, the Near and Middle East, North Africa, and most of the Pacific). Cases have been recorded among homosexual and bisexual males and IVDUs; heterosexual transmission and transmission through blood/blood products have also occurred. Until recently, reported AIDS cases were mostly among persons who had traveled to higher prevalence areas; if not, they were in individuals exposed to those who had been to such areas. Endogenous transmission is rising, and HIV infections are increasingly being documented among persons with high-risk behaviors in Pattern III areas. The current epidemic among IVDUs in Bangkok, Thailand demonstrates the potential for relatively rapid increase in HIV infections (Vanichseni *et al.*, 1989). As of the mid-1990, WHO estimates that in Pattern III areas about close to half a million people were HIV-infected, with an estimated cumulative total of over 1000 adult AIDS cases (Table I).

III. Epidemiologic Data Needed for HIV/AIDS Prevention and Control

For the prevention and control of HIV/AIDS the epidemiologic data listed in Table II are considered essential; these data needs are discussed in further detail below.

A. Estimation of the Total Number of Individuals HIV-Infected; Monitoring Trends in HIV Infection over Time

Estimates of the prevalence of HIV infection in different populations by serological surveys can serve to provide data on (a) the estimated current prevalence of HIV infections, as well as for (b) monitoring the trends in HIV infection over time. Such data are essential for the planning, monitoring, and evaluation of HIV/AIDS prevention and control programs. Yet the HIV sero-survey data obtained by the majority of studies across the world to date must be interpreted and compared with caution, due to wide differences in survey methodologies. However, it should be recalled that during the recognition phase of the pandemic, simple and rapid epidemiologic situation assessment was a priority, to determine whether a problem with HIV infection existed, as well as to obtain crude estimates of its overall dimensions.

There is now increasingly a need world wide for serological surveys of HIV infection conducted serially over time and across different sites using standardized methodologies. Such an approach would facilitate the establishment of valid inferences concerning the changes in HIV prevalence over time and place, The epidemiological framework for meeting these objectives of such HIV sero-surveillance is given in Table III. In conducting surveys of HIV infection and risk behavior, careful attention must continue to be given to confidentiality and related ethical and legal issues. If confidentiality of HIV test results is not sufficiently taken into account, substantial and unpredictable participation bias may occur (Hull *et al.*, 1988; Moye *et al.*, 1989; WHO, unpublished). Participation bias, in this context, is the distortion of collected HIV prevalence estimates resulting from the likelihood of being HIV-infected differing between those individuals or groups of individuals

Table II
Epidemiologic Data Needed for the Prevention and Control of HIV/AIDS

A. An estimate of the total number of individuals HIV-infected.
B. An estimate of the total number of persons at risk of HIV infection.
C. Trends in HIV infection over time.
D. An estimate of the current and future burden of HIV-related disease, including AIDS.

Table III

Essential Epidemiological Framework for HIV Surveillance, Including Sero-Surveillance

1. Serological surveys serially conducted over time at different sites.
2. Consistent and standardized sero-survey protocols.
3. A clear and consistently defined population to be sampled, and attention to obtaining a representative sample of that population.
4. Attention to participation bias (distortion of data as a result of individuals participating in HIV testing being more or less likely to be infected than those who do not participate).

electing to participate in a sero-survey, and those who choose not to participate. Breaches of confidentiality may also seriously undermine public health efforts motivate changes in risk behaviors based on voluntary HIV testing and counseling (WHO, unpublished).

B. Estimation of the Number of Persons at Risk of HIV Infection

Estimation of the total numbers of persons at risk of HIV infection remains difficult. Not all IVDU or homosexual men are at equal risk of HIV infection since this risk depends on the frequency of at-risk behaviors or practices, such as sharing injection equipment with multiple individuals, or sexual intercourse with multiple sexual partners. Behavioral studies measuring such variables as knowledge, attitudes, beliefs, and practices (KABP) in different populations world wide will aid the development of such estimates of populations at risk.

C. Estimation of the Current and Future Burden of HIV-Related Disease

Estimates of the current and future burden of HIV-related disease, including AIDS, are essential for allocation of adequate resources to the health care and social services sectors. Such estimates can be derived from reported AIDS cases, with projections of the number of future AIDS cases derived from extrapolation. However, cases of AIDS reported to national public health authorities are often subject to distortion as a result of under-recognition, under-reporting, as well as delays in reporting of cases. Therefore, alternative approaches must be considered in such situations.

One alternative approach for estimating the current AIDS burden, and over the next 4–5 years, involves the use of estimates of the number of HIV-infected persons combined with estimates of the annual rate of progression from infection to AIDS (Chin and Mann, 1989). In the short-term, such estimates are essentially independent of future trends in HIV incidence, as the median time from infection to the development of AIDS is estimated to be about 10 years (Moss and Bacchetti, 1989). Thus the majority of AIDS cases which can be projected within the next 5 years would be expected to occur even if HIV transmission were to cease completely. However, these projections do not take into consideration the potential effect of treatments which may delay the development of AIDS, or otherwise prolong survival of HIV-infected persons. Such treatments would, however, be expected to increase the prevalence of AIDS and HIV infection. Hence, projections which do not account for the potential lengthening of survial of those HIV-infected, or with AIDS, can be considered a relatively conservative indication of future trends.

IV. Global Estimates of the Number of HIV-Infected Individuals

As of mid-1990 the number of people world wide who are infected with HIV is estimated to be 8–10 million. This estimate refines previous WHO estimates of 5–10 million HIV-infected globally, and is based largely on the more recent estimates of HIV prevalence made for the United States (Dondero et al., 1989) and Europe (WHO, 1988) as well as an analysis of available HIV sero-survey data for each country in sub-Saharan Africa (Sato et al., 1989). Estimates of the number of individuals HIV-infected have generally been revised downward as more data have accumulated. For example, it was initially estimated in 1988 that there were between 24,000 and 80,000 individuals HIV-infected in the United Kingdom by the end of 1987 (WHO, 1988). By late 1989, these initial estimates had been revised downwards to 20,000–50,000 (Department of Health, Welsh Office, 1988). However, relatively wide confidence intervals remain around such estimates.

V. Global AIDS Projections

Projections of the number of AIDS cases that may occur in the future are important for the global and regional coordination of HIV/AIDS prevention and control. WHO has developed a projection model that makes use of what is known of the rate of progression from infection to AIDS, along with estimates of the current prevalence of HIV infection world wide (WHO, unpublished). Using such a model, WHO has projected that at least 3 million additional AIDS cases in adults can be expected world wide over the next decade from the over 8 million persons who are estimated to have *already* been HIV-infected as of mid-1990.

Estimation of the current and future extent of pediatric AIDS is more difficult, for several reasons. First, although improved diagnostic methods are being developed, at present the definitive diagnosis of HIV infection cannot be made before the child reaches 15–18 months of age, in the absence of specific clinical or immunological findings. Prior to that age the presence of passively acquired maternal antibody to HIV confounds the diagnosis

(Centers for Disease Control, 1987). The second fact is that child mortality and morbidity are high in those countries where the majority of HIV infections in childhood occur. Under such conditions, the clinical features of pediatric AIDS are often difficult to distinguish from the clinical manifestations of the other severe diseases of childhood. Available sero-survey data from sub-Saharan Africa, where the majority of HIV infections in childhood are thought to occur, suggest that as of 1987, there was a cumulative total of about 70,000–80,000 HIV-infected infants. This was projected to approach 150,000 by 1990 (Sato *et al.*, 1989; Chin *et al.*, 1989a).

It is difficult to predict the longer-term (10 or more years) dimensions of the HIV/AIDS pandemic. Nevertheless, some view of the future in the longer term, however limited and imperfect, is needed. Therefore, the WHO in early 1989 conducted a Delphi survey (Levine, 1984) of individuals selected for their knowledge of the epidemiology of HIV/AIDS on a global basis (Chin *et al.*, 1989b) to project the potential future course of the AIDS pandemic up to the year 2000. These Delphi HIV projections combined with the WHO model to project AIDS cases suggest that approximately nine

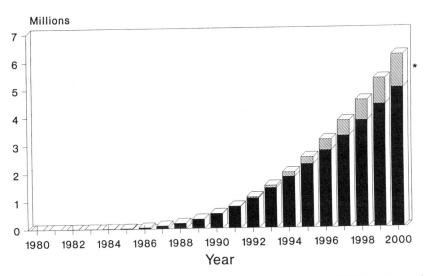

Figure 2. Projection of the future course of the pandemic of AIDS and HIV infection to the year 2000 by the Delphi questionnaire survey method. It can clearly be seen that, world wide, only a small proportion of the cumulative adult AIDS cases projected to occur by the end of the coming decade will have occurred prior to 1990. About nine times more AIDS cases in adults were predicted for the 1990s, when compared to the 1980s. Over half of the adult AIDS cases occurring in the 1990s were projected to result from individuals already infected by the middle of 1988. More than a third of the adult AIDS cases projected to arise from HIV infections occurring in the remaining years to the year 2000 were considered preventable by a globally and regionally coordinated HIV prevention and control effort, in concert with national programs. *Cases considered preventable.

Table IV
The Objectives of the Global AIDS Strategy

1. Prevention of HIV infection.
2. Reduction of the personal and social impact of HIV/AIDS.
3. Unification of national and international efforts against HIV/AIDS.

times more AIDS cases may occur in adults in the 1990s than had occurred during the 1980s. The total of AIDS cases in adults was projected to reach cumulatively about 5–6 million by the year 2000. Of the AIDS cases among adults in the 1990s, more than half would occur in persons already infected by mid-1988. Of the AIDS cases which were projected to arise from HIV infections in the future, over one-third were considered preventable by a globally and regionally coordinated effort, in conjunction with national HIV/AIDS prevention and control programs (Fig. 2).

VI. The Global AIDS Strategy

The World Health Organization has the constitutional responsibility to direct and coordinate international health work. In late 1985, as the global dimensions of HIV infection and AIDS became apparent, the WHO initiated the development of a global strategy for the prevention and control of AIDS. This Global AIDS Strategy was approved and adopted by the World Health Assembly (1987), the Economic and Social Council of the United Nations (1987), the General Assembly of the United Nations (1987), and the World Summit of Ministers of Health on Programmes for AIDS Prevention (1988).

The Global AIDS Strategy has three objectives (Table IV), based on several underlying principles (Table V). To implement this strategy, national

Table V
Underlying Principles of the Global AIDS Strategy

1. Protection of the public's health.
2. Respect for human rights, and the prevention of discrimination against people with AIDS.
3. AIDS can be prevented now, even in the absence of an HIV vaccine.
4. The key to AIDS prevention is information, education, and communication (IEC). HIV transmission can be prevented through informed and responsible behavior.
5. The prevention and control of AIDS and HIV infection will require a sustained social and political commitment.
6. All countries need a comprehensive national AIDS prevention and control program integrated into the national health infrastructure and linked within a global network.
7. Systematic surveillance, monitoring, and evaluation will ensure that the global strategy can adapt and grow stronger with time.

HIV/AIDS prevention and control programs and international coordination are recognized to be essential. The objectives of the Global AIDS Strategy are discussed in further detail below.

A. Prevention of HIV Infections

Currently, the single most important component of national HIV/AIDS prevention and control programs remains information, education, and communication (IEC). In the absence of an HIV vaccine or curative therapy, the prevention of HIV infection can be achieved through the adoption by individuals of informed and responsible behaviors and practices. HIV transmission essentially requires the active participation of two persons; thus, the chain of transmission can be broke by a change in the behavior of either the infected or noninfected individual.

B. Reduction of the Personal and Social Impact of HIV/AIDS

An underlying principle that needs particular emphasis is that the prevention and control of HIV/AIDS, and thus the attainment of the objectives of the Global AIDS Strategy, must continue to be based on respect for human rights and prevention of discrimination against HIV-infected individuals, including people with AIDS. This need to minimize the personal and social impact of HIV infection highlights the importance of health and social services for those who are HIV-infected, as well as for those living with such individuals. Thus the control of the social impact of the disease will continue to require a broad social and political commitment to prevention and control, recognizing that this impact extends far beyond the limits of the health care sector.

C. Unification of National and International Efforts

The unification of national and international efforts against HIV/AIDS pandemic is the third main objective of the Global AIDS Strategy. This effort is well underway, and particularly over the past two years, an unprecedented mobilization at the global and national level against AIDS and HIV infection has been achieved. As of 1 September 1989 the WHO Global Programme on AIDS is working with over 155 countries, and providing financial support of over US$60 million to 127 countries across the globe. In addition, over 1000 consultant and expert missions for the planning, training, and implementation of national HIV/AIDS prevention and control programs have been conducted. Such national programs now exist in most countries across the globe, and should be established in all of the world's countries and states by the end of 1989.

VII. Conclusions

The global mobilization for the prevention and control of AIDS and HIV infections is well underway. Implementation of the objectives of the Global AIDS Strategy will present a continuing challenge nationally, regionally, and internationally. Part of this continuing challenge will be the collection of data on the estimated current HIV prevalence, potential future extent (estimate of population at risk) of HIV infection, the trends in HIV infection over time, and estimation of the current and future burden of HIV-related disease including AIDS. Such data continue to be essential for the rational planning, monitoring, and evaluation of HIV/AIDS prevention and control programs world wide.

The pandemic of AIDS and HIV infection is dynamic, and has shown changes over time. One example is Latin America, where earlier infections and cases largely involved homosexual or bisexual men, while infections in recent years have increasingly resulted from heterosexual transmission. The overall magnitude of HIV-related disease including AIDS will only begin to become fully evident during the 1990s. Based on estimates of the number of individuals already HIV-infected as of mid-1990 alone, and without accounting for new HIV infections in future years, at least three million additional AIDS cases are likely to occur in adults over the next decade. Health and social services world wide need to be alerted and strengthened in anticipation of this rapidly growing number of HIV-related illnesses and deaths.

References

Centers for Disease Control (1987). *Morbid. Mortal. Weekly Rep.* **36,** 225–236.

Chin, J., and Mann, J. M. (1989). Bull. Wld. Hlth. Org **67**(1), 1–7.

Chin, J., Sankaran, G., and Mann, J. M. (1989). *In* "Maternal and Child Care in Developing Countries" (E. Kessel, and A. K. Awan, eds). Ott Publishers, (Thun, Switzerland).

Chin, J., Sato, P. A., and Mann, J. M. (1990). *Bull. Wld. Hlth. Org.* **68**(1), 1–11.

Department of Health, Welsh Office, Great Britain (1988). "Short-term Prediction of HIV Infection and AIDS in England and Wales. Report of a Working Group (The Cox Report)." Her Majesty's Stationery Office, London.

Dondero, T. J., St. Louis, M., Anderson, J., Peterson, L., and Pappaioanou, M. (1989). *Abstracts of the V International Conference on AIDS, Montreal, 1989,* MAO.4.

Economic and Social Council of the United Nations (1987). "Prevention and Control of AIDS." Resolution 1987/75.

Fortieth World Health Assembly (1987). "Global Strategy for the Prevention and Control of AIDS." Resolution WHA40.26.

General Assembly of the United Nations (1987). "Prevention and Control of Acquired Immune Deficiency Syndrome. Resolution 42/8.

Hull, H. F., Bettinger, C. J., Gallaher, M. M., Keller, N. M. *et al.* (1988). *JAMA,* **260,** 935–938.

Levine, A. S. (1984). *Wld. Hlth. Statist. Q.* **37,** 306–317.

Moss, A. R., and Bacchetti, P. (1989). *AIDS* **3,** 55–61.

Moye, J., Bracci, P., Kappes, R., Kunches, L., *et al.* (1989). *Abstracts of the V International Conference on AIDS, Montreal 1989,* MAO. 25.

Sato, P., A., Chin, J., Lwanga, S., Beausoleil, E. G., and Mann, J. M. (1989). *Abstracts of the V International Conference on AIDS, Montreal, 1989,* (TGP. 7).

Vanichseni, S., Sonchai, W., Plangsringarm, K., Akarasewi, P. *et al.* (1989). *Abstracts of the V International Conference on AIDS, Montreal, 1989,* (TGO. 23).

World Health Organization (1988). Consultation on information support for the AIDS surveillance system, Tatry-Poprad/Strbske Plezo, Czechoslovakia, 17–19 February, 1988.

World Health Organization Global Programme on AIDS. Unpublished WHO document GPA/SFI/89.3.

World Summit of Ministers of Health on Programmes for AIDS Prevention (1988). London Declaration on AIDS prevention. Unpublished WHO document WHO/GPA/INF/88.6.

Part IV

Animal Models

Human T-cell Leukemia Virus Type I (HTLV-I): Studies of Disease Mechanism in a Transgenic Mouse System

Steven H. Hinrichs, Lian-sheng Chen, Joseph Fontes, and Gilbert Jay

I. Introduction

Human T-cell leukemia virus type I (HTLV-I) has been associated with a wide range of human disease including adult T-cell leukemia (ATL) (Poiesz *et al.*, 1980; Yoshida *et al.*, 1982), tropical spastic paraparesis (TSP) (Gessain *et al.*, 1985; Rodgers-Johnson *et al.*, 1985), polymyositis (Wiley *et al.*, 1989), and possibly multiple sclerosis (Ohta *et al.*, 1986; Reddy *et al.*, 1989). It is unclear how a single infectious agent can induce such a bewildering array of disease including cellular proliferation such as occurs in adult T-cell leukemia and, in a different setting, atrophic or degenerative changes such as those occurring in tropical spastic paraparesis. In addition, certain clinical features of HTLV-I infection such as the development of hypercalcemia must be considered in terms of possible secondary effects (Jaffe, 1985). It appears that each of these disorders presents certain unique issues for studying the underlying mechanisms through which HTLV-I induces disease.

II. Current Problems and Molecular Considerations

The initial association of HTLV-I with adult T-cell leukemia was made through epidemiologic observations (Hinuma, 1981; Robert-Guroff et al., 1982). However, the same epidemiologic data also raised several important questions (Ito, 1985; Gallo, 1987). (1) Why do less than 0.1% of individuals infected by the virus ever develop disease? (2) Why does malignancy develop characteristically after a long latency period? (3) Why do leukemic cells not uniformly express detectable viral antigens? As a possible answer to these questions, it has been suggested that in addition to the action of HTLV-I an independent second genetic event is required in order to induce the malignant disease. In order to further investigate these issues we sought to develop a transgenic mouse model of HTLV-I-associated disorders.

Experiments in transgenic animals have provided valuable insights into gene expression and function that have direct implications for human biology and disease. Through the selective use of DNA constructs which express viral genes in transgenic animals, it may be possible to investigate whether the virus acts through a direct or an indirect mechanism in inducing pathogenesis. Since HTLV-I is associated with diverse effects, both proliferative and atrophic, it may be possible to investigate whether expression of viral genes in different cell types will lead to pathological changes similar to those seen in humans.

The restriction of viral infections has been shown to occur through different mechanisms, such as (1) presence of a tissue-specific receptor for the viral envelope and (2) existence of a limited range of cell types in which the viral promoter is active. The suspected gene which encodes the cellular receptor to which the HTLV-I envelope binds has been localized to chromosome 17 (Sommerfelt et al., 1988). In vitro experiments have shown that the virus may be transmitted by cell fusion, not only to T cells but to a wide range of cell types including those of endothelial, fibroblastic, osteoid, and muscle origin (Ho et al., 1984; Nagy et al., 1983). The significance of these observations in an in vivo setting remains to be determined.

The long terminal repeat (LTR) located at each end of the proviral DNA contains sequences responsible for viral gene expression and thereby contributes to a large extent to the tissue tropism and disease spectrum of HTLV-I. The LTR contains both the transcriptional enhancer and promoter sequences recognized by the biosynthetic machinery of the host cell. Since a transgene may be constructed that incorporates the retroviral long terminal repeat, it is possible to observe in vivo those cell types that are capable of supporting gene expression via the HTLV-I LTR. The restriction imposed by the cell surface receptor for virus uptake will be bypassed in transgenic animals since the transgene becomes integrated into the genome of one-cell stage embryos and will be present in all of the cells in the resulting animal.

III. Value of a Transgenic Approach to the Study of HTLV Disease

Viral pathogenesis is the consequence of perturbed cellular functions and cell to cell interactions. Transgenic animals provide an *in vivo* setting for the study of HTLV-I disease beyond what tissue culture methods have to offer. Since it is difficult to recreate proper cell–cell interactions *in vitro*, observations made in tissue culture systems are not necessarily predictive of results from *in vivo* experiments with transgenic mice. For example, it has previously been shown that the transcriptional promoter from SV40, a simian papovavirus, while showing permissivity to a broad range of cell types *in vitro*, is capable of expression in only a very restricted range of tissue types *in vivo* (Brinster *et al.*, 1984). In addition, transgenic animals provide important advantages such as the continual presence of an intact immune system critical for assuring the appropriate T- and B-cell interactions which may be responsible for the pathological changes associated with HTLV-I infection. Functional deficits and abnormalities such as those that affect the exceedingly complex neural system may also not be apparent in tissue culture cells.

These are important considerations since the pathogenic consequences of a viral infection are likely to be the perturbed "functional" end points of alteration by the virus at specific steps in complex and interactive "biochemical" pathways. The ability to reconstruct such phenotypic alterations in a transgenic mouse through germline insertion of the viral genome not only proves that a disease is caused by specific viral sequences but also provides an experimental system to uncover the underlying molecular mechanism through which the virus acts. Once a transgenic line of mice has been shown to faithfully reproduce the disorders that are characteristic of the viral infection in humans, therapeutic interventions may also be considered. Such animals may prove particularly useful for testing molecular strategies or for design of pharmacologics either to reverse or to prevent the development of disease.

Therefore, the use of transgenic animals provides a unique approach to the investigation of the pathogenesis of HTLV-I and will allow us to answer many of the puzzling questions that have been raised.

IV. Effects of the HTLV *tax* Gene in Transgenic Mice

In terms of genome organization, HTLV-I is a relatively simple nondefective retrovirus that contains two well-defined nonstructural genes, designated *tax* and *rex*. The Tax and the Rex proteins act as *trans*-regulators of viral gene expression in cultured cells (Sodroski *et al.*, 1984). Specific DNA elements in the LTR have been identified through which Tax functions although Tax does not bind DNA directly (Paskalis *et al.*, 1986). What remains to be determined is whether these viral gene products are directly or indirectly responsible for some of the clinical manifestations associated with HTLV-I infection. As suggested by its role in transcriptional activation, the Tax protein has been shown to be predominantly localized to the nucleus (Goh *et al.*, 1985). Cytoplasmic expression and retention of the protein has been proposed to result in a different spectrum of pathology from nuclear localization of Tax (Nerenberg and Wiley, 1989).

Based upon the premise that viral *trans*-activators do not act on their own to control viral gene expression but, rather, through interactions with the host's biosynthetic machineries, there is increasing interest in the possibility that these viral regulatory proteins may also act to alter cellular gene expression that will lead to pathological changes in the host.

The *tax* gene has an established role in the *trans*-activation of several cellular genes that encode growth factors and growth factor receptors, including c-*fos* (Fujii *et al.*, 1988), granulocyte-macrophage colony-stimulating factor (GM-CSF)(Chen *et al.*, 1985), and the interleukin-2 (IL-2) receptor (Inoue *et al.*, 1986). Hence, we have initially focused our efforts on this gene because of the likelihood of its direct participation in inducing disorders in HTLV-I-infected individuals. We have derived a construct with the *tax* coding sequence placed under the control of the HTLV-I LTR. This construct, designated pHTLV LTR *tax*, should encode an authentic 40-kDa Tax protein, identical to that derived from the double-spliced viral mRNA (Nerenberg *et al.*, 1987). It will not encode the 27-kDa Rex protein. The exact cell type in which expression would occur with a subsequent disturbance in cell function would be anticipated to be determined by (1) the range of cell types in which the HTLV-I-LTR is active, and (2) the presence of a pathway through which the Tax protein functions in each specific cell type.

We have derived a total of 18 transgenic F_0 mice carrying different numbers of copies of the injected DNA fragment. Of those that were analyzed, all expressed the integrated *tax* gene. Over time, three distinct phenotypes were observed in these mice. They involve the thymus, the salivary gland, and the nerve sheath. Through the study of these transgenic mice, we have sought to investigate the underlying mechanisms responsible for the generation of each of the observed phenotypes.

V. Effects on the Thymus

We postulated that HTLV-I induces adult T-cell leukemia in humans through the action of its *trans*-activating *tax* gene. Consistent with this suggestion are the findings that the HTLV-I LTR is active in T cells and the Tax protein can induce a proliferative effect in these same cells (Grassmann *et al.*, 1989). Therefore, in our transgenic animals we expected to observe a hyperplastic disorder involving either circulating lymphocytes or lymphoid tissues such as the spleen and lymph nodes.

Of the 18 transgenic *tax* mice that were generated, ten of them died between 3 and 6 weeks of age. At the time of death, all of these mice (~60%) were runts and were no more than half the weight of the other transgenic littermates which developed to sexual maturity. Necropsies of these small mice showed extensive involution of the thymus and secondary infections as the probable cause of death (Nerenberg *et al.*, 1987). While a normal thymus has a thick and highly cellular cortex surrounding a less cellular medulla, the thymus of this subset of transgenic *tax* mice invariably showed disproportionate thinning of the cortex and blurring of the corticomedullary junctions (Fig. 1A). This observation has also been made in studies by other laboratories (Furuta *et al.*, 1989).

Since the T lymphocytes were an anticipated target of the Tax protein we looked for expression of the transgene in lymphoid as well as in nonlymphoid tissues. Analysis of extracts from different tissues of mice with the small phenotype showed high levels of expression of the Tax protein in thymus, muscle, and salivary gland. Interestingly, F_1 mice from those transgenic animals that developed to maturity did not develop thymic atrophy and showed *tax* expression predominantly in muscle and salivary gland, with sparing of the thymus. We interpret this observation as consistent with thymic expression of the *tax* gene being responsible for the observed thymic involution and early death.

The relationship of this observation to human disease is not clear. Although the affected cell type is similar between mice and humans, the observed atrophic change in the transgenic animals appears completely opposite from the anticipated proliferative changes seen in ATL patients. Most likely the difference is determined by the age in each case at which time the *tax* gene is expressed. Since immune competence in mice is acquired after birth, expression of the Tax protein during early development may interfere with the maturation of immunological functions and result in early death from opportunistic infections. Since HTLV-I transmission is mostly horizontal in humans (Tajima *et al.*, 1982), at a stage where immunological maturation has taken place, there is a better opportunity for the affected T cells to develop into a malignant phenotype. Alternatively, a phenotype similar to that seen in the transgenic mice may be as yet undetected in humans since death may occur before antibodies would identify the affected child. Our transgenic animal studies may have revealed a disorder that is induced only at an early developmental stage. Since vertical transmission occurs in humans it is important to search for thymic changes in such cases. A third explanation of this phenotype may be that developmental specific pathways through which the Tax protein acts or through which the viral LTR is activated are missing in young humans.

It is difficult to study this issue further because the small mice with thymic atrophy die before maturity and, therefore, have to be regenerated each time from microinjected embryos. In order to test the hypothesis that

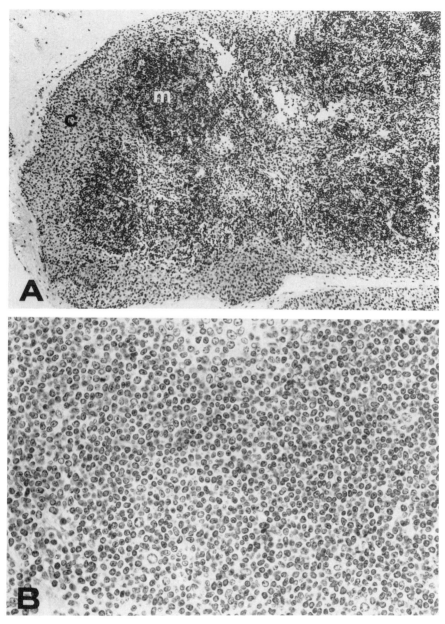

Figure 1. A. Histologic appearance of the thymus from a transgenic mouse with the runted phenotype. The cortex (c) is hypocellular in comparison with that of normal thymus from age-matched mice. M, Medulla. The tissue was formalin-fixed, paraffin-embedded and stained with hematoxylin and eosin (H&E). Magnification ×27. B. Histologic appearance of lymph node from a transgenic mouse with salivary gland epithelial hyperplasia and lymphoid infiltration. The lymph nodes adjacent to the salivary gland show hyperplasia with large numbers of plasmacytoid cells in the parafollicular and sinusoidal regions. Note the characteristic nuclear appearance with peripherally clumped chromatin and prominent single nucleolus. (H&E, magnification ×168.)

the observed phenotype is due to age-specific expression of the *tax* gene as directed by the HTLV-I LTR, it would be necessary to substitute the viral LTR with a regulatory element capable of expression only in mature T lymphocytes. One candidate is the transcriptional enhancer which controls the gene for the IL-2 receptor. The IL-2 receptor is expressed in only activated T cells and may prove useful in targeting *tax* gene expression in transgenic animals (Leonard *et al.*, 1985). It is likely that delayed expression of the Tax protein in mature T cells in an otherwise immunocompetent mouse will lead to the development of T-cell leukemia.

VI. Effects on the Salivary Gland

The remaining 40% of the F_0 transgenic mice were able to develop to sexual maturity and three of them have been selected to be bred out into established lines. Unlike mice with thymic atrophy and the associated lethality, this latter subset of mice do not have any expression of the Tax protein in the thymus. The reason for the disparity between the two groups of animals, both having received the same transgene, remains unknown. The fact that the thymic phenotype has been repeatedly observed in different F_0 transgenic mice rules out the possibility that the site of integration plays a role. One may speculate that thymic expression of the *tax* gene during development provides selective pressure on the fetus or on the neonate. Those that respond by repressing the *tax* gene will survive and continue to develop, while those that cannot will die. Similar epigenetic mechanisms in response to developmental choices have been described.

One of the phenotypes observed in each of the transgenic mice from the three established lines involved the salivary gland (Green *et al.*, 1989a). It showed features similar to a human disease that has been under consideration as potentially caused by retrovirus infection and referred to as Sjögren's syndrome. Sjögren's syndrome is thought to be a disease of autoimmune etiology (Daniels and Talal, 1987). The exact triggering mechanism for the development of the disease is unknown. Clinically, a patient experiences dry eyes and dry mouth which may occur in association with other autoimmune disorders including polymyositis, systemic lupus erythematosus, and vasculitis. Morphologically, the earliest finding is the presence of lymphocytes around the ducts of the salivary gland. The infiltrating cells consist predominantly of T cells in addition to B cells and plasma cells. Antibodies with different specificities have been detected in patients with this syndrome, particularly those directed against ribonucleoprotein antigens designated SS-A and SS-B (Provost, 1987).

The salivary gland changes in our transgenic mice are the result of expression of the *tax* gene, primarily in the submandibular and minor salivary glands, and to a lesser degree the parotid and sublingual glands. The initial alteration was the observation of ductal epithelial cell proliferation resulting in dilation of the ducts (Fig. 2A). This proliferation is multifocal and occurs simultaneously in multiple ducts throughout the gland, leading presumably to obstructions. The proliferation was observed in mice as young as 4 weeks of age with dramatic progression and subsequent distortion of the glandular architecture. The degree of salivary gland pathology in these transgenic mice correlates with the level of expression of the *tax* gene.

Low levels of the Tax protein were detected in 4-week-old mice with progressively increasing levels of Tax as the mice matured. This finding correlated with the increasing severity of the ductal cell proliferation in older mice and is consistent with cells expressing Tax becoming an increasingly larger proportion of the overall tissue mass. Immunohistologic studies of the salivary gland showed positive staining of the Tax protein in both the nucleus as well as the cytoplasm of the proliferating ductal cells. This disorder is also seen in mice with thymic depletion, where retrospective analysis revealed identical histopathology in the salivary gland. We have extended our analysis to other glandular tissues in these mice, and have observed similar hyperplastic lesions on occasion in lacrimal glands but not in mammary glands. It is noteworthy that the hyperplasia in the salivary gland has at times been observed to progress to neoplasia.

Lymphocytic infiltration of the salivary glands appears to occur after proliferation of the ductal epithelium (Fig. 2B). Lymphocytes are first seen adjacent to the proliferating nests of cells with subsequent infiltration of the epithelial islands. Mice which survive to 8 months of age exhibit extensive epithelial proliferation with lymphocytic and plasmolytic infiltration resulting in destruction of the glandular architecture and finally hyalinization, fibrosis, and thickening of basement membranes (Fig. 2C). These terminal features are similar to those seen in human Sjögren's syndrome, and cause destruction of the exocrine function leading to dry eyes and dry mouth. Pronounced lymphoid hyperplasia is frequently seen in the lymph node situated between the parotid and submandibular glands. The parafollicular zone contains an increased number of plasma cells and immunoblasts (Fig. 1B).

Several aspects of these observations may have clinical and causal relevance. An association between HTLV-I and Sjögren's syndrome has been suggested based on case reports of HTLV-I-infected patients with tropical spastic paraparesis who developed Sjögren's-like features (Gessain et al., 1985). Some of these features have also been reported in intravenous drug abusers (IVDA) who are at risk for HTLV-I infection (Smith et al., 1988). Salivary gland enlargement and dry mouth have also been reported in individuals infected with human immunodeficiency virus (HIV)-1 In addition, Sjögren's syndrome has been associated with a spectrum of lymphoproliferative disorders including lymphoid hyperplasia and lymphoma (Fox, 1984). While lymphoid hyperplasia is frequently seen in the *tax* transgenic mice, no lymphoid malignancies have been found to date. Since IVDA are at risk of infection by HTLV and HIV, it is important to search for antibodies to both viruses in patients with salivary gland pathology.

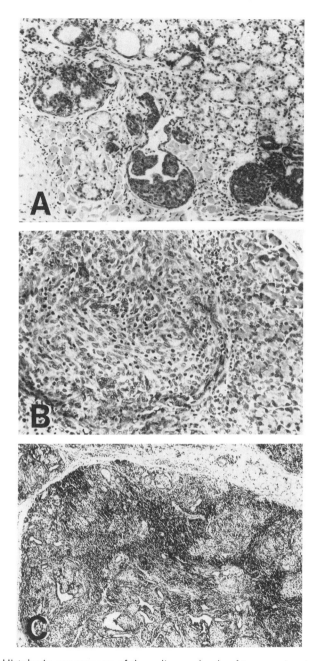

Figure 2. Histologic appearance of the salivary glands of transgenic mice at different stages of disease. A. Early ductal cell proliferation resulting in dilation of the ducts and filling of the lumen by proliferating cells. (H&E, magnification ×29.) B. Distortion of normal glandular architecture and infiltration of proliferating cells by lymphocytes. Very few changes are noted in the acinus. Note change in cytologic appearance of proliferating cells from round to elongate or spindle shape. (H&E, magnification ×180.) C. Late features of the salivary gland include loss of acinar tissue, dilation of ducts, and extensive infiltration by lymphocytes. Also present is hyalinization of connective tissue. (H&E, magnification ×18.)

Since this phenotype appeared in multiple lines of transgenic mice, it is unlikely to be due to a site-specific integration phenomenon such as the activation of cellular genes at or near the site of integration of the viral LTR *tax* gene. Similarly, in humans, integration of HTLV-I is not site-specific (Seiki *et al.*, 1983). The transgenic model demonstrates that the HTLV-I LTR is active in salivary gland and that expression of the Tax protein results in proliferation which is most likely the primary molecular event leading to the development of exocrinopathy. However, since this sequence of events differs from those thought to occur in Sjögren's syndrome where lymphocytic infiltration appears to precede ductal cell proliferation, it may be important to review cases of Sjögren's syndrome in patients infected with HTLV-I to search for a similar progression of events. Admittedly, it may be difficult to follow the development of this syndrome in patients who almost invariably are diagnosed at a fairly advanced stage of the disease. However, it may be that a subset of patients while having clinical symptoms similar to the autoimmune type of Sjögren's have a sequence of events which is more like that seen in the transgenic animals.

VII. Effects on the Nerve Sheath

Another phenotype appearing in the transgenic HTLV-I *tax* mice was totally unpredicted, but has resulted in several observations relevant to the study of human disease (Hinrichs *et al.*, 1987). All three independent lines of mice which have been studied developed tumors which first appear between 3 and 4 months of age on the extremities. These tumors are characteristically found arising from the sheath of peripheral nerves and grossly show the plexiform appearance characteristic of the human genetic disease, neurofibromatosis (NF) (Fig. 3A). The appearance of this phenotype in multiple founder lines is again consistent with tumor development being the consequence of the expression of the transgene. In fact, retrospective microscopic examination of tissues from transgenic mice that died of thymic involution also revealed identical hyperplastic lesions. That the *tax* gene can induce such tumors in mice has now been confirmed by studies from other laboratories (Felber *et al.*, 1989).

Histologically, the tumors were composed of spindle-shaped cells which formed small discrete nodules located on the ears, nose, legs, and tail (Fig. 3B). Although the majority of the tumors appeared benign, in some cases they demonstrated a high mitotic index and extended into the surrounding connective tissue, features which have been used to indicate malignant transformation (Fig. 3C). Infiltration by polymorphonuclear leukocytes is a consistent finding appearing in all tumors despite little evidence of tumor necrosis. Electron microscopic studies of peripheral nerves with increased diameters showed the proliferating cells to arise from the endoneurium and perineurium. Very few differentiating features were seen except for a discontinuous external lamina. Consistent with their lack of differentiation, the

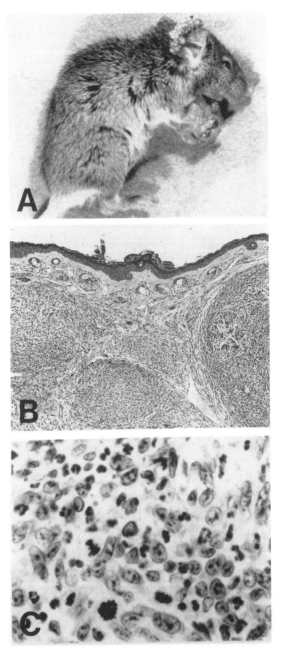

Figure 3. A. Multinodular peripheral tumors with a beadlike appearance involving the ears, feet, and nose of a transgenic mouse. B. Histologic section through a plexiform tumor showing distinctive nodules beneath the skin. (H&E, magnification ×29.) C. Highpower microscopic appearance of tumor with frequent mitotic figures and numerous polymorphonuclear leukocytes. (H&E, magnification ×180.)

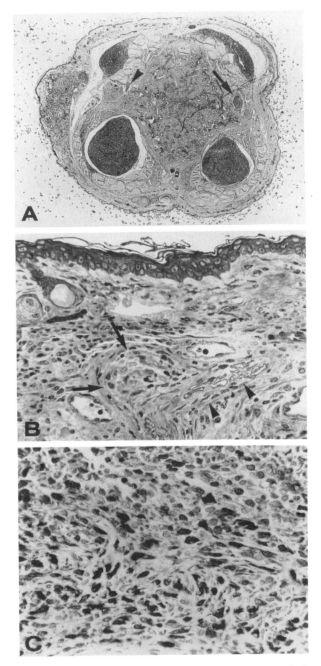

Figure 4. A. Cross-section through the tail of a transgenic mouse with dorsal artery and vein at bottom. Nerves are identified by arrowhead and nerve tumors by arrows. In addition to involvement of major nerve trunks the smaller subcutaneous nerve branches also show replacement by tumor (H&E, magnification ×11.) B. High-resolution light microscopic study

tumor cells are negative for the S-100 protein. Also noted in the earliest lesions identified by electron microscopy were infiltrating granulocytes.

Analysis of protein extracts from tumors demonstrated levels of Tax that were substantially higher than that seen in muscle tissue. Tumor cells removed from transgenic animals and grown in tissue culture were shown by indirect immunofluorescence to express the Tax protein which localized in the nucleus. These tissue culture cells continue to produce the Tax protein after extended passage and are also capable of inducing tumors in nude mice. Further examination of the transgenic mice resulted in additional evidence associating tumor development with nerve tissues. Transverse sections through the tail demonstrated a clear progression of tumors from peripheral nerves (Fig. 4A). Necropsy of animals without tumors on the extremities revealed tumors arising from the cranial nerves or nerve ganglion. Studies on peripheral nerves showed that all large tumors were in continuity with nerve trunks. However, smaller tumors frequently required histologic examination to identify the associated nerve twig. In epon-embedded tissues it was possible to demonstrate early proliferation of perineural cells of peripheral nerves (Fig. 4B). More recently we have also noted tumors involving the adrenal gland, as well as discrete spindle cell proliferation in the iris and corneal nerves. Immunohistohemistry using an antibody directed against the Tax protein has confirmed expression of the *tax* gene in the proliferating cells from each of the lesions described (Fig. 4C). These findings show that the *tax* gene under the control of the HTLV-I LTR has neurotropic activity and is capable of reproducibly generating morphologically similar tumors in widely separated locations.

Neurofibromatosis is one of the most common genetic syndromes affecting the nervous system and has an estimated incidence of approximately 1 in 3000 live births. In humans it is thought to be inherited as an autosomal dominant disorder (Riccardi and Eichner, 1986). However, most large studies have reported a spontaneous case rate of approximately 50%. If this were a single gene disorder, it would represent the highest gene mutation rate for any genetic disease (Crowe *et al.*, 1956). Alternatively, environmental factors such as viral infection may also be responsible for this high incidence of sporadic cases.

There is a wide diversity in the clinical symptoms, which have primarily

---

of early lesions. Epon-embedded ear tissue sectioned at 1 μm and stained with Methylene Blue Azure II. Small myelinated nerve is identified by arrowhead with a tumor nodule (arrows) extending from the perineurium. The same nerve twig may have small tumors developing at multiple sites. (Magnification ×112.) C. Immunohistochemical study of nerve sheath tumor using antibody against HTLV-I Tax. Tissues were fixed in methanol:acetic acid and paraffin-embedded. Dark granular staining is seen primarily in the nucleus of tumor cells. Also note many negative cells within the same tumor. (Magnification ×175.)

been categorized into two main forms. The classical Von Recklinghausen's neurofibromatosis is the prototypic form of the disease with multiple cafe au lait spots and dermal neurofibromas. When these dermal tumors have a plexiform appearance, it is generally diagnostic of the genetic disease. Interestingly, mast cells are frequently found infiltrating the tumors suggesting a chemotactic function or the activation of a growth factor. Central NF refers to a less common syndrome with frequent bilateral tumor involvement of the vestibulocochlear nerve and occasionally development of brain or spinal cord tumors. Studies of the genetic basis for human neurofibromatosis have resulted in mapping the suspected gene to chromosome 17 (Barker *et al.*, 1987). The cloning of the NF gene remains a project of great interest with several groups reporting progress. It is tempting to speculate that the suspected NF gene may have been expressed in the transgenic mice through *trans*-activation by the HTLV-I *tax* gene.

The neurotropic properties of HTLV-I and its effect on the development of neurological disease have only recently been recognized (Rodgers-Johnson *et al.*, 1985). The latency period between infection and development of nervous system disorders may also be lengthy. As demonstrated in the transgenic mouse system, HTLV-I has a tropism not only for the lymphoid system, but the nervous system as well. The underlying mechanisms for tropical spastic paraparesis, a degenerative disorder of the central nervous system, is unknown (Ceroni *et al.*, 1988). We suspect that specific cell types in both the central nervous system as well as peripheral nervous system will be shown to be capable of supporting HTLV-I gene expression. So far, there has been no epidemiologic or molecular studies to link HTLV-I infection to the development of NF. While it is unlikely that the familial cases of NF have a viral etiology, it remains to be shown whether the sporadic cases are induced by an infectious agent. The transgenic animal system in this case has further shown how an unexpected finding may lead to the discovery of basic mechanisms of cell proliferation.

VIII. Summary

It is clear from these studies that the HTLV-I LTR, in the presence of its transcriptional *trans*-activator (the Tax protein), is active in only a small subset of cell types in the transgenic mice. They include cells in the salivary gland, in nerve sheaths, and in muscle. It is not clear, however, whether the lack of expression in most other cell types is due to the methylation of *cis*-acting elements within the LTR, the presence of LTR-responsive negative regulatory factors, or the absence of LTR-responsive positive *trans*-acting factors. Our animal model provides an opportunity for us to address this issue. On the basis of our findings, the suggestion that HTLV-I may be transmitted to diverse cell types in infected individuals through cell fusion may not have broad implications if viral gene expression does not occur in a majority of those cells.

The most striking observation with our transgenic mice is that expression of a single gene from HTLV-I, the *tax* gene, will suffice to induce tumors at different target sites. Benign tumors developed from ductal cells in the salivary gland and neurofibroblasts in the nerve sheath. It has to be pointed out, however, that expression of the Tax protein in a cell does not by itself suffice to induce tumors in all cases. Despite consistent expression of the *tax* gene in muscle cells, proliferative disorders originating from these cells have not so far been detected (Nerenberg and Wiley, 1989). It appears that only select cell types may have subcellular targets for the HTLV-I Tax protein. The fact that numerous tumor lesions arise simultaneously at widely separated body sites in each animal at a very early age is indicative that the Tax protein can act "directly" to induce cell proliferation and hyperplastic growth in the involved tissues. The occasional finding of a malignant tumor, and then only in older mice, would suggest the involvement of a second genetic event in the progression of disease. This would imply that the development of adult T-cell leukemia in patients is likely not the sole action of HTLV-I, a conclusion that would account for the long latency and low penetrance of the disease in infected individuals. Our transgenic mice provide an ideal model system to identify potential secondary genetic events and to delineate their mechanisms of action.

It has long been suspected that the *tax* gene can activate the expression of growth factors and growth factor receptors in cells in culture. The gene for (GM-CSF) is one of those that have been shown to be *trans*-activated (Chan *et al.*, 1986). It is interesting that this observation can be extended to our transgenic mice. Tumor cells derived from the neurofibroblasts synthesize and secrete GM-CSF (Green *et al.*, 1989b). The consequences of excessive GM-CSF production include peripheral granulocytosis, splenomegaly resulting from myeloid hyperplasia, and neutrophil infiltration into tumor tissues similar to that seen in human NF cases. It should be noted, however, that *tax*-induced proliferation does not always lead to the activation of GM-CSF or other cellular genes. The hyperplasia seen in the salivary gland is not accompanied by the expression of GM-CSF or the infiltration of polymorphonuclear leukocytes. One of the take home lessons from the study of our transgenic mice is that genes from HTLV-I not only induce functional changes in the infected cells but may also have profound effects on uninfected cells in the same individuals through the activation and secretion of growth-modulating factors. A systematic survey of such factors in our transgenic mice may prove rewarding.

The complexity of the action of HTLV-I has been suggested by its epidemiologic association with both a proliferative disease called ATL and a degenerative disorder termed TSP. How a single virus can induce such seemingly opposing disease phenotypes remains a mystery. Many lessons have already been learned through the study of our transgenic mice. Answers to many of the remaining questions await further dissection of our transgenic model.

Acknowledgments

This work was supported by USPHS grant CA49624 and 51779. We thank Ms. Lisa Ruiz for her assistance in the preparation of this manuscript.

References

Barker, D., Wright, E., Nguyen, K., Cannon, L., Fain, P., Goldgar, D., Bishop, D. T., Carey, J., Baty, B., Kivlin, J., Willard, H., Waye, J. S., Greig, G., Leinwand, L., Nakamura, Y., O'Connell, P., Leppert, M., Lalouel, J.-M., White, R., and Skolnick, M. (1987). *Science* **236**, 1100–1102.
Brinster, R. L., Chen, H. Y., Messing, A., van Dyke, T., Levine, A. J., and Palmiter, R. D. (1984). *Cell* **37**, 367–379.
Ceroni, M., Piccardo, P., Rodgers-Johnson, P., Mora, C., Asher, D. M., Gajdusek, C., and Gibbs, C. J. (1988). *Ann. Neurol.* **23**, 188–191.
Chan, J. Y., Slamon, D. J., Nimer, S. D., Golde, D. W., and Gasson, J. C. (1986). *Proc. Natl. Acad. Sci. U.S.A.* **83**, 8669–8673.
Chen, I. S., Slamon, D., Rosenblatt, J. D., Shah, P., Quan, S., and Wachsman, W. (1985). *Science* **229**, 54–58.
Crowe, F. W., Schull, W. J., and Neel, J. V. (1956). *In* "A Clinical, Pathological, and Genetic Study of Multiple Neurofibromatosis" (F. W. Crowe, W. J., Schull, and J. V. Neel, eds.), pp. 147–169. C. C. Thomas, Springfield, Illinois.
Daniels, T., and Talal, N. (1987). *In* "Sjögren's Syndrome, Clinical and Immunological Aspects" (N. Talal, H. M. Moutsopoulos, and S. S. Kassan, eds.), pp. 193–199. Springer-Verlag, New York.
Felber, B. K., Rosenberg, M. P., Grammatikakis, W., Ewald, J., Swing, D., Jenkins, N., Copeland, N., and Pavlakis, G. N. (1989). *J. Cell Biochem.* **13B**, 202.
Fox, R. I. (1984). *Semin. Arthr. Rheum.* **14**, 77–105.
Fujii, M., Sassone-Corsi, P., and Verma, I. M. (1988). *Proc. Natl. Acad. Sci. U.S.A.* **85**, 8526–8530.
Furuta, Y., Aizawa, S., Suda, A., Ikawa, Y., Kishimoto, H., Asano, Y., Tada, T., Hikikoshi, A., Yoshida, M., and Seiki, M. (1989). *J. Virol.* **63**, 3185–3189.
Gallo, R. C. (1987). *Sci. Am.* **255**, 88–98.
Gessain, A., Vernant, J.-C., Maurs, L., Barin, F., Gout, O., Calender, A., and De The, G. (1985). *Lancet* **ii**, 407–409.
Goh, W. C., Sodroski, J., Rosen, C., Essex, M., and Haseltine, W. A. (1985). *Science* **227**, 1227–1228.
Grassmann, R., Dangler, C., Muller-Blackenstein, I., Blackenstein, B., McGuire, K., Bokhelar, M. C., Sodroski, J. G., and Haseltine, W. A. (1989). *Proc. Natl. Acad. Sci. U.S.A.* **86**, 3351–3355.
Green, J. E., Hinrichs, S. H., Vogel, J., and Jay, G. (1989a). *Nature (London)* **341**, 72–74.
Green, J. E., Begley, C. G., Wagner, D. K., Waldmann, T. A., and Jay, G. (1989b). *Mol. Cell. Biol.* **9**, 4731–4737.
Hinrichs, S. H., Nerenberg, N., Reynolds, R. K., Khoury, G., and Jay, G. (1987). *Science* **237**, 1340–1343.
Hinuma, Y. (1981). *Nature (London)* **294**, 770–771.
Ho, D. D., Rota, R. R., and Hirsch, M. S. (1984). *Proc. Natl. Acad. Sci. U.S.A.* **81**, 7588–7590.
Inoue, J., Seiki, M., Taniguchi, T., Turu, S., and Yoshida, M. (1986). *EMBO J.* **5**, 2883–2888.
Ito, Y. (1985). *Curr. Top. Microbiol. Immunol.* **115**, 99–112.
Jaffe, E. S. (1985). *Cancer Res.* **45**, 4662–4664.

Leonard, W. J., Depper, J. M., Kanehisa, M., Kronke, M., Peffer, N. J., Svetil, P. B., Sullivan, M., and Greene, W. C. (1985). *Science* **230**, 633–639.

Nagy, K., Clapman, P., Cheingsongg-Popov R., and Weiss, R. (1983). *Int. J. Cancer* **32**, 321–328.

Nerenberg, M. I., and Wiley, C. A. (1989). *J. Pathol.* **135**, 1025–1033.

Nerenberg, M., Hinrichs, S. H., Reynolds, R. K., Khoury, G., and Jay, G. (1987). *Science* **237**, 1324–1329.

Ohta, M., Ohta, D., Mori, F., Nishitani, H., and Saida, T. (1986). *J. Immunol.* **137**, 3440–3443.

Paskalis, H., Felber, B. K., and Pavlakis, G. N. (1986). *Proc. Natl. Acad. Sci. U.S.A.* **83**, 6558–6562.

Poiesz, B. J., Ruscetti, F. M., Gazdar, A. F., Bunn, P. A., Minna, J. D., and Gallo, R. C. (1980). *Proc. Natl. Acad. Sci. U.S.A.* **77**, 7415–7419.

Provost, T. (1987). *Neurol. Clin.* **5**, 405–426.

Reddy, E. P., Sandberg-Wollheim, M., Mettus, R. V., Ray, P. E., DeFreitis, E., and Koprowski, H. (1989). *Science* **243**, 529–533.

Riccardi, V. M., and Eichner, J. E. (1986). In "Neurofibromatosis: Phenotype Natural History, and Pathogenesis" (V. M. Riccardi and J. E. Eichner, eds.), pp. 214–226. Johns Hopkins University Press, Maryland.

Robert-Guroff, M., Nakao, Y., Natake, K., Ito, Y., Sliski, A., and Gallo, R. C. (1982). *Science* **215**, 975–978.

Rodgers-Johnson, P., St. C. Morgan, O., Zaninovic, V., Sarin, P., and Graham, D. S. (1985). *Lancet* **ii**, 1247–1248.

Seiki, M., Hattori, S., Hirayama, Y., and Yoshida, M. (1983). *Proc. Natl. Acad. Sci. U.S.A.* **80**, 3618–3622.

Smith, F. B., Rajdeo, H., Panesar, N., Bhuta, K., and Stahl, R. (1988). *Arch. Pathol. Lab. Med.* **112**, 742–745.

Sodroski, J. G., Rosen, C. A., and Haseltine, W. A. (1984). *Science* **225**, 381–385.

Sommerfelt, M. A., Williams, B. P., Clapham, P. R., Solomon, E., Goodfellow, P. N., and Weiss, R. A. (1988). *Nature (London)* **340**, 1557–1558.

Tajima, K., Tominaga, S., and Suchi, T. (1982). *Gann* **73**, 891–893.

Wiley, C. A., Nerenberg, M., Cros, D., and Soto-Aguilar, M. C. (1989). *N. Engl. J. Med.* **320**, 992–995.

Yoshida, M., Miyoshi, I., and Hinuma, Y. (1982). *Proc. Natl. Acad. Sci. U.S.A.* **78**, 2031–2035.

Simian Immunodeficiency Virus (SIV) from Old World Monkeys

Christopher H. Contag, Stephen Dewhurst, Gregory A. Viglianti, and James I. Mullins

I. Introduction

A. Animal Models for Acquired Immunodeficiency Syndrome (AIDS)

The human retroviruses HIV-1 and HIV-2 (human immunodeficiency viruses type 1 and type 2) have been implicated as the causative agents of AIDS and are estimated to infect perhaps 10 million people worldwide (Mann, 1990). The mechanisms by which HIVs induce disease are, however, unclear and no satisfactory means for control, treatment, or immunization against these viruses exist at present. A major limitation in our understanding of HIV-1 pathogenesis, which impedes efforts to develop effective antiviral therapies and fully evaluate vaccine strategies, is the lack of an adequate animal model.

Several animal models for AIDS are under investigation, but the most desirable model naturally involves direct HIV infection of a nonhuman primate. Infection of chimpanzees with HIV-1 results in persistent infection, which appear to be transmissible from persistently infected chimpanzees to seronegative cohoused animals (Fultz et al., 1986b). However, chimpanzees are endangered and in such limited supply for research purposes that their use is generally limited to evaluation of a small number of vaccine candidates (Prince et al., 1988). Furthermore, the host–virus relationship in chimpanzees is clearly different from that in humans, since no HIV-1-infected chimpanzee has developed overt symptoms of AIDS after as many as 8 years post infection (Fultz et al., 1986b; Nara et al., 1987). Development of alternative nonhuman primate models is therefore necessary. In contrast to HIV-1, which may give rise to transient infections, some isolates of HIV-2 will persistently infect smaller primate species such as rhesus (*Macaca mulatta*) and pig-tailed (*Macaca nemestrina*) macaques, but thus far no clear induction of immunodeficiency disease has been reported in these animals (Dormont et al., 1989). Recently, promising murine models for HIV-1 infection in SCID (severe combined immune deficiency) mice reconstituted with human lymphoid systems have been reported and are being actively investigated (Namikawa et al., 1988); these studies are reviewed in Chapter 19 by Weissman and McCune.

The nonprimate animal models for AIDS, including the feline (Hardy and Essex, 1986), avian (Rup et al., 1982), and murine (Mosier et al., 1985) leukemia onco-retroviruses, as well as the feline (Pedersen et al., 1987), bovine (Gonda et al., 1987), ovine (Haase, 1986; Querat et al., 1990) and equine (Cheevers and McGuire, 1985) lenti-retroviruses have been reviewed recently (Haase, 1986; Gardner and Luciw, 1989) and will not be discussed here. The immunodeficiency disease induced by the simian type D retrovirus (SAIDS; Gardner and Marx, 1985; Thayer et al., 1987) also will not be discussed here. All of the above animal models have value for the understanding and treatment of AIDS; however, simian immunodeficiency virus (SIV) infection of Old World monkeys may provide some of the most illuminating insights into the pathogenesis and control of AIDS.

The virions of HIV and SIV are isomorphic and the biological effects that the two viruses have on susceptible cells in culture are quite similar. SIV and HIV virions demonstrate a characteristic tapered cylindrical core as seen by electron microscopy, exhibit C-type viral budding, and have Mg^{2+}-dependent reverse transcriptases (Kannagi et al., 1985; Daniel et al., 1985; Fultz et al., 1986a). SIV and HIV also share an affinity for growth in human $CD4^+$ cells, both utilize CD4 as at least a component of its cellular receptor (see Section III.B; Kannagi et al., 1985; Daniel et al., 1985; Fultz et al.,

1986a), and both induce cytopathic effects in human CD4$^+$ lymphocytes, characterized by syncytium formation and cell killing (Fultz *et al.*, 1986a; Ohta *et al.*, 1988). Genetically, SIV is more closely related to HIV than any of the other animal lentiviruses, with a genome organization very much like that of HIV (Section II.A). Moreover, SIV infections in some nonhuman primate species result in an immunodeficiency disease resembling that of HIV-infected humans. This chapter will, therefore, both highlight molecular aspects of SIV biology and discuss the significance of SIV as a model system for AIDS.

B. Early Isolates of SIV

SIV was initially discovered in a captive rhesus macaque at the New England Primate Center in the United States (Daniel *et al.*, 1985; Kanki *et al.*, 1985a). This was followed closely by seroepidemiologic studies of SIV infection in Old and New World monkeys (Kanki *et al.*, 1985b; Lowenstine *et al.*, 1986; Daniel *et al.*, 1988b; Ohta *et al.*, 1988). Evidence of SIV infection in rhesus monkeys was limited to a few animals that had died in primate centers of an immunodeficiency disease similar to human AIDS (Daniel *et al.*, 1985; Murphey Corb *et al.*, 1986), which we will refer to as "SIV disease". Subsequent to these early isolations from captive macaques, genetically distinct SIV isolates have been obtained from wild-caught African Green monkeys (agm; Ohta *et al.*, 1988; Fukasawa *et al.*, 1988), mandrills (mnd; Tsujimoto *et al.*, 1989), sooty mangabey monkeys (smm; Fultz *et al.* 1986a), Chimpanzee (cpz; Huet *et al.*, 1990) and from various captive primate species (Table I). The initial isolates of the simian viruses were referred

Table I

Nonhuman Primate Lentiviruses

Species	Common name	Virus
Macaca:		
M. mulatta	Rhesus monkey	SIV$_{mac}$
M. nemestrina	Pig-tailed macaque	SIV$_{mne}$
M. arctoides	Stump-tailed macaque	SIV$_{stm}$
M. fascicularis	Cynomolgus monkey	SIV$_{cyn}$
Cercocebus:		
C. atys	Sooty mangabey	SIV$_{smm}$
Cercopithecus:		
C. aethiops	African Green monkey	SIV$_{agm}$
subspecies	Grivet, Vervet, Tantalus[a]	
Papio:		
P. sphinx	Mandrill	SIV$_{mnd}$
Pan:		
P. troglodytes	Chimpanzee	SIV$_{cpz}$

[a] The three-letter subscript designation for isolates from subspecies is underlined.

to as simian T-cell leukemia virus (STLV)-III to indicate parallels with human T-cell leukemia virus (HTLV)-III, one of the early terms for HIV. Currently, retroviruses isolated from nonhuman primates that demonstrate the typical lentivirus genomic structure are referred to as SIV, with a three letter subscript designating the species of origin and additional designations to indicate the origin of the strain and/or molecular clone (e.g., $SIV_{smm}PBj4.41$ is clone 4.41 of the PBj strain of an SIV isolated from a sooty mangabey monkey).

Early seroepidemiologic studies of people living in central and Western Africa revealed that some West Africans had antibodies which reacted with both SIV and HIV, and which recognized SIV antigens preferentially (Barin et al., 1985; Kanki et al., 1986). These studies suggested that West Africans may have been infected with a nonhuman primate virus or that the simian and human viruses may have recently evolved from a common progenitor. The relationship among the viruses from each host species was confused for some time. However, extensive genetic characterization of the initial viral isolates from African Green monkeys ($STLV-III_{agm}$) and West African humans (HTLV-IV) revealed contamination with, and evidently overgrowth by, the original SIV_{mac} virus isolate (Kornfeld et al., 1987; Kestler et al., 1988; and discussed by Mulder, 1988). It is now clear that viruses present in African Green monkeys, macaques, and West African humans are all quite distinct genetically (see next section). Consequently, the West African human viruses initially referred to as HTLV-IV and LAV-2 (Kanki et al., 1986; Clavel et al., 1986; Clavel, 1987) are presently known as HIV-2.

C. Evolutionary Relationships among SIV and HIV Isolates

Viral genetic diversity among SIVs within infected monkeys and HIV-2 in infected humans appears to be at least as great as, if not greater than, that found in people infected with HIV-1 (Zagury et al., 1988; Li et al., 1989; Johnson et al., 1990). In addition, each nonhuman primate species from which SIV has been isolated has given rise to SIV strains with unique biological, genetic and serological properties. At the time of writing, SIV isolates from eight species have been analyzed (Table I). Overall, SIV isolates are more closely related to HIV-2 than to HIV-1. SIV_{mac} displays approximately 75% DNA sequence similarity to HIV-2, versus 55% similarity between HIV-1 and HIV-2. Predicted amino acid similarity between SIV_{mac} and other lentiviruses are listed for each protein in Table II. Furthermore, comparison of the predicted amino acid composition of the *pol* gene products of these viruses is shown in Table III. HIV-2 and the SIV family are approximately 56% identical to HIV-1 in this region, whereas HIV-2, SIV_{mac}, and SIV_{smm} are considerably more closely related (>80% amino acid identity), while SIV_{agm} and SIV_{mnd} are approximately equidistant from one another and also from other primate lentiviruses. Comparison of the predicted amino acid sequences of *gag* product supports the proposed evolutionary relationships

Table II

Comparison of Primate Lentiviral Amino Acid Sequences
to $SIV_{mac}BK28$ Sequences

			Env[a]		Accessory genes					
$SIV_{mac}BK28$ vs:	Gag	Pol	SU	TM[b]	Vif	Vpx	Vpr	Tat	Rev	Nef
HIV-1BRU[c]	55[d]	56	29	43	30	—	53	35	33	39
HIV-2ROD	85	80	71	71	73	83	70	58	59	55
$SIV_{smm}PBJ$	91	92	83	79	82	90	89	77	75	70
$SIV_{agm}TYO-1$	57	57	46	44	35	35	—	36	37	43
$SIV_{mnd}GB1$	51	57	29	39	37	—	32	38	30	40

[a] The two Env proteins are compared separately.
[b] TM protein is prematurely truncated in many HIV-2, SIV isolates.
[c] Three-letter clone designations are given for each viral sequence.
[d] Values indicate percentage amino acid identity when sequences are compared to $SIV_{mac}BK28$.

of these viruses (Fig. 1). The fact that HIV-2, SIV_{smm}, and SIV_{mac} possess roughly 80% amino acid identity in Pol (Li *et al.*, 1989) suggests that these viruses constitute a single group, a notion reinforced by the finding that the degree of variation in Pol among HIV-2 isolates obtained from humans approximates that observed in SIV_{smm} and SIV_{mac} (Zagury *et al.*, 1988).

Extensive serological surveys of nonhuman primates have shown that most macaques in the wild and in captivity are seronegative. In contrast, some sooty mangabeys (*Cercobus atys*) captured in West Africa, where they are endemic, and most sooty mangabeys in captivity are seropositive for SIV (Ohta *et al.*, 1988). This has led to the suggestion that SIV_{smm} had infected macaques in captivity by the late 1960s, which was the last time wild-caught mangabeys were added to U.S. primate colonies (Fultz *et al.*, 1986a; Murphey-Corb *et al.*, 1986; Hirsch *et al.*, 1989a). Furthermore, the co-localization of SIV-infected sooty mangabeys and HIV-2-infected humans has led to the proposal that SIV_{smm} or a SIV_{smm} progenitor virus also infected

Table III

Amino Acid Identity of Predicted *pol* Gene Products of Primate Lentiviruses

	HIV-1BRU	HIV-2ROD	$SIV_{mac}BK28$	$SIV_{smm}PBj$	$SIV_{agm}TYO-1$	$SIV_{mnd}GB1$
HIV-1BRU	—	56[a]	56	56	56	57
HIV-2ROD	56	—	80	82	56	55
$SIV_{mac}BK28$	56	80	—	92	57	57
$SIV_{smm}PBj$	57	82	92	—	57	57
$SIV_{agm}TYO-1$	56	56	57	57	—	58
$SIV_{mnd}GB1$	56	55	57	57	58	—

[a] Values expressed as percentage amino acid identity.

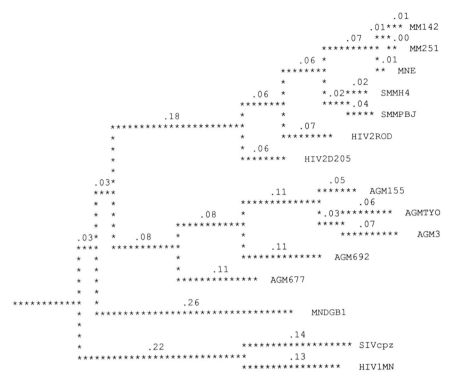

Figure 1. Computer-generated evolutionary tree of primate lentiviruses. The evolutionary relationships shown are for the primate lentiviruses that have been proposed by Myers *et al.* (1988 and unpublished). The numbers indicate the relative sequence divergence, and addition of the numbers along the path between viruses yields the extent of relatedness. Total sites = 1385; variable sites = 969; consistency index = 0.576. This figure kindly provided by Dr. G. Myers.

humans in West Africa, and is therefore the origin of both SIV_{mac} and HIV-2 (Hirsch *et al.*, 1989a; Dietrich *et al.*, 1989, and discussed by Doolittle, 1989). Intraspecies transmission is not without precedent among simian retroviruses. For example, a gibbon to woolly monkey transfer gave rise to the oncogenic simian sarcoma virus (Thielen *et al.*, 1971). It is formally possible that the lentiviral transmission went from a human to a simian host; however, this seems unlikely since SIV_{smm} has recently been isolated from a wild-caught sooty mangabey from a geographically distinct region, Liberia (Henderson *et al.*, 1989).

The computer-generated evolutionary relationships among primate lentiviruses illustrated in Fig. 1 are also consistent with SIV_{smm} being the progenitor of HIV-2 and SIV_{mac}; indeed it has been estimated that these viruses diverged no more than 20–40 years ago (Hirsch *et al.*, 1989a; Sharp and Li, 1988). These estimates for SIV and HIV evolution were determined by assuming that SIV_{mac} arose from a transspecies transmission of SIV from sooty mangabeys to cohoused macaques. The earliest possible time for this transmission was estimated and was correlated with sequence divergence which was then extrapolated to a simian to human transmission. In contrast to HIV-2, the natural history of HIV-1 is less clear since reasonable animal links to HIV-1 have only recently been established. The HIV-1-related virus recently isolated from a wild-caught chimpanzee in Gabon (Peeters *et al.*, 1989; Huet *et al.*, 1990) also suggests a simian origin for HIV-1. Moreover, 30–50% of feral African Green monkey sera are cross-reactive with SIV_{mac} and HIV-1, and the variation among SIVs isolated from African Green monkeys is considerable (Li *et al.*, 1989; Johnson *et al.*, 1990). Therefore, the possibility that a novel isolate of SIV_{agm} may yet be found that more closely resembles HIV-1 remains viable. Nonetheless, present data suggest that HIV-1 may have existed in human populations for a considerable length of time, possibly over one hundred years (Sharp and Li, 1988). It has also been suggested that until recenty, HIV-1 most likely existed only in humans within geographically isolated areas of central Africa where the virus may have diverged significantly from its simian progenitor (Saxinger *et al.*, 1985; Nzilambi *et al.*, 1988; Fukasawa *et al.*, 1988). Furthermore, it is considered unlikely that two viruses as similar as HIV-1 and HIV-2 evolved entirely independently, and it is estimated that these viruses may have diverged from a common parent perhaps as recently as 150 years ago (Sharp and Li, 1988).

Many animal species contain endogenous onco-retrovirus-like sequences related to exogenous viruses which infect these animal species (Coffin, 1984; Stoye and Coffin, 1985). These endogenous elements serve as reservoirs for acquisition of new genetic and pathogenic properties by exogenous agents through recombination (Overbaugh *et al.*, 1988b; discussed by Doolittle, 1989). However, since endogenous proviral sequences are replicated in the context of host chromosomal DNA, their sequences are relatively fixed, and this may tend to dampen the evolving genetic diversity of the virus pool (Sphaer and Mullins, 1990). In contrast to oncoviruses, no closely related endogenous counterparts of any lentivirus (or of the human or bovine oncoviruses) have been clearly identified, although endogenous sequences more distantly related to HIV have recently been identified in the human genome (Horwitz *et al.*, 1989). Thus, despite similar rates of mutation due to the high misincorporation rate of retroviral reverse transcriptases as compared to the cellular DNA replicative and error-corrective machinery, the observed rate of evolution of lentiviruses may be faster than that of oncoviruses.

II. Molecular Biology of SIV

A. Genomic Structure

The general structure of lentiviral genomes is similar to that of other retroviruses which possess three viral structural genes, *gag, pol,* and *env* bounded by long terminal repeats (LTR) containing the transcription control

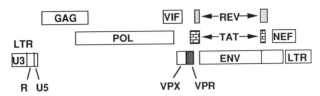

Figure 2. The genome organization of SIV_mac. The location and relative size of the open reading frames used for both the structural and potential accessory genes of SIV_mac are indicated as well as the LTR structure. The site of the TM stop codon is indicated by a vertical line within the *env* region.

elements (Fig. 2). In addition to the structural genes, primate lentiviruses also have small open reading frames (ORFs) which encode accessory proteins, some of which have been shown to perform viral regulatory functions in infected cells (Cann and Karn, 1989, and later this section). The genomic structure and regulation of HIV are reviewed in Chapters 4 and 5 by Wong-Staal and Haseltine respectively. The accessory proteins of primate lentiviruses are translated from multiply spliced messages, the pattern of which appears to be more complex for SIV than for HIV-1. The overall genomic organization of SIV and the HIVs are quite similar (Hirsch *et al.*, 1987); however, one notable difference is the presence or absence of the accessory genes *vpu* (present in HIV-1) and *vpx* (present in SIV and HIV-2; Chakrabarti *et al.*, 1987).

B. SIV Proteins

The similarities between SIV and HIV are further demonstrated at the protein level. Metabolic labeling of SIV and HIV viral proteins and immunoblots of virus from culture supernatants demonstrate cross-reactive proteins of similar but not identical molecular weight with the strongest cross-reactivity seen between *gag*- and *pol*-encoded proteins and weak reactivity between *nef*- and *env*-encoded proteins (Ohta *et al.*, 1988; Murphey-Corb *et al.*, 1986; Fultz *et al.*, 1986a). The structural gene products of SIV, as for all retroviruses, are initially synthesized as polyprotein precursors which are further processed to generate the major constituents of viral particles (Table IV). The present discussion focuses on SIV_mac proteins, for which the greatest amount of information is available. The SIV Gag and Pol proteins are probably encoded by unspliced, full length (9 kb) mRNAs which are synthesized late in infection. The Gag precursor of SIV (and HIV) undergoes proteolytic cleavage by the C-terminally associated viral-encoded protease to generate mature products. Evidence to support this has been obtained by purification and partial sequence analysis of six structural proteins obtained from SIV_mac (p28, p16, p8, p6, p2, and p1). These proteins derive from the Gag precursor, Pr57–60^{gag} and are ordered p16-p28-p2-p8-p1-p6 in the polyprotein (Henderson *et al.*, 1988a). SIV Gag p16 has been localized within the virion (Niedrig *et al.*, 1988) and further, by comparison to HIV-1

Table IV
Genes of Primate Lentiviruses

Gene (Protein)	Protein (kDa)		Comment
	Precursor	Final	
gag	p160; p57–60		
CA		p28	Capsid protein
MA		p16	Matrix protein
NC		p6	Nucleocapsid protein
pol	p160		
RT		p66; p52	Reverse transcriptase
IN		p31	Integrase
PR		p10	Protease
vif		p23	Unknown function in SIV (affects virion infectivity in HIV-1)
vpx		p14	Unknown function in SIV (alters infectivity of HIV-2 for primary cells) absent in SIV_{mnd}
vpr		p15	Unknown function in SIV (may affect virion infectivity in HIV-1) absent in SIV_{agm}
tat		p14	*Trans*-activator of transcription
rev		p19	Antirepressor of translation of viral mRNAs
env	gp140–160		
SU		gp115–125	External virion glycoprotein, binds CD4
TM		gp28–41	Transmembrane anchor of SU, mediates fusion with host cell
nef		p27	Unknown function in SIV (binds GTP; autophosphorylation in HIV-1) (may reduce viral expression in HIV-1)

proteins, the largest of the Gag proteins have tentatively been identified as capsid (p28), matrix (p16), and nucleocapsid (p6) proteins. These analyses also indicated that the Gag precursor is modified near the N-terminus by addition of myristic acid (Henderson *et al.*, 1988a). The importance of the myristoylation of the Gag precursor of SIV_{mac} has been demonstrated in a baculovirus-based expression system in insect cells (Delchambre *et al.*, 1989). In this system, Pr57–60^{gag} was found to spontaneously assemble at the plasma membrane, and to be released into the culture supernatant, by budding, as virus-like 100–120 nm particles. In contrast, an unmyristoylated mutant protein formed particles which accumulated inside cells, suggesting that myristoylation is necessary for extracellular release, budding, or association with the plasma membrane (Delchambre *et al.*, 1989).

Also transcribed from the genomic length mRNA is the Pol protein

(Table IV) which by analogy to HIV-1 is thought to be derived from a Gag/Pol fusion polyprotein (p160) which undergoes proteolysis by the viral protease to generate the mature products. These include the heterodimeric reverse transcriptase–RNaseH protein (p61/P52), as well as integrase/endonuclease (p31) and protease (p10) proteins (Henderson et al., 1988a; Cann and Karn, 1989). Pol gene products are the most highly conserved of all retroviral proteins, and the putative active site of the viral protease is conserved in all SIVs and HIVs (Pearl and Taylor, 1987). Indeed, the kinetic properties of reverse transcriptase from SIV_{mac} and HIV-1 are almost indistinguishable, as is their susceptibility to inhibition by nucleoside analogs such as 2′,3′-dideoxynucleosides, 3′-azido-3′-deoxythymidine, and their respective triphosphate metabolites (Mitsuya and Broder, 1988; Wu et al., 1988; Eriksson et al., 1989).

SIV envelope glycoproteins are synthesized from a singly spliced 4.2-kb mRNA species that, by analogy to HIV-1, is presumed to be expressed late in infection. The Env precursor of SIV_{mac} (gp140–160) is proteolytically cleaved by cellular proteases, initially to remove the N-terminal signal peptide and subsequently to give rise to mature products of 115–125 and 28–41 kDa, which correspond to the surface (SU) and transmembrane (TM) envelope glycoproteins, respectively (Table IV). The sites of proteolytic cleavage in the Env precursor of SIV_{mac} have been identified and are located between the cysteine residue at position 21 and the threonine residue at position 22, and between arginine 526 and glycine 527 (Veronese et al., 1989). The role of the SU protein of SIV_{mac}, like that of HIV-1, includes mediation of viral binding to the CD4 cellular receptor (Hoxie et al., 1988; Rey et al., 1990), while the TM product is involved in anchorage of the SU protein, viral entry into cells, and virus-mediated cell fusion (Bosch et al., 1989).

The SIV_{mac} Env polyprotein is a multimer, perhaps a homodimer (Rey et al., 1990); however, the Env proteins of HIV-1 appear to be trimers or tetramers (Schawaller et al., 1989; Pinter et al., 1989). Multimerization of the Env precursor of SIV_{mac} may be required for proteolytic processing, whereas the TM multimer may be essential to virion structure and infectivity (Rey et al., 1990). The tridimensional conformation of fusion peptides of all syncytia-inducing viruses in a lipid bilayer has been theoretically investigated (Brasseur et al., 1988), and all such peptides are predicted to adopt an asymmetric, amphipathic α-helical structure which is inserted obliquely into the lipid bilayer. The unusual oblique orientation is suspected to locally disorganize the lipid bilayer, which could be an initial step in membrane fusion. The SIV fusion peptide has been modified by site-directed mutagenesis in order to perturb this oblique orientation. Preliminary results indicate that alteration of the oblique orientation of the fusion peptide to a vertical orientation (parallel to the acyl chains of the phospholipids) significantly decreases the fusion properties of the envelope glycoproteins (M. Horth and A. Burny, personal communication).

Although the SIV *env* gene is the most divergent region of the viral genome, it consists of both variable and constant regions. Among the variable regions is a highly immunodominant domain near the N-terminus of the transmembrane glycoprotein; this region has been exploited for serological screening assays designed to distinguish HIV-1, HIV-2, and SIV (Kodama *et al.*, 1988). The evolution of the variable regions is likely influenced by pressure from the immune system, and, by analogy to other lentiviruses (Montelaro *et al.*, 1984), probably facilitates evasion of neutralization by pre-existing antibodies. It has further been hypothesized that variable regions in the SU protein serve as immunological "umbrellas" which cover essential and more highly conserved regions (Coffin, 1986). Computer-predicted protein structures for the SU protein place the hypervariable regions in exposed loops (Gallaher *et al.*, 1989), which is consistent with this hypothesis. The constant regions of the SU protein contain the conformationally important cysteine residues and potential N-linked glycosylation sites, most of which are conserved in all SIV and HIV isolates (Chakrabarti *et al.*, 1987; Franchini *et al.*, 1987; Guyader *et al.*, 1987; Hirsch *et al.*, 1987; Tschachler *et al.*, 1990).

The putative fusion domain at the N-terminus of the TM glycoprotein is also conserved in all SIV isolates, and the TM proteins of SIV and HIV are predicted to have a structure similar to the influenza HA2 TM protein based on the conservation of linear sequence motifs (Gallaher *et al.*, 1989). Site-directed mutagenesis of the region encoding the fusion domain of SIV_{mac} has demonstrated that the overall hydrophobicity of this domain is critical since mutations that introduced polar residues in this domain abolished fusogenic activity, whereas those that favored hydrophobicity promoted syncytium formation (Bosch *et al.*, 1989). Compared to the N-terminus, the C-terminus of the TM glycoprotein of SIV is poorly conserved (Hirsch *et al.*, 1987). Furthermore, a stop codon is found in many SIV and HIV-2 isolates near the junction between the conserved and variable regions. These stop codon mutations result in truncated transmembrane proteins in mature SIV and HIV-2 virions generating a size heterogeneity of TM proteins among various isolates. The presence of the variable domain in the SIV TM leads to greatly reduced viral infectivity in many human cell lines (Hirsch *et al.*, 1989b; Kodama *et al.*, 1989; Chakrabarti *et al.*, 1989) and truncation of the protein apparently represents an adaptation through mutation and selection for improved growth in some cell types in culture.

The primate immunodeficiency viruses encode a number of proteins which are believed to be either positive or negative regulators of viral expression or infectivity. These include Tat, Rev, Vif, Vpr, and Nef, and either Vpx (SIV and HIV-2) or Vpu (HIV-1). The function(s) of most of these proteins are still poorly understood and what is known has been primarily determined for the HIV-1 proteins. However, when the functions of the SIV and HIV-1 proteins have been compared, they have been shown to be homologous but not always interchangeable. Comparisons of the regulatory proteins of primate lentiviruses may be important in evaluating both the life cycle of SIV and its utility as a model for human AIDS. At least two genes (*tat* and *rev*) are obviously involved in the regulation of SIV gene expression and will be discussed in detail in the following two sections.

C. Tat and Tat-mediated *Trans*-activation

The Tat proteins of SIV_{mac}, SIV_{agm}, and HIV-1 are required for viral replication, and *trans*-activate viral gene expression through interaction with a *trans*-activation responsive (TAR) RNA element within the R region of the viral LTR (Arya *et al.*, 1985; Sodroski *et al.*, 1985a,b; Rosen *et al.*, 1985). TAR is positioned downstream from the start of transcription and is therefore present at the 5′ ends of viral mRNAs (Rosen *et al.*, 1985; Muesing *et al.*, 1987; Jakobovits *et al.* 1988; Frankel *et al.*, 1988). Tat seems to act at two different levels of gene expression. First, it apparently increases transcriptional activity of the LTR and may either directly or indirectly act by initiating and/or elongating transcription complexes (Hauber *et al.*, 1987; Jakobovits *et al.*, 1988; Jeang *et al.*, 1988; Rice and Mathews, 1988; Laspia *et al.*, 1989). Second, it apparently increases the translational efficiency of TAR-containing mRNAs (Cullen, 1986; Wright *et al.*, 1986). Both of these functions of Tat require an intact TAR element. The Tat protein in HIV-1 is known to be composed of several functionally distinct domains (Green and Lowenstein, 1989) including the following: (i) a cysteine-rich region involved in metal binding and dimerization of the protein, (ii) a *trans*-activation domain that is rich in hydroxylated residues, (iii) a highly basic domain involved in nucleic acid binding and/or nuclear localization, and (iv) a dispensable region encoded by the second exon of the *tat* gene (Green and Lowenstein, 1989; Hauber *et al.*, 1989; Rice and Carlotti, 1990). Mutational analysis of the cysteine-rich region (Rice and Carlotti, 1990) and the conserved basic domain (Hauber *et al.*, 1989) of HIV indicate that both of these regions are required for full *trans*-activating function. Regions homologous to the HIV *tat* domains also exist in the *tat* gene product of SIV_{mac}, and it is tempting to speculate that they fulfill similar functions.

It should be pointed out that differences between Tat proteins of HIV and SIV do exist, the most obvious being that exon 2 of SIV_{mac} *tat* is not dispensable for function (Viglianti and Mullins, 1988). In addition, the SIV_{mac} Tat protein is 130 amino acids in length while the HIV-1 protein is 86 amino acids (Myers *et al.*, 1988). In those areas which overlap, these proteins share about 48% sequence identity (Hirsch *et al.*, 1987; Viglianti and Mullins, 1988). Most of this similarity is restricted to amino acids 40 though 75 of SIV_{mac} and 21 through 47 of HIV-1, where these proteins share 75% sequence identity. Significantly, this conserved area in HIV-1 contains a region which is believed to possess *trans*-activator activity. The Tat proteins of SIV_{mac} and HIV-1 are also functionally homologous but not interchangeable. SIV_{mac} and HIV-1 Tat are equally effective in stimulating expression from SIV_{mac} LTRs whereas SIV_{mac} Tat is significantly less effective than HIV-1 Tat in stimulating expression from the HIV-1 LTR (Arya et al., 1987; Arya, 1988; Viglianti and Mullins, 1988). The reason for this lack of reciprocal cross trans-activation is not clear and may be related to either the specificity of the Tat proteins for their target molecules or to the specificity of their trans-activating domains. Deletion of regions at each end of the HIV-2 tat, those sequences that do not overlap with HIV-1 tat, did not alter the specificity of the HIV-2 Tat for its respective TAR element (Berkhout et al., 1990). The sequences that determine the differential trans-activation capabilities for HIV and SIV Tat proteins are, therefore, likely contained within the shared regions.

In HIV-1, TAR RNA adopts a hairpin stem loop structure (Muesing et al., 1987). Mutations that disrupt the stability of this structure also disrupt trans-activation while compensatory mutations that restore the structure also restore trans-activation (Feng and Holland, 1988; Berkhout and Jeang, 1989; Berkhout et al., 1990). The sequence 5'-CUGGGX-3' is contained within a single-stranded loop at the end of the hairpin. Mutation of the individual nucleotides within this loop lower trans-activation to 5–50% of wild-type (Feng and Holland, 1988; Berkhout and Jeang, 1989). The other important single-stranded region of the HIV-1 TAR structure is a three-nucleotide bulge loop, 5'-UCU-3' which is located five nucleotides before the terminal hairpin loop. Mutation of the nucleotides within this bulge lower trans-activation to 12–49% of wild-type (Berkhout and Jeang, 1989; Berkhout et al., 1990). The TAR element of SIV has only about 65% sequence identity with HIV-1 TAR (Myers et al., 1988). Furthermore, the SIV TAR RNA element is larger and instead of containing a single stem loop structure, it is predicted to contain three stem loops (Arya, 1988). Perhaps significantly, the two 5' most loops conserve the single-stranded sequence 5'-CUGGGX-3'. The importance of these single-stranded regions in SIV TAR has not yet been determined.

RNA splicing removes an intron from the 5' ends of about 10% of SIV_{mac} mRNAs in chronically infected CEM and HuT78 cells (Viglianti et al., 1990). This intron (nucleotides +61 through +205) is contained within the R and U5 regions of the LTR and overlaps the 3' end of the predicted SIV_{mac} TAR structure. Therefore, splicing removes the two 3'-most stem loops from SIV_{mac} TAR. Transient trans-activation assays have demonstrated that truncation of the SIV_{mac} LTR at the splice donor site of this intron reduced trans-activation response to about 30% of wild-type levels. Therefore sequences contained within this intron constitute a portion of the TAR element. However, it is not yet known whether this requirement is at a transcriptional or post-transcriptional level. The splice sites for this intron are also conserved in SIV_{smm} and HIV-2 (although their use has not been demonstrated); however, they are not conserved in HIV-1, implying that splicing of viral mRNAs may differ between SIVs, HIV-2, and HIV-1.

D. Rev and Rev-mediated Accumulation of mRNA

Rev is a positive regulator that was originally identified in HIV-1 as an antirepressor of translation (Sodroski et al., 1986b) and later shown to affect the generation and/or translocation of mRNA species encoding structural proteins (Feinberg et al., 1986; Felber et al., 1989; Malim et al., 1989b; Emerman et al., 1989; Hammarskjöld et al., 1989). SIV_{mac} Rev proteins are 108 amino acids in length while the HIV-1 Rev protein contains 117 amino acids (Myers et al., 1988) and these proteins share about 34% amino acid sequence identity. The Rev protein has a molecular weight of approximately 19 kDa in SIV_{mac}, but functional mapping of the protein remains to be performed. Nonetheless, comparison to the HIV-1 protein offers some clues. HIV-1 Rev contains at least two functional domains, including an arginine-rich N-terminal region important for nucleolar translocation (and possibly also for nucleic acid binding), as well as a C-terminal domain involved in Rev activity (Malim et al., 1989a). These structural features are also conserved in SIV Rev proteins and may therefore fulfill similar functions (Malim et al., 1989a).

The Rev protein is believed to facilitate the nuclear export of viral mRNAs containing a Rev-responsive element (RRE; Felber et al., 1989; Malim et al., 1989b; Emerman et al., 1989; Hammarskjöld et al., 1989; Lewis et al., 1990). The RRE is located within the common intron separating the two coding exons of tat and rev and is therefore contained within the unspliced gag and pol mRNAs and the singly spliced env mRNAs. In the absence of rev, these mRNAs are preferentially localized within the nucleus (Felber et al., 1989; Malim et al., 1989b). Rev apparently acts by directly binding to the RRE within these mRNAs and mediating their transport to the cytoplasm (Zapp and Green, 1989; Daly et al., 1989; Malim et al., 1990; Olsen et al., 1990; Heaphy et al., 1990). The precise mechanism of Rev action is not known but may in part involve release of the mRNAs from splicing complexes (Chang and Sharp, 1989). Functional comparison of the activities of the SIV_{mac} and HIV-1 Rev proteins indicates that, like Tat, they are functionally homologous but not interchangeable (Malim et al., 1989a; Sakai et al., 1990). The HIV-1 Rev protein is able to efficiently mediate the cytoplasmic accumulation of mRNAs containing either the HIV-1 or SIV_{mac} RRE (Malim et al., 1989a). However, the SIV_{mac} Rev protein is able to mediate the cytoplasmic accumulation of only mRNAs containing the SIV_{mac} RRE. The lack of reciprocal activity may reflect differences in the specificity of the HIV-1 and SIV_{mac} Rev proteins for their respective RREs.

Like the TAR element, the RRE is predicted to adopt an RNA secondary structure. The HIV-1 RRE is believed to be about 230 nucleotides in length and to be organized into five stem loops (Malim *et al.*, 1989b, 1990; Dayton *et al.*, 1989; Olsen *et al.*, 1990; Heaphy *et al.*, 1990). In HIV-1, mutations which are predicted to disrupt the secondary structure of the RRE disrupt both Rev binding and Rev-mediated nuclear transport (Malim *et al.*, 1990; Olsen *et al.*, 1990; Heaphy *et al.*, 1990). A complex RNA secondary structure is also predicted for sequences contained within the *tat/rev* intron of SIV_{mac}. This structure is 216 nucleotides long and like that of HIV-1 contains five stem loops; comparison of the nucleotide sequences of the SIV_{mac} and HIV-1 RREs indicates that they share about 69% identity (Malim *et al.*, 1989a). Initial experiments indicate that a deletion that removes a portion of the predicted SIV_{mac} RRE disrupts Rev-mediated nuclear transport (Malim *et al.*, 1989a).

E. Additional Accessory Genes

The remaining accessory gene products of SIV (Vpr, Vpx, Vif, and Nef) are less well studied, though insights into their function can again be gained by comparison to their homologs in HIV-1 and HIV-2. The *nef* product of HIV-1 is a 27-kDa myristoylated protein with GTPase and GTP-binding activities (Guy *et al.*, 1987) that was initially identified as a negative regulator of virus expression (Luciw *et al.*, 1987; Ahmad and Venkatesan, 1988). More recently, however, the negative effects of Nef have come into question (Kim *et al.*, 1989b; Hammes *et al.*, 1989). In the case of SIV_{agm} and HIV-2, there is currently no evidence for a negative regulatory function since *nef*-defective mutants of these viruses replicated no more efficiently than wild-type virus in the only published study (Shibata *et al.*, 1990b).

The *vif* gene product of HIV-1 is a 23-kDa protein which is required for cell-free infectivity of HIV-1 virions, and *vif*-defective virions have been shown to be roughly 1000 times less infectious than wild-type, whereas cell-to-cell spread of mutant virus was essentially normal (Shibata *et al.*, 1990b; Strebel *et al.*, 1987). In contrast, *vif*-defective mutants of SIV_{agm} or HIV-2 exhibited unchanged cell-free infectivity as compared to wild-type virus (Shibata *et al.*, 1990a,b).

Vpr, vpx and/or *vpu* genes are dispensable for replication of HIV-1, HIV-2, and SIV_{agm} in cell culture (Shibata *et al.*, 1990a), but may play a role in determining the functional integrity of viral particles. *vpr* is absent in SIV_{agm} but is presenting HIV-1, SIV_{mac}, and HIV-2, and encodes a 15-kDa protein. Mutation of this gene in HIV-1 results in the formation of cytopathic virus with somewhat slowed kinetics of growth (Ogawa *et al.*, 1989) but the mechanism of delayed kinetics and its function in replication remain unclear. Recent data suggest that most isolates of HIV-1 have a truncated *vpr* gene and that the extended *vpr* gene of HIV-1 encodes a protein which functions as a transcriptional enhancer and significantly increases viral cytopathicity (Cohen *et al.*, 1990). SIV isolates which contain inactivated *vpr* genes, as observed in many HIV-1 isolates, have not been reported. *Vpx* on the other hand is found exclusively in SIVs and HIV-2 (though it is missing in SIV$_{mnd}$) and encodes a product of 14 and 16 kDa in size in SIV$_{mac}$ and HIV-2, respectively (Yu *et al.*, 1988; Henderson *et al.*, 1988b). *Vpx*-defective mutants of HIV-2 exhibit greatly reduced infectivity for peripheral blood lymphocytes, but not for established cell lines (Guyader *et al.*, 1989).

In HIV-1, Vpu is a nonvirion associated 16-kDa phosphoprotein that shows a similar amino acid composition to some small membrane-associated proteins of the orthomyxo- and paramyxovirus families, leading to the suggestion that it may function as a matrix protein at the level of virus assembly (Cohen *et al.*, 1988). This view is consistent with biochemical studies which have demonstrated that the infectivity of *vpu*-defective virions is markedly reduced relative to wild-type virus (Strebel *et al.*, 1988), likely as a result of delayed virus release with concomitant intracellular accumulation of viral proteins (Klimkait *et al.*, 1990). As expected then, *vpu*-expressing cells *trans*-complement such *vpu*-defective HIV-1 constructs, increasing their infectivity, but they do not do so for infectious clones in SIV$_{mac}$ (Terwilliger *et al.*, 1989) which normally lack a *vpu* gene.

Whether or not these findings indicate a fundamental difference in the maturation and assembly of SIV virions as compared to those of HIV-1 is unclear. Terwilliger and colleagues (1989) have speculated that Vpu may increase the transmission efficiency of HIV-1 relative to SIV and HIV-2, thereby accounting for the worldwide distribution of HIV-1 as compared to the more geographically restricted distribution of HIV-2 (Mann, 1990). However, it is difficult to evaluate this hypothesis in the absence of epidemiological data concerning the relative transmission efficiencies of primate lentiviruses.

F. RNA Splicing Patterns of SIV

The SIV regulatory proteins are expressed from a complex set of spliced mRNAs (Fig. 3; Viglianti *et al.*, 1990). In SIV$_{mac}$, mRNAs have been identified that use splice acceptor sites 5' to the initiator codons for each of the regulatory proteins. In all cases, these are the first initiator codons present in the mRNA, suggesting that each protein can be encoded by a separate message. Furthermore, some proteins, for example Tat and Rev, can be encoded by more than one message. Rev mRNAs can exist in at least seven forms and Tat mRNAs can exist in at least nine forms. These forms differ by the presence of various combinations of untranslated exons positioned upstream of the coding exons and by alternative use of different splice acceptor sites at the start of coding exon 2 (Viglianti *et al.*, 1990). The splicing patterns of the mRNAs of the SIV regulatory genes is thus more complex than those of HIV

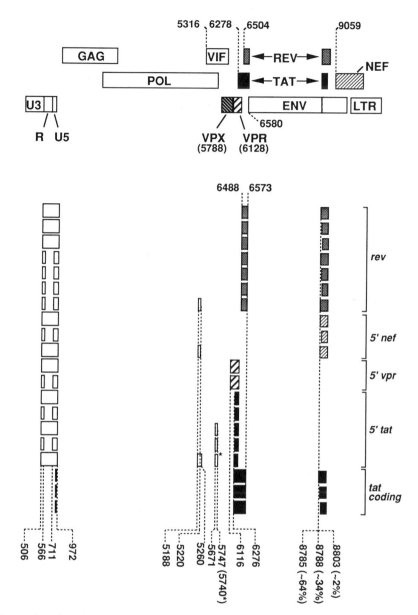

Figure 3. Complex splicing patterns of SIV_mac mRNAs (Viglianti et al., 1990). The upper portion of the figure shows the genome organization of SIV_mac. The numbers indicate the positions of the initiator codons for various genes. The lower portion of the figure shows the structures of a variety of 5′ cDNA clones isolated from HuT78 T cells chronically infected with SIV_mac. The boxes represent untranslated exons while filled boxes represent coding exons. The numbers indicate the positions of the exon borders. Each cDNA is identified as corresponding to a particular protein based in the first initiator codon present. Three different splice acceptor sites were used at the beginning of *tat/rev* coding exon 2 and their relative frequencies are indicated.

(Colombini *et al.*, 1989; Viglianti *et al.*, 1990), but the importance of this complexity remains unknown.

In HIV-1-infected cells, spliced and unspliced mRNAs accumulate differentially throughout the viral replication cycle (Kim *et al.*, 1989a). During the early stages of HIV-1 infection of H9 cells, there is a preferential accumulation of small multiply spliced mRNAs presumably encoding the regulatory proteins. At later stages there is a transition to the preferential accumulation of singly spliced and unspliced mRNAs encoding the virion-associated proteins, Gag, Pol, and Env. In light of these observations, it seems important to assess the role that the temporal accumulation of alternatively spliced mRNAs might play in the life cycle of SIV.

G. Cellular Transcription Factor Binding Sites in Lentiviral LTRs

Located within the LTRs of retroviruses are canonical sequences for cellular transcription factors. The LTRs of primate lentiviruses contain one or multiple copies of consensus sequences for NFκB and Sp1 transcription factors clustered just upstream from the TATA boxes. Minimally pathogenic SIV$_{mac}$ clones contain a single NFκB site and two to four possible Sp1 sites. Of the Sp1 sites in SIV clones, the central two copies (Jones *et al.*, 1986) closely match the consensus sequence while all others are poorly matching copies. The two 5' sites are the most critical for transcriptional activation of HIV-1 in cell culture (Jones *et al.*, 1986). HIV, SIV$_{agm}$, and SIV$_{mnd}$GB1 also have two NFκB sites but only three Sp1 sites. The highly virulent SIV$_{smm}$ molecular clone PBj4.41 contains a duplication which encompasses the NFκB site and thus contains two NFκB sites, in contrast to other SIV$_{smm}$ clones which have only one NFκB site (Dewhurst *et al.*, 1990). The weakly pathogenic SIV$_{mac}$ BK28 clone is also missing one of the well-conserved Sp1 sites. The possibility of cell-type-specific, increased transcriptional activity of the PBj LTR due to its constellation of factor binding sites has been implicated as a potentially critical difference between low virulent isolates and the super virulence of the PBj isolate (Dewhurst *et al.*, 1990).

III. Properties of SIV in Cell Culture

A. Cell Tropism and Viral Selection in Culture

SIVs have been isolated by cocultivation of virus-infected monkey tissues or lymphocytes with established human T-cell lines such as HuT78 and H9 (SIV$_{mne}$; Benveniste *et al.*, 1986), MT-4 and Molt-4 clone 8 (SIV$_{agm}$; Ohta *et al.*, 1988; Daniel *et al.*, 1988a and SIV$_{mnd}$; Tsujimoto *et al.*, 1988) or with lectin-stimulated human peripheral blood mononuclear cells (SIV$_{mac}$; Daniel *et al.*, 1985; Murphey-Corb *et al.*, 1986; SIV$_{smm}$; Fultz *et al.*, 1986a). SIVs, like HIVs, are likely to be in a constant state of genetic flux due to selective pressures on the viral genome. The forces are different and less complex in cell culture than they are *in vivo* and as a result, SIV populations in cell culture may diverge rapidly from those found *in vivo*. One consequence of differential selection is that isolates may develop strong host cell preferences in cell culture, and that highly pathogenic viral quasispecies may be rapidly selected against if they have cytolytic characteristics or are replication defective (Overbaugh et al., 1988a).

Perhaps the best studied example of this phenomenon is the premature truncation of the transmembrane protein that is observed in most SIV isolates (Hirsch et al., 1989b; Kodama et al., 1989; Chakrabarti et al., 1989). The TM protein in SIV-infected primates is likely to be full length since this form is selected over an otherwise isogenic SIV when propagated *in vivo* and in rhesus cells. However, when the virus is propagated in certain human cell lines it becomes truncated as a result of strong selective pressures against the stable expression of the C-terminal region of the protein. The precise site of the truncation is variable, giving rise to TMs of divergent size in different SIV isolates. The origin of this instability appears to be that expression of the C-terminus of the TM protein renders the protein highly susceptible to digestion by cellular proteases (Khiroya, Edmonson, Hirsch, Arbeille, and Mullins, unpublished results).

B. CD4 Receptor Utilization and Use of Accessory or Alternative Receptors

Receptor-mediated viral entry into the host cell is a crucial aspect of the infection process for SIV_{mac} since nonsusceptible host cells will support viral replication if the SIV_{mac} genome is introduced via transfection (Koenig et al., 1989). SIVs preferentially infect CD4$^+$ cells in culture, and studies of the SIV_{mac} Env polyprotein have established that the SU protein of the virus interacts specifically with the CD4 molecule, as is also the case for HIV-1 (Koenig et al., 1989). Furthermore, (i) anti-CD4 monoclonal antibodies prevent infection of cultured cells by SIV_{mac} and SIV_{agm} (Kornfeld et al., 1987; Allan et al., 1990), (ii) a CD4–*Pseudomonas*-exotoxin hybrid protein inhibits viral spread in cell culture (Berger et al., 1989), and (iii) soluble recombinant CD4 protein inhibits SIV_{mac}, SIV_{smm}, and SIV_{agm} infection of some cell types in culture and retards the spread of SIV_{mac} *in vivo* (Clapham et al., 1989; Watanabe et al., 1989). These data suggest that the CD4 molecule is likely to represent at least a major component of the primary cellular receptor for SIV.

However, it appears that CD4 is not the sole component of the SIV receptor. SIV_{mac} fails to replicate in HeLa cells stably transfected with and expressing CD4 (Koenig et al., 1989; Kornfeld and Mullins, unpublished results). SIV also appears to replicate preferentially in a subset of CD4$^+$ lymphocytes which express high levels of CD44 on the surface of the cell

(Gallatin *et al.*, 1989). These studies suggest that CD4 is not sufficient for the infection of lymphocytes, and other studies demonstrating low-level infection of CD4$^-$ cells suggest that it is also not absolutely required for infection (Tsubota *et al.*, 1989). It is interesting to note that SIV$_{mac}$ does not infect either CEM cells (a CD4$^+$ human T-cell line) or B721.174 cells (a human B-cell line), but that the virus does infect a hybrid cell line derived from these two cell types (CEM × 174 cells; Hoxie *et al.*, 1988). In addition, soluble CD4 actually enhances SIV$_{agm}$ infection of Molt-4 clone 8 cells, suggesting that SIV$_{agm}$ infection may proceed by a mechanism, in some cells, that is different from HIV-1 (Allan *et al.*, 1990). Furthermore, SIV$_{mac}$ infects only peripheral blood lymphocytes of Old World monkeys (and humans) but not those of New World primates such as the Squirrel monkey and cotton-top tamarin, even though these cells express a CD4 receptor analog (Kannagi *et al.*, 1985). These species of New World monkey may have subtle changes in the CD4 molecule, or lack additional proteins that may be required for SIV infection.

The current data therefore suggest that accessory molecules may be involved in the interaction of the SIV SU protein with the cellular CD4 molecule. Accessory proteins also appear to be required for HIV infection of cells. The nature of these cellular proteins is unclear, but antibodies to the leukocyte adhesion receptor, LFA-1, have been shown to prevent HIV-1-induced syncytium formation in cell culture (Hildreth and Orentas, 1989). It is possible that accessory receptors of this nature (e.g., CD44) could play a role in SIV infection by stabilizing weak interactions between the virus and its host cell, thereby permitting fusion of the viral envelope with the host cell membrane and allowing viral RNA to enter the cell (Hildreth and Orentas, 1989; Gallatin *et al.*, 1989). SIV may also employ alternative receptors for infection, since SIV$_{mac}$ has been reported to infect CD8$^+$ CD4$^-$ cells (Tsubota *et al.*, 1989). This is likely to be important in lentiviral infection of, and virus-induced damage to, the primate brain and gastrointestinal tract, since reports indicate that HIV-1 can infect brain-derived cells in a CD4-independent manner (Kunsch *et al.*, 1989), as well as infecting nonlymphoid cells of the gastrointestinal tract (Nelson *et al.*, 1988).

C. Cytopathic Effects

SIV infection of cultured cells is typically accompanied by cytopathology similar to that produced by HIV-1 (i.e., syncytium formation and cell killing). In some host cell/virus combinations, however, gross cytopathic effects (CPE) are not observed. SIV$_{mac}$ infection of SupT1 cells for example causes no apparent CPE nor down-regulation of CD4 from the cell surface, in contrast to HIV-1 infection of this cell line (Hoxie *et al.*, 1988). More subtle changes in differentiated functions of cells may, however, occur. SIV$_{mac}$ indirectly causes inhibition of bone marrow hematopoietic progenitor cell growth through infection of macrophages, which may alter cytokine production thereby leading to cell loss (Watanabe *et al.*, 1990). This concept of viral-induced loss of cellular "luxury" functions has been well documented in other viral systems (discussed by Southern and Oldstone, 1986), and may have a significant role in lentiviral pathogenesis. Other physiological changes have also been noted in SIV-infected cells. For example, H9 cells persistently infected with SIV_{mac} have been shown to exhibit increased membrane polarization and alterations in intracellular calcium levels (Wright and Olsen, 1989).

It has been suggested that the observed properties of SIV in cell culture may be correlated with *in vivo* pathogenicity. Studies by Fultz and colleagues indicate that cytolysis of CD4$^+$ mangabey lymphocytes may correlate with the capacity of particular isolates of SIV_{smm} to induce acute disease *in vivo* (Dewhurst *et al.*, 1990). Clearly, the identification of such predictive correlates of disease will greatly facilitate the identification of genetic determinants of virulence and the mechanisms of pathogenesis of SIV, and may potentially provide analogous correlates applicable for distinguishing HIV isolates.

IV. Pathogenesis of SIV Infections

Various SIV isolates result in different courses of disease in different host species. Certain SIV_{smm} isolates, for example, were originally isolated from an asymptomatic sooty mangabey and yet are some of the most virulent isolates in pig-tailed macaques and rhesus monkey (Fultz *et al.*, 1989). SIV infection of macaques results in an immunodeficiency syndrome which demonstrates remarkable similarities to human AIDS. The typical course of SIV disease is characterized by persistent infection with the causative retrovirus, depletion of CD4$^+$ T cells, impaired cellular and humoral immunity, development of opportunistic infections, and neuropathology (Letvin *et al.*, 1985). This constellation of symptoms is typically part of a protracted disease course reminiscent of HIV infections of humans. However, one SIV isolate has been reported to induce disease with survival times as short as one week ($SIV_{smm}PBj$, discussed below) with disease characterized by diarrhea, weight loss, and dehydration similar to that observed in FeLV-induced immunodeficiency disease (Hoover *et al.*, 1987) and many cases of human AIDS. SIV disease generally differs from human AIDS in the prevalence of anemia and the absence of Kaposi's sarcoma-like syndrome (Arthur, 1989).

Assuming that the evolutionary relationships and epidemiological studies are accurate in their prediction of transspecies transmission as the origin of HIV infections in humans, several observations from other viral systems can be extended to human AIDS. A typical observation of transspecies transmission of viruses is that the infection of the primary host is generally asymptomatic, whereas infection of the heterologous host results in overt disease. This phenomenon has been documented for the transspecies transmission of other simian viruses. For example, there is no evidence of pathogenicity in the natural hosts of herpes saimiri and herpes ateles, namely squirrel monkeys and spider monkeys respectively, but these viruses are extremely pathogenic when other primate species are experimentally or naturally infected. This is also reflected in SIV infections; in the case of SIV_{smm}, infected mangabeys usually remain healthy whereas infected macaques develop immunodeficiency disease (Murphey-Corb et al., 1986) as do humans infected with the SIV-related virus, HIV-2 (Clavel et al., 1987). Similarly, SIV_{agm} does not induce disease in infected African Green monkeys and SIV_{mnd} fails to induce disease in infected mandrills; however, their pathogenicity in other host species remains to be determined.

A. SIV-Induced Immunodeficiency

Early studies on naturally occurring immunodeficiency in captive macaque monkeys revealed symptoms that included anemia, neutropenia, lymphopenia, variable thrombocytopenia, monocytosis, hypogammaglobinemia, reduced blastogenic responsiveness of T-lymphocytes, and reduced CD4/CD8 lymphocyte ratio (Zetvin and Hunt, 1984). More recently a thorough study involving experimental infection of juvenile rhesus macaques with at least two field isolates of SIV more clearly defined the pathology and disease course with a duration of 5–12 months characterized by wasting, diarrhea, thymic atrophy, depletion of $CD4^+$ lymphocytes, opportunistic infections (i.e., adenovirus, candida, cryptosporidia, pneumocystis, cytomegalovirus and trichosmoniasis) and central nervous system degeneration (Baskin et al., 1988), all of which are typical of AIDS in humans.

Inoculation of African macaques with SIV usually results in asymptomatic infections whereas in Asian macaques SIV infection results in immunodeficiency disease. The SIV_{smm} isolates appear to not only be among the most virulent but also demonstrate the broadest range of tissue involvement. SIV_{smm} induces persistent lymphadenopathy and reduced numbers of $CD4^+$ lymphocytes in cynomologus and rhesus monkeys (Putkonen et al., 1990; Fultz et al., 1990), and both neurotropic and enterotropic viral isolates have been described (Fultz et al., 1989; Sharer et al., 1988). Recently immunodeficiency disease has been induced by molecular clones of SIV (Dewhurst et al., 1990; Kestler et al., 1990). In these studies the symptoms and histopathology produced by inoculation of molecularly cloned virus into rhesus monkeys, although the courses of disease are different from each other, are identical to those observed in animals inoculated with each of the respective original viral isolates. The reports by Dewhurst et al. (1990) and Kestler et al. (1990) are initial studies in the molecular disection of the pathogenic determinants of SIV disease. In the study by Dewhurst et al. (1990) the regulatory regions in the viral LTR are implicated as major determinants of the highly virulent phenotype of the $SIV_{smm}PBj$ isolate (see later this section).

SIV infections can result in a variety of disease manifestations and affect multiple organ systems even when the inoculating virus has been biologically or molecularly cloned (Fultz et al., 1989; Dewhurst et al., 1990; Kestler et al., 1990). When rhesus monkeys were infected with SIV_{smm}/Delta (a virus isolate originally isolated at the Delta Primate Research Center and originating, presumably, from an asymptomatic sooty mangabey monkey, Murphey-Corb et al., 1986), nearly all animals presented with lymphoid and thymic atrophy but other disease manifestations were variable. Infections by enteropathogenic opportunists (*Shigella, Campylobacter, Cryptosporidum,* and Giardia) were often found to be associated with persistent or recurring diarrhea, a common cause of death in SIV-infected rhesus monkeys, but there were also instances where no pathogen other than SIV could be identified and in these cases enteropathogenesis of SIV is implied (see below). The variability in the disease course may reflect viral heterogeneity, efficiency of the host defenses, the extent of the host-mediated damage or pathogenesis, and the presence or absence of a variety of pathogens and opportunists which may contribute additional symptoms or potentiate the pathogenesis of SIV and HIV.

Infectious molecular clones of SIV have been used in attempts to define more clearly this model system. The first molecularly cloned viruses of the SIV/HIV-2 group which were studied *in vivo* were replication-competent clones derived from viral isolates that were extensively passaged in cell culture. These clones differ in their ability to induce viremia in macaques, but like HIV-1 infection in chimpanzees, have failed to induce rapid onset immunodeficiency (in some cases by more than 4 years, post infection; Murphey-Corb and Mullins, unpublished results; Naidu et al., 1988; Baier et al., 1989). As described earlier in this section, the possibility that pathogenic SIV_{mac} variants exist and are selected against by standard virus isolation techniques requires further evaluation.

An enteropathogenic mechanism of SIV disease is supported by the study of the PBj isolate of SIV_{smm} (Fultz et al., 1989; Dewhurst et al., 1990) as molecular clones of SIV_{smm}PBj14 have been shown to reproduce the severe intestinal dysfunction that is observed with the uncloned isolate. The fatal disease induced by PBj often has a rapid onset, suggesting that SIV can be directly enteropathogenic. Some of the animals survive the acute phase of the infection yet have persistent diarrhea accompanied by severe progressive weight loss, and marked lymphopenia with decreased number of total T lymphocytes, as well as CD4$^+$ lymphocytes, indicating the immunodeficiency potential of the enteropathogenic isolate. In contrast to SIV_{smm}PBj infections in pig-tailed macaques, SIV_{smm}PBj infections of rhesus macaques sometimes results in no overt symptoms of immunodeficiency (Fultz et al., 1989; Dewhurst et al., 1990). It has been suggested that rhesus macaques are innately resistant to infection by SIV_{smm}PBj14 (Fultz et al., 1989), but the nature of this resistance is unclear. Moreover, prior infection of pig-tailed macaques with SIV_{smm9} (the parental virus PBj14, which is not acutely pathogenic) prevents the induction of acute disease by PBj infection (Dewhurst *et al.*, 1990). This observation is encouraging for vaccine development against immunodeficiency-inducing viruses, since it indicates that prior infection with an attenuated virus may prevent development of acute disease upon challenge with a virulent variant.

Depending upon the infecting viral isolate and the host species a spectrum of antiviral humoral immune responses can be observed in SIV-infected animals. At the extremes of this spectrum, no detectable antibody (and no detectable virus) was observed in $SIV_{mm}PBj$ infection of some rhesus macaques (Fultz *et al.*, 1989). In contrast, reciprocal fluctuations of SIV-specific antigens and anti-Gag antibodies were observed in SIV/Delta-infected rhesus macaques (Zhang *et al.*, 1988), and SIV_{smm}-infected cynomolgus monkeys (Putkonen *et al.*, 1989). These reciprocal fluxes parallel the response typically seen in HIV-infected humans.

B. SIV Infections of the Central Nervous System (CNS)

HIV infection of humans often leads to a progressive dementia resulting from HIV infection of cells in the CNS (AIDS dementia complex). CNS complications occur in up to 70% of HIV-infected individuals (Navia *et al.*, 1986b; Sutjipto *et al.*, 1990), and can be especially severe in pediatric cases (Epstein *et al.*, 1985). SIV also appears to infect and induce histological lesions in the CNS in at least 60% of infected animals (Ringler *et al.*, 1988; Baskin *et al.*, 1988) with especially severe symptoms in the immature host (Sharer *et al.*, 1988). The CNS disease associated with these viruses is characterized by multinucleated giant cells, which are found in the gray and white matter of the brain and primarily the gray matter of the spinal cord, and by perivascular infiltrates of macrophages (Ringler *et al.*, 1988; Navia *et al.*, 1986a; Price *et al.*, 1988). In humans, white matter myelin pallor is also commonly observed in the brain (Navia *et al.*, 1986b; Sutjipto *et al.*, 1990).

The hallmark of HIV and SIV infections of the CNS is the presence of virus-infected syncytial cells in inflammatory lesions (Ward *et al.*, 1987; Ringler *et al.*, 1988; Baskin *et al.*, 1988; Sharer *et al.*, 1988). In the case of SIV these CNS lesions consist of lymphocytic infiltrates of the leptomeninges and perivascular infiltration in the gray and white matter with foamy macrophages mixed with occasional small mononuclear cells (Ringler *et al.*, 1988). To date, however, CNS infections by SIV have been reported by only two groups and consequently data pertaining to the neurovirulence of SIV have been obtained for essentially only two viral isolates. Thus, it will be important to study other isolates, especially since the neurovirulence of SIV isolates appears to be variable (Benveniste *et al.*, 1988; Sharer *et al.*, 1988) and also because SIV infections of the CNS do not parallel the human disease in all aspects (i.e., myelin pallor present in HIV-infected humans, and SIV involvement of the leptomeninges).

V. Vaccines against SIV

The ability to immunize against SIV infections appeared, at the onset, to be unlikely or difficult due to: (i) viral persistence and induction of disease despite strong humoral and cellular immune responses in naturally infected animals, and (ii) failure to immunize chimpanzees against HIV using subunit and vaccinia recombinant vaccines (Berman *et al.*, 1988; Hu *et al.*, 1987). Several attempts to immunize macaques against subsequent challenge with SIV have been reported (Desrosiers *et al.*, 1989; Murphey-Corb *et al.*, 1989; Sutjipto *et al.*, 1990). These studies have revealed mixed results but reports of protection conferred by vaccination with formalin-inactivated whole virus are encouraging. Macaques vaccinated with formalin-inactivated SIV developed antiviral antibodies with weak yet detectable neutralizing activity (Desrosiers *et al.*, 1989; Murphey-Corb *et al.*, 1989) and complete protection was observed in one study (Murphey-Corb *et al.*, 1989). Quantitation of neutralizing antibodies in vaccinated animals has been used as a relatively convenient assay to assess the success or failure of immunization. However, in these studies the neutralizing antibody titers in vaccinated animals were low and the significance of neutralizing antibodies in protecting vaccinated animals from virus infection has not been determined. Neutralizing antibodies have been shown to modulate the disease course of other viral infections without interfering with the replication of the virus or conferring protection against infection. Studies by Harty *et al.* (1987) and Harty and Plagemann (1990) are consistent with the notion that protection against viral infections often requires multifactorial immune responses. The results from the SIV vaccine studies also support this notion by invoking mechanisms of protection in addition to simply the elimination of input virus by neutralizing antibodies.

Vaccination of primates with detergent-disrupted SIV resulted in partial protection (Desrosiers *et al.*, 1989). In those vaccinated macaques that were not protected from subsequent challenge, viral RNA levels in the lymph nodes, as detected by *in situ* hybridization, were found to be considerably lower than those in unvaccinated controls, and the resulting disease appeared to be less severe in vaccinated animals (Desrosiers *et al.*, 1989). This partial protection suggests that the vaccine elicited responses comprising only some aspects of a complex anti-viral immune response, or that all possible responses were only partially primed.

The efficacy of SIV vaccines has also been attributed to the amount and purity of the virus in the pool used for the preparation of the inactivated immunogen, as well as to the nature of the challenge virus. To date, all SIV vaccine studies have employed genetically identical immunizing and challenge viruses. Hence, although protection from a homologous challenge is encouraging, the genetic variability in natural infections requires that vaccination also protects against heterologous virus challenge, and such studies are urgently needed. It can be anticipated that the future of the vaccine studies for SIV will have a profound influence on the development of vaccines against HIV in humans.

References

Ahmad, N., and Venkatesan S. (1988). *Science* **241**, 1481–1485.

Allan, J. S., Strauss, J., and Buck, D. W. (1990). *Science* **247**, 1084–1088.

Arthur, L. O. (1989). *PAHO Bull.* **23**, 215–234

Arya, S. K. (1988). *AIDS Res. Hum. Retroviruses* **4**, 175–186.

Arya, S. K., Guo, C., Josephs, S. F., and Wong-Staal, F. (1985). *Science* **229**, 69–73.

Arya, S. K., Beaver, B., Jagodzinski, L., Ensoli, B., Kanki, P. J., Albert, J., Fenyo, E. M., Biberfeld, G., Zagury, J. F., Laure, F., Essex, M., Norrby, E., Wong-Staal, F., and Gallo, R. C. (1987). *Nature (London)* **328**, 548–551.

Baier, M., Werner, A., Cichutek, K., Garber, C., Muller, C., Kraus, G., Ferdinand, F. J., Hartung, S., Papas, T. S., and Kurth, R. (1989). *J. Virol.* **63**, 5119–5123.

Barin, F., Boup, S. M., Denis, F., Kanki, P., Allan, J. S., Lee, T. H., and Essex, M. (1985). *Lancet* i, 1387–1389.

Baskin, G. B., Murphey-Corb, M., Watson, E. A., and Martin, L. N. (1988). *Vet. Pathol.* **25**, 456–467.

Benveniste, R. E., Arthur, L. O., Tsai, C., Sowder, R., Copeland, T. D., Henderson, L. E., and Oroszlan, S. (1986). *J. Virol.* **60**, 483–490.

Benveniste, R. E., Morton, W. R., Clark, E. A., Tsai, C. C., Ochs, H. D., Ward, J. M., Kuller, L., Knott, W. B., Hill, R. W., Gale, M. J., and Thouless, M. E. (1988). *J. Virol.* **62**, 2091–2101.

Berger, E. A., Clouse, K. A., Chaudhary, V. K., Chakrabarti, S., FitzGerald, D. J., Pastan, I., and Moss, B. (1989). *Proc. Natl. Acad. Sci. U.S.A.* **86**, 9539–9543.

Berkhout, B., and Jeang, K. T. (1989). *J. Virol.* **63**, 5501–5504.

Berkhout, B., Gatignol, A., Silver, J., and Jeang, K. (1990). *Nucl. Acids Res.* **18**, 1839–1846.

Berman, P. W., Groopman, J. E., Gregory, T., Clapham, P. R., Weiss, R. A., Ferriani, R., Riddle, L., Shimasaki, C., Lucas, C., Lasky, L. A., and Eichberg, J. W. (1988). *Proc. Natl. Acad. Sci. U.S.A.* **85**, 5200–5204.

Bosch, M. L., Earl, P. L., Fargnoli, K., Picciafuoco, S., Giombini, F., Wong-Staal, F., and Franchini, G. (1989). *Science* **244**, 694–697.

Brasseur, R., Cornet, B., Burny, A., Vandenbranden, M., and Ruysschaert, J. M. (1988). *AIDS Res. Hum. Retroviruses* **4**, 83–90.

Cann, A. J., and Karn, J. (1989). *AIDS* **3**, S19–S34.

Chakrabarti, L., Guyader, M., Alizon, M., Daniel, M. D., Desrosiers, R. C., Tiollais P., and Sonigo, P. (1987). *Nature (London)* **328**, 543–547.

Chakrabarti, L., Emerman, M., Tiollais, P., and Sonigo, P. (1989). *J. Virol* **63**, 4395–4403.

Chang, D. D., and Sharp, P. A. (1989). *Cell* **59**, 789–795.

Cheevers, W. P., and McGuire, T. C. (1985). *Rev. Infect. Dis.* **7**, 83–88.

Clapham, P. R., Weber, J. N., Whitby, D., McIntosh, K., Dalgleish, A. G., Maddon, P. J., Deen, K. C., Sweet, R. W., and Weiss, R. A. (1989). *Nature (London)* **337**, 368–370.

Clavel, F. (1987). *AIDS* **1**, 135–140.

Clavel, F., Guetard, D., Brun-Vezinet, F., Chamaret, S., Rey, M. A., Santos-Ferreira, M. O., Laurent, A. G., Dauguet, C., Katlama, C., Rouzioux, C., Klatzmann, D., Champalimaud, J. L., and Montagnier, L. (1986). *Science* **233**, 343–346.

Clavel, F., Mansingo, K., Chamaret, S., Guetard, D., Favier, V., Nina, J., Santos-Ferreira, M. O., Champalimaud, J. L., and Montagnier, L. (1987). *N. Engl. J. Med.* **316**, 1180–1185.

Coffin, J. (1984). *In* "RNA Tumor Viruses (R. Weiss, N. Teich, H. Varmus, and J. Coffin, eds.), pp. 17–74. Cold Spring Harbor Laboratory, Cold Spring Harbor, New York.

Coffin, J. M. (1986). *Cell* **46**, 1–4.

Cohen, E. A., Terwilliger, E. F., Sodroski, J. G., and Haseltine, W. A. (1988). *Nature (London)* **334**, 532–534.

Cohen, E. A., Terwilliger, E. F., Jalinoos, Y., Proulx, J., Sodroski, J. G., and Haseltine, W. A. (1990). *J. AIDS* **3**, 11–18.

Colombini, S., Arya, S. K., Reitz, M. S., Jagodzinski, L., Beaver, B., and Wong-Staal, F. (1989). *Proc. Natl. Acad. Sci. U.S.A.* **86**, 4813–4817.

Cullen, B. R. (1986). *Cell* **46**, 973–982.

Daly, T. J., Cook, K. S., Malone, T. E., and Rusche, J. R. (1989). *Nature (London)* **342**, 714–716.

Daniel, M. D., Letvin, N. L., King, N. W., Kannagi, M., Sehgal, P. K., Hunt, R. D., Kanki, P. J., Essex, M., and Desrosiers, R. C. (1985). *Science* **228**, 1201–1204.

Daniel, M. D., Li, Y., Naidu, Y. M., Durda, P. J., Schmidt, D. K., Troup, C. D., Silva, D. P., MacKey, J. J., Kestler, H. W., III, Sehgal, P. K., King, N. W., Ohta, Y., Hayami, M., and Desrosiers, R. C. (1988a). *J. Virol.* **62**, 4123–4128.

Daniel, M. D., Letvin, N. L., Sehgal, P. K., Schmidt, D. K., Silva, D. P., Solomon, K. R., Hodi, F. S., Jr., Ringler, D. J., Hunt, R. D., King, N. W., and Desrosiers, R. C. (1988b). *Int. J. Cancer* **41**, 601–608.

Dayton, E. T., Powell, D. M., and Dayton, A. (1989). *Science* **246**, 1625–1629.

Delchambre, M., Gheysen, D., Thines, D., Thiriart, C., Jacobs, E., Verdin, E., Horth, M., Burny, A., and Bex, F. (1989). *EMBO J.* **8**, 2653–2660.

Desrosiers, R. C., Wyand, M. S., Kodama, T., Ringler, D. J., Arthur, L. O., Sehgal, P. K., Letvin, N. L., King, N. W., and Daniel, M. D. (1989). *Proc. Natl. Acad. Sci. U.S.A.* **86**, 6353–6357.

Dewhurst, S., Embretson, J. E., Anderson, D. C., Mullins, J. I., and Fultz, P. N. (1990). *Nature (London)* **345**, 636–640.

Dietrich, U., Adamski, M., Kreutz, R., Seipp, A., and Rubsamen-Waigmann, H. (1989). *Nature (London)* **342**, 948–950.

Doolittle, R. F. (1989). *Nature (London)* **339**, 338–339.

Dormont, D., Livartowski, J., Chamaret, S., Guetard, D., Henin, D., Levagueresse, R., van de Moortelle, P. F., Larke, B., Gourmelon, P., Vazeux, R., Metivier, H., Flageat, J., Court, L., Hauw, J. J., and Montagnier, L. (1989). *Intervirology* **30**, 59S–65S.

Emerman, M., Vazeux, R., and Peden, K. (1989). *Cell* **57**, 1155–1165.

Epstein, L. G., Sharer, L. R., Joshi, V. V., Fojas, M. M., Koenigsberger, M. R., and Oleske, J. M. (1985). *Ann. Neurol.* **17**, 488–496.

Eriksson, B. F. H., Chu, C. K., and Schinazi, R. F. (1989). *Antimicrob. Ag. Chemother.* **33**, 1729–1734.

Felber, B. K., Hadzopoulou-Cladaras, M., Cladaras, C., Copeland, T., and Pavlakis, G. N. (1989). *Proc. Natl. Acad. Sci. U.S.A.* **86**, 1495–1499.

Feinberg, M. B., Jarrett, R. F., Aldovini, A., Gallo, R. C., and Wong-Staal, F. (1986). *Cell* **46**, 807–817.

Feng, S., and Holland, E. C. (1988). *Nature (London)* **334**, 165–167.

Franchini, G., Gurgo, C., Guo, H. G., Gallo, R. C., Collalti, E., Fargnoli, K. A., Hall, L. F., Wong-Staal, F., and Reitz, M. S. (1987). *Nature (London)* **328**, 539–543.

Frankel, A. D., Bredt, D. S., and Pabo, C. O. (1988). *Science* **240**, 70–73.

Fukasawa, M., Miura, T., Hasegawa, A., Morikawa, S., Tsujimoto, H., Miki, K., Kitamura, T., and Hayami, M. (1988). *Nature (London)* **333**, 457–461.

Fultz, P. N., McClure, H. M., Anderson, D. C., Swenson, R. B., Anand, R., and Srinivasan, A. (1986a). *Proc. Natl. Acad. Sci. U.S.A.* **83**, 5286–5290.

Fultz, P. N., McClure, H. M., Swenson, R. B., McGrath, C. R., Brodie, A., Getchell, J. P., Jensen, F. C., Anderson, D. C., Broderson, J. R., and Francis, D. P. (1986b). *J. Virol.* **58**, 116–124.

Fultz, P. N., McClure, H. M., Anderson, D. C., and Switzer, W. M. (1989). *AIDS Res. Hum. Retroviruses,* **5**, 397–409.

Fultz, P. N., Stricker, R. B., McClure, H. M., Anderson, D. C., Switzer, W. M., and Horaist, C. (1990). *J. AIDS* **3**, 319–329.

Gallaher, W. R., Ball, J. M., Garry, R. F., Griffin, M. C., and Montelaro, R. C. (1989). *AIDS Res. Hum. Retroviruses* **5**, 431–440.

Gallatin, W. M., Gale, M. J., Jr., Hoffman, P. A., Willerford, D. M., Draves, K. E., Benveniste, R. E., Morton, W. R., and Clark, E. A. (1989). *Proc. Natl. Acad. Sci. U.S.A.* **86**, 3301–3305.

Gardner, M. L., and Luciw, P. A. (1989). *FASEB J.* **3**, 2593–2606.

Gardner, M. B., and Marx, P. A. (1985). *In* "Advances in Viral Oncology," (G. Klein, ed.), pp. 57–81. Raven Press, New York.

Gonda, M. A., Braun, M. J., Carter, S. G., Kost, T. A., Bess, J. W., Arthur, L. O., and Van Der Maaten, M. J. (1987). *Nature (London)* **330**, 388–391.

Green, M., and Loewenstein, P. M. (1989). *Cell* **55**, 1179–1188.

Guy, B., Kieny, M. P., Riviere, Y., Le Peuch, C., Dott, K., Girard, M., Montagnier, L., and Lecocq, J. P. (1987). *Nature (London)* **330**, 266–269.

Guyader, M., Emerman, M., Sonigo, P., Clavel, F., Montagnier, L., and Alizon, M. (1987). *Nature (London)* **326**, 662–669.

Guyader, M., Emerman, M., Montagnier, L., and Peden, K. (1989). *EMBO J.* **8**, 1169–1175.

Haase, A. T. (1986). *Nature (London)* **322**, 130–136.

Hammarskjold, M.-L., Heimer, J. Hammarskjold, B., Sangwan, I., Albert, L., and Rekosh, D. (1989). *J. Virol.* **63**, 1959–1966.

Hammes, S. R., Dixon, E. P., Malim, M. H., Cullen, B. R., and Greene, W. C., (1989). *Proc. Natl. Acad. Sci. U.S.A.* **86**, 9549–9553.

Hardy, W. D., and Essex, M. (1986). *Prog. Allergy* **37**, 353–376.

Harty, J. T., and Plagemann, P. G. W. (1990). *J. Virol.* **64**, 6257–6262.

Harty, J. T., Chan, S. P. K., Contag, C. H., and Plagemann, P. G. W. (1987). *J. Neuroimmunol.* **15**, 195–206.

Hauber, J., Perkins, A., Heimer, E. P., and Cullen, B. R. (1987). *Proc. Natl. Acad. Sci. U.S.A.* **84**, 6364–6368

Hauber, J., Malim, M. H., and Cullen, B. R. (1989). *J. Virol.* **63**, 1181–1187.

Heaphy, S., Dingwall, C., Ernberg, I., Gait, M. J., Green, S. M., Karn, J., Lowe, A. D., Singh, M., and Skinner, M. A. (1990). *Cell* **60**, 685–693.

Henderson, L. E., Benveniste, R. E., Sowder, R., Copeland, T. D., Schultz, A. M., and Oroszlan, S. (1988a). *J. Virol.* **62**, 2587–2595.

Henderson, L. E., Sowder, R. C., Copeland, T. D., Benveniste, R. E., and Oroszlan, S. (1988b). *Science* **241**, 199–201.

Henderson, L., Sowder, R., Copeland, T., Orozlan, S., and Benveniste, R. E. (1989). *J. Med. Primatol.* **18**, 287–303.

Hildreth, J. E. K., and Orentas, R. J. (1989). *Science* **244**, 1075–1078.

Hirsch, V. H., Riedel, N., and Mullins, J. I. (1987). *Cell* **49**, 309–317.

Hirsch, V. M., Olmsted, R. A., Murphey-Corb, M., Purcell, R. H., and Johnson, P. R. (1989a). *Nature (London)* **339**, 389–392.

Hirsch, V. M., Murphey-Corb, M., Edmonson, P. E., Arbeille, B., and Mullins, J. I. (1989b). *Nature (London)* **341**, 573–574.

Hoover, E. A., Mullins, J. I., Quackenbush, S. L., and Gasper, P. W. (1987). *Blood* **70**, 1880–1892.

Horwitz, M. S., Boyce-Jacino, M., and Faras, A. J. (1989). *Abstract, RNA Tumor Virus Symposium,* Cold Spring Harbor Laboratory, Cold Spring Harbor, New York

Hoxie, J. A., Haggarty, B. S., Bonser, S. E., Rackowski, J. J., Shan, H., and Kanki, P. J. (1988). *J. Virol* **62**, 2557–2568.

Hu, S. L., Fultz, P. N., McClure, H. M., Eichberg, J. W., Thomas, E. K., Zarling, J., Singhal, M. C., Kosowski, S. G., Swenson, R. B., Anderson, D. C., and Todaro, G. (1987). *Nature (London)* **328**, 721–723.

Huet, T., Cheynier, R., Meyerhans, A., Roelants, G., and Wain-Hobson, S. (1990). *Nature (London)* **345**, 356–359.

Jakobovits, A., Smith, D. H., Jakobovits, E. B., and Capon, D. J. (1988). *Mol. Cell. Biol.* **8**, 2555–2561.

Jeang, K.-T., Shank, P. R., and Kumar, A. (1988). *Proc. Natl. Acad. Sci. U.S.A.* **85**, 8291–8295.

Johnson, P. R., Fomsgaard, A., Allan, J., Gravell, M., London., W. T., Olmsted, R. A., and Hirsch, V. M. (1990). *J. Virol.* **64**, 1086–1092.

Jones, K. A., Kadonaga, J. T., Luciw, P. A., and Tjian, R. (1986). *Science* **232**, 755–759.

Kanki, P. J., McLane, M. F., King, N. W. Jr., Letvin, N. L., Hunt, R. D., Sehgal, P., Daniel, M. D., Desrosiers, R. C., and Essex, M. (1985a). *Science* **228**, 1199–1201.

Kanki, P. J., Kurth, R., Becker, W., Dreesman, G., McLane, M. F., and Essex, M. (1985b). *Lancet* **i**, 1330–1332.

Kanki, P. J., Barin, F., M'Boup, S., Allan, J. S., Romet-Lemonne, J. L., Marlink, R., McLane, M. F., Lee, T. H., Arbeille, B., Denis, F., and Essex, M. (1986). *Science* **232**, 238–243.

Kannagi, M., Yetz, J. M., and Letvin, N.J. (1985). *Proc. Natl. Acad. Sci. U.S.A.* **82**, 7053–7057.

Kestler, H. W., Li, Y., Naidu, Y. M., Butler, C. V., Ochs, M. F., Jaenel, G., King, N. W., Daniel, M. D., and Desrosiers, R. C. (1988). *Nature (London)* **331**, 619–622.

Kestler, H., Kodama, T., Ringler, D., Marthas, M., Pederson, N., Lackner, A., Regier, D., Sehgal, P., Daniel, M., Kind, N., and Descrosiers, R. (1990). *Science* **248**, 1109–1112.

Kim, S., Byrn, R., Groopman, J., and Baltimore, D. (1989a). *J. Virol.* **63**, 3708–3713.

Kim, S., Ikeuchi, K., Byrn, R., Groopman, J., and Baltimore, D. (1989b). *Proc. Natl. Acad. Sci. U.S.A.* **86**, 9544–9548.

Klimkait, T., Strebel, K., Hoggan, M. D., Martin, M. A., and Orenstein, J. M. (1990). *J. Virol.* **64**, 621–629.

Kodama, T., Ohta, Y., Masuda, T., Ishikawa, K.-I., Tsujimoto, H., Isahakia, M., Hayami, M. (1988). *J. Virol.* **62**, 4782–4785.

Kodama, T., Wooley, D. P., Naidu, Y. M., Kestler, H. W., Daniel, M. D., and Desrosiers, R. C. (1989). *J. Virol.* **63**, 4709–4714.

Koenig, S., Hirsch, V. M., Olmsted, R. A., Powell, D., Maury, W., Rabson, A., Fauci, A. S., Purcell, R. H., and Johnson, P. R. (1989). *Proc. Natl. Acad. Sci. U.S.A.* **86**, 2443–2447.

Kornfeld, H., Riedel, N., Viglianti, G., Hirsch, V., and Mullins, J. I. (1987). *Nature (London)* **326**, 610–613.

Kunsch, C., Hartle, H. T., and Wigdahl, B. (1989). *J. Virol.* **63**, 5054–5061.

Laspia, M. F., Rice, A. P., and Mathews, M. B. (1989). *Cell* **59**, 283–292.

Letvin, N. L., and Hunt, R. D. (1984). *In* "Human T-cell Leukemia/Lymphoma Virus" (R. C. Gallo, M.E. Essex, and L. Gross, eds.), pp 347–354. Cold Spring Harbor Laboratory, Cold Spring Harbor, New York.

Letvin, N. L., Daniel, M. D., Sehgal, P. K., Desrosiers, R. C., Hunt, R. D., Waldron, L. M., MacKey, J. J., Schmidt, D. K., Chalifoux, L. V., and King, N. W. (1985). *Science* **230**, 71–73.

Lewis, N., Williams, J., Rekosh, D., and Hammarskjöld, M. L. (1990). *J. Virol.* **64**, 1690–1697.

Li, Y., Naidu, Y. M., Daniel, M. D., and Desrosiers, R. C. (1989). *J. Virol.* **63**, 1800–1802.

Lowenstine, L. J., Pederson, N. C., Higgins, J., Pallis, K. C., Uyeda, A., Marx, P., Lerche, N. W., Munn, R. J., and Gardner M. B. (1986). *Int. J. Cancer* **38**, 563–574.

Luciw, P. A., Cheng-Meyer, C., and Levy, J. A. (1987). *Proc. Natl. Acad. Sci. U.S.A.* **84**, 1434–1438.

Malim, M. H., Bohnlein, S., Fenrick, R., Le, S., Maizel, J. V., and Cullen, B. R. (1989a). *Proc. Natl. Acad. Sci. U.S.A.* **86**, 8222–8226.

Malim, M. H., Hauber, J., Le, S. Y., Maizel, J. V., and Cullen, B. R. (1989b). *Nature (London)* **338**, 254–257.

Malim, M. H., Tiley, L. S., McCarn, D. F., Rusche, J. R., Hauber, J., and Cullen, B. R. (1990). *Cell* **60,** 675–683.

Mann, J. M. (1990) *J. AIDS* **3,** 438–442.

Mitsuya, H., and Broder, S. (1988). *AIDS Res. Hum. Retroviruses* **4,** 107–113.

Montelaro, R. C., Parekh, B., Orrego, A., and Issel, C. J. (1984). *J. Biol. Chem.* **259,** 10539–10544.

Mosier, D. E., Yetter, R. A., and Morse, H. C. III (1985). *J. Exp. Med.* **161,** 766–784.

Muesing, M. A., Smith, D. H., and Capon, D. J. (1987). *Cell* **48,** 691–701.

Mulder, C. (1988). *Nature (London)* **333,** 396.

Murphey-Corb, M., Martin, L. N., Rangan, S. R., Baskin, G. B., Gormus, B. J., Wolf, R. H., Andes, W. A., West, M., and Montelaro, R. C. (1986). *Nature (London)* **321,** 435–437.

Murphey-Corb, M., Martin, L. N., Davidson-Fairburn, B., Montelaro, R. C., Miller, M., West, M., Ohkawa, S., Baskin, G. B., Zhang, J., Putney, S. D., Allison, A. C., and Epstein, D. A. (1989). *Science* **246,** 1293–1297.

Myers, G., Josephs, S. F., Rabson, A. B., Smith, T. F., and Wong-Staal, F. (1988). Los Alamos: Theoretical Biology and Biophysics Group, Los Alamos National Laboratory.

Naidu, Y. M., Kestler III, H. W., Li, Y., Butler, C. V., Silva, D. P., Schmidt, D. K., Troup, C. D., Sehgal, P. K., Sonigo, P., Daniel, M. D., and Desrosiers, R. C. (1988). *J. Virol.* **62,** 4691–4696.

Namikawa, R., Kaneshima, H., Lieverman, M., Weissman, I. L., and McCune, J. M. (1988). *Science* **242,** 1684–1686.

Nara, P. L., Robey, W. G., Arthur, L. O., Asher, D. M., Wolff, A. V., Gibbs, C. J. Jr., Gajdusek, D. C., and Fischinger, P. J. (1987). *J. Virol.* **61,** 3173–3180.

Navia, B.A., Cho, E., Petito, C.K., and Price, R.W. (1986a). *Ann. Neurol.* **19,** 525–535.

Navia, B. A., Jordan, B. D., and Price, R. W. (1986b). *Ann. Neurol.* **19,** 517–524.

Nelson, J. A., Wiley, C. A., Reynolds-Kohler, C., Reese, C. E., Margaretten, W., and Levy, J. A. (1988), *Lancet* **i,** 259–262.

Niedrig, M., Rabanus, J. P., L'age Stehr, J., Gelderblom, H. R., and Pauli, G. (1988). *J. Gen. Virol.* **69,** 2109–2114.

Nzilambi, N., De Cock, K. M., Forthal, D. N., Francis, H., Ryder, R. W., Malebe, I., Getchell, J., Laga, M., Piot, P., and McCormick, J. B. (1988). *N. Engl. J. Med.* **318,** 276–279.

Ogawa, K., Shibata, R., Kiyomasu, T., Higuchi, I., Kishida, Y., Ishimoto, A., and Adachi, A. (1989). *J. Virol* **63,** 4110–4114.

Ohta, Y., Masuda, T., Tsujimoto, H., Ishikawa, K., Kodama, T., Morikawa, S., Nakai, M., Honjo, S., and Hayami, M. (1988). *Int. J. Cancer* **41,** 115–122.

Olsen, H. S., Nelbock, R., Cochrane, A. W., and Rosen, C. A. (1990). *Science* **247,** 845–848.

Overbaugh, J., Donahue, P. R., Quackenbush, S. L., Hoover, E. A., and Mullins, J. I. (1988a). *Science* **239,** 906–910.

Overbaugh, J., Riedel, N., Hoover, E. A., and Mullins, J. I. (1988b). *Nature (London)* **332,** 731–734.

Pearl, L. H., and Taylor, W. R. (1987). *Nature (London)* **329,** 351–354.

Pedersen, N. C., Ho, E. W., Brown, M. L., and Yamamoto, J. K. (1987). *Science* **235,** 790–793.

Peeters, M., Honore, C., Huet, T., Bedjabaga, L., Ossari, S., Bussi, P., Cooper, R., and Delaporte, E. (1989). *AIDS* **3,** 625–630.

Pinter, A., Honnen, A., and Tilley, W. J. (1989). *J. Virol.* **63,** 2674–2679.

Price, R. W., Brew, B., Sidtis, J., Rosenblum, M., Scheck, A. C., and Cleary, P. (1988). *Science* **239,** 586–592.

Prince, A. M., Moor-Janowski, J., Eichberg, J. W., Schellens, H., Mauler, R. F., Girard, M., and Goodall, J. (1988). *Nature (London)* **333,** 513.

Putkonen, P., Warstedt, K., Thorstensseon, R., Benthin, R., Albert, J., Lundgren, B., Oberg, B., Norrby, E., and Biberfeld, G. (1989). *J. AIDS* **2,** 359–365.

Querat, G., Audoly, G., Sonigo, P., and Vigne, R. (1990). *Virology* **175,** 434–447.

Rey, M. A., Laurent, A. G., McClure, J., Prust, B., Montagnier, L., and Hovanessian, A. G. (1990). *J. Virol.* **64**, 922–926.

Rice, A. P., and Carlotti, F. (1990). *J. Virol.* **64**, 1864–1868.

Rice, A. P., and Mathews, M. B. (1988). *Nature (London)* **332**, 551–553.

Ringler, D. J., Hunt, R. D., Desrosiers, R. C., Daniel, M. D., Chalifoux, L. V., and King, N. W. (1988). *Ann. Neurol.* **23**, S101-S107.

Rosen, C. A., Sodroski, J. G., and Haseltine, W. A. (1985). *Cell* **41**, 813–823.

Rup, B. J., Hoelzer, J. D., and Bose, H. R. Jr. (1982). *Virology* **116**, 61–71.

Sakai, H., Shibata, R., Miura, T., Hayami, M., Ogawa, K., Kiyomasu, T., Ishimoto, A., and Adachi, A. (1990). *J. Virol.* **64**, 2202–2207.

Saxinger, W. C., Levine, P. H., Dean, A. G., de The, G., Lange, M., Wantzin, G., Moghissi, J., Laurent, F., Hoh, M., Sarngadharan, M. G., and Gallo, R. C. (1985). *Science* **227**, 1036–1038.

Schawaller, M., Smith, G. E., Skehel, J. J., and Wiley, D. C. (1989). *Virology* **172**, 367–369.

Sharer, L. R., Baskin, G. B., Cho, E., Murphey-Corb, M., Blumberg, B. M., and Epstein, L. G. (1988). *Ann. Neurol.* **23**, S108-S112.

Sharp, P. M., and Li, W-H. (1988). *Nature (London)* **336**, 315.

Shibata, R., Miura, T., Hayami, M., Sakai, H., Ogawa, K., Kiyomasu, T., Ishimoto, A., and Adachi, A. (1990a). *J. Virol.* **64**, 307–312.

Shibata, R., Miura, T., Hayami, M., Ogawa, K., Sakai, H., Kiyomasu, T., Ishimoto, A., and Adachi, A. (1990b). *J. Virol.* **64**, 742–747.

Shpaer, E. G., and Mullins, J. I. (1990). *Nucl. Acids Res.* **18**, 5793–5797.

Sodroski, J., Patarca, R., Rosen, C., Wong-Staal, F., and Haseltine, W. (1985a). *Science* **229**, 74–77.

Sodroski, J., Rosen, C., Wong-Staal, F., Salahuddin, S. Z., Popovic, M., Arya, S., Gallo, R. C., and Haseltine, W. A. (1985b). *Science* **227**, 171–173.

Sodroski, J., Goh, W. C., Rosen, C., Dayton, A., Terwilliger, E., and Haseltine, W. (1986b). *Nature (London)* **321**, 412–417.

Southern, P., and Oldstone, M. B. A. (1986) *N. Engl. J. Med.* **314**, 359–367.

Stoye, J., and Coffin, J. M. (1985). *In* "RNA Tumor Viruses" Vol II (R. Weiss, N. Teich, H. Varmus and J. Coffin, eds.), pp. 357–404. Cold Spring Harbor Laboratory, Cold Spring Harbor, New York.

Strebel, K., Daugherty, D., Clouse, K., Cohen, D., Folks, T., and Martin, M. A. (1987). *Nature (London)* **328**, 728–730.

Strebel, K., Klimkait, T., and Martin, M. A. (1988). *Science* **241**, 1221–1223.

Sutjipto, S., Pedersen, N. C., Miller, C. J., Gardner, M. B., Hanson, C. V., Gettie, A., Jennings, M., Higgins, J., and Marx, P. A. (1990). *J. Virol.* **64**, 2290–2297.

Terwilliger, E. F., Cohen, E. A., Lu, Y., Sodroski, J. G., and Haseltine, W. A. (1989). *Proc. Natl. Acad. Sci. U.S.A.* **86**, 5163–5167.

Thayer, R. M., Power, M. D., Bryant, M. L., Gardner, M. B., Barr, P. S., and Luciw, P. A. (1987). *Virology* **157**, 317–329.

Thielen, G. H., Gould, D., Fowler, M., and Dungworth, D. J. (1971). *J. Natl. Cancer Inst.* **47**, 881–889.

Tschachler, E., Buchow, H., Gallow, R. C., and Reitz, M. S. Jr. (1990). *J. Virol.* **64**, 2250–2259.

Tsubota, H., Ringler, D. J., Kannagi, M., King, N. W., Solomon, K. R., MacKey, J. J., Walsh, D. G., and Letvin, N. L. (1989). *J. Immunol.* **143**, 858–863.

Tsujimoto, H., Cooper, R. W., Kodama, T., Fukasawa, M., Miura, T., Ohta, Y., Ishikawa, K.-I., Nakai, M., Frost, E., Roelants, G. E., Roffi, J., and Hayami, M. (1988). *J. Virol.* **62**, 4044–4050.

Tsujimoto, H., Hasegawa, A., Maki, N., Fukasawa, M., Miura, T., Speidel, S., Cooper, R. W., Moriyama, E. N., Gojobori, T., and Hayami, M. (1989). *Nature (London)* **341**, 539–541.

Veronese, F. D., Joseph, B., Copeland, T. D., Oroszlan, S., Gallo, R. C., and Sarngadharan, M. G. (1989). *J. Virol.* **63**, 1416–1419.

Viglianti, G. A., and Mullins, J. I. (1988). *J. Virol.* **62,** 4523–4532.

Viglianti, G. A., Sharma, P., and Mullins, J. I. (1990). *J. Virol.* **64,** 4207–4216

Ward, J. M., O'Leary, T. J., Baskin, G. B., Benveniste, R., Harris, C. A., Nara, P.L., and Rhodes, R.H. (1987) *Am. J. Pathol.* **127,** 199–205.

Watanabe, M., Reimann, K. A., DeLong, P. A., Liu, T., Fisher, R. A., and Letvin, N. L. (1989). *Nature (London)* **337,** 267–270.

Watanabe, M., Ringler, D. J., Nakamura, M., DeLong, P. A., and Letvin, N. L. (1990). *J. Virol.* **64,** 656–663.

Wright, K. A., and Olsen, R. G. (1989). *Int. J. Cancer* **44,** 753–756.

Wright, C., Felber, B., Paskalis, H., and Pavlakis, G. (1986). *Science* **234,** 988.

Wu, J. C., Chernow, M., Boehme, R. E., Suttmann, R. T., McRoberts, M. J., Prisbe, E. J., Matthews, T. R., Marx, P. A., Chuang, R. Y., and Chen, M. S. (1988). *Antimicrob. Ag. Chemother.* **32,** 1887–1890.

Yu, X. F., Ito, S., Essex, M., and Lee, T. H. (1988). *Nature (London)* **335,** 262–265.

Zagury, J. F., Franchini, G., Reitz, M., Collalti, E., Starcich, B., Hall, L., Fargnoli, K., Jagodzinski, L., Guo, H.-G., Laure, F., Arya, S. K., Josephs, S., Zagury, D., Wong-Staal, F., and Gallo, R. C. (1988). *Proc. Natl. Acad. Sci. U.S.A.* **85,** 5941–5945.

Zapp, M. L., and Green, M. R. (1989). *Nature (London)* **342,** 1959–1966.

Zhang, J., Martin, L. N., Watson, E. A., Montelaro, R. C., West, M., Epstein, L., and Murphey-Corb, M. (1988). *J. Inf. Dis.* **158,** 1277–1286.

Transgenic Mouse Model of Human Immunodeficiency Virus (HIV)-Induced Kaposi's Sarcoma

Jonathan Vogel, Jonathan A. Rhim, Desmond B. Jay, and Gilbert Jay

I. Introduction

The acquired immune deficiency syndrome (AIDS) encompasses many disorders, involving in particular cells of the immune and neural systems and resulting in both atrophic and proliferative responses (Gallo, 1987; Fauci, 1988). While the HIV plays an essential role in the development of AIDS, it is not clear which of the disorders that HIV-infected individuals present are the direct consequence of the expression of viral genes. To better understand the etiology of this complex syndrome, one would need experimental models which reflect the human disease.

While *in vitro* approaches, such as the use of tissue culture cells, have been rewarding in the study of viral gene expression and replication, they invariably do not allow or mimic the interactions between different cell types which are indispensable for our understanding of viral pathogenesis. This deficiency has called for the use of *in vivo* approaches which promise to offer a more accurate view of the whole picture rather than a somewhat distorted view of part of the picture. Early attempts to derive animal models for the study of AIDS have focused on identifying animal species which are susceptible to infection with HIV. While chimpanzees, gibbons, and rabbits have been successfully infected, they do not in general develop disorders similar to those found in humans (Gajdusek *et al.*, 1984; Alter *et al.*, 1984; Fultz *et al.*, 1986; McClure *et al.*, 1987; Kulaga *et al.*; 1988; Filice *et al.*, 1988). Explanations for this discrepancy include possible infection of different subsets of target cells and differences in host immune response to the infecting virus. These findings have in a way led to a shift of interest to the identification of related viruses in other animal species, including Old World monkeys and cats (Letvin *et al.*, 1985; Mullins *et al.*, 1986; Pedersen *et al.*, 1987; Schneider and Hunsmann, 1988). This latter approach has turned out to be more encouraging. Macaques infected with the HIV-related simian immunodeficiency virus (SIV) develop AIDS. Such an animal model may prove useful in the development of therapeutic agents and vaccines that may be relevant to human AIDS.

An alternative approach which has become more and more attractive over the past couple of years is to derive animal models by placing the viral genome or segments of it directly in the germline of animals (Brinster *et al.*, 1984; Small *et al.*, 1986b; Reynolds *et al.*; 1988). Briefly, a linear fragment of DNA carrying the gene(s) of interest is microinjected into fertilized single-cell embryos obtained from superovulated female mice (Brinster *et al.*, 1985). These manipulated embryos are implanted back into the oviducts of pseudopregnant females which are then allowed to come to term. Progeny mice are screened for the presence of the microinjected gene(s) and individual transgenic mice are then selected to be bred into lines.

II. Transgenic Approach to the Study of Viral Pathogenesis

The use of transgenic mice to study viral pathogenesis has several unique advantages. First, it circumvents the complex process of viral infection, including variability in mode of virus uptake, virus titer, virus receptor interaction, cell type variation and susceptibility to infection, that are seen during natural infection of individuals. Since the viral sequence is directly integrated into the genome of every cell in the transgenic animal, it has the potential to be expressed in every permissive cell type and to induce disorders in all those tissues. Second, specific transcriptional control elements may be chosen to target viral gene expression to particular tissues of interest or to broaden the repertoire of tissue-specific expression beyond what is commonly seen in natural infections. This latter approach may help to elucidate syndromes that are only sporadically seen in patients. Third, the resulting animal models will provide biochemical and cellular reagents which are essential for a molecular definition of the etiology of the disorders. Such reagents will be reliable, reproducible, and available in practically unlimited quantities.

Transgenic animal models are unique in the sense that there is a precise knowledge, in most cases, of the gene(s) whose expression has led to the observed phenotype, and which will serve as a molecular handle to facilitate further analysis at a biochemical level. This contrasts with conventional animal models where the lack of knowledge of the gene(s) involved often precludes a molecular dissection of the phenotype. The justification for using a murine system to study a complex human virus is based upon the recurring observation that human genes when properly inserted in the germline of mice will be expressed in a tissue-restricted manner as they would in humans. Selective erythroid expression of the human β-globin gene (Grosveld et al., 1986), restricted pancreatic expression of the human insulin gene (Dusbois et al., 1986), and regulated hepatic expression of the human C-reactive protein gene (Ciliberto et al., 1987), all in transgenic mice, are but a few of the many examples that have been reported. In addition, it has repeatedly been demonstrated that human genes encode protein products that function appropriately in transgenic mice, just as if they were in human cells. Expression of the human low-density lipoprotein (LDL) receptor in transgenic mice resulted in rapid clearance of apoproteins B-100 and E from their circulation (Hofmann et al., 1988) and expression of the human β-globin gene effectively corrected β-thalassemia in transgenic mice (Constantini et al., 1986). It is, therefore, quite appropriate and attractive to study the expression and the function of human genes, or genes derived from human pathogens, in mice.

The transgenic approach is particularly appropriate for the study of the human immunodeficiency virus. Because of the multitude of clinical manifestations that HIV- infected individuals present and the fact that this virus encodes large numbers of proteins that regulate viral gene expression (Haseltine and Wong-Staal, 1988), it is anticipated that some of these viral 'accessory' genes may, either singly or in combinations, be involved in inducing disorders that are seen in patients. At least four of these nonstructural proteins have been shown to be *trans*-regulators of viral gene expression in tissue culture cells. In the case of Tat, Rev, and Nef, these proteins act either directly or indirectly on specific *cis*- recognition elements within the viral genome to up- or down-regulate expression of other viral genes (Haseltine and Wong-Staal, 1988; Varmus, 1988). There is now increasing evidence that these viral *trans*-activators act not singly but in concert with cellular proteins that may exist only in specific cell types or are induced by other environmental signals (Nabel and Baltimore, 1987; Franza et al., 1987; Wu et al., 1988). These findings have suggested that the viral control proteins have 'coevolved' with the host cell's biosynthetic machineries, and have further led to the speculation that these viral proteins not only can regulate viral gene expression but may also perturb cellular gene expression as well. Interference with differentiated functions of a cell often leads to uncontrolled proliferation or transformation. This has been demonstrated with *trans*-activator genes of many human viruses, including adenovirus, JC virus, BK virus, and the human T-lymphotropic virus (Small *et al.*, 1986a; Hinrichs *et al.*, 1987; Nerenberg *et al.*, 1987; Green *et al.*, 1989; Koike *et al.*, 1989). Such alterations in cell growth may well result in the manifestation of disorders, particularly malignancies, in infected individuals and warrant more definitive investigations.

III. Derivation of Transgenic Mice with the HIV *tat* Gene

From the outset, we have decided not to begin our studies by introducing the entire HIV proviral DNA into the germline of mice, as this would result in a complexity of phenotypes that could not subsequently be dissected easily. In addition, such a multigenic construct would likely disrupt development and induce lethality in the recipient mice. The first chimeric construct that we derived for microinjection into single-cell embryos contained the HIV *tat* gene under the control of the viral long terminal repeat (LTR). Since the Tat protein is such a potent trans-activator of HIV gene expression (Arya *et al.*, 1985; Sodroski *et al.*, 1985; Haseltine and Wong-Staal, 1988), it is the most likely candidate to have an effect on cellular functions. The choice of the HIV LTR as the transcriptional element to drive the *tat*. gene was based upon our desire to mimic as closely as possible viral gene expression in natural infections and thus avoid potential artifacts. The LTR sequence used not only included the positive regulatory element (PRE) and the negative regulatory element (NRE) in the U3 region, but also the *trans*-acting responsive (TAR) sequence located within the R region (Ayra *et al.*, 1985; Sodroski *et al.* 1985; Luciw *et al.*, 1987). This will support *tat trans*-activation of the LTR, as has been shown in murine cells (Khillan *et al.*, 1988), leading hopefully to higher levels of *tat* expression and, perhaps, a better chance of perturbing cellular functions.

Despite the presence of *tat* gene in all cell types in the resulting transgenic mice, *tat* mRNA was detected only in the skin of these animals (Vogel *et al.*, 1988). Why there is selective expression of the viral LTR in this but not other tissues is by itself intriguing. At least in human AIDS patients, expression of HIV mRNAs or proteins has been detected albeit rarely in a variety of cell types, including both T and B cells, cells in the macrophage- monocyte lineage, neural cells such as neurons, oligodendrocytes, astrocytes and microglial cells, and possibly endothelial cells as well (Koening *et al.*, 1986; Wiley *et al.*, 1986; Gartner *et al.*, 1986, Ho *et al.*, 1986).

While we are attempting to type the cell in the skin that is supporting *tat* expression, it is possible that the cellular targets in the transgenic mice are the Langerhans cells which have previously been suggested to serve as a major reservoir for HIV in infected humans (Levy *et al.*, 1985; Tschachler *et al.*, 1987). If they are Langerhans cells, it is curious that expression is not extended to other macrophage-derived cells like monocytes in the peripheral blood and Kupffer cells in the liver. At least in humans, monocytes appear to be an easy target for certain HIV isolates (Gartner *et al.*, 1986; Salahuddin *et al.*, 1986). This raises the possibility that preferential expression of the LTR—*tat* gene in the skin of the transgenic mice is the consequence of certain activation processes, such as repeated exposure to ultraviolet light or physical trauma. There is evidence that the HIV LTR is responsive to such inducing agents and can turn on viral gene expression as a result (Valeria *et al.*, 1988). These possibilities are attractive and are currently being tested using this transgenic animal model.

IV. Development of Skin Lesions in the Transgenic *tat* Mice

The restricted expression of the *tat* gene has prompted us to look for histological changes in the skin of these mice. Microscopic examination of skin biopsies from the back revealed characteristic focal changes (Vogel *et al.*, 1988). Among the distinguishing features observed was the presence of a hyperplastic epidermis at focal areas which, instead of being three-cell layers thick and made up of well-differentiated cells (Fig. 1A), appeared thickened and poorly differentiated (Figs. 1B,C). Even more prominent was an increase in cellularity in the underlying dermis, frequently in the superficial layer and sometimes in the deep dermis. The extent of involvement varied from a subtle patch of spindle cells commonly seen in young animals (Fig. 1B) to a well-defined plaque observed in many older mice (Fig. 1C). The fact that changes are seen in both the epidermis and dermis may suggest that one compartment of the skin is responding reactively to changes occurring in the other compartment. In fact, the hyperplastic areas in the epidermis appeared not only earlier but also were clearly more uniform morphologically than those in the dermis. On the basis of these observations, mediators elicited in the epidermis may be responsible for the marked changes in the dermis of these transgenic mice.

With advancing age, tumors appear in some of the animals (Fig. 2A). The nodules were frequently highly erythematous and were present mostly on the flank and the back. Histologically, these tumors were found underlying the hyperplastic or ulcerated epidermis and were composed of interweaving bands of spindle cells and numerous extravasated erythrocytes (Fig. 2B).

In transgenic mice, a given phenotype may be a consequence either of expression of the transgene, or of interruption of an essential cellular gene at the specific site of integration. The phenotypic alteration in the former case is reproducible in multiple independent founder lines of mice, while in the latter it is isolated. We have observed restricted expression of the *tat* gene in

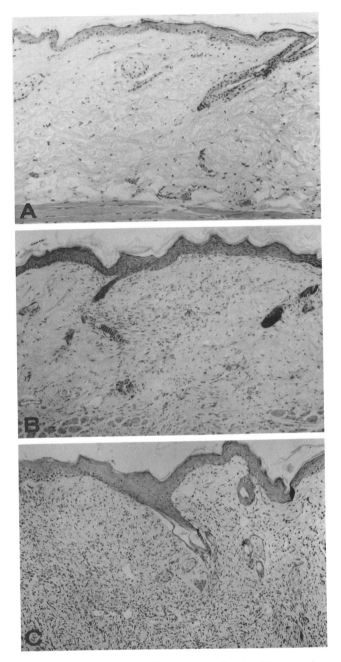

Figure 1. Progressive changes in the skin of transgenic mice. Microscopic examination of the back skin of transgenic *tat* mice at different ages. Panel A, negative male mouse at 4 months. Panel B, transgenic male mouse at 4 months. Panel C, transgenic male mouse at 12 months. Paraffin-embedded sections of fresh biopsies were stained by hematoxylin and eosin.

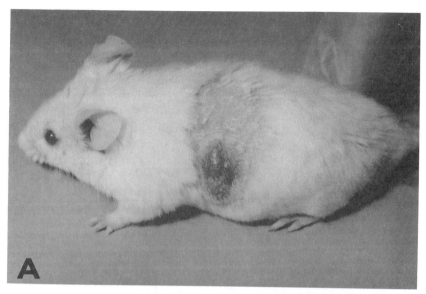

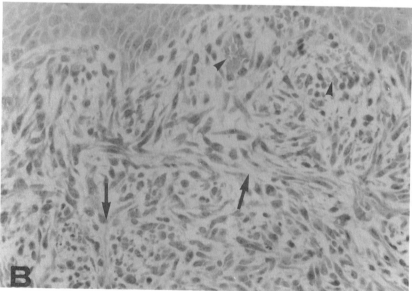

Figure 2. Characterization of tumors in the transgenic mice. Panel A, gross examination of a 15-month-old male transgenic mouse with a highly erythematous tumor nodule overlying a shaved area on the left flank. Panel B, microscopic examination of a tumor from a 12-month-old male transgenic mouse showing fascicles of interweaving spindle cells (arrows), being cut either along the long axis or on end, with extravasated red blood cells (arrowheads) scattered throughout.

the skin and the development of the characteristic skin lesions in mice from each of our transgenic mouse lines. The changes we observed were caused by the HIV *tat* gene.

While there is overwhelming evidence for an increased incidence of different forms of cancers among patients with AIDS, the potential role of HIV in inducing these malignancies has always been overshadowed by the general perception that they are opportunistic neoplasms arising as a consequence of the pronounced immunodeficiency of HIV-infected individuals (Marmor *et al.* 1982; Reichert *et al.*, 1983; Welch *et al.*, 1984; Niedt and Schinella, 1985; Safai *et al.*, 1985). Unfortunately, this cannot account for the extreme predilection of AIDS patients to develop one rare form of neoplasm called Kaposi's sarcoma, sometimes even without detectable immunodeficiency in the individuals (Safai *et al.*, 1985). Results with our transgenic mice carrying the *tat* gene clearly show that a single gene from HIV will suffice to induce tumors in animals and that this virus has to be considered a cancer-causing virus. Even more interesting is the finding that the tumors which appeared in the transgenic *tat* mice have a marked resemblance to Kaposi's sarcoma (Hofmann *et al.*, 1988).

V. Association between HIV Infection and Kaposi's Sarcoma

Kaposi's sarcoma (KS), formerly a rare disorder, has come into prominence with the AIDS epidemic (Templeton, 1981; Gottlieb and Ackerman, 1982; Ziegler *et al.* 1984; Francis *et al.*, 1986; Jones *et al.*, 1986; Volberding, 1986; Mitsuyasu, 1988). While clearly a proliferative disorder related to the vascular system, the nature of the proliferating spindle cells has so far evaded identification. Indeed it is even debatable as to whether it is truly a malignancy or a mere benign dermatosis. Whatever the case, KS has become a major component of AIDS and its etiology has elicited much interest.

Three major forms of KS have been described (Templeton, 1981; Gottlieb and Ackerman, 1982; Ziegler *et al.*, 1984). The classical form is seen sporadically among older individuals of Mediterranean descent, the endemic African form among individuals of all ages, and the epidemic AIDS-related form among HIV-infected male homosexuals. While they share certain diagnostic features such as characteristic histologic patterns, substantial male preponderance, and frequent multifocal involvement, they differ considerably in other clinical features (Ziegler *et al.*, 1984). The skin lesions are restricted predominantly to the extremities in both the classical and African forms, but are widely dispersed in the AIDS-related form with common mucosal (oral and anal) and lymph node involvements. While the classical and African forms frequently follow an indolent course with excellent response to treatment, the AIDS-related form is generally more aggressive with poor response to therapy. On the basis of these clinical findings, one might speculate that the three forms of KS result from similar host cell responses but to different initiating etiologies.

VI. Similarities between the Murine Skin Lesions and Human Kaposi's Sarcoma

The skin lesions we observed in the transgenic *tat* mice have a marked resemblance to those of Kaposi's sarcoma in humans. Since KS has not previously been reported in mice and its diagnosis in humans is not always unambiguous, a review of the findings of our animal model in comparison to the human disease appears appropriate.

The development of the skin lesions in the transgenic mice appears to progress through distinct stages, starting initially with small focal areas of maybe a hundred spindle-shaped cells scattered throughout the dermis (Fig. 1B). These lesions were more frequently observed in skin biopsies from the back of a mouse rather than from the front or the tail. This stage of the disease is seen in most male mice as young as 3–4 months of age. Characteristically, no matter how minor the involvement, these nests of spindle cells in the dermis were always underlying a less differentiated and somewhat hyperplastic epidermis, changes which are always multifocal in distribution. Histologically, these lesions are virtually indistinguishable from the earliest "patch" form of human KS which is characterized by the presence of subtle patches of spindle cells with round to oval nuclei that are frequently found surrounding pre-existing vascular plexuses (Templeton, 1981; Francis *et al.*, 1986).

With older male mice between 4 and 5 months old, many but not all showed more extensive involvement with an increased number of spindle cells between collagen bundles in the dermis (Fig. 1C). Widely dilated, thin-walled, bizarre-shaped vessels lined with thin endothelial cells are found scattered throughout. A sparse lymphocytic infiltration is frequently observed. Again, epidermal hyperplasia is invariably seen overlying these involved areas of the dermis. Virtually identical features characterize the "plaque" form of human KS (Templeton, 1981; Francis *et al.*, 1986). Morphologically, these lesions although extensive suggest a benign and reactive process.

Given time, when the mice reached the age of 15–24 months, about one out of every six males developed distinct nodules. Particularly noticeable was the erythematous nature of these tumors (Fig. 2A). On histological examination, they showed short or elongated fascicles of interweaving bands of spindle cells, forming a defined border and underlying either an intact but hyperplastic or an ulcerated epidermis (Fig. 2B). Extravasated erythrocytes were present either singly or in small collections in the spaces between the spindle cells. Lymphocytic or lymphoplasmacytic infiltration was present in most lesions but not a prominent feature. In some large tumors, foci of necrosis were seen. These descriptions befit a diagnosis of the "nodular" form of human KS (Templeton, 1981; Francis *et al.*, 1986). The apparent progression of disease observed over time may suggest the stepwise involvement of one or more successive genetic events occurring subsequent to the expression of the HIV *tat* gene in the transgenic mice. This will be important in any consideration of how KS arises.

Characteristic of KS are the presence of many slit-like spaces each lined by a thin endothelium interspersed among the interweaving spindle cells (Fig. 3A). Most but not all tumors had a low mitotic index which reflects the slow-growing nature of most of these tumors (cf. Figs. 3A and B). Sometimes, the nodules showed a monocellular proliferation of spindle cells (Fig. 3A). In these cases, nuclear atypia were generally not detected. Differentiation from fibrosarcoma was made frequently by the presence of vascular slit-like spaces and extravasated red blood cells. With some of the faster-growing tumors, mixed cellularity either within the same tumor section or between neighboring regions of the same nodule has been observed (Fig. 3B). They comprised a mixture of spindle cells and abnormal endothelial cells. Vessels lined with enlarged and atypical endothelial cells may be seen among neoplastic cells with similar pleomorphic nuclear morphology (Fig. 3B). These fast-growing tumors frequently had a high mitotic index. Interestingly, these are some of the diagnostic features of human KS (Gottlieb and Ackerman, 1982; Ziegler *et al.*, 1984; Francis *et al.*, 1986; Jones *et al.*, 1986; Volberding, 1986; Mitsuyasu, 1988).

In humans, KS is generally thought of as a multifocal disease with patients having multiple nodules at disparate sites but no evidence of metastasis (Templeton, 1981; Gottlieb and Ackerman, 1982). In the transgenic mice, despite the fact that long latency in the development of nodules gave rise to a low penetrance of disease, where only one out of every six mice had tumors, those animals that showed tumors frequently had nodules at more than one site. This observation not only reflects the human disease, but is particularly germane to our proposed model of the molecular mechanism of etiology of this tumor.

Another similarity found between AIDS patients with KS and the transgenic *tat* mice is the tendency to undergo possibly trauma-induced hemorrhaging under the skin. This may reflect the abnormal vascular structures in the dermis and the fact that KS is a disorder which has its origin in cells related to the vascular system (Templeton, 1981).

A distinguishing feature seen in the transgenic mice which parallels the human disease is the male preponderance in the development of the skin lesions. It is well recognized that both the classical Mediterranean form and the endemic African form of KS have a male preponderance of 3- or 4-to-1 in the former and 15-to-1 in the latter (Templeton, 1981). In the transgenic mice, despite the expression of a similar level of the *tat* gene in both sexes,

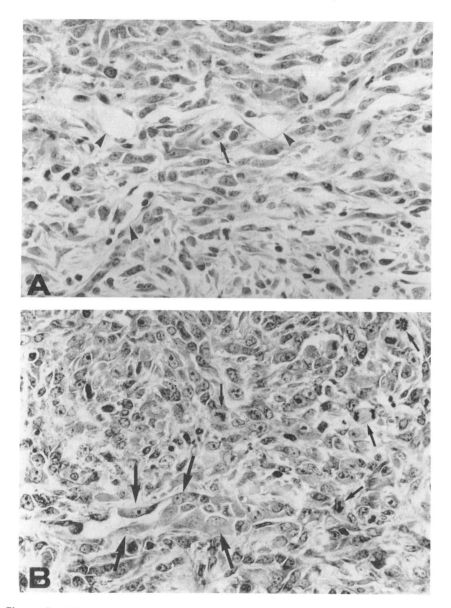

Figure 3. Microscopic examination of tumors in the transgenic mice. Panel A, high magnification of a tumor section from a 12-month-old male mouse showing many vascular slitlike spaces (arrowheads) lined by thin endothelial cells among spindle-shaped tumor cells with an infrequent mitotic flgure (arrow). Panel B, high magnification of a tumor section from a 14-month-old male transgenic mouse showing abnormally large pleomorphic endothelial cells (large arrows) and multiple mitotic figures (small arrows).

only male mice developed either the early or late stages of the disease. On the basis of these observations, it is tempting to conclude that the skin abnormalities detected in the male transgenic *tat* mice are indeed KS lesions. There are subtle differences between the human and mouse disease. While the sex difference observed in humans reflects a preponderance, that in mice is absolute. The skin lesions in the Mediterranean and African forms are frequently localized to the extremities, but those in mice are restricted to the trunk. At least the AIDS-related form of KS is frequently reported to have lymph node and mucosal involvement, including the gastrointestinal tract and pulmonary system, which has not been observed in mice. The dissection of collagen by irregular small branching vascular spaces in the involved areas of the dermis while clearly present in mice are not as impressive as those in humans. In all fairness, the composition of the skin is understandably different between men and mice. One would not expect every aspect of the disease to be identical between the two species. Suffice it to say, the overall similarities far exceed the minor discrepancies.

VII. Suggested Mechanisms for the Development of Kaposi's Sarcoma

In any discussion of the development of a disease, defining the cell of origin is essential. In the case of human KS, the culprit responsible for initiating the entire series of events has so far eluded identification. Potential candidates include endothelial cells of blood or lymphatic vessels, vascular pericytes, smooth muscle cells, myofibroblasts, and fibroblasts (Templeton, 1981; Jones *et al.*, 1986; Nickoloff and Griffith, 1989). The difficulty in naming the normal progenitor cell by morphological criteria or marker analysis may well reflect the dynamic or pluripotent nature of the mesodermal cells. Even more complex is the suggestion that the KS lesions may be made up of tumor cells of diverse genealogies (Templeton, 1981). It appears that the only consensus at the present time is that the proliferation is related to the vasculature and must involve cellular constituents of vessels.

On the basis of the clinical observations in human KS patients, there have been numerous suggestions as to its possible etiology. Perhaps the most gravitating piece of information available is the sex difference in the development of the disease (Templeton, 1981). However, the possibility that the presence of a Y chromosome in males confers predisposition, or two X chromosomes in females provides resistance may not be tenable because of the observation that the incidence of disease in children is similar between boys and girls. This leads to the possibility that development of KS in adults has a hormonal base. It is possible that androgens have a stimulatory activity or estrogens an inhibitory effect. It is worth pointing out, however, that estrogen therapy has no effect in the treatment of established disease. Given all these inconsistencies, it is likely that hormonal influence plays only a contributory rather than a causative role in the development of human KS.

The suggestion that KS may have a genetic origin is derived mainly from the strong variation in the instance of disease among different racial groups living under similar geographical environments and sporadic reports of close relatives coming down with the disease (Templeton, 1981). However, it has also been noted that emigrants from high to low risk areas do not maintain a high incidence of disease. It appears that genetics may confer predisposition but clearly is insufficient in causing the disease.

A feature associated with KS patients is immune deficiency. The disease appears to develop frequently among individuals who have acquired an immunocompromised state either from other disorders, like Hodgkin's disease and AIDS, or from therapy, including organ transplantation (Templeton, 1981). Since the associated disorders and the development of immunodeficiency may either precede or follow the onset of KS, it is not clear whether KS arises as a consequence of immune deficiency.

If indeed immunosuppression induces the development of KS, opportunistic infections have to be considered. The fact that developing KS lesions show close resemblance to inflammatory disease and the detection of regressive KS lesions are suspicious for the involvement of an infectious agent. In this regard, cytomegalovirus (CMV) has been a candidate but its role if any has been difficult to prove (Harwood et al., 1979).

HIV, while present in all patients with AIDS, has not been seriously considered the direct etiological agent of AIDS-associated KS. The best argument against direct involvement of the virus is the observation that KS appears to be restricted to only one subset of AIDS patients, namely the male homosexuals. This has led to the suggestion that environmental factors may have more than a contributory role in the development of KS.

VIII. A Proposed Model for the Etiology of Kaposi's Sarcoma

Our inability to identify the cell of origin and to define the mechanism of etiology of human KS clearly emphasizes the need of an appropriate animal model. The derivation of transgenic animals bearing skin lesions closely resembling KS offers an opportunity for us to dissect this proliferative disease at the molecular level.

Perhaps the most puzzling but intriguing observation we have made with our transgenic mice is that tumors which developed did not express the transgenic *tat* gene (Vogel et al., 1988). This implies that the Tat protein is not required for the maintenance of neoplastic growth. While one can argue that the level of expression required may be below our ability to detect or that expression has been turned off at that particular stage of tumor development, a more likely explanation is that the *tat* gene is not expressed at any time in the cell that grew into the tumor. Consistent with this suggestion is the finding that while the tumors clearly arose from the dermis, expression of the *tat* gene was restricted to cells in the epidermis. While we are actively engaged in defining the exact cell type that supports expression of the *tat* gene, it is tempting to speculate that *tat* gene expression in the epidermis can induce cells in the dermis to proliferate.

The restricted expression of the *tat* gene in the epidermis is not without morphological consequences. Focal areas of thickened epidermis consisting of poorly differentiated cells are seen even when the mice are at a very young age. It is likely that this hyperplastic growth reflects an underlying biochemical change resulting from the action of the Tat protein in the involved cells. This epidermal abnormality is often accompanied by the proliferation of spindle cells in the adjoining dermis.

These observations have led us to propose that expression of the Tat protein in specific cells in the epidermis alters their differentiated functions which may induce them to proliferate, thus giving rise to focal areas of hyperplasia. These involved areas may on occasions become extensive, resembling keratosis or even psoriatic lesions, but have not been observed to acquire neoplastic features. In response to *trans*-activation by the Tat protein, the hyperplastic epidermal cells may release signals, most likely specific cytokines, which can induce locally a subset of cells in the underlying dermis to proliferate. This reactive process is initially mild (the "patch" form) but may progress to be severe (the "plaque" or "nodular" forms). Since the dermal proliferation is in response to extracellular signals from the overlying epidermis, a withdrawal of signals for whatever reason would be expected to result in regression of the dermal lesions. A sustained provision of signals, however, will not only induce extensive dermal hyperplasia, but will also increase the probability of selecting a secondary alteration which will give rise to a malignant tumor. This latter change is a genetic event which results from phenotypic selection of variants within the population of proliferating spindle cells and is consistent with the multistep nature of the development of many human cancers.

Such a mechanism of action would explain the many intricacies of human KS. First, it is consistent with the absence of HIV genomic sequences in KS tissues (Bovi *et al.*, 1986). Second, a "diffusible" factor from the epidermis that acts simultaneously to induce multiple dermal lesions is consistent with the multifocal nature of the development of human KS. Third, the spontaneous regression frequently reported for human KS (Costa and Rabson, 1983; Janier *et al.*, 1985; Real and Krown, 1985; Brooks, 1986) may reflect the termination of epidermal signals, for one reason or another, which are responsible for inducing and sustaining dermal proliferation. Fourth, the release of such signals or cytokines may concomitantly or successively induce responses from not one but multiple cell types in the dermis and thus result in the mixed involvement of cells in human KS lesions. Fifth, the paracrine effect of epidermal signals may not only be localized to target cells in the dermis but may also affect responsive targets in distant tissues and organs, and thus account in part for the detection of other malignancies in AIDS patients. Indeed, we have noted a high incidence of hepatocellular carcinoma in our transgenic *tat* mice and have again failed to detect the expression of the *tat* gene in the involved tissue either before or after the development of malignancy. Lastly, the extensive involvement of the dermis in certain cases in response to sustained epidermal signals may provide a better opportunity for the proliferating cells to select secondary genetic alterations which will allow acquisition of a true malignant phenotype, as occasionally seen in patients with KS.

The release of growth signals by cells in the skin is by no means reserved for cells that express the Tat protein. The skin is not a mere physical integument but a center of many biological activities (Shimada and Katz, 1988; Morhenn, 1988). There is now overwhelming evidence for the existence of a dynamic interplay between cells in the various compartments of the skin which is orchestrated by a consortium of growth mediators. This cross-talk between cells in the skin is important not only for normal growth and development, but also for wound healing, local inflammation, and the pathogenesis of an array of skin diseases. The dynamic nature of the interactions between cell types in the skin appears vital and complex. Indeed the normal process of wound healing involves the proliferation of multiple cell types, including epidermal keratinocytes, fibroblasts, and endothelial cells. Surely there is a sequence of signals and cellular interactions which normally coordinate this series of proliferative events. KS may in fact be the result of the disequilibrium of these physiologic pathways.

In the epidermis, the major cell type is keratinocytes. Apart from the role in the production of keratin, keratinocytes are involved in the secretion of a large number of modulators such as basic fibroblast growth factor (bFGF), interleukin 1 (IL-1), interferon-α (IFN-α), transforming growth factors α and β (TGF-α and -β), granulocyte-macrophage colony-stimulating factor (GM-CSF), and prostaglandin E_2 (PGE$_2$) (Shimada and Katz, 1988; Morhenn, 1988; Halaban *et al.*, 1988). Some of these factors act in an autocrine manner while others in a paracrine fashion to control functions in the skin. While TGF-α can enhance keratinocyte growth and migration, TGF-β most likely acts reciprocally to inhibit their proliferation. Growth of melanocytes is sustained by bFGF produced by the keratinocytes. PGE$_2$ may be responsible for turning off IL-2 production by infiltrating T cells and thus inhibit their proliferation when not required. Langerhans' cells, while a minor component in the epidermis, play a pivotal role in immune recognition by acting as antigenpresenting cells. They are bone marrow-derived and appear to have retained the capacity to undergo mitosis within the epidermis. Perhaps best known is their ability to secrete IL-1 which, because of its chemotactic activity, acts to attract T cells to the skin. In turn, these infiltrating T cells secrete interferon γ (IFN-γ) which has been suggested to be responsible for controlling keratinocyte growth in the epidermis and for stimulating fibroblast proliferation in the dermis. Also present in the epidermis are Merkel's cells and Thy-1[+] dendritic cells; the latter has been implicated in immunologic and inflammatory responses in the skin as well.

The selective expression of the *tat* gene in the epidermis of transgenic mice implies the involvement of one or more of the epidermal cells. In this regard, it is interesting to note that Langerhans cells have been shown to support HIV replication and suspected to serve as a major reservoir for the virus (Tschachler *et al.*, 1987). Given the central role Langerhans cells play in controlling key activities in the epidermis and dermis, perturbation of their differentiated functions by the *trans*-acting Tat protein may have profound consequences in the various compartments of the skin. We have not ruled out keratinocytes as the cells expressing the *tat* gene under the control of the viral LTR. Altering the activity of this subset of cells will likewise suffice to induce the multiplicity of morphologic changes that have been observed in the transgenic mice.

IX. Conclusion

Kaposi's sarcoma may represent communication between epidermis and dermis that has gone astray. An imbalance in the release of certain cytokines by specific cells in the skin may suffice to induce inappropriate responses by different target cells which lead to the development of disease. It is interesting to note that cells obtained from human KS lesions and maintained in culture have indeed been found to release a variety of growth modulators (Salahuddin *et al.*, 1988; Ensoli *et al.*, 1989). A better definition of the mediators that may be involved is essential to our understanding of the etiology of KS. We believe that our transgenic animal model is ideally suited for testing our proposed mechanism of the development of this disease.

Acknowledgments

This work was supported by NIH grants CA53633 and CA52408. We thank Ms. Lisa Ruiz for her assistance in the preparation of this manuscript.

References

Alter, H. J., Eichberg, J. W., Masur, H., Saxinger, W. C., Gallo, R., Macher, A. M., Lane, H. C., and Fauci, A. S. (1984). *Science* **226,** 549–552.

Arya, S. K., Guo, C. X., Josephs, S. F., and Wong-Staal, F. (1985). *Science* **229,** 69–73.

Bovi, P. D., Donti, E., Knowles, D. M., Friedman-Kien, A., Luciw, P. A., Dina, D., Dalla-Favera, R., and Basilico, C. (1986). *Cancer Res.* **46,** 6333–6338.

Brinster, R. L., Chen, H. Y., Messing, A., van Dyke, T., Levine, A. J., and Palmiter, R. D. (1984). *Cell* **37,** 367–379.

Brinster, R. L., Chen, H. Y., Trumbauer, M. E., Yagle, M. K., and Palmiter, R. D. (1985). *Proc. Natl. Acad. Sci. U.S.A* **82**, 4438–4442.

Brooks, J. J. (1986). *Lancet* **ii**, 1309–1311.

Ciliberto, G., Arconi, R., Wagner, E. F., and Rugher, U. (1987). *EMBO J.* **6**, 4017–4022.

Constantini, R., Chada, K., and Magram, J. (1986). *Science* **233**, 1192–1194.

Costa, J., Rabson, A. S. (1983). *Lancet* **i**, 58.

Dusbois, P., Lores, P., Monthioux, E., Absil, J., Lepesant, J. A., Pitet, R., and Jami, J. (1986). *Proc. Natl. Acad. Sci. U.S.A.* **83**, 2511–2515.

Ensoli B., Nakamura, S., Salahuddin, S. Z., Biberfeld, P., Larsson, L., Beaver, B., Wong-Staal, F., and Gallo, R. C. (1989). *Science* **243**, 223–226.

Fauci, A. S. (1988). *Science* **239**, 617–622.

Filice, G., Cereda, P. M., and Varnier, O. E. (1988). *Nature (London)* **335**, 366–369.

Francis, N. D., Parkin, J. M., Weber, J., and Boylston, A. W. (1986). *J. Clin. Pathol.* **39**, 469–474.

Franza, Jr., B. R., Josephs, S. F., Gilman, M. Z., Ryan, W., and Clarkson, B. (1987). *Nature (London)* **330**, 391–395.

Fultz, P. N., McClure, H. M., Swenson, R. B., McGrath, C.R., Brodie A., Getchell, J. P., Jensen, F. C., Anderson, D. C., Broderson, J. R., and Francis, D. P. (1986). *J Virol.* **58**, 116–124.

Gajdusek, D. C., Amyx, H. L., Gibbs, C. J., Jr., Asher, D. M., Yanagihara, R. T., Rodgers-Johnson, P., Brown, P. W., Sarin, P. S., Gallo, R. C., Jr., Maluish, A., Arthur, L. O., Gilden, R. V., Montagnier, L., Cherman, J.-C., Barré Sinoussi, F., Mildran, D., Mathur, U., and Leavitt, R. (1984). *Lancet* **i**, 1415–1416.

Gallo, R. C. (1987). *Sci. Am.* **256**, 46–56.

Gartner, S., Markovits, P., Markovits, D. M., Kaplan, M. H., Gallo, R. C., and Popovic, M. (1986). *Science* **233**, 215–219.

Gottlieb, G. J., and Ackerman, A. B. (1982). *Hum. Pathol.* **13**, 882–892.

Green, J., Hinrichs, S. H., Vogel, J., and Jay, G. (1989). *Nature (London)* **341**, 72–74.

Grosveld, F., van Assendelft, G. B., Greaves, D. R., and Kollias, G. (1986). *Proc. Natl. Acad. Sci. U.S.A.* **83**, 2511–2515.

Halaban, R., Langdon, R., Birchall, N., Cuono, C., Baird, A., Scott, G., Moellmann, G., and McGuire, J. (1988). *J Cell Biol.* **107**, 1611–1619.

Harwood, A. R., Osoba, D., and Hofstader, S. L. (1979). *Am J. Med.* **67**, 759–765.

Haseltine, W. A., and Wong-Staal, F. (1988). *Sci. Am.* **259**, 52–62.

Hinrichs, S. H., Nerenberg, M., Reynolds, R. K., Khoury, G., and Jay, G. (1987). *Science* **237**, 1340–1343.

Ho, D. D., Rota, T. R., and Hirsch, M. S. (1986). *J. Clin. Invest.* **77**, 1712–1715.

Hofmann, S. L., Russell, D. W., Brown, M. S., Goldstein, J. L., and Hammer, R. E. (1988). *Science* **239**, 1277–1281.

Janier, M., Vignon, M. D., and Cottenot, F. (1985). *N. Engl. J. Med.* **312**, 1638–1639.

Jones, R. R., Spaull, J., Spry, C., and Jones, E. W. (1986). *J.Clin. Pathol.* **39**, 742–749.

Khillan, J. S., Deen, K. C., Yu, S.-H., Sweet, R. H., Rosenberg, M., and Westphal, H. (1988). *Nucl. Acids Res.* **16**, 1423–1430.

Koening, S., Gendelman, H. E., Orenstein, J. M., Dal Canto, M. C., Pezeshkpour, H., Yungbluth, M., Janotta, F., Akasamit, A., Martin, M. A., and Fauci, A. S. (1986). *Science* **233**, 1089–1093.

Koike, K., Hinrichs, S. H., Isselbacher, K. J., and Jay, G. (1989). *Proc. Natl. Acad. Sci. U.S.A.* **86**, 5615–5619.

Kulaga, H., Folks, T. M., Rutledge, R., and Kindt, T. J. (1988). *Proc. Natl. Acad. Sci. U.S.A.* **85**, 4455–4499.

Letvin, N. L., Daniel, M. D., Sehgal, P. K., Desrosiers, R. C., Hunt, R. D., Waldron, L. M., MacKey, J. J., Schmidt, D. K., Chalifoux, L. V., and King, N. W. (1985). *Science* **230**, 71–73.

Levy, J. A., Shimabukura, J., McHugh, T., Casavant, C., Stites, D., and Oshirol, L. (1985). *Virology* **147**, 441–448.

Luciw, P. A., Chen-Mayer, C., and Levy, J. A. (1987). *Proc. Natl. Acad. Sci. U.S.A.* **84**, 1434–1438.

Marmor, M., Friedman, A. E., Laubenstein, L., Byrum, R. D., William, D. C., O'Onofrio, and Dubin, N. (1982). *Lancet* **i**, 1003–1007.

McClure, M. O., Sattentau, Q. J., Beverley, P. C. L., Hearn, J. P., Fitzgerald, A. K., Zuckerman, A. J., and Weiss, R. A. (1987). *Nature (London)* **330**, 487–489.

Mitsuyasu, R. T. (1988). *Blood Rev.* **2**, 222–231.

Morhenn, V. B. (1988). *Immunol. Today* **9**, 104–107.

Mullins, J. I., Chen, C. S., and Hoover, E. A. (1986). *Nature (London)* **319**, 333–336.

Nabel, G., and Baltimore, D., (1987). *Nature (London)* **326**, 711–713.

Nerenberg, M., Hinrichs, S. H., Reynolds, R. K., Khoury, G., and Jay, G. (1987). *Science* **237**, 1324–1329.

Nickoloff, B. J., and Griffith, C. E. M. (1989). *Science* **243**, 1736–1737.

Niedt, G. W., and Schinella, R. A. (1985). *Arch. Pathol. Lab. Med.* **109**, 727–734.

Pedersen, N. C. Ho, E. W., Brown, M. L., and Yamamoto, J. K. (1987). *Science* **235**, 790–793.

Real, F. X., and Krown, S. E. (1985). *N. Engl. J. Med.* **313**, 1659.

Reichert, C. M., O'Leary, T. J., Levens, D. L., Simrell, C. R., and Macher, A. B. (1983). *Am. J. Pathol.* **112**, 357–382.

Reynolds, R. K., Hoekzema, G. S., Vogel, J., Hinrichs, S. H., and Jay, G. (1988). *Proc. Natl. Acad. Sci. U.S.A* **85**, 3135–3139.

Safai, B., Johnson, K. G., Myskowski, P. L., Koziner, B., Yang, Y., Cunninghan-Rundles, S., Godbold, J. H., and Dupont, B. (1985). *Ann. Int. Med.* **103**, 744–750.

Salahuddin, S. Z., Rose, R. M., Groopman, J. E., Markham, P. D., and Gallo, R. C. (1986). *Blood* 68, 281–284.

Salahuddin, S. Z., Nakamura, S. S., Biberfeld, P., Kaplan, M. H., Markham, P. D., Larsson, L., and Gallo, R. C. (1988). *Science* **242**, 430–433.

Schneider, J., and Hunsmann, G. (1988). *AIDS* **2**, 1–9.

Shimada, S., and Katz, S. I. (1988). *Arch. Pathol. Lab. Med.* **112**, 231–234.

Small, J. A., Khoury, G., Jay, G., Howley, P., and Scangos, G. A. (1986a). *Proc. Natl. Acad. Sci. U.S.A.* **83**, 8288–8292.

Small, J. A., Scangos, G., Cork, L., Jay, G., and Khoury, G. (1986b). *Cell* **46**, 13–18.

Sodroski, J. G., Rosen, C., Wong-Staal, F., Salahuddin, Z. S., Popovic, M., Arya, S., Gallo, R., and Haseltine, W.A. (1985). *Science* **227**, 171–173.

Templeton, A. C. (1981). *In* "Pathology Annual" (S. C. Sommers and P. P. Rosen eds), p. 315–336. Appleton Century Crofts, New York.

Tschachler, E., Groh, V., Popovic, M., Mann, D. L., Konrad, K., Safai, B., Eron, L., Veronese, F.-d., Wolff, K., and Stingl, G. (1987). *J. Invest. Dermatol.* **88** 233–237.

Valeria, K., Delers, A., Bruck, C., Thiriart, C., Rosenberg, H., Debouck, C., and Rosenberg, M. (1988). *Nature (London)* **333**, 78–81.

Varmus, H. (1988). *Genes Develop.* **2**, 1055–1962.

Vogel, J., Hinrichs, S. H., Reynolds, R. K., Luciw, P. A., and Jay, G. (1988). *Nature (London)* **335**, 606–611.

Volberding, P. A. (1986). *Med. Clin. North Am.* **70**, 665–675.

Welch, K., Finkbeiner, W., Alpers, C. E., Blumenfeld, W., Davis, R. L., Smuckler, E. A., and Beckstead, J. H. (1984). *JAMA* **252**, 1152–1159.

Wiley, C. A., Schrier, R. D., Nelson, J. A., Lampert, P. W., and Oldstone, M. B. A. (1986). *Proc. Natl. Acad. Sci. U.S.A.* **83**, 7089–7093.

Wu, F. K., Farcia, J. A., Harrish, D., and Gaynor, R. B. (1988). *EMBO J.* **7**, 2117–2129.

Ziegler, J. L., Templeton, A. C., and Vogel, C. L. (1984). *Semin. Oncol.* **11**, 47–52.

The SCID-hu Mouse as a Model for Human Immunodeficiency Virus (HIV) Infection

J. M. Mc Cune

I. Introduction

Multiple lines of evidence suggest that the acquired immune deficiency syndrome (AIDS) is etiologically associated with the human immunodeficiency virus (HIV) (Barre-Sinoussi *et al.*, 1983; Popovic *et al.*, 1984). Infection by this virus is predicated by interactions between its envelope protein (gp120/41) and the human cell surface receptor, CD4 (Dalgleish *et al.*, 1984, Klatzmann *et al.*, 1984). This receptor is found on a number of cell populations, including helper/inducer T cells, monocytes/macrophages, Langerhans/dendritic cells, eosinophils, and certain cells of the central nervous system. To the extent that these cells are responsible for recognition of foreign antigen in association with Class II major histocompatibility complex (MHC) antigens (Engleman *et al.*, 1981), their destruction and/or dysfunction results in severe deficits in immune function. AIDS patients are accordingly susceptible to a variety of rare opportunistic infections and uncommon neoplasms.

Although much has been learned from studies of the interactions between HIV and CD4$^+$ cells *in vitro*, this knowledge is insufficient to understand the pathophysiology of infection *in vivo* (for review, see Mc Cune, 1991). Indeed, little is known about the events that occur immediately after infection which lead to eventual immunodeficiency and death. Given the long period of time between infection and the development of symptoms, most HIV-infected individuals do not know if and when they are infected. When symptoms consistent with the clinical definition of AIDS supervene, the orderly evaluation of disease process and treatment options is complicated in the setting of overburdened medical institutions. As a result, it is difficult to study the disease process induced by HIV in man; it is also difficult to systematically analyze large numbers of antiviral compounds, alone or in combination, so that optimal treatment strategies might be devised.

A major obstacle to research has been the lack of an adequate animal model for HIV infection. Clearly, studies of the human immune response cannot easily use humans as an experimental model. This is true both in the setting of studies related to the analysis of the human immune response to acute infection as well as to the analysis of immune responses during the progression to AIDS. Even after diagnosis, it is only possible to study the effects of virus on the immune system indirectly by sampling peripheral blood; the interactions which occur between HIV and cells differentiating through fixed microenvironments (e.g., bone marrow, thymus, lymph node, spleen, skin) can only be inferred. Thus, in the absence of a suitable experimental model system for the analysis of HIV infection *in vivo*, many questions are raised without the chance for an answer.

A number of animal models have been advanced. All are approximations with severe limitations: either an animal retrovirus other than HIV is used to infect a host other than man (e.g., Cheevers and McGuire, 1985; Daniel *et al.*, 1985, Haase, 1986, Mullins *et al.*, 1986) or HIV is used in a context beyond that of its natural tropism (e.g., Alter *et al.*, 1984; Francis *et al.*, 1984; Letvin *et al.*, 1987; Felice *et al.*, 1988; Kulaga *et al.*, 1989). At this time, no model precisely fits the pattern of selective CD4$^+$ T-cell deficiency and of the development of opportunistic infections, maligancies, and encephalopathies associated with HIV infection.

II. The Optimal Animal Model for HIV Infection

If the urgent need for an animal model system to study HIV pathogenesis is clear, it should also be apparent that the construction of any model will be met, *a priori*, with a host of confounding obstacles. The work to be described in this review was started in early 1987 with a simple outline for an "optimal small animal model." The following characteristics were thought to be of importance.

1. The animal model would re-create the pathogenesis of HIV infection in a genetically defined animal that is available in large numbers, that is easy to breed, and that displays known markers for MHC and subsets of cells involved in the immune response. Given the wealth of information already available about the murine immune system, the obvious first choice would be a mouse.

2. The virus to be used for infection would be HIV, not a surrogate virus from another species (e.g., a small animal lentiretrovirus) or a recombinant virus of uncertain relevance. Preferably, the HIV challenge would be an infectious molecularly cloned isolate which was not adapted to growth conditions *in vitro*. Isolates which have been so passaged might have been selected in both phenotype and genotype to differ markedly from those HIV isolates that are pathogenic *in vivo* (Meyerhans *et al.*, 1989; Hirsch *et al.*, 1989).

3. There should be no known restriction to the replication and spread of HIV infection; the cells, therefore, that are initially infected must be human and there should be uninfected human cells around to support further cycles of infection. Given the tissue, cell type, and species tropism so common amongst all retroviruses, any other situation would likely present multiple levels of restriction. For instance, the mere presentation of human CD4 on mouse cells is not sufficient to permit viral entry and/or intracellular replication; other human-specific factors may also be required (Maddon *et al.*, 1986).

4. Finally, the cells that are human should be those that are likely to be infected *in vivo* in man, namely, the cells of the hematolymphoid organs (e.g., the fetal liver, bone marrow, thymus, lymph node, spleen, and skin). Although most studies of HIV infection in man have focused on surrogate markers in the peripheral blood, it is clear from histologic analyses that these organs are also infected (Ioachim *et al.*, 1983; Armstrong *et al.*, 1984; Biberfeld *et al.*, 1986; Le Tourneau *et al.*, 1986; Pekovic *et al.*, 1987). Since they are critical to the regeneration, maintenance, and coordination of hematopoietic cells, observation of infection within them may be of even greater relevance than studies done on peripheral blood cells. Of the subpopulations of cells present in these organs, many are interactive, most cannot be grown *in vitro*, and all are likely to be affected, directly or indirectly, by HIV infection.

In short, it appeared optimal to create a situation in which HIV can be observed to infect human hematolymphoid organ systems within the confines of a mouse.

III. The SCID-hu Mouse

The SCID-hu mouse (Mc Cune *et al.*, 1988) was constructed precisely for this purpose. The rationale for the experiment was based upon several key concepts in immunology. First, during fetal life, tolerance to "self" is predicated by the environment in which the developing immune system resides. Thus, in the classic experiments of Billingham, Brent, and Medawar, the introduction of A strain tissue into the developing embryos of CBA mice permitted the resultant offspring to be tolerant of CBA as well as A strain skin grafts (Billingham *et al.*, 1953). Secondly, self—nonself education of T cells occurs primarily in the thymus. Thus, MHC restriction of the developing T cell lineage is dictated by the thymic microenvironment in which T progenitor cells initially differentiate (Zinkernagel *et al.*, 1978a,b). Finally, after T cells are mature, self recognition is fixed: MHC-disparate cells are rejected.

We reasoned that human T cell differentiation might proceed in a physiologic fashion, if human stem cells and human thymus were used and if they were grown within a mouse unable to reject them. We therefore sought to use a mutant mouse strain which bears a specific defect in T-cell maturation, in hopes that the absence of murine T cells would facilitate the engraftment of human cells.

The C.B-17 *scid/scid* stock lacks functional T cells and B cells (Bosma *et al.*, 1983; Dorshkind *et al.*, 1984; Custer *et al.*, 1985). Homozygotes have severe combined immunodeficiency (SCID): an inability to mount an effective cellular response or an effective antibody response against foreign antigens. This defect resides in a qualitative and/or quantitative defect in a recombinase-associated function: T and B cell antigen receptors do not rearrange as they normally would during differentiation (Schuler *et al.*, 1986; Malynn *et al.*, 1988; Hendrickson *et al.*, 1988). Although they are otherwise replete with all other hematopoietic lineages (Czitrom *et al.*, 1985; Dorshkind *et al.*, 1985), SCID mice die at an early age. Often, they succumb to the same opportunistic infection which afflicts many patients with AIDS, namely *Pneumocystis carinii* pneumonia (Shultz and Sidman, 1987; Shultz *et al.*, 1989). The hematopoietic microenvironment of the SCID mouse (including the rudimentary thymic stroma) is intact. Engrafted with syngeneic bone marrow, reconstitution with functional T cells and B cells can occur. We reasoned that these mice would accept hematolymphoid organs from man and that human progenitor cells might thereafter give rise to mature human T and B cells circulating in the periphery. We wondered whether such cells would functionally complement the defect of the SCID and thereby protect the mice from the acquisition of opportunistic infection.

To reconstitute SCID mice with human hematopoietic cells, a source of progenitor cells is required. After careful subfractionation, 0.05% of mouse bone marrow cells represent a pluripotent stem cell source. After intravenous injection of 20–30 of these cells, 50% of lethally irradiated mice are reconstituted (Spangrude *et al.*, 1988). In humans, donor bone marrow depleted of mature T cells may also be used to reconstitute immunodeficient recipients; in the absence of such depletion, graft-versus-host (GvH) reactions can occur. The human fetal liver is a major site of hematopoiesis between 8 and 24 weeks of gestation (Namikawa *et al.*, 1986) and thus represents an alternative source of hematopoietic stem cells. All lineages are derived from these stem cells, including the erythroid, the myelomonocytic, and the lymphoid. Most significantly, since the T-cell progenitors have not migrated to and through the fetal thymus, they are not yet committed to a given pattern of self–nonself recognition. Therefore, GvH reactions are less likely to occur upon transfer to a genetically disparate recipient (Touraine, 1983).

Thus, an animal model for the evaluation of human hematolymphoid differentiation is formed by the engraftment of human fetal liver, human fetal thymus, and human fetal lymph node into the C.B-17 *scid/scid* mouse, producing the hematochimeric SCID-hu mouse.

IV. Construction of the SCID-hu Mouse

This work was started with the intent to create a model which is easily and reproducibly constructed and in which data can be gathered in a statistically meaningful fashion. For some applications (e.g., analysis of antiviral compounds against HIV; see Section VI,B), that goal has been achieved; for others (e.g., implantation of a fully functioning human hematopoietic system), there is still much to be done. It was clear from the outset that there were many variables associated with the construction of SCID-hu mice and many potential artifacts in their analysis. Some of these can be controlled; others cannot. Therefore, the key to any use of the SCID-hu mouse is that a reproducible and sensitive assay be identified for the purpose of observing a given biologic function within many animals at once.

Although it is beyond the scope of this review to provide detailed methods for the construction of SCID-hu mice (see Mc Cune *et al.*, 1988), the following points are of importance.

A. Animal Husbandry

The C.B-17 *scid/scid* strain of mice is susceptible to a variety of opportunistic infections, including that caused by *Pneumocystis carinii*. Under optimal conditions, breeding pairs are obtained by cesarean derivation and maintained in germ-free conditions. Alternatively, animals may be treated with prophylactic trimethoprim-sulfamethoxasole, either in the drinking water or within specially-formulated food pellets (Mc Cune *et al.*, 1988; Shultz *et al.*, 1989). The mice must be at least 5–6 weeks old before use in SCID-hu procedures: prior to that age, the renal capsule is too fragile to support implants.

B. Tissue Collection

Although human fetal tissue has long been used for research and commercial purposes (e.g, vaccine development), extraordinary care must be taken to ensure that its collection is performed in an ethical manner. Only after the decision to have an abortion has been made should consent for donation be obtained. This signed consent must acknowledge that the tissue will be used for research purposes and that it may be tested for infectious agents. Where appropriate, the consent should also include permission to use the tissue for research which might later have commercial applications. The donation and collection of tissue should not in any way change the way in which the procedure is performed or the standard practices of the clinic.

Once obtained, the tissue should be coded so that its source cannot be identified later. Usually, the gestational age of the tissue is recorded under this code; all other information about the tissue is destroyed. The tissues should be processed as quickly as possible. Fragments of tissue may be stored overnight (or even frozen) and used successfully later; it appears, however, that fresh tissue is superior. In general, the less preparation the better: all efforts should be made to preserve organ structure prior to and during implantation.

Given the fact that human fetal tissue could be infected with any number of infectious agents, including HIV, universal precautions should be in force at all times and all personnel should be vaccinated with the hepatitis B vaccine. It is also prudent to test all specimens for HIV using a sensitive and reproducible assay, e.g., the polymerase chain reaction (PCR). If test results are positive, the tissue should be quarantined into a Biosafety Level 3 (BSL3) facility or destroyed.

C. Analysis

After implantation, the success of the procedure may be determined by a variety of techniques. The implantation site can be biopsied for histologic study; peripheral blood can be prepared for analysis by flow cytometry (FACS) or PCR; serum can be assayed for human proteins, e.g., immunoglobulins. In all cases, multiple layers of positive and negative controls must be applied to rule out the possibility of artifact. This is especially the case for analyses using antibodies against human cell-surface proteins: many of these antibodies have never been assayed for specificity on mixed populations of murine and human cells. Indeed, attempts to reproduce the original published results of Mosier (1988) have run into this problem: after intraperitoneal injection of human peripheral blood cells into SCID mice, most of the cells thought to be human were in fact murine (Bankert *et al.*, 1989; Mosier *et al.*, 1989).

The most precise analysis of human cells in the peripheral circulation of the SCID-hu mouse is made by triple-color flow cytometry using directly conjugated antibodies. Fluorescein isothiocyanate-conjugated HLe-1 (anti-CD45) can be used to identify all nucleated human peripheral cells; Texas Red-conjugated OKT4 (anti-CD4) will identify the fraction of $CD4^+$ T cells within the gated population of $CD45^+$ cells; nonviable cells can be gated out with propidium iodide. Murine cells should not, of course, stain with these reagents, but the following potential artifacts should be kept in mind: (1) "leaky" SCID mice with peripheral murine B cells will show reactions with fluorochrome-conjugated anti-mouse Ig (second-stage) reagents; and (2) immune complexes or aggregates in the staining reagents may be internalized or bound by subpopulations of murine cells. To avoid these problems, it is best to always have a normal SCID control, isotype-matched antibody controls, and well-defined antibody reagents. In the final analysis, a second layer of nonimmunologic proof is comforting, e.g., a nucleic acid probe specific for human DNA (the Alu repeat) or PCR oligo primers that are human specific (Mc Cune et al., 1988; Mc Cune et al., 1990c).

D. Assay for Function

Different assays will be required for different purposes. In their design, the following points should be remembered. First, potential variables (source of tissue, age of tissue, extent of reconstitution, etc.) must be controlled. Populations of animals or of cells must be analyzed with statistical precision before statements of confidence are declared. Secondly, there may be limiting numbers of human cells that could take part in some functions (e.g., clonally dispersed B cells capable of generating antibody against a given epitope). Therefore, sensitive assays must be used. In general, it is best to initiate experiments with large numbers of animals, covering as many potential variables as possible. Once a qualitative result is obtained, efforts can be expended to make the associated techniques reproducible. Then, and only then, is it possible to quantitate the events.

E. Maintenance of Human Organ Structures and Hematopoiesis

The concept of the SCID-hu is relatively simple. If human organ structures can be implanted in such a fashion that they are vascularized, then they may grow. With growth, the intricate relationships which normally occur between cells therein may be preserved. If organs that are normally related (e.g., bone marrow, thymus; thymus, lymph node) can be implanted successfully into the same animal, then movement of cells between organs may also occur. If obtained, this situation is vastly different from a tissue culture flask: organs (and most eukaryotic cells) do not easily grow in vitro.

A number of human hematolymphoid organs have been surgically implanted into the SCID mouse, in a variety of locations and with a variety of techniques. It is clear from the experience of the past several years that a multiplicity of variables contribute to subtle and sometimes major differences in the success rate of implantation, vascularization, and eventual function. The less controllable variables are inclusive of the gestational age of the tissue, the condition of the tissue at the time of collection, and the duration of time between collection and implantation. Empiric experimentation has offered some insight into the control of other variables. It must be stressed that these methods will not yield 100% success rates; therefore, sufficiently large numbers of animals must be analyzed in any experiment so that statistically meaningful statements might be made. In general, the following observations hold true for the implantation of specific organs.

1. Thymus Introduction of fragments of thymus beneath the fibrous kidney capsule is a time-honored technique, borrowed in initial constructions of the SCID-hu mouse (Mc Cune *et al.*, 1988). SCID-hu mice have been constructed with human thymus ranging in gestational age from 9 to 22 weeks. Once transplanted, the organ is found to be vascularized and to grow with great reproducibility (80–90% take rates). Some 4–8 weeks after engraftment, biopsy sections show a morphology which is quite similar to that of normal human fetal thymus; immunohistochemical stains reveal a normal distribution of cells expressing the markers CD3, CD4, CD8, CDR2, and MD1. Using flow cytometry, subpopulations present during normal thymic development are also found in the SCID-hu thymus, including cells double-positive, double-negative, and single-positive for CD4 and CD8. Histologic stains for human MHC Class I antigens reveal high expression on medullary thymocytes. Cortical thymocytes do not stain with high intensity, but MHC Class I$^+$ interdigitating cells are delineated in the cortex. These cells are of bone-marrow origin and myelomonocytic lineage. To date, the only difference observed between the SCID-hu thymus and a normal human fetal thymus relates to the murine counterpart of this latter cell type: interdigitating cells which stain with antibodies against murine MHC (H-2d) are also found in the cortex of the engrafted human thymus. If these cells are involved in the processes by which developing T cells become tolerant of or restricted to self-MHC antigens, it is possible that the human T cells of the SCID-hu are tolerized/restricted not simply to human, but also to mouse (H-2d) MHC alloantigens (see Section VII,A).

When human fetal thymus is implanted beneath the kidney capsule and no other manipulations are performed, the organ frequently becomes stromal in nature, particularly in the cortical compartment. It might be inferred that early T cell progenitors in the cortex are following their normal developmental program, differentiating through to the medulla and then out. In the absence of a source of replenishment, the cortex and medulla eventually become depleted of thymocytes. Less frequently, the isolated thymus continues to grow and to maintain a richly populated cortical compartment.

2. Progenitor Cells When human fetal liver is teased apart, pluripotent hematopoietic stem cells can be isolated in a heterogeneous fraction of mononuclear cells ("fetal liver cells"); the stromal microenvironment which supports hematopoiesis is left behind intact. In early experiments, dispersed fetal liver cells were injected intravenously (i.v.) into SCID-hu mice that had been previously engrafted with human fetal thymus. Thereafter, in a time-dependent and time-limited fashion, circulating human T cells were observed in the periphery. We reasoned that the fetal liver mononuclear cell preparation might give rise to (and/or contain) T-restricted progenitors, that these cells might home to and differentiate through the thymus and thereafter circulate as mature CD4$^+$ or CD8$^+$ T cells. In support of this hypothesis, donor fetal liver cells were typed for MHC Class I alleles and administered to SCID-hu recipients implanted with MHC-disparate thymic implants. Using the MHC marker to trace cell movement, both T- and myelomonocytic-lineage cells could be directly shown to differentiate from progenitor cells in the SCID-hu mouse (Mc Cune *et al.*, 1988). Human T cells were found in the peripheral blood; mature myelomonocytic cells were found in the parenchyma of the human thymus. After 3–4 months, the human T cells had usually disappeared. At such a time, however, the human organ systems (e.g., the thymus) of the mouse were still intact and functioning: a second administration of fetal liver cells resulted in a second (transient) wave of human T cells in the periphery.

It stood to reason that human T cells in the peripheral circulation might disappear because (a) they were being destroyed, (b) they were not being self-renewed, and/or (c) they were simply moving out of the peripheral circulation and going elsewhere. All of these possibilities may be operative coincidentally; all are being explored (Mc Cune *et al.*, 1989). Reasoning that the stromal cells of the fetal liver might be necessary for hematopoiesis, we have attempted to enhance self-renewal of progenitor cells in SCID-hu mice by implanting whole fetal liver. Experiments of this design began several years ago and have been recently analyzed (Namikawa *et al.*, 1990a; Weilbaecher *et al.*, 1990). It has been found that long-term multilineage human hematopoiesis can occur in the SCID-hu mouse when fragments of human fetal liver are transplanted adjacent to fragments of human fetal thymus ("Thy/Liv" implants). More than half the animals transplanted in this fashion have circulating human T cells for 5–11 (and sometimes, as long as 15) months; at the same time, the cortical (immature) thymocyte compartment is maintained. These observations are consistent with long-term survival and differentiation of human T cell progenitors. For the same period of time, Thy/Liv implants are found to contain progenitor cells to the myeloid and erythroid lineages. In some cases, areas indistinguishable from normal human bone marrow are found to form *in vivo*, containing blast cells, immature and mature myeloid cells, and megakaryocytes. In sum, it is now possible to readily implant combined human organ structures in such a way that multilineage hematopoietic human stem cells can be maintained and allowed to differentiate within a mouse.

3. Lymph Node Early studies demonstrated that human fetal lymph nodes could be implanted beneath the kidney capsule of the SCID mouse. The take rate for such implants was considerably lower than that for thymic implants (15–20% versus 80–90%). When biopsied, growing lymph nodes demonstrated a morphology devoid of primary follicles or T-cell areas. Histologic stains for human immunoglobulin did, however, demonstrate the presence of plasma cells within the lymph nodes; in parallel, high levels of circulating IgG and IgM could be found in the plasma.

Given the importance of the lymph node to immune functions and to the analysis of infection by HIV, an extensive series of experiments was carried out to establish conditions for optimal lymph node growth. The results of these experiments indicate that reproducible transplantation of human lymph node into SCID mice is possible. Take rates exceed 95%. In many cases, subanatomic structures such as primary follicles develop. A current limitation is that the nodes do not last for indefinite periods of time.

V. Infection of the SCID-hu Mouse with HIV

A. General Considerations

To date, the analysis of HIV infection in the laboratory setting has been largely confined to infection of either continuously growing cell lines or of mitogen-activated peripheral blood cells in tissue culture flasks. These conditions ensure that infectious virions will enter, integrate, and replicate within cells that are moving through the cell cycle. Indeed, most assays for HIV infection (e.g., syncytium formation, reverse transcriptase assays, supernatant p24 levels) only detect virus which is actively replicating. The virus isolates which have been used most extensively are generally infectious molecular clones (e.g., HTLV-IIIB, LAV, ARV) which have been propagated in lymphoid cell lines for years. In general, these isolates replicate to high levels in such lines and are cytopathic; it is these characteristics which brought them to notice in the first place.

There are three problems inherent with such analyses (Mc Cune *et al.*, 1991).

1. Many cell types which can be shown to be infected by HIV *in vivo* (e.g., monocytes, antigen-presenting cells, neurons) can only be grown with difficulty, if at all, *in vitro*.

2. Some cells (e.g., CD4$^+$ T cells) can be grown and infected by HIV *in vivo* and *in vitro*, but the analysis is skewed depending upon where one

looks. In vivo, most CD4$^+$ T cells are resting in G_0; *in vitro*, by definition of the culture conditions, most are cycling. Therefore, although steps associated with viral replication can be observed *in vitro*, those associated with viral latency (e.g., in a resting cell) cannot.

3. The infectious molecular clones studied *in vitro* have been selected, in part, because they replicate *in vitro*. It is clear that retroviruses mutate. By some estimates, the error rate in reverse transcription approaches 1 base/10(4) (Dougherty and Temin,1988; Leider *et al.*, 1988); in other words, for the 10 kB HIV genome, a distinct genome may arise with each cycle of viral infection. It is also obvious that the selective pressures of tissue culture conditions *in vitro* are likely to differ from those operative *in vivo*. Formally, therefore, it is possible that the molecular clones studied *in vitro* are different from those HIV isolates which cause disease *in vivo* (Mayerhans *et al.*, 1989; Hirsch *et al.*, 1989).

The rationale of the SCID-hu mouse is that these pitfalls may be minimized if human hematolymphoid organ structures are instead infected with isolates of HIV which have not been adapted to growth *in vitro*. These organ structures, when vascularized, support a vast array of different cell types, in different cell lineages and in different states of cell activation. In the thymus, for instance, HIV may infect a CD4$^+$ CD8$^+$ cortical thymocyte, actively cycling and en route to a differentiative step; thereafter, the integrated genome might be found in resting (G_0) CD4$^+$ medullary thymocytes, progeny of the first infected cell. Notably, neither the parental nor the progeny cell type can be maintained or observed to differentiate *in vitro*.

HIV has been shown to infect bone marrow cells, thymus, lymph node, and skin of man; as detailed above, all of these organs can be grown in the SCID-hu. Therefore, on one level, the SCID-hu may be used immediately to analyze the process by which HIV infects these organs. All cell subpopulations within them can be observed, not just a subset of those that can be grown *in vitro*. In this sense, the SCID-hu can be properly viewed as an *in vivo* system for the vascularization and long-term maintenance of interactive human hematolymphoid organs. This makes it quite distinct from the "hu-PBL-SCID" mouse (Mosier *et al.*, 1988), in which human peripheral blood cells are injected into the peritoneal cavity of a SCID mouse without the supportive stroma of human lymphoid organs.

On another level, the human hematolymphoid organs engrafted into the SCID-hu normally interact with one another to generate mature lymphocytes from immature progenitors and to orchestrate an immune response against antigen. If such interactions can be formed in the SCID-hu, then the analysis of infection with HIV may be even more comprehensive. The ability of the virus to interfere with hematopoiesis (in the bone marrow), with thymocyte maturation (in the thymus), and with antigen presentation (in the lymph node or skin) might be explored. The pathophysiology of HIV infection known to exist in man could then be systematically studied in a mouse. Further, if the human lymphocytes of the SCID-hu could be shown to be functional, immune responses to HIV could be examined. Studies of this sort might shed light on the immunopathogenesis of HIV infection; they may also contribute to an analysis of vaccines against HIV.

Our initial experiments have sought to define conditions in which infection of SCID-hu mice with HIV can be reproducibly obtained. We have therefore developed a set of experimental conditions in which certain questions related to HIV infection can now be approached *in vivo*.

B. Virus Isolates

Our studies have initially focused on two isolates of HIV, JR-CSF, and JR-FL (Koyanagi *et al.*, 1987). These isolates were derived from the cerebrospinal fluid or frontal lobe of a patient with AIDS encephalopathy and Kaposi's sarcoma. They were cultivated for a period of 2 weeks in phytohemagglutinin (PHA)-activated T cell blasts and then directly cloned into phage lambda. They have not been adapted to long-term culture *in vitro*. JR-CSF has been reduced to an infectious molecular clone; it is T-tropic and nonlytic. JR-FL is a biologic isolate; it is T and monocytotropic and also nonlytic.

These isolates are prepared as large stocks under standardized conditions, quantitated with respect to p24 levels and biologic activity *in vitro* (using tissue culture infectious dose [TCID]$_{50}$ determinations in PHA blasts) and then frozen as aliquots for future use in SCID-hu mice. It is important that when used, standardized procedures are also employed: free virus at room temperature has a finite half-life. Therefore, stock virus should be thawed and virus infected in a similar fashion each time. Under current practice, all procedures are carried out under BSL3 precautions

C. Systems for Infection

Initially, SCID-hu mice (implanted with human fetal liver, thymus, and lymph node) were infected with varying doses of JR-CSF, either as free virus or cell-associated, by a variety of routes: intravenously, intraperitoneally, and by direct intraorgan injection. Evidence of infection was later sought with a broad range of techniques. Biopsy specimens from infected human organs were analyzed by histology, immunohistochemistry, *in situ* hybridization, and PCR (for viral RNA or DNA). Peripheral blood cells were assayed for virus with the same techniques. Virus in plasma was additionally screened using RNA PCR and p24 enzyme-linked immunosorbent assay (ELISA). Finally, all tissue and blood specimens were cocultivated with PHA blasts *in vitro* in an attempt to isolate virus. In general, the cumulative data presented two distinct systems for analysis of HIV infection in SCID-hu mice.

1. Intrathymic Injection   Intrathymic or intranodal injection of free virus results in viral replication and spread (Namikawa *et al.*, 1988). Cells producing virus can be visualized in both the cortex and the medulla of the thymus with immunohistochemical stains for viral structural proteins; progeny are also detectable by p24 ELISA and by subsequent isolation *in vitro*. By in situ hybridization, a 35S-labeled probe to the 3' end of the viral genomic transcript (which hybridizes to all viral mRNAs) detects cortical and medullary thymocytes. Interestingly, when immunohistochemical stains for viral protein are combined with *in situ* hybridization for viral RNA, some cells (especially those found in the cortex) are found to harbor viral RNA species in the absence of detectable viral structural proteins. This observation is possibly related to a quantitative phenomenon: some cells may have more viral structural protein than others. Alternatively, it could reflect a qualitative phenomenon associated with "latency." In some cells, for instance, viral transcripts detectable by *in situ* hybridization may be associated with down-regulation of the production of viral structural proteins.

The above events are reproducible, time-dependent, and dose-dependent. Two weeks after the intrathymic injection of 1600 $TCID_{50}$s of JR-CSF or JR-FL, 100% of thymic biopsy specimens show evidence of infection by HIV. As time proceeds, progressively more cells are positive by *in situ* hybridization. As the dose is decreased, fewer cells are infected at a given time point until, finally, at a dose of 16 $TCID_{50}$s, no sign of infection is later observed.

2. Intravenous Injection   When SCID-hu mice are injected with HIV by the intravenous or intraperitoneal route, the engrafted human lymph node is infected; the human thymus implant is not (Kaneshima *et al.*, 1990, 1991). In biopsy specimens taken from the lymph node, infected cells can be detected by immunohistochemistry, *in situ* hybridization, and p24 ELISA; infectious virus can also be recovered by cocultivation with PHA-blasts *in vitro*. The cells that are infected in the human lymphoid organs include CD4$^+$ T as well as myeloid cells (Namikawa *et al.*, 1990b). Human organ implants without CD4$^+$ cells as well as mouse organs are not infected by HIV in the SCID-hu mouse. Therefore, at least within the limits of detection by PCR, expansion of host range as a result of phenotypic mixing or pseudotype formation with endogenous murine retroviruses does not occur (Lusso *et al.*, 1990; Mc Cune *et al.*, 1990a).

After infection with HIV, SCID-hu mice become viremic in a time-dependent manner. Three days after infection, none are viremic; 2 weeks after infection, greater than 95% are viremic. In the small number of animals that are not viremic, the human lymphoid organs *are* infected. Using an

RNA PCR assay to quantitate free virus, the circulating load in HIV-infected SCID-hu mice is not widely varying (Shih *et al.*, 1990).

Like the assay using intrathymic injection, intravenous infection of SCID-hu mice with HIV is dose-dependent. For example, 120,000 $TCID_{50}$s of JR-CSF are required to result in infection of 100% of SCID-hu mice within 2 weeks; for other fresh patient isolates, input doses as low as 1000 $TCID_{50}$s achieve the same result. The variation in dose from one fresh patient isolate to another suggests that there are differences in infectivity that are intrinsic to the virus. In all cases, relatively high input doses are used for the following reasons: (a) the virus is rapidly cleared after intravenous injection (e.g., by passage through the murine lung, liver, and spleen); (b) the "target size" for HIV infection in the SCID-hu mouse is small (several cubic millimeters of human lymphoid organ implant); and (c) for the purpose of general utility (see Section VI,B), we demand a stringent end point: 100% infection after 2 weeks.

Infection of the SCID-hu mouse with HIV is not only time- and dose-dependent but also strain dependent. Thus, a striking dichotomy has been found between isolates of HIV taken directly from patients and those which have been adapted to tissue culture growth *in vitro* (Rabin *et al.*, 1990). All of the "patient isolates" (including JR-CSF and JR-FL) have been observed to replicate in the human organs of the SCID-hu. In contrast, none of the lab isolates (including IIIb, MN, and RF) do so, even when inoculated at doses as high as $10(6)$ $TCID_{50}$s intravenously or directly into the human lymphoid grafts. It is possible that these viruses, like others that have been propagated *in vitro* for long periods of time, have met selective pressures that are different from those which exist *in vivo*. Those which have survived *in vitro* may be dissimilar to those which are pathogenic in humans.

In sum, intravenous infection of SCID-hu mice with HIV provides a novel system for the analysis of HIV. Infection of human cells can be observed *in vivo*, with a tropism that is apparently unchanged. Infectious isolates have not been adapted to tissue culture growth. Many features of the virus infectious cycle are encompassed, from entry into the bloodstream, to passage across endothelia into the parenchyma of a lymphoid organ, to infection of a permissive $CD4^+$ cell therein, to integration, replication, and eventual viral release. Finally, the assay system is highly reproducible, easily constructed, and safely used.

VI. Current Applications of the SCID-hu Mouse to the Analysis of HIV Infection in Humans

The observations cited above describe the construction of several forms of SCID-hu mice, one with implanted thymus and one with implanted lymph node, each of which results in reproducible levels of HIV infection (100% of animals) after a short period of time (2 weeks). These systems permit the evaluation of questions which could not be previously approached.

A. Tropism

Many isolates of HIV have been described by *in vitro* analysis to be tropic for T cells and/or monocytes; there is also evidence to suggest that some isolates of HIV may be tropic for cells of different lineages as well. Often, such studies are dependent upon the use of cell lines which grow *in vitro*. Acute infection of the SCID-hu mouse with these isolates permits comparison of tropism as it is defined *in vitro* to tropism as it is reflected within the heterogeneous cell populations which exist in the engrafted human organs. After infection, the cell types permissive for infection by a given isolate might be identified by simultaneous immunohistochemistry (e.g., for a cell surface marker)/*in situ* hybridization (probing for HIV mRNA); they may also be isolated by multiparameter flow cytometry and subsequently examined for signs of HIV infection.

Perhaps most important for an understanding of the pathogenesis of HIV infection, the SCID-hu will provide an experimental construct in which to evaluate the tropism of HIV for human hematopoietic progenitor cells (Weilbaecher *et al.*, 1990). These cells, inclusive of both pluripotent as well as T-restricted cells, cannot be studied elsewhere now.

B. Analysis of Antiviral Compounds

The standard pathway for development of a therapeutic compound involves early discovery work *in vitro* followed by progressive steps through animal models to man. Usually, moving a compound from the laboratory to the clinic necessitates a series of decision trees of ever-increasing importance: is the drug toxic? is it efficacious? in what dosing regimen? is a related congener more efficacious, or less toxic, or both? For HIV (and other viruses causing disease in man), the absence of suitable animal models often means that compounds move quickly through to clinical trials, usually before these questions have adequate preliminary answers. As a result, some compounds are later found to be toxic in man, others are found not to be efficacious, and perhaps most importantly, some potentially useful compounds are left at the bench and never tried at all.

The reproducible systems for intrathymic and intravenous infection of SCID-hu mice with HIV can be easily tapped for the rational analysis of experimental antiviral compounds *in vivo* (Mc Cune, 1990b). This analysis can proceed in the setting of HIV infection of human organs, within a small animal. Large numbers of animals can therefore be studied at once to understand multiple and independent attributes of a given antiviral compound, including: bioavailability, distribution, the ability to enter human lymphoid organs, and efficacy against HIV *in vivo*. Careful dose–response experiments can also be constructed to gain an appreciation for the appropriate dose that might be efficacious against HIV infection in man. This last point is of particular importance when comparing the SCID-hu system to other animal models for HIV infection. The nucleoside analogues AZT and ddI are, for instance, metabolized inside human cells and must be triphosphorylated prior to the time that they can be active against HIV. In the SCID-hu, it will not only be possible to ask whether a given human cell is infected by HIV, but also whether or not it might appropriately metabolize such antiviral compounds.

We have initiated experiments of this type using AZT in the setting of SCID-hu mice infected intrathymically with HIV (Mc Cune *et al.*, 1990c). Two weeks after infection with HIV, 100% (40/40) of untreated animals were found to be positive for HIV by DNA PCR analysis of infected thymic tissue. When treated first with AZT (200 mg/kg/day in the drinking water, for 1 day prior to infection and for 2 weeks thereafter), no SCID-hu mice (0/17) were HIV-positive by this assay. Qualitatively, at least, suppression was complete; the known efficacy of AZT against HIV in man was reproduced in the SCID-hu. On closer analysis, however, infection was not completely blocked. When thymus sections of AZT-treated, HIV-infected SCID-hu mice were observed by *in situ* hybridization, rare infected cells could be seen 2 weeks post-infection. If AZT treatment was discontinued at that time, spread of infection could be demonstrated 4 weeks later by DNA PCR assays for HIV and by *in situ* hybridization. The nature of those cells that are infected in the presence of AZT is currently under investigation. It is possible that higher doses of AZT or, alternatively, another antiviral agent might inhibit HIV infection of them. If so, the SCID-hu might be a useful system in which to demonstrate improvement of antiviral efficacy.

Using the assay system for intravenous infection, these studies have been expanded to include other antiviral compounds (Kaneshima *et al.*, 1990, 1991; Mc Cune *et al.*, 1990d). In a dose-dependent manner, both AZT and dideoxyinosine have been shown to suppress HIV infection of human lymph nodes in the SCID-hu mouse when given 1 day prior to infection. Strikingly, the doses of these drugs that are found to be effective in the SCID-hu are very similar to equivalent doses used in the clinic in humans. Given the fact that the end points to analysis occur at 2 weeks postinfection, statistically meaningful data can be quickly derived. It is possible that this system will prove useful for the rapid screening of multiple experimental antiviral compounds, for estimation of efficacious dose ranges, and perhaps more importantly, for the evaluation of synergy in various combination therapies.

In addition, the SCID-hu mouse has recently been used to evaluate the efficacy of postexposure administration of AZT (Shih *et al.*, 1991). These studies, which are difficult if not impossible to carry out in the clinic, indicate that HIV replication can be suppressed *in vivo* if AZT is administered promptly after exposure.

C. Analysis of Vaccine Candidates

The formulation of a vaccine against a given infectious agent requires two basic ingredients: first, a knowledge of those epitopes which should be included, and secondly, information about the adjuvant, route, dose, and timing of boosts necessary to provoke a protective set of T-cell and/or B-cell responses against those epitopes. The first of these tasks can now be carried out in the HIV-infected SCID-hu mouse. In the same fashion that AZT can be given prophylactically and shown to suppress infection, putative neutralizing antibodies might also be given to determine whether they block or alter the course of infection *in vivo*. The neutralizing B cell epitopes of HIV *in vivo* might thereby be mapped. It is also clear that cytolytic T lymphocytes (CTL) can be elicited during the course of HIV infection; these cells are MHC-restricted and specific for given viral epitopes. Using adoptive transfer of CTL clones, it is now possible to construct experiments with MHC-defined, HIV-infected SCID-hu mice to ask the questions: can CTLs inhibit or suppress infection? If so, which epitopes do they recognize in the setting of which MHC types? These experiments might contribute to the identification of T-cell epitopes recognized by a protective CTL response.

VII. Potential Improvements of the SCID-hu Mouse for the Analysis of HIV Infection in Humans

A. Demonstration of Human Immune Function

The applications cited above represent important problems that can now be addressed using the HIV-infected SCID-hu mouse. Our goal, however, is to create a SCID-hu mouse in which the pathogenesis of HIV infection can be fully dissected. It is important to be able to understand the time course of events after infection, the immune response to infection (which responses are protective and which are not?), patterns of viral replication in the face of an immune response (does latency ensue? are ever-more virulent strains selected?), and the conditions that ultimately lead to the collapse of the immune system. Once understood, greater insight might be gained enabling development of ever more effective treatment modalities in man.

In large part, this goal is dependent upon improvement of two limiting features of the SCID-hu mouse. On the one hand, the extent of repopulation of the mouse by human hematopoietic cells must be higher, more consistent, and of longer duration. Progress is being made in this direction (see above, Section IV,E,2). On the other hand, the function of the human immune system within the animal must be understood and amenable to manipulation.

At the outset of these experiments, 21 mice were obtained from a breeding colony at Jackson Laboratory. Ten were kept in a clean room as controls. Within 4 months, all of these mice had died of a wasting disease consistent with *Pneumocystis carinii* pneumonia. Eleven mice received implants of human fetal thymus and an intravenous bolus of human fetal liver cells. All of these animals survived. This observation suggested that the provision of certain components of the human immune system to the SCID mouse might protect it, directly or indirectly, from the acquisition of opportunistic infection.

Since that time, the following observations have been made about immune function in the SCID-hu mouse.

1. No sign of graft-versus-host disease is seen: the CD4/CD8 ratio in the peripheral blood is normal; the human T cells do not express activation antigens (e.g., CD25, CD69) (Krowka et al., 1990, 1991); human T cells are not observed in the murine skin, near the murine hepatobiliary tree, or in the murine lymph node or thymus (Namikawa et al., 1990).

2. The T cells circulating in the SCID-hu are, however, phenotypically and functionally similar to those found in humans (Krowka et al., 1990, 1991). They express the heterodimeric alpha/beta T cell receptor and are responsive to exogenous signals such as phytohemagglutinin, antibodies against CD3, and alloantigens.

3. It has long been known that the human B cells of the lymph node implants could differentiate and class switch to IgG-secreting plasma cells, presumably in response to unknown, environmental antigens (Mc Cune et al., 1988). Human lymph nodes can now be implanted with greater reproducibility and preliminary data indicate that human antibody responses can now be elicited against known antigens, introduced into the SCID-hu mouse intentionally.

In much the same way that empiric experimentation led to established protocols for immunizatin of rabbits, a multiple of variables is now being approached in the SCID-hu. It should be clear within a reasonable period of time whether or not human immune functions can be elicited in a practical manner. If it is possible to study normal function, then it is but a step to ask questions about active immunization against HIV (which B- and/or T-cell epitopes provoke protective responses? by which route should they be given? with which adjuvant? etc.) It should also be possible to study immunopathogenic events which occur after HIV infection.

B. Analysis of Multiple Viral Infections

The reality of the clinical setting is that patients with HIV infection are often infected with other viruses as well. In some cases (e.g., hepatitis, cytomegalovirus), these viruses may actually synergize with HIV to accelerate disease progression. Ultimately, it is not unlikely that multiple combinations of antiviral compounds will be administered against multiple combinations of viruses, all within a single patient. The SCID-hu mouse could potentially be a useful system in which to study (a) the infectious process of a single viral agent other than HIV, (b) the interaction, if any, between HIV and other viral pathogens, and (c) the efficacy of combinations of antiviral compounds to suppress coinfections. Viral agents now being pursued include HTLV-I, hepatitis B, papilloma viruses, cytomegalovirus, and Epstein-Barr virus. To the extent that the human target organs for these viruses can be maintained in the SCID-hu, it should be possible to produce small animal models for their infection.

C. Immunologic Standardization

It would be optimal to create large populations of SCID-hu mice in which functional populations of T, B, and antigen-presenting cells are all restricted to the same complement of MHC gene products. In a sense, such mice would represent inbred strains of animals, isogenic for human MHC alleles. These cohorts would permit extensive analysis of immune function in the animal model, particularly as it relates to MHC-restricted responses.

To this end, we have already observed that human fetal thymus can not only be grown in SCID mice but also transferred after growth into naive secondary recipients for further growth. In practice, one 16-gestational-week human fetal thymus might be sufficient for the production of 100 SCID-hu mice; after growth *in vivo*, the human thymus of a single SCID-hu might then be transferred to an additional cohort of animals. In this fashion, an exponentially expanding colony of SCID-hus isogenic for MHC alleles could be produced.

D. Creation of Safe Systems for Analysis

In the formulation of the SCID-hu as a small animal surrogate for human disease states, a decision has been made up front that agents infectious to man will be studied in the laboratory. Most of these agents, e.g. HIV, are not thought to be spread by aerosol and are therefore ones which could be properly studied under BSL2 conditions. Given the exigencies of the AIDS epidemic and the fact that animals infected with HIV are being used, the general consensus has been that these and similar experiments should be conducted under BSL3 (and, in some instances, BSL4 conditions). To the extent that these experiments work, their continuation will then become rapidly untenable: there are simply not enough BSL3 facilities available and little in the way of future research funds in sight to build more.

The solution to this problem will be to find better methods, tools, cages, and spaces, tailored to the practical considerations of the experiments at hand. In particular, in the setting of HIV-infected SCID-hu mice, it is likely that it will be possible to construct better cages and access devices, so that animals can be infected and assayed in the absence of sharp instruments and/or frequent handling. Thereafter, samples drawn for assay can be analyzed under standard BSL2 practice. Such improvements would dramatically increase the availability of this animal model for extensive experimentation by other investigators in multiple sites.

Acknowledgments

The author would like to thank Hideto Kaneshima, Reiko Namikawa, Linda Rabin, Chu-Chih Shih, Joe Barisse, Jane Fedor, Tom O'Toole, and Suzan Salimi for their contributions to studies of HIV infection in SCID-hu mice; John Krowka, Susan Mayo, Reina Mebius, Henry Outzen, Sujata Sarin, and Phil Streeter for their contributions to analysis of immune function in SCID-hu mice; Edwin Yee and members of the Animal Production Group at SyStemix, Inc., for their skill and care in making SCID-hu mice; and Miriam Lieberman, Irving Weissman (Stanford University), and Leonard Shultz (Jackson Laboratories) for their initial contributions to the development of SCID-hu technology.

References

Alter, J. H., Eichberg, J. W., Masur, H., Saxinger, W. C., Gallo, R., Macher, A. M., Lane, H. C., and Fauci, A. S. (1984). *Science* **226,** 549–552.

Armstrong, J. A., and Horne, R., (1984). *Lancet* **ii,** 370–372.

Bankert, R. B., Umemoto, T., Sugiyama, Y., Chen, E. A., Repasky, E., and Yokota, S. (1989). *In* "The Scid Mouse: Characterization and Potential Uses" (M. J. Bosma, R. A. Phillips, and W. Schuler, eds.), pp. 201–210. Springer-Verlag, Berlin.

Barre-Sinoussi, E., Chermann, J. V., Rey, E., Nugeyre, M. T., Chamaret, S., Gruest, J., Dauguet, C., Axler-Blin, F., Rouzioux, C., Rozenbaum, W., and Montagnier, L. (1983). *Science* **220,** 868–871.

Biberfeld, P., Chayt, K. J., Marselle, L. M., Biberfeld, G., Gallo, R. C., and Harper, M. E. (1986). *Am. J. Pathol.* **125,** 436–442.

Billingham, R. E., Brent, L., and Medawar, P. B. (1953). *Nature* (*London*) **172,** 603–605.

Bosma, G. C., Custer, R. P., and Bosma, M. J. (1983). *Nature* (*London*) **301,** 527–530.

Cheevers, W. A., and McGuire, T. (1985). *Rev. Infect Dis.* **7,** 83–101.

Custer, R. P., Bosma, G. C., and Bosma, M. J. (1985). *Am. J. Pathol* **120,** 464–477.

Czitrom, A. A., Edwards, S., Phillips, R. A., Bosma, M. J., Marrack, P., and Kappler, J. W. (1985). *J. Immunol.* **134,** 2276–2280.

Dalgleish, A. G., Beverly, P. C. L., Clapham, P. R., Crawford, D. H., Greaves, M. F., and Weiss, R. A. (1984). *Nature* (*London*) **312,** 763–767.

Daniel, M. D., Letvin, N. L., King, N. W., Kannagi, M., Sehgal, P. K., Hunt, R. D., Kanki, P. J., Essex, M., and Desrosiers, R. C. (1985). *Science* **228,** 1201–1204.

Dorshkind K., Keller, G. M., Phillips, R. A., Miller, R. G., Bosma, G. C., O'Toole, M., and Bosma, M. J. (1984). *J. Immunol.* **132,** 1804–1808.

Dorshkind, K., Pollack, S. B., Bosma, M. J., and Phillips, R. A. (1985). *J. Immunol.* **134,** 3798–3801.

Dougherty, J. P., and Temin, H. M. (1988). *J. Virol.* **62,** 2817–2822.

Engleman, E. G., Benike, C. J., Grumet, C., and Evans, R. L. (1981). *J. Immunol.* **127,** 2124–2129.

Felice, G., Cereda, P. M., and Varnier, O. E. (1988). *Nature* (*London*) **335,** 366–369.

Francis, D. P., Feorino, P. M., Broderson, J. R., Mc Clure, H. M., Getchell, J. P., Mc Grath, C. R., Swenson, B., McDougal, J. S., Palmer, E. L., Harrison, A. K., Barre-Sinoussi, F., Chermann, J.-C., Montagnier, L., Curran, J. W., Cabradilla, C. D., and Kalyanaraman, V. S. (1984). *Lancet* **ii,** 1276–1277.

Haase, A. T. (1986). *Nature* (*London*) **322,** 130–136.

Hendrickson, E. A., Schatz, D. G., and Weaver, D. T. (1988). *Genes and Dev.* **2,** 817–829.

Hirsch, V. M., Edmondson, P., Murphey-Corb, M., Arbeille, B., Johnson, P. R., and Mullins, J. I. (1989). *Nature* (*London*) **341,** 574–575.

Ioachim, H. L., Lerner, C. W., and Tapper, M. L. (1983). *Am J. Surg. Pathol.* **7,** 543–553.

Kaneshima, H., Namikawa, R., Shih, C.-C., Rabin, L., and Mc Cune, J. M. (1990). *6th Intl. Conf. AIDS* **3**, 99.

Kaneshima, H., Shih, C.-C., Namikawa, R., Rabin, L., Machado, S., Outzen, H., and Mc Cune, J. M. (1991). *Proc. Natl. Acad. Sci. U.S.A.* (in press).

Klatzmann, D., Champagne, E., Chamaret, S., Gruest, J., Guetard, D., Hercend, T., Gluckman, J.-C., and Montagnier, L. (1984). *Nature (London)* **312**, 767–768.

Koyanagi, Y., Miles, S., Mitsuyasu, R. T., Merrill, J. E., Vinters, H. V., and Chen, I. S. Y. (1987). *Science* **236**, 819–822.

Krowka, J., Sarin, S., Namikawa, R., Mc Cune, J. M., and Kaneshima, H. (1990). *6th Intl. Conf. AIDS* **1**, 194.

Krowka, J., Sarin, S., Namikawa, R., Mc Cune, J. M., and Kaneshima, H. (1991). *J. Immunol.* (in press).

Kulaga, H., Folks T., Rutledge, R., Truckenmiller, M. E., Gugel, E., and Kindt, T. J. (1989). *J. Exp. Med.* **16**, 321–326.

Leider, J. M., Palese, P., and Smith, F. I. (1988). *J. Virol.* **62**, 3084–3091.

Letvin, N. L., Daniel, M. A., Sehgal, P. K., Yetz, J. M., Solomon, K. R., Kannagi, M., Schmidt, D. K., Silva, D. P., Montagnier, L., and Desrosiers, R. C. (1987). *J. Inf. Dis.* **156**, 406–407.

Le Tourneau, A., Audouin, J., Diebold, J. Marche, D., Tricottet, V., and Reynes M. (1986). *Hum. Pathol.* **17**, 1047–1053.

Lusso, P., Veronese, B., Ensoli, G., Franchini, V., Jemma, C., DeRocco, S. E., Kalyanaraman, V. S., and Gallo, R. C. (1990). *Science* **247**, 848-851.

Maddon, P. J., Dalgleish, A. G., McDougal, J. S., Clapham, P. R., Weiss, R. A., and Axel, R. (1986). *Cell* **47**, 333–348.

Malynn, B. A., Blackwell, T. K., Fulop, G. M., Rathbun, G. A., Furley, A. J. W., Ferrier, P., Heinke, L. B., Phillips, R. A., Yancopoulos, G. D., and Alt, F. W. (1988). *Cell* **54**, 453–460.

Mc Cune, J. M., Namikawa, R., Kaneshima, H., Shultz, L. D., Lieberman, M., and Weissman, I. L. (1988). *Science* **241**, 1632–1639.

Mc Cune, J. M., Kaneshima, H., Lieberman, M., Weissman, I. L., and Namikawa, R. (1989a). *In* "The Scid Mouse: Characterization and Potential Uses" (M. J. Bosma, R. A. Phillips, and W. Schuler, eds.), pp. 183–194. Springer-Verlag, Berlin.

Mc Cune, J. M., Namikawa, R., Shih, C.-C., Rabin, L., and Kaneshima, H. (1990a). *Science* **250**, 1152–1153.

Mc Cune, J. M. (1990b). *Sem. Virol.* **1**, 229–235.

Mc Cune, J. M., Namikawa, R., Shih, C. C., Rabin, L., and Kaneshima, H. (1990c). *Science* **247**, 564–567.

Mc Cune, J. M., Shih, C.-C., Namikawa, R., Rabin, L., and Kaneshima, H., (1990d). *6th Intl. Conf. AIDS* **2**, 107.

Mc Cune, J. M. (1991). *Cell* **64**, 351–363.

Meyerhans, A., Cheynier, R., Albert, J., Seth, M., Kwok, S., Sninsky, J., Morfeldt-Manson, L., Asjo, B., and Wain-Hobson, S. (1989). *Cell* **58**, 901–910.

Mosier, D. E., Gulizia, R. J., Baird, S., and Wilson, D. B. (1988). *Nature (London)* **335**, 256–259.

Mosier, D. E., Gulizia, R. J., Baird, S., and Wilson, D. B. (1989). *Nature (London)* **338**, 211.

Mullins, J. I., Chen, C. S., and Hoover, E. A. (1986). *Science* **319**, 333–336.

Namikawa, R., Mizuno, T., Matsuoka, H., Fukami, H., Ueda, R., Itoh, G., Matsuyama, M., and Takahashi, T. (1986). *Immunology* **57**, 61–69.

Namikawa, R., Kaneshima, H., Lieberman, M., Weissman, I. L., and Mc Cune, J. M. (1988). *Science* **242**, 1684–1686.

Namikawa, R., Weilbaecher, K. N., Kaneshima, H., Yee, E. J., and Mc Cune, J. M. (1990a). *J. Exp. Med.* **172**, 1055–1063.

Namikawa, R., Fedor, J., Kaneshima, H., and Mc Cune, J. M. (1990b) *6th Intl. Conf. AIDS* **1,** 194.

Pekovic, D. D., Gornitsky, M., Ajdukovic, D., Dupuy, J.-M., Chausseau, J. P., Michaud, J., Lapointe, N., Gilmore, N., Tsoukas, C., Zwadlo, G., and Popovic, M. (1987). *J. Pathol.* **152,** 31–35.

Popovic, M., Sarngadharan, M. G., Read, E., and Gallo, R. C. (1984). *Science* **224,** 497–500.

Rabin, L., Kaneshima, H., Shih, C.-C., and Mc Cune, J. M. (1990). *6th Intl. Conf. AIDS* **3,** 113.

Schuler, W., Weiler, I. J., Schuler, A., Phillips, R. A., Rosenberg, N., Mak, T., Kearney, J. F., Perry, R. P., and Bosma, M. J. (1986). *Cell* **46,** 963–972.

Shultz, L. D., and Sidman, C. L. (1987). *Annu. Rev. Immunol.* **5,** 367–403.

Shultz, L. D., Schweitzer, P. A., Hall, E. J., Sundberg, J. P., Taylor, S., and Walzer, P. D. (1989). *In* "The Scid Mouse: Characterization and Potential Uses" (M. J. Bosma, R. A. Phillips, and W. Schuler, eds), pp. 243–249. Springer-Verlag, Berlin.

Shih, C.-C., O'Toole, T., Salimi, S., Kaneshima, H., Namikawa, R., Rabin, L., and Mc Cune, J. M. (1990). *6th Intl. Conf. AIDS* **1,** 184.

Shih, C.-C., Kaneshima, H., Rabin, L., Namikawa, R., Sager, P., McGowan, J., and Mc Cune, J. M. (1991). *J. Inf. Dis.* (in press).

Spangrude, G. J., Heimfeld, S., and Weissman, I. L. (1988). *Science* **241,** 58–62.

Touraine, J. L. (1983). *Immunol. Rev.* **71,** 103–121.

Weilbaecher, K., Namikawa, R., and Mc Cune, J. M. (1990). *6th Intl. Conf. AIDS* **1,** 193.

Zinkernagel, R. M., Callahan, G. N., Althage, A., Cooper, S., Klein, P. A., and Klein, J. (1978a). *J. Exp. Med.* **147,** 882–896.

Zinkernagel, R. M., Callahan, G. N., Althage, A., Cooper, S., Streilein, J. W., and Klein, J. (1978b). *J. Exp. Med.* **147,** 897–911.

Part V

Therapy

Adult T-cell Leukemia: Prospects for Immunotherapy

Thomas A. Waldmann

I. Introduction

The field of clinical immunology began almost two centuries ago when Jenner introduced a form of immune intervention, vaccination, with cowpox as a means of protecting against smallpox. This form of immune intervention directed at prevention of disease culminated in the complete elimination for the first time of a major human disease, smallpox. Similarly, with the development of radioimmunoassays, enzyme-linked immunoassays, monoclonal antibodies, now microfluorometry, and modern molecular immunogenetics, immunological approaches are playing a major role in diagnosis.

Immune intervention is also playing an increasing role in therapy. However, we have clearly not achieved the potential provided by the great specificity of the immune system. Most of the medical therapy for cancer has focused on chemotherapeutic agents acting within the cell. The hybridoma technique of Köhler and Milstein (1975) rekindled interest in the use of antibodies targeted to the cell surface as agents to treat cancer patients. However, such monoclonal antibodies have to date been relatively ineffective, with only 23 partial and 3 complete remissions reported in the initial 185 patients studied among 25 clinical trials (Catane and Longo, 1989). There have been a number of explanations for this observed low therapeutic efficacy. One of the factors is that the murine monoclonal antibodies are immunogenic. An even more critical factor is that most of the monoclonal antibodies employed are neither cytocidal nor cytostatic agents against human neoplastic cells. Furthermore, in most cases, the antibodies are not directed against a vital structure present on the surface of malignant cells, such as a growth factor receptor required for tumor cell proliferation. We have readdressed this issue utilizing the interleukin-2 receptor (IL-2R) as a target for immune intervention. The scientific basis for this approach is that resting normal cells do not express the IL-2R, in contrast to the abnormal T cells in select patients with autoimmune disorders, in individuals rejecting allografts, and in patients with lymphoid neoplasms, especially the human T-cell leukemia virus (HTLV)-I-associated adult T-cell leukemia (ATL) (Waldmann 1986, 1989a, b). To exploit this difference in IL-2R expression, we have initiated therapeutic trials using unmodified antibodies to the IL-2R (anti-Tac), conjugates of anti- Tac with mutated forms of a *Pseudomonas* exotoxin (PE40 and PE66[4]-Glu), IL-2-truncated toxin fusion proteins, and α- and β-emitting isotopic chelates of anti-Tac. Humanized hyperchimeric anti-Tac molecules have been prepared by genetic engineering in which the molecule is entirely human IgG_1 except for the small complementarity-determining regions that are retained from the mouse antibody (Queen *et al.*, 1989). This "humanized" antibody manifested the ability to perform antibody-dependent cell-mediated cytotoxicity (ADCC) which was absent in the original mouse monoclonal (Junghans *et al.*, 1990). The clinical implication of IL-2R-directed therapy in HTLV-I-associated ATL represents a new perspective for the treatment of this neoplastic disease.

II. Structure and Function of the Multisubunit IL-2R

The sequence of events involved in the activation of T cells begins when a foreign pathogen encounters the antigen-specific receptor on the surface of resting T cells. The antigen-stimulated activation of these resting T cells induces the synthesis of the 15.5-kDa IL-2 lymphokine molecule. To exert its biological effect, IL-2 must interact with specific high-affinity membrane receptors. Resting cells do not express high-affinity IL-2R, but they are rapidly expressed on T cells after activation with antigen or mitogen. Progress in analyzing the structure, function, and expression of human IL-2R was greatly facilitated by our production of an IgG_{2a} mouse monoclonal antibody (anti-Tac) that blocks the binding of IL-2 to one of its receptor peptides, preventing T-cell proliferation (Uchiyama *et al.*, 1981a, b). The IL-2-binding receptor peptide identified by the anti-Tac monoclonal on phytohemagglutinin (PHA)-activated normal lymphocytes is a 55-kDa glycoprotein.

A series of issues were difficult to resolve when only the 55-kDa Tac peptide was considered Specifically, most cells display two classes of receptors, one with an affinity of 10^{-11} M and another with an affinity of 10^{-8} M. The Tac peptide was shown to participate in both the high- and low-affinity forms of the IL-2R. Isolation of cDNAs encoding the Tac peptide did not provide an explanation for the great difference in affinity between high and low-affinity receptors. In addition, the amino acid sequence of the Tac peptide deduced from the cloned cDNA revealed a very short (13 amino acid) cytoplasmic domain that was too short to independently transduce receptor signals to the nucleus (Leonard et al., 1984; Nikaido et al., 1984). Furthermore, it had been shown that certain cells not expressing the Tac peptide, including large granular lymphocytes (LGL), which are precursors of activated natural killer (NK) and lymphokine-activated killer (LAK) cells, could be activated by IL-2 to become efficient killers (Grimm and Rosenberg, 1984; Ortaldo et al., 1984). Finally, the cell line MLA-144 did not express the Tac peptide, yet manifested 4000 IL-2-binding sites with intermediate affinity. These observations led us to consider the possibility that the high-affinity IL-2R was not a single peptide but rather a receptor complex that included the Tac peptide as well as novel non Tac peptides. In studies initially presented at the Sixth International Immunology Congress in July 1986, we used radiolabeled IL-2 in cross-linking to define the size of the IL-2R peptide on MLA-144 (Waldmann et al., 1986; Tsudo et al., 1986, 1987). We identified a 70/75-kDa IL-2-binding protein on the MLA-144 cell line. The binding of IL-2 to this peptide was blocked by excess unlabeled IL-2 but not by the anti-Tac antibody, confirming the presence of a novel 70/75-kDa IL-2-binding protein. When a series of cell lines were subjected to IL-2 cross-linking studies, it was shown that cell lines bearing either the p55 Tac or the p75 peptide alone manifested low- or intermediate-affinity IL-2 binding, whereas cell lines bearing both peptides manifested both high- and low-affinity receptors (Tsudo et al., 1987). In light of these observations, we proposed a multi-chain model for the high-affinity IL-2R. In this model, an independently existing p55 or p75 peptide would create low- or intermediate-affinity receptors, whereas high-affinity receptors would be created when both receptors were expressed and noncovalently associated in a receptor complex. In independent studies, Sharon and co-workers (1986) proposed a similar model.

Kinetic binding studies with IL-2 have provided an interesting perspective on how the two separate IL-2-binding chains cooperate to form the high-affinity receptor. Each chain reacts very differently with IL-2, with distinct kinetic and equilibrium binding constants. The on-and-off rate for IL-2 binding to the Tac protein is rapid (5–10 sec), while the on-and-off rate for IL-2 binding to the p70/75 protein is markedly slower (>40 min) (Lowenthal and Greene, 1987; Wang and Smith, 1988). The kinetic binding data obtained when the high-affinity receptors are analyzed show that the association rate of this receptor depends on the fast-reacting p55 Tac or α chain, whereas the dissociation rate is derived from the slow-reacting p75 or β chain. Because the affinity of binding at equilibrium is determined by the ratio of the dissociation constant and the association rate constant, this kinetic cooperation between the low- and intermediate-affinity ligand binding sites results in a receptor with a high affinity for IL-2.

There are a number of features that suggest a more complex relationship between IL-2 and the p55 and p75 peptides than implied by the preformed binary complex model just discussed. To examine this issue, Saito and co-workers (1988) carried out kinetic studies on IL-2 binding to the high-affinity IL-2R on T lymphocytes expressing various numbers of the p55 chains and a relatively constant number of the p75 chains. They found that the expression of a larger number of p55 chains accelerated the association of IL-2 to the high-affinity receptor. These results were not compatible with the binary complex model that assumes a fixed number of high-affinity sites determined by the numbers of a limiting chain. Instead, the results were consistent with the prediction of an affinity conversion model that assumes that the association of IL-2 to the p55 chain is the first step of the ternary complex formation, and they indicate that the possible role of excess p55 chains is to accelerate the formation of the ternary complex. Support for an affinity conversion model implying an initial interaction between IL-2 and the p55 Tac peptide emerged from the studies of Tsudo and co-workers (1989) utilizing a monoclonal antibody, Mik-β1, that inhibits IL-2 binding to the p75 or β chain. Mik-β1 completely abolished the high-affinity IL-2 binding to the IL-2-dependent T-cell line Kit225, which supported the view that the β p75 chain is an indispensable component of the high-affinity receptor. A surprising observation in this study, which has been reproduced in other laboratories is that the addition of Mik-β1 did not inhibit IL-2-induced proliferation of Kit225 cells at any of the IL-2 concentrations used. The addition of the anti-Tac antibody increased the IL-2 concentrations required for growth 20-fold. When both antibodies were present, the proliferation was almost completely inhibited. The fact that the IL-2 signal through the high-affinity receptors is blocked by Mik-β1 only when the anti-Tac antibody is simultaneously present suggests that the p55 α chain is involved in the initial events of IL-2 interaction and, ultimately, signal transduction through the β chain.

Hatakeyama and co-workers (1989) have molecularly cloned the gene encoding the p75 β chain of the IL-2R and have not only provided further information concerning this peptide but have also provided evidence suggesting a more complex subunit structure that involves peptides in addition to the p55 and p75 IL-2-binding peptides. D'Andrea and co-workers (1989), on cloning the erythropoietin receptor, have pointed out several structural features shared by the erythropoietin receptor and the IL-2R β chain, including significant sequence homology. Furthermore, a common Trp-Ser-X-Trp-

Ser motif has been identified in the extracellular region just beyond the membrane-spanning domain, not only of the erythropoietin and the IL-2 β receptor but also the cellular receptors for IL-4, IL-6, and GM-CSF. Hatakeyama and co-workers (1989) examined the IL-2-binding properties of the cDNA encoded IL-2R β chain expressed in various cell types. The cDNA encoded β have bound and internalized IL-2 when expressed on T lymphoid cells but not fibroblast cell lines. These results suggest the involvement of either a cell type-specific processing mechanism or, more likely, an additional cellular component for the functional IL-2R β chain expression.

Evidence thus suggests a more complex subunit structure that involves peptides in addition to the p55 and p75 IL-2-binding peptides. With the use of coprecipitation analysis, radiolabeled IL-2 cross-linking procedures, binding by insolubilized IL-2, and flow cytometric resonance energy-transfer measurements, a series of peptides of molecular weight 22,000, 35,000, 40,000, 75,000 (non-IL-2 binding), 95–105,000, and 180,000 have been associated with the two IL-2-binding peptides (Szöllösi et al., 1987; Saragovi and Malek, 1987, 1989; Herrmann and Diamantstein, 1987). We have utilized a flow cytometric energy-transfer technique to demonstrate a close, nonrandom proximity between the p55 Tac and the 95-kDa peptide ICAM-1 (Szöllösi et at., 1987). Furthermore, Saragovi and Malek (1989) provided evidence indicating that the p75 IL-2-binding peptide of mice exists as a p75/p22–30 heterodimer. The p75 peptide, precipitated by immobilized IL-2 or coprecipitated with anti-p55 antibodies, yielded an M_r-105,000 species when analyzed under nonreducing conditions. Under reducing conditions and in diagonal nonreduced versus reduced SDS-PAGE analysis, 75- and 22-kDa species were identified from the appropriately internally mdiolabeled cells. Finally, evidence suggests that there are two independent M_r-70,000 and M_r-75,000 peptides in the high-affinity receptor. Although a series of antibodies have been generated that completely inhibit IL-2 binding to the p75 IL-2-binding peptide, none has agonist activity for activated T cells. In contrast, other monoclonal antibodies have been generated that recognize a 70/75-kDa peptide on human activated T cells and LGLs that have agonist activity. However, these monoclonal antibodies do not fulfill the criteria for an antibody to an IL-2-binding protein. The most extensively studied monoclonal antibodies of this group, YTA-1 and YTA-2, described by Nakamura and co-workers (1989), recognize 75-kDa molecules on human LGLs. However, these antibodies do not block IL-2 binding to p75 IL-2R expressed on LGLs, and they do not precipitate IL-2 cross-linked to the prototypical p75 IL-2-binding peptide. Nevertheless, these monoclonal antibodies down-regulate/modulate high-affinity IL-2R from peripheral blood mononuclear cells. Furthermore, YTA-1 and YTA-2 were mitogenic and were different from other mitogenic monoclonal antibodies. Clearly, the p75 peptide identified by monoclonal antibodies such as YTA-1 may not be a participant in the high-affinity IL-2R. Alternatively, this peptide may be a non-IL-2-binding peptide that plays a role in the transduction of the signal from the two IL-2-binding subunits p55 and p75.

In summary, with the use of coprecipitation analysis and radiolabeled IL-2 cross-linking procedures, as well as of other techniques, a series of peptides of M_r 22,000, 35,000, 40,000, 75,000, 95,000–105,000, and 180,000 have been associated with the two IL-2-binding peptides. Further studies, including molecular cloning followed by expression of the peptides in concert in cells not expressing functional IL-2R, will be required to define which of these peptides plays a meaningful role in the multisubunit IL-2R.

III. Disorders of IL-2R Expression in Malignant and Autoimmune Diseases

In contrast to the lack of Tac peptide expression in normal resting mononuclear cells, this receptor peptide is expressed by a proportion of the abnormal cells in certain forms of lymphoid neoplasia, in select autoimmune diseases, and in individuals rejecting allografts (Waldmann, 1989a, b; Diamantstein and Osawa, 1986; Williams et al., 1988). That is, a proportion of the abnormal cells in these diseases express the p55 peptide on their cell surface. Furthermore, the serum concentration of the soluble form of the Tac peptide is elevated. Rubin and co-workers (1985) demonstrated that activated peripheral blood mononuclear cells and certain cell lines of T- or B-cell origin release a soluble 45- kDa form of the IL-2R Tac peptide into the culture medium. Using an enzyme-linked immunosorbent assay with two monoclonal antibodies that recognize distinct epitopes on the human IL-2 Tac receptor, they showed that normal individuals have measurable amounts of IL-2R in their plasma and that certain lymphoreticular malignancies, autoimmune disorders, and allograft reactions are associated with elevated plasma levels of this receptor (Rubin et al., 1985; Nelson et al., 1986). In terms of neoplasia, certain T-cell, B-cell, monocytic, and even granulocytic leukemias express the Tac antigen. Specifically, virtually all of the abnormal cells of patients with HTLV-I-associated ATL express the Tac antigen (Waldmann et al., 1984; Uchiyama et al., 1985). Similarly, a proportion of patients with cutaneous T-cell lymphomas, including the Sézary syndrome and mycosis fungoides, express the Tac peptide (Waldmann et al., 1984; Schwarting et al., 1985). Furthermore, the malignant B cells of virtually all patients with hairy cell leukemia and a proportion of patients with large and mixed cell diffuse lymphomas are Tac expressing (Korsmeyer et al., 1983). The Tac antigen is also expressed on the Reed-Sternberg cells of patients with Hodgkin's disease and on the malignant cells of patients with true histiocytic lymphoma (Schwarting et al., 1985). Finally, a proportion of the leukemic cells of patients with chronic and acute myelogenous leukemia are Tac positive. In addition to these Tac-expressing leukemias and lymphomas, there are certain leukemias (e.g., acute lymphoblastic leukemia and LGL leukemia) that do not express the Tac peptide but do express the p75 β peptide of the IL-2R. Autoimmune diseases may also be associated with disorders of Tac antigen expression (Diamantstein and Osawa, 1986). Some mononuclear cells in the involved tissues express the Tac antigen, and the serum concentration of the soluble form of the Tac peptide is elevated. Such evidence for T-cell activation and disorders of Tac antigen expression appear in certain patients with rheumatoid arthritis and systemic lupus, select patients with aplastic anemia, and more particularly, individuals with HTLV-I-associated tropical spastic paraparesis. Finally, the T cells that recognize foreign histocompatibility antigens following allograft implantation as well as those involved in graft versus host disease express the Tac antigen.

IV. Disorders of IL-2R Expression in HTLV-I-Associated ATL

A distinct form of mature T-cell leukemia was defined by Uchiyama *et al.* (1977) and termed "adult T-cell leukemia" (ATL). HTLV-I has been shown to be the primary etiologic agent in ATL (Poiesz *et al.*, 1980). ATL is a malignant proliferation of mature CD3/CD4-expressing T cells that tend to infiltrate the skin, lung, and liver. Cases of ATL are associated with hypercalcemia and an immunodeficiency state that usually has a very aggressive course. Although the leukemic cells are usually of the $CD4^+$ phenotype, they do not manifest functional T helper activity but function as suppressor effectors and inhibit the immunoglobulin synthesis of cocultured pokeweed mitogen-stimulated mononuclear target populations (Waldmann *et al.*, 1984). The leukemic cells that we and others have examined from patients with HTLV-I-associated ATL expressed high- and low-affinity IL-2R, including the Tac peptide (Waldmann *et al.*, 1984; Uchiyama *et al.*, 1985). An analysis of HTLV-I and its protein products suggests a potential mechanism for this association between HTLV-I and constitutive IL-2R expression. In addition to the presence of long terminal repeats (LTRs), *gag, pol,* and *env* genes, retroviral gene sequences common to other groups of retroviruses, HTLV-I, and HTLV-II were shown to contain a genomic region between *env* and the 3' LTR referred to as pX. This region encodes at least three peptides of 21, 27, and 42 kDa (Seiki *et al.*, 1983). One of these, a 42-kDa protein now termed "Tax" is essential for viral replication (Sodroski *et al.*, 1984). The mRNA for this protein is produced by a double splicing event. To stimulate viral transcription, the tax protein requires the presence of three 21base pair (bp) enhancer-like repeats within the LTR of HTLV-I (Shimotohno *et al.*, 1985).

Studies involving the transfection of cDNA encoding the Tax product of HTLV-I into Jurkat T cells showed that the Tax protein plays a central role in increasing the transcription of host genes governing human IL-2 and IL-2 p55 α-chain receptor peptide expression (Shimotohno *et al.*, 1985; Inoue *et al.*, 1986; Cross *et al.*, 1987; Maruyama *et al.*, 1987; Siekewitz *et al.*, 1987; Suzuki *et al.*, 1987; Ruben *et al.*, 1988). In these transient expression studies, the Tax protein stimulated an increase in IL-2R-promoter activity. A direct interaction of the Tax protein with specific DNA sequences in the 5′ region of the IL-2R appears unlikely, since this promoter region does not share strong sequence homologies with the 21-bp enhancer-like regions of the LTR of HTLV-I. Ruben and co-workers (1988) demonstrated that the 12 base sequence motif in the −255 to −267 5′ region of the IL-2R *tac* gene is required for activation by the *tax* gene. This is the site of action of the nuclear factor NFκB. Ruben and co-workers (1988) suggested that activation of the IL-2R *tac* gene expressed by HTLV-I tax protein occurs through an interaction with, or activation of, a host transcription factor with properties similar to those of NFκB. Others using mutational analysis have suggested that the NFκB binding site is not required for *tax trans*-activation. In addition, Fujii and co-workers (1988) pointed out that a consensus sequence C-C(A+T-rich)-G-G is a protein-binding site shared by the IL-2R and by c-*fos*. This region is required for c-*fos* gene activation by Tax. As noted previously, an uncontrolled autocrine T-cell growth model has been proposed for the early events of HTLV-I-induced T-cell transformation. One limitation of the hypothesis suggesting a role for *tax* expression in HTLV-I-associated ATL is that it has not been possible to demonstrate mRNA expression for HTLV-I in general and the *tax* region in particular in circulating ATL cells examined by Northern blot analysis. In order to determine whether this reflects the insensitivity of the Northern blot analysis or the examination of cells from the wrong body compartment, we have applied a modification of gene amplification by the polymerase chain reaction (PCR) to this question. Our strategy exploits the limited efficiency of PCR for sequences greater than 2000 bp and the reduction in distance that occurs in doubly spliced tax mRNA when compared with genomic DNA. Using primers chosen to be at great distances in the DNA but at a dramatically reduced distance in the spliced mRNA for the pX region, we were able to detect pX mRNA expression in the peripheral blood lymphocytes of all patients examined with HTLV-I-associated tropical spastic paraparesis (Tendler *et al.*, 1989). Furthermore, pX mRNA could be demonstrated in the peripheral blood lymphocytes of certain patients who had antibodies to HTLV-I without ATL. In contrast, the peripheral blood leukemic cells from patients with the aggressive form of ATL did not express pX message. Such message was, however, demonstrable in the lymph node cells of approximately one-half of the patients. Thus, the tax protein may play an important role in HTLV-I-associated tropical spastic paraparesis and in the early phases of HTLV-I-induced ATL by deregulating the expression of the cellular genes encoding IL-2 and the IL-2R that are involved in the normal control of T-cell proliferation.

V. IL-2R as a Target for Therapy in Patients with HTLV-I-Associated ATL

A. Unmodified Anti-Tac Monoclonal Antibody

HTLV-I-induced ATL cells constitutively express IL-2R identified by the anti-Tac monoclonal antibody, whereas normal resting cells do not. This observation provided the scientific basis for IL-2R-directed immunotherapy. Such agents could theoretically eliminate Tac-expressing leukemic cells or activated T cells involved in other disease states, while retaining the Tac-nonexpressing mature normal T cells and their precursors that express the full repertoire of antigen receptors for T-cell immune responses. The agents that we have developed include: (a) unmodified anti-Tac monoclonal antibody; (b) toxin conjugates of anti-Tac [e.g., PE, truncated PE (PE40), and mutated PE (PE66^4-Glu)]; (c) IL-2 truncated toxin fusion proteins (e.g., IL-2 PE40); (d) α- and β-emitting isotopic (e.g., ^{212}Bi and ^{90}Y) chelates of anti-Tac; (e) bifunctional antibodies (e.g., anti-Tac-anti-CD3 or anti-Tac-anti-CD16 heteroconjugates); and (f) hybrid "humanized" anti-Tac with mouse light and heavy chain hypervariable regions joined to the human constant κ light chain and IgG$_1$ heavy chain regions.

In the initial studies, we have performed a trial of intravenous anti-Tac in the treatment of patients with ATL (Waldmann et al., 1985, 1988). The patients did not suffer untoward reactions and did not have a reduction in the normal formed elements of the blood, and only one produced antibodies to the anti-Tac monoclonal antibody. Five of 15 patients had transient mixed (one), partial (one), or complete remissions (three) lasting from 1 to over 8 months following anti-Tac therapy, as assessed by routine hematologic tests, immunofluorescent analysis of circulating cells, and molecular genetic analysis of HTLV-I proviral integration and of the T-cell receptor gene rearrangement (Waldmann et al., 1988; T. A. Waldmann, unpublished). Thus, the use of a monoclonal antibody that prevents the interaction of IL-2 with its growth factor receptor on ATL cells provides a rational approach for the treatment of this malignancy. Indeed, Maeda and co-workers (1987) have presented evidence for the IL-2-dependent expansion of leukemic cells in ATL in approximately 20% of cases. In the majority of cases of the aggressive phase of ATL, however, the leukemic cells no longer produce IL-2 or require IL-2 for their proliferation. In this phase of the disease, patients are not responsive to unmodified anti-Tac.

B. Recombinant Toxins Directed toward the IL-2R

The second IL-2R-directed approach to the therapy of ATL was based on the observation that ATL cells continue to express the IL-2R during the late stages of the disease when they no longer require IL-2 and thus are no longer responsive to therapy with unmodified anti-Tac. In this approach, anti-Tac or IL-2 was used to deliver toxins to IL-2R-expressing cells. In these cases, the toxin was modified so that it no longer indiscriminately killed all cells. The toxin that has been primarily used has been *Pseudomonas* exotoxin, a 66,000-M_r enzyme that catalyzes the ADP-ribosylation and inactivation of elongation factor 2, thus inhibiting protein synthesis and killing the cell. The addition of anti-Tac antibody coupled to PE inhibited protein synthesis by Tac-expressing HUT-102 cells, but not by the acute T-cell leukemia line MOLT 4, which does not express the Tac antigen FitzGerald *et al.*, 1984). However, we found that we could only give a few milligrams of this agent to patients with ATL without producing undesirable liver damage. This result was obtained because the toxin had not been sufficiently changed so that it would no longer bind to normal cells, such as liver cells, which are very sensitive to PE. Functional analysis of deletion mutants of the PE structural gene has shown that domain Ia of the 66-kDa PE molecule is responsible for cell binding, analogous to the recognition portion of a polypeptide hormone (Hwang *et al.*, 1987). Domain II had the properties of a cell-penetrating protein and helped the toxin get across the endosomic membrane into the cytosol. Domain III was responsible for ADP-ribosylation of elongation factor 2, the step actually responsible for cell death. PE molecule from which domain I had been deleted (PE40) has virtually full ADP-ribosylating activity but extremely low cell-killing activity when used alone because of the loss of the cell recognition domain. The PE40 was produced in *Escherichia coli*, purified, and conjugated to anti-Tac. The anti-Tac–PE conjugates inhibited the protein synthesis of Tac-expressing T-cell lines but not that of lines not expressing Tac. However, immunotoxins made by chemically attaching a toxin to an intact antibody generate a product that is heterogeneous, with yields that are often poor. To circumvent this problem, Chudhary and co-workers (1989) have produced a recombinant immunotoxin consisting of anti-Tac variable light and heavy chain domains fused to PE40. This work followed the observation that active single chain Fv fragments of antibodies can be produced in *E. coli* by attaching the light and heavy chain variable domains together with a peptide linker. A single-chain antibody toxin fusion protein [anti-Tac(Fv)-PE40] in which the variable regions of anti-Tac are joined in peptide linkage to PE40 was constructed and expressed in *E. coli*. Anti-Tac(Fv)-PE40 was very cytotoxic to two IL-2R-bearing human cell lines but was not cytotoxic to receptor-negative cells.

IL-2-PE40, a chimeric protein composed of human IL-2 genetically fused to the N-terminus of the modified form of PE40, was constructed to provide an alternative (lymphokine-mediated) method of delivering PE40 to the surface of IL-2R Tac-expressing cells (Hwang *et al.*, 1987; Lorberboum-Galski *et al.*, 1988). The IL-2-PE40, a cytotoxic protein, was produced by fusion of a cDNA encoding the human IL-2 gene to the 5' end of a modified PE40 gene that lacks sequences encoding the cell-recognition domain. The addition of IL-2-PE40 led to the inhibition of protein synthesis by the toxin moiety of IL-2-PE40 when added to human cell lines expressing either p55 or p75 or both IL-2R subunits. The receptor internalization was much more efficient when high-affinity receptors composed of both units were present. IL-2-PE40 has been effective in preventing the rejection of cardiac allografts in mice, in the amelioration of experimental arthritis in rats, and in the inhibition of tumor development following the transfer of IL-2R-expressing tumor cells to mice. Although IL-2-PE40 was also very effective in inhibiting protein synthesis of Tac-expressing HUT-102 cells, it was much less effective in inhibiting the protein synthesis of primate T cells undergoing a mixed leukocyte reaction. An alternative lymphokine-mediated toxin fusion protein, IL-2-PE66^4-Glu, has been generated that is effective in this regard. In this construct, all domains of the PE are retained, but amino acid positions 57, 247, 248, and 249 have been mutated from a basic amino acid to amino acid glutamate. This molecule is relatively nontoxic to IL-2-nonexpressing cells but is quite effective in inhibiting the protein synthesis of human IL-2R-expressing cell lines, as well as in inhibiting protein synthesis in PHA- and mixed leukocyte reaction-induced blasts of monkeys and humans. The toxicity and efficacy in preventing allograft rejection in cynomolgus monkeys of the different *Pseudomonas* toxin-antibodies and interleukin fusion molecules are being studied prior to their application to the therapy of IL-2R-expressing ATL.

Bacha and associates (1988) have performed parallel studies in which the portion of the diphtheria toxin gene that encodes the receptor-binding domain has been genetically replaced with the cDNAs encoding IL-2. Using this agent, they have achieved prolongation of allograft survival and suppression of delayed-type hypersensitivity. Furthermore, the diphtheria toxin-IL-2 fusion protein at a concentration of 10^{-8} M inhibited protein synthesis by 60–98% in lymph-node ATL cells, whereas protein synthesis in peripheral blood ATL cells was inhibited by 20–57% in acute-type, and by 3–13% in chronic ATL (Kiyokawa et al., 1989). In contrast, the IL-2 toxin had no measurable effects on T cells from either patients with the smoldering type of ATL or normal controls. Taken together with the IL-2-PE toxin studies, these findings suggest that the high-affinity IL-2R on acute- and lymphoma-type ATL cells may serve as a target for therapy with recombinant chimeric toxins.

C. α- and β-Emitting Isotopic Chelates of Anti-Tac

The action of toxin conjugates of monoclonal antibodies and lymphokines depends on their ability to be internalized by the cell and released into the cytoplasm. In fact, the toxin conjugates do not pass easily from the endosome to the cytosol as required for their action on elongation factor 2. Furthermore, the toxins are quite immunogenic and thus provide only a short window for therapy prior to the development of antibodies directed toward the toxin. To circumvent these limitations, alternative cytotoxic reagents were developed that were not immunogenic, that could be conjugated to anti-Tac, and that were effective when bound to the surface of Tac-expressing cells. In collaboration with Dr. Otto Gansow, α- and β-emitting radionuclides were conjugated to anti-Tac by the use of bifunctional chelates (Kozak et al., 1986, 1989). In initial studies, a series of chelating agents were developed that were evaluated in terms of their ability to fulfill the following criteria for suitability: (a) the chelating agent coupled to the monoclonal antibody should not compromise antibody specificity; (b) the chelation and radiolabeling procedure should not alter the metabolism of the monoclonal-antibody; and (c) the chelating agent should not permit elution and thus premature release of the radiometal in vivo. The chelate 1(2)-methyl-4-(P-isothiocyanatobenzyl)diethylenetriaminepentaacetic acid fulfills these requirements. We have shown with in vito studies that bismuth-212 (^{212}Bi), an α-emitting radionuclide conjugated to anti-Tac by the use of a bifunctional chelate, was well suited for immunotherapy. Activity levels of 0.5 μCi or the equivalent of 12 rad/ml of α radiation targeted by ^{212}Bi-labeled anti-Tac eliminated greater than 98% of the proliferative capacity of the HUT-102 cells, with only a modest effect on IL-2R-negative lines. This specific cytotoxicity was blocked by excess unlabeled anti-Tac but not by control IgG. Parallel studies were performed for the β-emitting yttrium-90 (^{90}Y) chelated to anti-Tac, using a chelation agent that did not permit elution of the radiolabeled yttrium from the monoclonal antibody. This agent has been studied for efficacy and toxicity in a primate xenograft organ transplantation model. Rhesus monkeys receiving a xenograft of a cynomolgus monkey heart showed a marked prolongation (with a mean graft survival in treated animals of 40 days compared with 7 days in controls) of xenograft survival following the administration of ^{90}Y-labeled anti-Tac (Cooper et al., 1988). No prolongation of xenograft survival was observed following the administration of unmodified anti-Tac, and only a modest increase in survival occurred following the administration of ^{90}Y on an irrelevant monoclonal antibody. Thus, ^{212}Bi-labeled anti-Tac and ^{90}Y-labeled anti-Tac are potentially effective and specific immunocytotoxic agents for the elimination of Tac-expressing cells.

D. Bifunctional Antibodies with Specificity against Tac Tumor Targets and against CD3 or CD16 Effector Cells

Two major problems confront the use of unmodified mouse monoclonal antibodies in anti-tumor therapy: (a) neutralization with anti-mouse antibodies; and (b) poor recruitment of host effector mechanisms. Mouse monoclonals in general and the anti-Tac monoclonal antibody in particular do not participate in ADCC with human mononuclear cells. In one approach to increase the efficacy of anti-Tac therapy, bifunctional antibody products were prepared by chemical cross-linking, disulfide exchange, and by the production of hybrid hybridomas with specificity against Tac-expressing tumor targets on the one hand and against CD3 or CD16 effectors on the other. The anti-Tac CD3 bifunctional agents showed potent killing of targets with cytotoxic lymphocytes. In parallel, the bifunctional antibodies with specificities against the Tac peptide and the Fc receptors of LGLs (CD16) showed potent killing of Tac-expressing HUT-102 cells by LGLs, which was enhanced by IL-2 activation.

E. Anti-Tac-H, A Humanized Antibody That Binds to the IL-2R

As noted previously, our prior attempts to use the anti-Tac monoclonal antibody in humans have been limited by weak recruitment of effector functions and neutralization by antibodies to mouse immunoglobulins. To circumvent these difficulties, in conjunction with Dr. Cary Queen, we have constructed a "humanized" antibody, anti-Tac-H, by combining the complementarity-determining regions (CDRs) of the anti-Tac antibody with human framework and constant regions (Queen et al., 1989). The human framework regions were chosen to maximize homology with the anti-Tac antibody sequence. In addition, a computer model of murine anti-Tac was used to identify several amino acids which, while outside the CDRs, are likely to interact with the CDRs or antigen. These mouse amino acids were also retained in the humanized antibody. It is hoped that the humanized anti-Tac antibodies will lack the T helper cell recognition units required for the production of anti-monoclonal antibodies. The hyperchimeric antibodies to different antigens studied by Hale (1988) and by LoBuglio (1989) and their co-workers elicited either no or only modest anti-globulin responses. The chimeric anti-Tac antibodies maintained high binding affinity (3×10^{-9} M) for Tac-expressing cells and preserved their abilities to inhibit antigen and mixed leukocyte-induced T-cell proliferation. Furthermore, the IgG_1 hyperchimeric form of anti-Tac manifested a new activity of ADCC with human mononuclear cells that was absent in the parental mouse anti-Tac (Junghans et al., 1990). With this new ADCC activity it is hoped that there will be a substantial improvement in the performance of the antibody in vivo which should translate into an increase in therapeutic efficacy.

VI. Summary

We have proposed a multichain model for the high-affinity IL-2R involving two IL-2-binding peptides, a 70/75-kDa β and a 55-kDa α peptide, which are associated in a receptor complex. A series of additional peptides of M_r 22,000, 35,000, 40,000, 75,000 (non-IL-2 binding), 95,000–105,000, and

180,000 have been associated with the two IL-2-binding peptides. In contrast to resting T cells, the abnormal T cells of patients with select autoimmune disorders, individuals rejecting allografts, and of patients with HTLV-I associated tropical spastic paraparesis or ATL express the Tac peptide of the IL-2R. To exploit this difference in Tac expression, we have initiated therapeutic trials using unmodified anti-Tac, anti-Tac variable region truncated toxin fusion proteins [anti-Tac(Fv)-PE40], IL-2 mutated toxin fusion proteins (IL-2-PE66^4-Glu), and α- and β-emitting isotopic chelates of anti-Tac. Humanized hyperchimeric anti-Tac molecules (anti-Tac-H) have been prepared by genetic engineering in which the molecule is entirely human IgG$_1$ except for the small complementarity-determining regions that are retained from the mouse antibody. The IgG$_1$ hyperchimerics manifested a new activity of ADCC with human mononuclear cells that was absent in the parental mouse anti-Tac. Thus, our present understanding of the IL-2/IL-2R system opens the possibility for more specific immune intervention strategies. The IL-2R may prove to be an extraordinarily versatile therapeutic target. The clinical applications of anti-IL-2R-directed therapy represent a new perspective for the treatment of HTLV-1-associated neurological disoilers such as tropical spastic paraparesis, and for the treatment of adult T-cell leukemia which is caused by this retrovirus.

References

Bacha, P., Williams, D. P., Waters, C., Williams, J. R., Murphy, J. R., and Strom, T. B. (1988). *J. Exp. Med.* **167**, 612–622.

Catane, R., and Longo, D. L. (1989). *Isr. J. Med. Sci.* **24**, 471–476.

Chudhary, V. K., Queen, C., Junghans, R. P., Waldmann, T. A., FitzGerald, D. J., and Pastan, I. (1989). *Nature (London)* **339**, 394–397.

Cooper M. M., Robbins, R. C., Waldmann, T. A., Gansow, O. A., and Clark, R. E. (1988). *Surg. Forum* **39**, 353–355.

Cross, S. L., Feinberg, M. D., Wolf, J. B., Holbrook, N. J., Wong-Staal, F., and Leonard, W. J. (1987). *Cell* **49**, 47–56.

D'Andrea, A. D., Fasman, G. D., and Lodish, H. F. (1989). *Cell* **58**, 1023–1024.

Diamantstein, T., and Osawa, H. (1986). *Immunol. Rev.* **92**, 5–27.

FitzGerald, D., Waldmann, T. A., Willingham, M. C., and Pastan, I. (1984). *J. Clin. Invest.* **74**, 966–971.

Fujii, M., Sassone-Corsi, P., and Verma, I. M. (1988). *Proc. Natl. Acad. Sci. U.S.A.* **85**, 8526–8530.

Grimm, E. A., and Rosenberg, S. A. (1984). *In* "Lymphokine" (E. Pick, ed.), Vol. IX, p. 279. Academic Press, New York.

Hale, G., Dyer, M. J. S., Clark, M. R., Phillips, J. M., Marcus, R., Riechmann, L., Winter, G., and Waldmann, H. (1988). *Lancet* ii, 1394–1399.

Hatakeyama, M., Tsudo, M., Minamoto, S., Kono, T., Doi, T., Miyata, T., Miyasaka, M., and Tomiguchi, T. (1989). *Science* **244**, 551–556.

Herrmann, F., and Diamantstein, T. (1987). *Immunobiology* **175**, 145–158.

Hwang, J., FitzGerald, D. J. T., Adya, S., and Pastan, I. (1987). *Cell* **48**, 129–136.

Inoue, J., Seiki, M., Taniguchi, T., Tsuru, S., and Yoshida, M. (1986). *EMBO J.* **5**, 2883–2888.

Junghans, R. P., Landolfi, N. F., Avdalovic, N. M., Schneider, W. P., Waldmann, T. A., and Queen, C. (1990). *Cancer Res.* **50**, 1495–1502.

Kiyokawa, T. Shirono, K., Hattori, T., Nishimura, H., Yamaguchi, K., Nichols, J. C., Strom, T. M., Murphy, J. R., and Takatsuki, K. (1989). *Cancer Res.* **49**, 4042–4046.

Köhler, G., and Milstein, C. (1975). *Nature (London)* **256**, 495–497.

Korsmeyer, S. J., Greene, W. C., Cossman, J., Hsu, S. M., Jensen, J. P., Neckers, L. M., Marshall, S. L., Bakhshi, A., Depper, J. M., Leonard, W. J., Jaffe, E. S., and Waldmann, T. A. (1983). *Proc. Natl. Acad. Sci. U.S.A.* **80**, 4522–4526.

Kozak, R. W., Atcher, R. W., Gansow, O. A., Friedman, A. M., Hines, J. J., and Waldmann, T. A. (1986). *Proc. Natl. Acad. Sci. U.S.A.* **83**, 474–478.

Kozak, R. W., Raubitschek, A., Mirzadeh, S., Brechbiel, M. W., Junghans, R., Gansow, O. A., and Waldmann, T. A. (1989). *Cancer Res.* **49**, 2639–2644.

Leonard, W. J., Depper, J. M., Crabtree, G. R., Rudikoff, S., Pumphrey, J., Robb, R. J., Krönke, M., Svetlik, P. B., Peffer, N. J., Waldmann, T. A., and Greene, W. C. (1984). *Nature (London)* **311**, 626–631.

LoBuglio, A. F., Wheeler, R. H., Trang, J., Haynes, A., Rogers, K., Harvey, E. B., Sun, L., Ghrayeb, J., and Khazaeli, M. B. (1989). *Proc. Natl. Acad. Sci. U.S.A.* **86**, 4220–4224.

Lorberboum-Galski, H., Kozak, R., Waldmann, T., Bailon, P., FitzGerald, D., and Pastan, I. (1988). *J. Biol. Chem.* **263**, 18650–18656.

Lowenthal, J. L., and Greene, W. C. (1987). *J. Exp. Med.* **166**, 1156–1161.

Maeda, M., Arima, N., Daitoku, Y., Kashihara, M., Okamoto, H., Uchiyama, T., Shirono, K., Matsuoka, M., Hattori, T., Takatsuki, K., Ikuta, K., Shimizu, A., Honjo, T., and Yodoi, J. (1987). *Blood* **70**, 1407–1411.

Maruyama, M., Shibuya, H., Harada, H., Hatakeyama, M., Seiki, M., Fujita, T., Inoue, J-I., Yoshida, M., and Taniguchi, T. (1987). *Cell* **48**, 343–350.

Nakamura, Y., Inamoto, T., Sugie, K., Masutani, H., Shendo, T., Tagaya, Y., Yamaguchi, A., Ozawa, K., and Yodoi, J. (1989). *Proc. Natl. Acad. Sci. U.S.A.* **86**, 1318–1322.

Nelson, D. L., Rubin, L. A., Kurman, C. C., Fritz, M. E., and Boutin, B. (1986). *J. Clin. Immunol.* **6**, 114–120.

Nikaido, T., Shimizu, N., Ishida, N., Sabe, H., Teshigawara, K., Maeda, M., Uchiyama, T, Yodor, S., and Honjo, T. (1984). *Nature (London)* **311**, 631–635.

Ortaldo, J. R., Mason, A. T., Gerard, J. P., Henderson, L. E., Farrer, W., Hopkins, W. F., Herberman, R. B., and Rabin, H. (1984). *J. Immunol.* **133**, 779–783.

Poiesz, B. J., Ruscetti, F. W., Gazdar, A. F., Bunn, P. A., Minna, J. D., and Gallo, R. C. (1980). *Proc. Natl. Acad. Sci. U.S.A.* **77**, 7415–7419.

Queen, C., Schneider, W. P., Selick, H. E., Payne, P. W., Landolfi, N. F., Duncan, J. F., Avdalovic, N. M., Levitt, M., Junghans, R. P., and Waldmann, T. A. (1989). *Proc. Natl. Acad. Sci. U.S.A.* **86**, 10029–10033.

Ruben, S., Poteat, H., Tan, T-H, Kawakame, K., Roeder, R., Haseltine, W., and Rosen, C. A. (1988). *Science* **241**, 89–92.

Rubin, L. A., Kurman, C. C., Biddison, W. E., Goldman, N. D., and Nelson, D. L. (1985). *Hybridoma* **4**, 91–102.

Saito, Y., Sabe, H., Suzuki, N., Kondo, S., Ogura, T., Shimuzu, A., and Honjo, T. (1988). *J. Exp. Med.* **168**, 1563–1572.

Saragovi, H., and Malek, T. R. (1987). *J. Immunol.* **139**, 1918–1926.

Saragovi, H., and Malek, T. R. (1990). *Proc. Natl. Acad. Sci. U.S.A.* (in press)

Schwarting R., Gerdes, J., and Stein, H. (1985). *J. Clin. Pathol.* **38**, 1196–1197.

Seiki, M., Hattori, S., Hirayama, Y., and Yoshida, M. (1983). *Proc. Natl. Acad. Sci. U.S.A.* **80**, 3618–3622.

Sharon, M., Klausner, R. D., Cullen, B. R., Chizzonite, R., and Leonard, W. J. (1986). *Science* **234**, 859–863.

Shimotohno, K., Miwa, M., Slamon, D. J., Chen, I.S.Y., Hiro-o, H., Takano, M., Fujino, M., and Sugimura, T. (1985). *Proc. Natl. Acad. Sci. U.S.A.* **82**, 302–306.

Siekewitz, M., Feinberg, M. B., Holbrook, N., Wong-Staal, F., and Greene, W. (1987). *Proc. Natl. Acad. Sci. U.S.A.* **84**, 5389–5393.

Sodroski, J. G., Rosen, C. A., and Haseltine, W. A. (1984). *Science* **225**, 381–385.

Suzuki, N., Matsunami, N., Kanamori, H., Ishida, N., and Shimizu, A. (1987). *J. Biol. Chem.* **262**, 5079–5086.

Szöllösi, J., Damjanovich, S., Goldman, C. K., Fulwyler, M., Aszalos, A. A., Goldstein, G., Rao, P., Talle, M. A., and Waldmann, T. A. (1987). *Proc. Natl. Acad. Sci. U.S.A.* **84**, 7246–7251.

Tendler, C., Greenberg, S., Blattner, W., Manns, A., and Waldmann, T. A. (1989). *Blood* **74**, 205a.

Tsudo, M., Kozak, R. W., Goldman, C. K., and Waldmann, T. A. (1986). *Proc. Natl. Acad. Sci. U.S.A.* **83**, 9694–9698.

Tsudo, M., Kozak, R. W., Goldman, C. K., and Waldmann, T. A. (1987). *Proc. Natl. Acad. Sci. U.S.A.* **84**, 4215–4218.

Tsudo, M., Kitamura, F., and Migasaka, M. (1989). *Proc. Natl. Acad. Sci. U.S.A.* **86**, 1982–1986.

Uchiyama, T., Yodoi, J., Sagawa, K., Takasuki, K., and Uchino, H. (1977). *Blood* **50**, 481–492.

Uchiyama, T., Broder, S., and Waldmann, T. A. (1981a). *J. Immunol.* **126**, 1393–1397.

Uchiyama, T., Nelson, D. L., Fleischer, T. A., and Waldmann, T. A. (1981b). *J. Immunol.* **126**, 1398–1403.

Uchiyama, T., Hori, T., Tsudo, M., Wano, Y., Umadome, H., Tamori, S., Yodoi, J., Maeda, H., Sawami, H., and Uchino, H. (1985). *J. Clin. Invest.* **76**, 446–453.

Waldmann, T. A. (1986). *Science* **232**, 727–732.

Waldmann, T. A. (1989a). *J. Natl. Cancer Inst.* **81**, 914–922.

Waldmann, T. A. (1989b). *Annu. Rev. Biochem.* **58**, 875–911.

Waldmann, T. A., Greene, W. C., Sarin, P. S., Saxinger, C., Blayney, W., Blattner, W. A., Goldman, C. K., Bongiovanni, K., Sharrow, S., Depper, J. M., Leonard W., Uchiyama, T., and Gallo, R. C. (1984). *J. Clin. Invest.* **73**, 1711–1718.

Waldmann, T. A., Longo, D. L., Leonard, W. J., Depper, J. M., Thompson, C. B., Krönke, M. Goldman, C. K., Sharrow, S., Bongiovanni, K., and Greene, W. C. (1985). *Cancer Res.* **45**, 4559s–4562s.

Waldmann, T. A., Kozak, R. W., Tsudo, M., Oh-ishi, T. Bongiovanni K. F., and Goldman, C. K. (1986). *In* "Progress in Immunology" (B. Cinidar and R. G. Miller, eds.), Vol. VI, p. 553. Academic Press, Orlando.

Waldmann, T. A., Goldman, C. K., Bongiovanni, K. F., Sharrow, S. O., Davey, M. P., Cease, K. B., Greenberg, S. J., and Longo, D. (1988). *Blood* **72**, 1805–1816.

Wang, H-M., and Smith, K. A. (1988). *J. Exp. Med.* **166**, 1055–1069.

Williams, J. M., Kelley, V. E., Kirkman, R. L., Tilney, N. L., Shapiro E. Murphy, J. R., and Strom, T. B, (1988). *Immunol. Invest.* **16**, 687–723.

Toward the Rational Design of Antiretroviral Therapy for Human Immunodeficiency Virus (HIV) Infection

Hiroaki Mitsuya and Samuel Broder

I. Introduction

The acquired immunodeficiency syndrome (AIDS) remains a significant and worsening medical problem since it was first described in 1981 as a new clinical entity (Gottlieb *et al.*, 1981; Masur *et al.*, 1981; Siegal *et al.*, 1981). It is estimated that in the United States alone more than one million people may have been infected by the causative agent, human immunodeficiency virus, or HIV (Barré-Sinoussi *et al.*, 1983; Popovic *et al.*, 1984) and over 10 million people may harbor this virus worldwide. Most of these people may finally develop a serious clinical illness related to their HIV infection. However, we can say that in the past several years substantial progress has been made in the chemotherapy of HIV infection; such progress is likely to exert a major effect against the HIV epidemic in the coming decade.

It has been shown that the replication of HIV can be suppressed in patients with HIV-related diseases by several antiretroviral drugs (Mitsuya *et al.*, 1984; Mitsuya and Broder, 1986, 1987; Yarchoan *et al.*, 1986, 1988a, b, 1989a, b, 1990; Yarchoan and Broder, 1987; Fischl *et al.*, 1987; Hirsch and Kaplan, 1987; Merigan *et al.* 1989). One such drug, 3'-azido-2',3'-dideoxy-nucleoside (AZT or zidovudine), has been formally shown to confer prolonged survival and improved quality of life in patients with advanced HIV infection (Mitsuya and Broder, 1987). More recently, it has also been shown that the administration of AZT to symptomatic or asymptomatic HIV-infected individuals can in certain circumstances delay the short-term progression to AIDS or symptomatic AIDS-related complex (ARC). Furthermore, recent data suggest that another drug, called 2',3'-dideoxyinosine (ddI) (Mitsuya and Broder, 1986, 1987), has an antiretroviral activity against HIV in patients with AIDS or ARC at doses that can be well tolerated (Yarchoan *et al.*, 1989a, 1990; Cooley *et al.*, 1990; Lambert *et al.*, 1990). The development of such drugs has clearly ushered in a new era of therapy against retroviral diseases.

In theory, antiviral drugs exert their effects by interacting with viral structural components, virally encoded enzymes, viral genomes or proteins, cellular receptors, cellular enzymes, or factors required for viral replication. The discovery of new modalities of therapy for HIV infection is not an exception. For example, in the development of antiretroviral therapy, the virally encoded reverse transcriptase has been one of the most attractive targets (Mitsuya and Broder; 1986, 1987, 1990; Yarchoan and Broder, 1987; Hirsch and Kaplan, 1987; Yarchoan *et al.,* 1989b). A number of drugs targeting this enzyme have been shown to be active against various animal and human retroviruses including HIV (Ruprecht *et al.*, 1986; Dahlberg *et al.*, 1987; Matsushita *et al.*, 1987; Sharp *et al.*, 1987; Travares *et al.*, 1987; Mitsuya and Broder, 1988; Tsai *et al.*, 1988). Recombinant soluble CD4 (rCD4), a soluble form of the receptor for HIV, to be discussed later, is also among the first of what might be called rationally designed therapies for HIV infection (Smith *et al.*, 1987; Fisher *et al.*, 1988; Hussey *et al.*, 1988; Deen *et al.*, 1988; Traunecker *et al.*, 1988). Some selected stages which may be targeted for therapeutic interventions of HIV are illustrated in Fig. 1.

The development of antiviral therapy against AIDS has been an area of intense research effort since the seminal discovery of the causative virus (Barré-Sinoussi *et al.*, 1983; Popovic *et al.*, 1984). In this chapter, we will highlight advances in the antiretroviral therapy against AIDS and its related

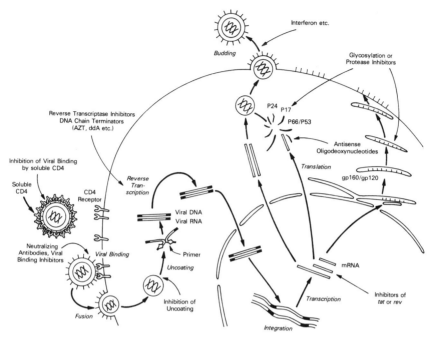

Figure 1. Some stages of the HIV replicative cycle that may theoretically be targeted for therapeutic intervention.

diseases in the past six years. We will also discuss selected antiretroviral drugs, whose antiviral actions have been studied at a molecular level.

II. Genetic Organization of HIV

HIV, an enveloped single-stranded RNA virus, which exists as a dimer of identical RNA molecules, is perhaps the most complex retrovirus thus far studied. HIV has at least nine known genes. These genes are flanked at their ends by sequences called long terminal repeats, or LTRs, which do not code for any protein but serve to initiate the expression of the viral genes (Fig. 2). A newly proposed nomenclature of these genes (Gallo *et al.*, 1988) is illustrated in Table I. Before human pathogenic retroviruses were identified, retroviruses were known to contain a set of three genes, designated *gag*, *pol*, and *env*, as basic components of a replicating genome. In addition to this standard set of genes, HIV has at least six more extra genes, none of which is homologous to mammalian cellular genes (i.e., not *onc* genes) and whose origin is as yet unknown. HIV, in common with previously known animal retroviruses, has as its major structural components: a core of genomic RNA; Gag proteins which play a role both in the structure of the core and

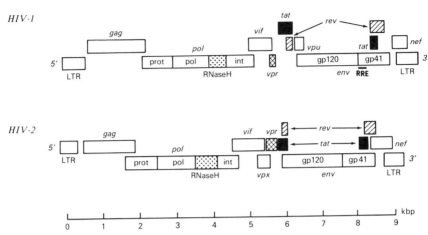

Figure 2. Genetic organization of HIV-1 and HIV-2. Three lanes in each panel represent three different reading frames where each gene is located. LTR, long terminal repeat; *gag*, group-specific antigen; *pol*, reverse transcriptase; prot, protease; int, integrase (endonuclease); *vif*, viral infectivity factor; *vpr*, viral protein R: *vpu*, viral protein U; *tat*, transactivating protein; *rev*, regulator of expression of virion proteins; *nef*, negative regulator factor; *vpx*, viral protein X. The minimal RRE (*rev*-responsive element) (positions 7,361–7,569), as defined in Malim *et al.* (1989), is indicated. TAR (*trans*-acting responsive element) is located between nucleotides −17 and +80 in the 5′ LTR (Rosen *et al.*, 1985).

assembly of virion in the membrane of the host cell; a lipid bilayer; a transmembrane envelope glycoprotein (gp41) spanning the lipid layer; and an outside envelope glycoprotein (gp120).

The specific functions of several of these genes are as yet incompletely understood, and much more research is needed to learn if they could be the target of novel therapies. One such gene, *vpu*, has been recently suggested to have a role in assembly or maturation of progeny viruses, and thus to facilitate virus release from infected cells (Terwillinger *et al.*, 1988; Strebel *et al.*, 1989). However, it is still unclear whether this gene, encoding a 16-kDa protein, is essential for the replication of HIV. Indeed, the *vpu* gene distinguishes HIV-1 isolates from HIV-2 and simian immunodeficiency virus (SIV); the latter do not encode a similar protein (Strebel *et al.*, 1989). Instead both HIV-2 and SIV carry the *vpx* (viral protein X) gene, which encodes a protein capable of binding nucleic acids (Kappes *et al.*, 1988). The role of Vpx protein also remains obscure; however, the fact that HIV-1 entirely lacks this gene suggests that the *vpx* gene expression is not directly required for induction of the immunodeficiency state observed in HIV-1 infection. Another gene, *vif* (virion infectivity protein), has been linked to the ability of HIV to replicate by a pathway of cell-free virion infection (Fisher *et al.*, 1987). The *vif* gene encodes a 23-kDa protein that appears to play a crucial role in the efficient generation of infectious virus particles. The exact mechanism of the *vif* function is still to be determined. However, it is conceivable

Table I

Nomenclature of Genes of HIV-1 and HIV-2

Proposed name	Old name	Product	Function
ma	*gag*	p17	Matrix
ca		p24	Capsid
nc		p9	Nucleocapsid
pr	*pol*	p10	Protease
rt		p66, 51	Reverse transcriptase
in		p32	Endonuclease (integrase)
su	*env*	gp120	Surface envelope
tm		gp41	Transmembrane envelope
vif	*sor*(Q, P', orfA)	p23	Virion infectivity protein
vpr	*R*		Viral protein R (dispensable?)
vpu	*U*	p16	Viral protein U (release of mature virions?)
tat	*tat*-III	p15	*trans*-activation protein
rev	*art/trs*		Regulator of expression of virion protein
nef	3'*orf*(F, E', orfB)	p25	Negative regulatory factor
vpx	*X* (HIV-2 and SIV)		Viral protein X (function unknown; capable of binding nucleic acids)

that drugs or biologics could be developed to interfere with the *vif* gene and thereby attenuate the pathogenicity of HIV infection. The *vpr* gene (viral protein R) has been shown to be conserved in HIV-1 and HIV-2. Mutations within the *vpr* gene do not affect the viral replication kinetics and cytopathogenicity, suggesting that the *vpr* gene is dispensable at least in cultured lymphoid cells (Dedera *et al.*, 1989; Ogawa *et al.*, 1989). However, the significance of the presence of this gene and all the above genes as well both in HIV-1 and HIV-2 needs to be further studied. Three interesting regulatory genes, *tat*, *rev*, and *nef*, and several important viral and cellular elements linked to HIV replication such as TAR (*trans*-acting responsive sequence) in LTR and NFκB, will be discussed later.

III. Possible Targets for Therapeutic Intervention

A. Inhibition of Viral Binding to Target Cells

Since the interaction of a virus with the receptor on the target cell plays a pivotal role in the initiation and spread of infection, viral receptors on the cells have long been principal targets for antiviral strategies. The first molecule that HIV encounters (via the gp120 moiety of its envelope) as it begins its infection of a target cell is the CD4 molecule (Dalgleish et al., 1984; Klatzman et al., 1984) although CD4 does not seem to be the only target molecule for HIV infection (Clapham et al., 1989; Tateno et al., 1989; Miller et al., 1989a). The CD4 protein is a member of the immunoglobulin supergene family (White and Littman, 1989). Three of its four external domains resemble the constant regions of immunoglobulins. The site to which the HIV gp120 molecule binds lies in the outermost domain. In late 1987 and early 1988, several groups reported that recombinant soluble CD4 (rCD4), lacking the transmembrane and cytoplasmic portions, could inhibit the HIV binding to and infection of $CD4^+$ target cells (Smith et al., 1987; Fisher et al., 1988; Hussey et al., 1988; Deen et al., 1988). A potential advantage of this approach is that soluble CD4 is likely to inhibit diverse HIV isolates, since both HIV-1 and HIV-2 use CD4 as the receptor. However, our group and others have recently found that HIV-2 viruses are less susceptible to the inhibitory effects of soluble CD4 than are HIV-1 viruses (Clapham et al., 1989; Looney et al., 1990). It was subsequently noted that certain fresh primary HIV-1 isolates were also markedly less susceptible to rCD4 than certain viral prototypes such as HTLV-III$_B$ (Daar et al., 1990). The mechanism of the differential sensitivity of HIV to soluble CD4 has not been completely defined; however, the higher density of envelope glycoprotein on the virion surface of HIV-2, as compared to HIV-1, appears to contribute at least to some extent, to the functional resistance of HIV-2 to soluble CD4 (Looney et al., 1990).

The approach of using genetically created, truncated soluble CD4 as a therapeutic agent against HIV infection is still evolving. The plasma half-life of soluble CD4 averages only 15 min in rabbits (Capon et al., 1989) and 35–40 min in humans (Yarchoan et al., 1989c). To overcome this problem, hybrid molecules have been created by genetically combining the CD4 protein and constant heavy chain domains of human or murine immunoglobulins (Capon et al., 1989; Traunecker et al., 1989) (Fig. 3). We have reported that these hybrid CD4 molecules retain an antiviral activity against HIV-1 comparable to soluble CD4 in T cells and monocytes/macrophages (Capon et al., 1989). In some systems hybrid CD4 molecules exhibit a more potent antiviral activity than unmodified soluble CD4 (Traunecker et al., 1989). Such chimeric molecules also gain other desirable properties, such as a longer plasma half-life (Capon et al., 1989) and a capacity to bind Fc receptors with high affinity (Capon et al., 1989; Traunecker et al., 1989). It is not yet clear to what extent the addition of immunoglobulin components can provide advantages in a clinical setting. However, if a good steady-state level of such modified soluble CD4 products is attained in vivo, effective tissue concentrations and transplacental transfer could be achieved, and antiretroviral activity against HIV might be efficiently obtained.

Soluble CD4 can also be combined toxins, such as pseudomonas endotoxin (Chaudhary et al., 1988) or ricin, or with a radionuclide. Thus, one

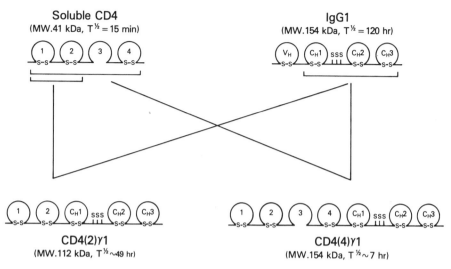

Figure 3. Structure of hybrid molecules of CD4 and immunolglobulin G. Two distal or all four domains of soluble CD4 were genetically combined to the Fc portion of human immunoglobulin G_1 (IgG$_1$), generating CD4(2)γ1 and CD4(4)γ1. Plasma half-life of each molecule was obtained in rabbits (Capon *et al.*, 1989).

could specifically destroy and eliminate chronically HIV-infected cells such as certain types of B cells (Montagnier *et al.*, 1984) and monocytes/macrophages (Gartner *et al.*, 1986; Koenig *et al.*, 1986; Ho *et al.*, 1986) that otherwise constitutively produce HIV without being killed by the virus. Another approach to block the binding of gp120 to CD4$^+$ target cells is the use of synthetic CD4 peptides (Hayashi *et al.*, 1989). However, to date, the *in vitro* anti-HIV activity of short CD4 peptides appears to be significantly less than that of recombinant CD4 molecules. More studies of these modified soluble CD4 proteins or "second generation" products are presently underway. Once the structural and clinical requirements for optimal effectiveness are defined, soluble CD4 species could be potent antiviral agents against AIDS and related diseases, alone or in combination with other therapies. As discussed earlier, HIV can also infect CD4$^-$ cells including neural cells; muscle cells, or fibroblastoid cells *in vitro* (Clapham *et al.*, 1989; Tateno *et al.*, 1989; Miller *et al.*, 1989a). Although the mechanism(s) and significance of the CD4-independent transmission remain to be determined, such a pathway may have implications concerning the efficacy of this therapeutic approach. The current clinical trials on soluble CD4 species might provide more evidence for or against this mechanism in a clinical setting.

Recently, low-molecular-weight dextran sulfate (7000–8000 Da) was shown to block the binding of HIV virions to CD4$^+$ target cells, inhibit virally induced syncytia formation, and exert a potent inhibitory effect against HIV *in vitro* (Ueno and Kuno, 1987; Ito *et al.*, 1987; Nakashima *et al.*, 1987; Mitsuya *et al.*, 1988a; Baba *et al.*, 1988). However, a phase I/II trial of orally administered dextran sulfate has suggested that this compound had little toxicity but also little clinical effect (Abrmas *et al.*, 1989). Subsequent studies showed that dextran sulfate was very poorly absorbed when given orally using changes in partial thromboplastin time as a marker (Lorentsen *et al.*, 1989). More recent data of ours using rats as a model have suggested that this anionic polysaccharide is almost totally degraded in the gastrointestinal tract when given orally, and sufficient inhibitory doses are not achieved in the plasma (Hartman *et al.*, 1990). Further *in vitro* analyses of the antiviral activity of dextran sulfate against HIV-1 in our laboratory have indicated that dextran sulfate with molecular weight less than 2300 had no antiviral effect (Hartman *et al.*, 1990). Furthermore, it has been shown that the higher the concentration of human serum, the higher the concentration of dextran sulfate required for antiviral effect (Hartman *et al.*, 1990). Dextran sulfate might theoretically bring about coagulation abnormalities at such high antiviral concentrations in the plasma. Perhaps dextran sulfate could be viewed as a prototype for a large class of anionic polysaccharides of both natural and synthetic origin which can block viral replication by inhibiting viral binding or other early events. It should be stressed that cautious interpretation of *in vitro* antiretroviral data is required and issues such as absorption and bioavailability need to be addressed in considering clinical application (Mitsuya *et al.*, 1989; Hartman *et al.*, 1990).

Yet another target is the envelope glycoprotein of HIV. While considerable variation in this glycoprotein of HIV has been demonstrated, the range of variation in the CD4-binding site of the virus must be somewhat constrained because the virus needs to bind to the CD4 molecule (Maddon *et al.*, 1986). It is possible that an antibody directed against the CD4-binding site could bind to various HIV isolates and destroy gp120-expressing infected cells. Monoclonal antibodies specific for the envelope glycoprotein (Matsushita *et al.*, 1988a) can also be combined to a toxin. Thus, such toxin-conjugated monoclonal antibodies could also destroy the viral envelope-expressing cells (Matsushita *et al.*, 1988b). However, virally infected cells could theoretically transmit the virus without the need for CD4 binding by cell-to-cell contact. Therefore the cells might escape such attacks by soluble CD4 or humoral antibodies. Furthermore the notorious variability in the nucleotide sequences in the *env* gene has made these humoral approaches difficult. Another approach combining an antiretroviral drug and a monoclonal antibody specifically reactive with a viral antigen expressed on the cell surface may render the delivery of the drug to chronically infected cells highly specific. However, these studies are still at an early stage.

B. Inhibition of Later Stages of Viral Entry

The binding of gp120 to cellular CD4 is only the first step of viral entry into the cell, and later stages could also be targets for the therapy of AIDS; however, the later stages of the viral entry are less thoroughly understood.

According to one plausible model, the binding of gp120 to cellular CD4 induces a change in the conformation of the gp120, unmasking a specific site in the hydrophobic gp41 transmembrane portion of the envelope, which is called the fusogenic domain (Weiss et al., 1985; Peterson and Seed, 1988). This hydrophobic site of gp41 is thought to be normally hidden under the gp120 molecule. Once it is unmasked, the hydrophobic fusogenic domain interacts with the adjacent cell membrane and induces virion–cell or cell–cell fusion. Pharmacological agents which interfere with this step might eventually be developed in the future. Indeed, antibodies directed against gp41 have been shown to neutralize diverse strains of HIV, presumably by inhibiting the fusion process (Weiss et al., 1985, 1986).

Binding of HIV virions may subsequently bring about phosphorylation of the CD4 molecule, via cellular protein kinases. The process of viral entry cannot be functionally completed without such CD4 phosphorylation (Fields et al., 1988), suggesting that specific inhibition of the CD4 phosphorylation could be a potential target for future therapies. After HIV enters the cell, the virus loses the envelope coat (uncoats) and releases its functional RNA genome into the cytoplasm as a ribonucleoprotein complex. Therapeutic strategies to attack this stage may prove to be effective in a setting of HIV infection. In fact, several inhibitors (WIN51711 and WIN52084) of the uncoating stage have already been shown to be valuable for rhino (common cold) viruses and related picornaviruses (Smith et al., 1986). These drugs may inhibit the viral disassembly by preventing the collapse of the "hydrophobic pocket" of the viral structural capsid protein or by blocking the flow of ions into the virus interior (Smith et al., 1986). The uncoating processes of these viruses could follow a different mechanism from that of retroviruses. However, the prospect of freezing the retrovirus in its nucleocapsid state, and thereby preventing RNA release, is attractive and worthy of further research.

C. Reverse Transcriptase and Viral Protease as Targeted Molecules

Once uncoating occurs, viral RNA is used as a template for the proviral DNA synthesis catalyzed by reverse transcriptase using a lysine transfer RNA as a primer (Wain-Hobson et al., 1985; Ratner et al., 1985a). This process has proven to be one of the most successful targets for the development of antiretroviral agents (Mitsuya et al., 1985; Mitsuya and Broder, 1986, 1987, 1990; Yarchoan and Broder, 1987; Yarchoan et al., 1989b). Several drugs thus far shown to be effective in patients and a number of compounds active against HIV in vitro act at this stage (Mitsuya et al., 1985; Mitsuya and Broder, 1986, 1987, 1990, Yarchoan et al., 1986, 1988a, b, 1989a, b, 1990; Yarchoan and Broder, 1987; Fischl et al., 1987; Hirsh and Kaplan, 1987; Merigan et al., 1989). The reverse transcriptase is encoded by the pol gene. The pol gene (Fig. 2) generates three distinct proteins: viral protease (vide infra); the reverse transcriptase (which is associated with an

RNaseH at a different catalytic site); and integrase (in protein), i.e., an endonuclease involved in integrating the proviral DNA into the cellular genome.

The reverse transcriptase is manufactured in HIV-infected cells as a fusion protein involving the adjacent gag gene that lies in a different translation reading frame through an adroit translational event, ribosomal frame shifting (Jacks et al., 1988). In an animal retroviral model of ribosomal frame shifting, it has recently been shown that the frame shifting reaction is mediated by slippage of two adjacent tRNAs by a single nucleotide in the 5' direction (Jacks et al., 1989). Further definition of the frame shifting mechanism will likely lead to the development of novel types of antiretroviral agents. The gag/pol fusion protein then undergoes posttranslational cleavage events mediated by the viral protease to form active gag and pol products. Mutation of the active site in the HIV-1 protease damages processing of the gag/pol polyprotein, resulting in immature and noninfectious virions. The HIV protease has a symmetrical structure of two units and has a structural homology to the family of microbial aspartyl protease (Navia et al., 1989; Miller et al., 1989b; Blundel and Pearl, 1989). This structure suggests a mechanism for the autoproteolytic release of HIV protease. The self-assembly of two identical monomers into a symmetric structure should be efficient for creating an active enzyme while encoding a minimal amount of genetic information. This self-assembly of two units to create a functional enzyme may take place very late, even in virions that have already been extruded from an infected cell. X-ray crystallographic analyses of HIV-1 protease at 2.7 Å resolution has recently confirmed that this enzyme exists as a homodimer and also revealed the amino acid invariance among retroviral proteases (Lapatto et al., 1989). Most investigators now believe that the protease represents a crucial virus-specific target for new therapies. Studies using molecular modeling of the HIV-1 protease have predicted that it interacts with a seven-residue portion of the protein substrate (Weber et al., 1989). Indeed, Meek and colleagues (1989) have synthesized peptide analogs which acted in vitro as competitive inhibitors of the HIV-1 protease. More recent studies by Miller and her co-workers have provided a model of the interaction of synthesized inhibitors and substrates with HIV-1 protease by cocrystalizing HIV protease-based inhibitor (Miller et al., 1989a). Roberts and his co-workers (1990) have recently synthesized a series of peptide derivatives based on the same transition-state mimetic concept. Some peptide derivatives incorporating an active hydroxyethylamine moiety inhibit both HIV-1 and HIV-2 proteases with little effect against the structurally related human aspartic proteases and exert a potent and highly specific antiretroviral activity against HIV-1 in the nanomolar range in multiple cell systems. Two-fold (C_2) symmetric inhibitors of HIV protease have recently been designed by Erickson et al. (1990) based on the three-dimensional symmetry of the enzyme active site. These C_2 symmetric compounds inhibited both protease activity and acute HIV-1 infection in vitro. Some compounds were at least

10,000-fold more potent against HIV-1 protease than against related cellular enzymes (Erickson *et al.*, 1990). While further definition of the structure and function of the retroviral protease is important to be able to rationally design inhibitors more active against the HIV-1 protease than proteases normally present in cells, there appear to be enough data to proceed with appropriate preclinical and clinical trials using the protease as a target.

The mature HIV reverse transcriptase is found as two forms of protein, with molecular weights of 51,000 and 64,000 (Di Marzo Veronese *et al.*, 1986). The latter contains a 15,000-Da protein with ribonuclease H (RNaseH) activity (Hansen *et al.*, 1987) which specifically degrades the RNA template from an RNA–DNA hybrid to permit the synthesis of a double-standard viral DNA which is also catalyzed by the reverse transcriptase. In theory, inhibition of the RNaseH activity should suppress viral replication because an effective and orderly degradation of the template viral RNA is required for effective conversion of genomic RNA to proviral DNA. However, many of the already known RNaseH inhibitors are toxic or carcinogenic. Nevertheless, this is a crucial area for future research.

It has recently been shown that HIV-1 reverse transcriptase is highly error-prone (Roberts *et al.*, 1988; Preston *et al.*, 1988). The high error rate of HIV-1 reverse transcriptase produces approximately five to ten errors per HIV-1 genome per round of replication *in vivo* (Roberts *et al.*, 1988; Preston *et al.*, 1988). This high error rate of HIV reverse transcriptase may be, at least in part, responsible for the hypermutability of HIV. Thus, from a clinical point of view, the emergence of drug-resistant HIV variants must always be considered possible. The HIV drug-resistance issue will be discussed later.

D. Migration of the Proviral DNA and Its Integration

The synthesized double-stranded proviral DNA can then migrate into the nucleus by an as yet poorly characterized mechanism, although some proviral DNA can remain in the cytoplasm in an unintegrated linear form. Again, it is possible that this nuclear migration may be blocked by drugs or biologics in the future. Following migration into the nucleus, the linear proviral DNA may be circularized, although it is not proven that such circularized forms are important from a pathophysiological point of view. Indeed, it has recently been shown in murine leukemia virus infection that circularization of proviral DNA played no role in integration, and that the direct precursor to the integrated provirus was a linear molecule (Fujiwara and Mizuuchi, 1988; Bowerman *et al.*, 1989; Brown *et al.*, 1989). The initial step in the integration reaction catalyzed by the In protein (integrase) is perhaps a cleavage that removes the terminal bases from each 3' end of the proviral DNA. The resulting viral 3' ends are then joined to target cellular DNA to form the initial recombination intermediate (Bowerman *et al.*, 1989). Thus, this cleavage event depends on a virally encoded endonuclease, or integrase, which has long been known to be essential for this integration process. The integrase activity has been shown in infected cells as a nucleoprotein complex containing the viral capsid protein (Fujiwara and Mizuuchi, 1988; Bowerman et al., 1989). Further research of this integration process is required in order to design inhibitors of the integration reaction.

E. Transcription and Translation

As we discussed earlier, in addition to the three standard replicative genes, the HIV genome contains at least six extra genes, all of which appear to regulate viral replication. Some of these genes may enhance viral protein synthesis; other genes may down-regulate viral protein production. Following integration of the HIV genome into the cellular gene, perhaps upon activation of the infected cell by physiological signals such as antigens (Zagury et al., 1986) or regulatory interleukins (Perno et al., 1989), the proviral DNA is transcribed to mRNA (viral genomic RNA) using host cell RNA polymerases. Indeed, immunologic activation of T cells by antigenic or mitogenic stimuli has been correlated with induction of DNA-binding proteins which act on the viral LTR and increase HIV expression in synergy with another gene product (Tat) (Sodroski et al., 1985). Recent data have shown that a purified cellular factor NFκB activates transcription from the HIV-1 promoter through the interaction with the NFκB-binding site. This site contains two copies of the sequence GGGACTTTCC in the LTR, and functions as an enhancer element for HIV transcription (Nabel and Baltimore, 1987; Nabel et al., 1988). To date, several factors including EP-1 have been reported to bind to the same sequence of HIV-1 (Franza et al., 1987; Vano et al., 1987; Kawakami et al., 1988; Lenardo et al., 1988; Wu et al., 1988; Mackawa et al., 1989). Conceivably, mutant proteins capable of binding the NFκB-binding site without activation of HIV may be developed to block viral replication.

Proteins encoded by a variety of DNA viruses such as herpes simplex virus type I, adenovirus, or cytomegalovirus are also able to activate expression of HIV. It is theoretically possible that infection by such viruses (including the newly identified human herpes virus type 6) (Salahuddin et al., 1986) may also directly contribute to the activation of HIV by mechanisms independent of immune stimuli (Mosca et al., 1987; Nabel et al., 1988; Skolnik et al., 1988; Lusso et al., 1989). Recent data suggest that in the monocyte lineage, NFκB-binding activity is developmentally regulated and provides one signal for HIV activation in monocytes/macrophages as well (Nabel et al., 1988). It is also possible that human T-cell leukemia virus type I (HTLV-I) (and HTLV-II) produces transcriptional activating factors that potentiate transcription of HIV when both viruses coinfect the same cell. Coinfection by HIV-1 and HTLV-I in the same individuals has been reported (Harper et al., 1986; Robert-Guroff et al., 1986) although the consequences of the coinfection still remain to be determined. Whatever the mechanism of transcription initiation, viral RNA is subsequently translated to form viral proteins, again using the biochemical apparatus of the host cell.

1. The *tat* Gene   HIV has an essential regulatory gene, designated *tat*. The *tat* gene code for a diffusible protein that, through the LTR sequences of HIV, markedly enhances the expression of other viral genes and amplifies the production of new infectious virions at a transcriptional and/or post-transcriptional step (Sodroski *et al.*, 1985). The *tat* gene, like the *tax* gene (Seiki *et al.*, 1983; Gallo *et al.*, 1988) of the first two known human pathogenic retroviruses , HTLV-I and HTLV-II (Wong-Staal and Gallo, 1985), is so called because it mediates a *trans*-activation of transcription, that is, it works by a mechanism that affects the transcription of genes not in its direct proximity through a diffusible factor.

The Tat protein is small (86 amino acids) with a cluster of positively charged amino acids. The predicted amino acid sequence of the Tat protein contains two highly conserved domains which are rich in cysteine and arginine/lysine residues (Ratner *et al.*, 1985; Sodroski *et al.*, 1985). Data suggest that several functional domains of the Tat protein are involved in the transcriptional activation, with the cysteine-rich domain being required for complete activity of the Tat protein (Garcia *et al.*, 1988). More recently, Ruben and his co-workers (1989), using a site-directed mutagenesis technique, have identified three distinct domains which were essential for the *tat* activity but did not regulate subcellular localization. The Tat protein appears to form a dimer mediated by binding to a metal (cadmium or zinc), exerting primary effects in the cysteine-rich domain (Frankel *et al.*, 1988a, b). If dimerization of the Tat protein is required for *trans*-activation of HIV, the Tat function might be inactivated by forming abnormal heterodimers with truncated or inactive subunits, provided such subunits can penetrate into the target cell.

To exert its effects, the Tat protein interacts with a short nucleotide sequence designated TAR (*trans*-acting responsive sequence), which is located within the 5′ LTR and is included in the mRNA transcript of every HIV gene (Rosen *et al.*, 1985). The Tat protein may directly bind to and activate elongation of nascent TAR RNA (Dingwall *et al.*, 1989), or cellular RNA-binding proteins may play a major role in mediating the Tat-dependent LTR activation (Gatignol *et al.*, 1989). Although the direct interaction of the Tat protein with TAR requires more research, if simple inhibitors or strategies such as use of antisense oligonucleotides are developed which can block the binding of the Tat or TAR, viral replication may be efficiently inhibited. If "pseudo TAR" molecules are introduced into target cells, the Tat function might also be weakened or abrogated.

Frankel and Pabo (1988) and Green and Loewenstein (1988) have found that the Tat protein is taken up by cells and may subsequently *trans*-activate the viral promoter to enhance the expression of the virus. Conceivably, a nonactive but competitive Tat analog could be used as a drug. Thus, it may be possible in the future to develop chemicals or biologics that inhibit the production of the Tat protein, block its binding to nucleic acid, or alter other functional interaction of the Tat protein and other proteins, and thus inhibit the replication of the virus.

2. The *rev* Gene   The *rev* gene (formerly known as *art* or *trs*) encodes a small (116 amino acid) positively charged protein, which is thought to function as a second essential *trans*-acting factor in viral replication (Feinberg *et al.*, 1986; Sodroski *et al.*, 1986). In the absence of this second regulatory factor, *gag*- and *env*-encoded protein synthesis is severely diminished (Feinberg *et al.*, 1986; Sodroski *et al.*, 1986). The Rev protein binds to a *cis*-acting element, called the Rev-responsive element or RRE, located within the *env* region of HIV-1 (positions 7,361–7,569: Fig. 2) (Rosen *et al.*, 1988; Zapp and Green, 1989). There are data that the counterpart gene of HTLV-I (called *rex*) (Gallo *et al.*, 1988) might preserve the viral RNA in infected cells by affecting exit pathways for the viral RNA transcript out of the nucleus (Siomi *et al.*, 1988). The *rev* gene also appears to promote the transport of the unspliced viral mRNA containing RRE from the nucleus to the cytoplasm (Felber *et al.*, 1988; Malin *et al.*, 1989). Recent studies suggest that two other different *cis*-acting elements, although non-Rev-responsive, are located within the *gag* region and the *env* region and operate to decrease the level of HIV mRNA (Felber *et al.*, 1988), possibly by decreasing transport of RNA to the cytoplasm. In contrast, the *rev* gene product appears to increase the stability of unspliced and singly spliced viral mRNA species containing RRE (Felber *et al.*, 1988). The multiply spliced small mRNA species such as *tat*, *rev*, and *nef* RNA, unlike the unspliced and singly spliced viral mRNA species (full genome or *env* RNA), are stable and do not require the *rev* product (Felber *et al.*, 1988). The *rev* gene was originally thought to affect the splicing process of HIV RNA species (Feinberg *et al.*, 1986). However, a recent study using mutated proviral constructs producing mRNA that cannot be spliced suggests that the effect of the *rev* gene on the stability of viral RNA species is independent of splicing (Felber *et al.*, 1988).

While there is no consensus as to how *rev* works, there is a clear consensus that this gene is critical for effective viral expression. The *rev* gene is, therefore, an important target for future research. The specific interference by drugs or chemicals with the function of the *rev* gene or RRE function might lead to suppression of viral replication in chronically infected cells. Indeed, studies by Matsukura and his co-workers (1989) in our group have demonstrated that a nuclease-resistant phosphorothioate analog (an oligodeoxynucleotide that contains a sulfur atom in place of one of the two internucleotide nonbridging oxygens), in the form of a *rev* gene antisense construct, can suppress viral replication in chronically HIV-1-infected cells without inhibiting the growth of the target cells. We will discuss the antisense strategy in more detail later in this section.

3. The *nef* Gene   HIV has an additional regulatory gene, designated *nef* (negative regulatory factor), which has been reported to function as a transcriptional silencer which down-regulates the expression of the viral genome (Luciw *et al.*, 1987; Niederman *et al.*, 1989). Thus, the *nef* gene is thought to confer, in part, the ability of HIV to be dormant in the genome of the host cell, although this is an area for which a consensus has not been reached. Sequence similarities between the *nef* gene and the phosphorylation domain of human interleukin 2 receptor, the nucleotide-binding domains of a *ras* oncogene, epidermal growth factor, and insulin receptor have suggested (but not proved) a nucleotide-binding activity of the *nef* gene product (Samuel *et al.*, 1987). Indeed, it has been suggested that the Nef protein is a myristoylated GTP-binding phosphoprotein with features similar to the cellular *src* and *ras* oncogene products (Guy *et al.*, 1987), although this requires further research.

The sequence of the *nef* gene is highly variable (Ratner *et al.*, 1985b). In fact, the Nef protein is the most diverse protein among the different HIV-1 isolates with the exception of the *env* products. Within the *nef* gene, a termination codon has frequently been found (Ratner *et al.*, 1985a; Niederman *et al.*, 1989). Thus, the significance of the *nef* gene expression needs to be further explored; however, it is possible that if chemicals or biologics that can potentiate *nef* gene activity are developed, viral transcription may be more easily controlled with or without other modalities of antiviral strategies.

The three regulatory genes described above therefore form a complex network of regulatory interactions. This picture will be complicated even further when the function of other regulatory genes such as *vif*, *vpu*, and *vpr* are clarified in the future. However, the very complexity of HIV could make this virus more vulnerable to coordinated attacks by antiretroviral drugs at multiple stages in the viral life cycle.

4. Antiviral Agents Targeting the Stage of Transcription/Translation

An intriguing strategy of blocking the expression of retroviral genes by constructing negative strand (antisense) synthetic oligodeoxynucleotide was first proposed by Zamecnik and Stephenson (1978; Zamecknik *et al.*, 1986). However, several factors have complicated this area of research. First, unmodified oligodeoxynucleotides are subject to rapid hydrolysis by host nucleases. Second, certain chemically modified oligomers have poor solubility and require exceedingly high concentrations for biological effects (Smith *et al.*, 1986). If an antisense strategy can be operational, agents should affect the expression of specific viral genes; however, this remains a very controversial area. Many studies, particularly HIV-related studies, have not proven a strong effect at the level of viral expression in chronically HIV-infected cells. In this regard, Matsukura and his co-workers (1987) explored the possibility of using a phosphorothioate modification of oligodeoxynucleotides to make antiviral oligomers resistant to nucleases. A phosphorothioate analog of 28-mer homooligodeoxycytidine (S–dC$_{28}$) was found to be potent against HIV-1 *in vitro*. This compound was active against HIV-1

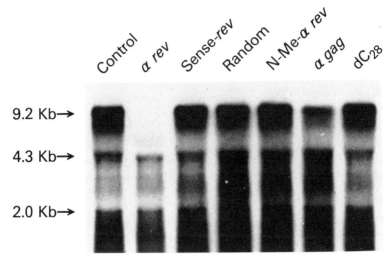

Figure 4. Sequence-specific anti-*rev* effects on the HIV RNA profile. Cytoplasmic RNA (10 μg) extracted from chronically HIV-1-infected H9 cells on day 5 in culture with no compound added (Control), or a phosphorothioate oligomer with an antisense construct against *rev* (α rev), with a sense-*rev* construct (Sense-*rev*), with a random sequence with the same composition as *rev* (Random), with a *rev* antisense sequence containing four N^3-methylthymidine residues (N-Me-α *rev*), with an antisense sequence against the initiation site of *gag* (α gag), or with a 28-mer homooligodeoxycytidine (dC$_{28}$) was subjected to Northern blot analysis. Note the remarkable change of RNA profile in the presence of α *rev*. (Reproduced from Matsukura *et al.* (1989) with permission of the Proceedings Office of the National Academy of Science, U.S.A.)

without regard to the nucleotide sequence. Its mechanism of action is non-specific and likely involves inhibition of the reverse transcriptase (Majumdar *et al.*, 1988) and the fusion process of HIV (Mitsuya *et al.*, 1988b). More recently, Matsukura and his co-workers (1989b) have identified a phos-phorothioate heterooligomer complementary to the initiation sequence of HIV-1 *rev* which had a significant and sequence-specific inhibitory effect on the production of virally encoded proteins in chronically HIV-1-infected T cells. This antisense oligomer also reduced the level of the genomic viral mRNA transcript, with relative sparing of smaller RNA species (Fig. 4) (Matsukura *et al.*, 1989). In contrast, the same antisense sequence with unmodified normal phosphodiester linkages; phosphorothioate oligomers containing the sense sequence, random or homopolymeric sequences; or an antisense sequence with N^3-methylthymidine residues (thus precluding appropriate hydrogen-bonding) did not have an inhibitory effect on the viral expression (Fig. 4). The altered HIV-1 mRNA profile induced by the *rev* antisense oligomer suggests that the mechanism for the inhibition of the observed viral expression is due to an interference with the regulatory *rev* gene and involves hybridization to relevant mRNA. These modified antisense oligodeoxynucleotides may yield important theoretical and clinical insights into the regulation of HIV expression and may lead to development of a new class of drugs that can control the expression of the virus in already infected cells.

F. Maturation, Transport, and Assembly of Viral Components

The very final stages in the replication of HIV involve crucial virus-specific secondary processing of certain viral proteins by a protease (one of the *pol* gene products) as we have already discussed. However, cellular enzymes such as glycosylating or myristoylating enzymes also play a key role. Mature proteins are transported to or near the cell membrane and there assembled into virus particles. Inhibitors of such critical enzymes or the migration of viral products from the cytoplasm to the membrane may also lead to suppression of the viral replication and infectivity.

1. Glycosidase Inhibitors   The gp120 of HIV is extremely heavily glyco-sylated, suggesting (but not proving) that *N*-linked sugars play a role in the functional interaction of the virus with target cell surface. Thus far, several trimming glycosidase inhibitors, including castanospermine (Gruters *et al.*, 1987; Walker *et al.*, 1987) and deoxynojirimycin (Gruters *et al.*, 1987) (both are plant alkaloids that modify glycosylation by inhibiting α-glucosidase) have been shown to block HIV-induced syncytia formation, and interfere with HIV infectivity *in vitro*. These compounds are thought to inhibit trim-ming glycosidases which are involved in the biosynthesis of the *N*-linked oligosaccharides of the envelope glycoprotein in HIV-infected cells. The data showing that virus production *per se* is not inhibited and that the virus produced in the presence of these inhibitors can still bind to the CD4 mole-cules (although the binding constant is as yet undetermined) suggest that these agents may disrupt the post gp120-CD4 binding step. The latter step is necessary for virion–cell and cell–cell fusion. More recently, *N*-butyldeoxy-nojirimycin, an analog of deoxynojirimycin, has been reported to be more potent in inhibiting HIV replication and reducing infectious virus titer by greater than five orders of magnitude at noncytotoxic concentrations (Kar-pas *et al.*, 1988). At this time, *N*-butyldeoxynojirimycin is under phase I clinical trial in the United States.

2. Myristoylation Inhibitors and Others   Celurenin, an inhibitor of *de novo* fatty acid biosynthesis, has been shown to inhibit the myristoylation and the proteolytic cleavage of a *gag*-encoded polyprotein to mature p24 Gag protein *in vitro* expression of membrane glycoproteins and several se-creting proteins (Pal *et al.*, 1988). A monovalent carboxylic ionophore, called monensin, is known to inhibit the transport and expression of mem-brane glycoproteins and several secretory proteins (Pal *et al.*, 1988). This agent has also been shown to block the proteolytic cleavage of the gp160 to gp120, leading to the accumulation of the precursor gp160 in the cytoplasm and the significant reduction of syncytia formation *in vitro* (Pal *et al.*, 1988). However, both celurenin and monensin were found to be highly toxic to the cells *in vitro* (Pal *et al.*, 1988), and their clinical application seems uncertain at this time.

A sequence located between the 5' LTR and the *gag* gene, designated ψ, and apparently encoding no proteins, is thought to be responsible for the efficient packaging of genomic RNA into viral particles (Lever *et al.*, 1989). The efficiency of an HIV-1 mutant with a deletion of 19 base pairs in producing infectious virions was less than 2% of that of the wild-type HIV (Lever *et al.*, 1989). This specific sequence could also be a target for drugs, biologics, or some molecular strategies for blocking the packaging process of the virus, to suppress the production of infectious viruses.

G. Budding Process

Finally, HIV virion particles are released by a process of viral budding, making the replicative cycle of HIV start over again. The retroviral budding process is not yet fully understood. Studies of murine retroviruses indicate that interferon α (IFN-α) affects the stage of virion release from the cell membrane (Reitz *et al.*, 1977; Pitha *et al.*, 1979). IFN-α has been shown to affect several other early events in the replicative cycle, such as transcription or translation during acute viral infection. This agent has been reported to be active against HIV *in vitro* during acute infection (Ho *et al.*, 1985). Poli and his co-workers (1989) have shown that IFN-α can also suppress the HIV expression in chronically infected cells. Their data suggest that this agent mainly affects posttranscriptional events in the viral replicative cycle but not in the production and assembly of the viral components. The mechanism of IFN-α activity against HIV could be an alteration of membrane fluidity, resulting in the reduction of the release of progeny virions, or inhibition of HIV gene expression. In this regard, the *vpu* gene, which may regulate the release of the virus as discussed earlier in this chapter, could be one of the targets of IFN-α (Poli *et al.*, 1989). Whatever the mechanism, IFN-α may be clinically useful in several ways. The agent has been reported to have an *in vivo* antiretroviral effect in certain patients (Kovacs *et al.*, 1989). Furthermore, reasonable *in vitro* synergistic antiviral activity has been reported when this agent was combined with other antiretroviral agents (Hartshorn *et al.*, 1987). IFN-α has also been shown to suppress Kaposi's sarcoma lesions in patients with AIDS (Krawn *et al.*, 1983; Groopman *et al.*, 1984).

In this regard, Nakamura, Salahuddin, and their co-workers have recently provided perhaps the first cell culture model for Kaposi's sarcoma (KS) (Nakamura *et al.*, 1988; Salahuddin *et al.*, 1988). They have identified new growth factors released by HTLV-II-infected CD4[+] cells which support the growth of KS-like cells with characteristic spindle-like morphology. These factors could support the temporary growth of normal vascular endothelial cells (Nakamura *et al.*, 1988). The inoculation of the growth factors into mice could also induce KS-like lesions (Salahuddin *et al.*, 1988). On the other hand, Vogel and his co-workers (1988) have provided the important finding that the HIV *tat* gene can induce KS-like lesions in male transgenic mice. Their data may suggest that humoral or genetic factors are associated with the development of KS in humans. These findings are intriguing because they might provide some clues for development of drugs for the specific treatment of KS. It is worth stressing that while HIV is the cause of AIDS, this virus may not necessarily be the proximate cause of Kaposi's sarcoma (Beral *et al.*, 1990). It is possible that this tumor is caused by several factors, including an immunodeficiency state *per se*.

It is as yet unknown whether IFN-α suppresses the KS lesion either directly or by mediation of cellular-gene-encoded regulatory proteins. Whatever the mechanism of the IFN-α activity, this agent may benefit certain patients with HIV infection by acting as both an antiretroviral drug and an antitumor agent in combinations. A pilot study has shown that AZT (which we will discuss further) and IFN-α can be coadministered safely to patients with AIDS-related KS and that the combination has an antitumor as well as an antiretroviral effect *in vivo* (Kovacs *et al.*, 1989). However, in this trial, apparently synergistic toxicities such as neutropenia, thrombocytopenia, and hepatic dysfunction were documented with a much greater frequency (Kovacs *et al.*, 1989). To determine the clinical usefulness of this regimen, additional large clinical trials are underway.

IV. DNA-Chain Terminators as Antiretroviral Agents

In 1977, Sanger and his co-workers established a new method for determining nucleotide sequences in DNA by using four 2',3'-dideoxynucleoside 5'-triphosphates as chain-terminating inhibitors. These 2',3'-dideoxynucleosides had been synthesized in the 1960s or earlier (Horwitz *et al.*, 1964, 1966; Robins and Robins, 1964) and pioneering studies of them had been accomplished over the past 30 years or so (Atkinson *et al.*, 1969; Toji and Cohen, 1970; Ono *et al.*, 1979; Furmanski *et al.*, 1980; Wagar *et al.*, 1984); however, these compounds had not been proved to be useful as anticancer drugs or antibiotics. Since the establishment of Sanger's method, these 2',3'-dideoxynucleosides have served as routine reagents in molecular laboratories, in the form of 5'-triphosphates. These DNA-chain terminators have now attracted attention as drugs active against retroviral infections, and in particular as candidate drugs for the therapy of AIDS. Such agents, following anabolic phosphorylation in target cells, can inhibit HIV reverse transcriptase by substrate competition and/or chain termination as discussed below.

A. Activity of 2',3'-Dideoxynucleosides against HIV

In 1985, we found that a broad family of nucleosides with a 2',3'-dideoxyribose moiety can inhibit the infectivity and replication of a divergent range of HIV-1 strains *in vitro* (Mitsuya *et al.*, 1985, 1987a; Mitsuya and Broder, 1986, 1987). Several 2′,3′-dideoxynucleoside analogs were found to inhibit the *de novo* replication and cytopathic effect of HIV at concentrations that are 10–20 fold lower than those that impair the function and proliferation of target helper T cells (Mitsuya *et al.*, 1985; Mitsuya and Broder, 1986). Among them were 3′-azido-2′,3′-dideoxythymidine (AZT or zidovudine), 2′,3′-dideoxycytidine (ddC), 2′,3′-dideoxyadenosine (ddA), 2′,3′-dideoxyinosine (ddI), and 2′,3′-dideoxyguanosine (ddG). It is worth noting that as a general rule, each drug is different, from both an activity and toxicity point of view. Therefore, in a sense each drug must be considered in its own right. During the past five years, a number of 2′,3′-dideoxynucleoside derivatives have been identified as agents that have antiretroviral activity, at least *in vitro* (Balzarini *et al.*, 1986, 1987a, b, 1989; Mitsuya *et al.*, 1987b; Lin *et al.*, 1987; Kim *et al.*, 1987; Hamamoto *et al.*, 1987; Marquez *et al.*, 1987, 1990; Schinazi *et al.*, 1987; Hartman *et al.*, 1987; Baba *et al.*, 1987a; Vince *et al.*, 1988; Webb *et al.*, 1988; Haertle *et al.*, 1988; Masood *et al.*, 1990) (Figs. 5–8). In 1986, it was shown that infected monocytes/macrophages play an important role in the pathogenesis of AIDS, in particular, in the traffic of HIV across the blood–brain barrier and as a reservoir of HIV-1 *in vivo* (Gartner *et al.*, 1986; Koenig *et al.*, 1986; Ho *et al.*, 1986). Perno and his co-workers (1988) in our group subsequently demonstrated that dideoxynucleoside analogs including AZT could also suppress the replication of HIV in monocytes/macrophages *in vitro*. More recently, it has been shown that granulocyte/macrophage colony-stimulating factor (GM-CSF) potentiates viral replication yet enhances the antiretroviral effect of AZT (Perno *et al.*, 1989). This might have a direct relevance to the therapy against HIV infection with AZT (Pluda *et al.*, 1989).

It should be noted that the search for new drugs relies on the availability of rapid and sensitive assay systems for determining the activity of a given compound. In 1984, immediately after HIV was formally shown to be the causative agent for AIDS (Barré-Sinoussi *et al.*, 1983; Popovic *et al.*, 1984), we developed *in vitro* assay systems and made a commitment to test drugs in large numbers for their potential antiretroviral effect against HIV (Mitsuya *et al.*, 1984, 1985, 1987b; Mitsuya and Broder, 1986, 1987). Within a few years, we had tested more than 800 different compounds in our laboratory including some 500 nucleoside analogs, and a great deal of knowledge in terms of structure/activity relationships has now emerged.

B. Structure/Activity Relationships in 2′,3′-Dideoxynucleoside Analogs

The 2′,3′-dideoxynucleosides are of interest, because they prove that a simple chemical modification in the sugar moiety can in some circumstances convert a normal substrate for nucleic acid synthesis into a compound with a potent ability to inhibit the infectivity and replication of HIV. The 2′,3′-dideoxynucleoside analogs are successively phosphorylated in the cyto-

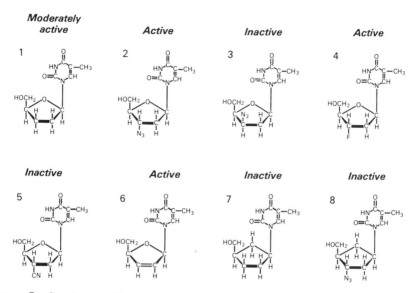

Figure 5. Structures and *in vitro* antiretroviral activity of 2',3'-dideoxythymidine analogs. Antiretroviral activity on each compound was assessed using ATH8 cells (Mitsuya *et al.*, 1985), H9 cells (Popovich *et al.*, 1984), normal clonal CD4[+] T cells (Mitsuya and Broder, 1988), MT-2 cells (Miyoshi *et al.*, 1981; Harada *et al.*, 1985) and/or normal unfractionated peripheral blood mononuclear cells, based on the inhibition of the cytopathic effect of HIV, suppression of Gag protein production, and/or suppression of HIV viral DNA or RNA synthesis (see Mitsuya *et al.*, 1987b). Compounds that can give a virtually complete inhibition (80–100%) of the infectivity and cytopathic effect of HIV at concentrations that do not significantly affect the growth of target cells are defined as *Active*. Compounds that can inhibit the infectivity and replication of HIV by 30–80% are defined as *Moderately active*. Compounds that give less than 30% inhibition are defined as *Inactive*. Unless otherwise stated, the *in vitro* antiviral activity of compounds against HIV defined here was determined by Mitsuya and Broder. *1,* 2',3'-dideoxythymidine (ddT or d2T); *2,* 3'-α-azido-2',3'-dideoxythymidine (AZT or *erythro*-AZT); *3,* 3'-β-azido-2',3',-dideoxythymidine (*threo*-AZT); *4,* 3'-α-fluoro-2',3'-dideoxythymidine; *5,* 3'-cyano-2',3'-dideoxythymidine (Greengrass *et al.*, 1989); *6,* 2',3'-didehydro-2',3'-dideoxythymidine (2',3'-dideoxythymidinene or d4T); *7,* carbocyclic-2',3'-dideoxythymidine; *8,* carbocyclic-3'-azido,-2',3'-dideoxythymidine.

plasm of a target cell to yield ultimately 2',3'-dideoxynucleoside 5'-triphosphate, although each drug may require a separate metabolic pathway (Mitsuya and Broder, 1986, 1987; Cooney *et al.*, 1986; Furman *et al.*, 1986; Ahluwalia *et al.*, 1987; Yarchoan and Broder, 1987; Mitsuya *et al.*, 1987b; Johnson *et al.*, 1988; Johnson and Fridland, 1989; Yarchoan *et al.*, 1989b). They then become analogs of the 2'-deoxynucleoside 5'-triphosphates that are the natural substrates for cellular DNA polymerases and the viral reverse transcriptase. It is worth stressing that, in general, the various dideoxynucleosides are not equivalent in either activity or toxicity profiles *in vivo*.

Several 2',3'-dideoxynucleoside 5'-triphosphates have been extensively

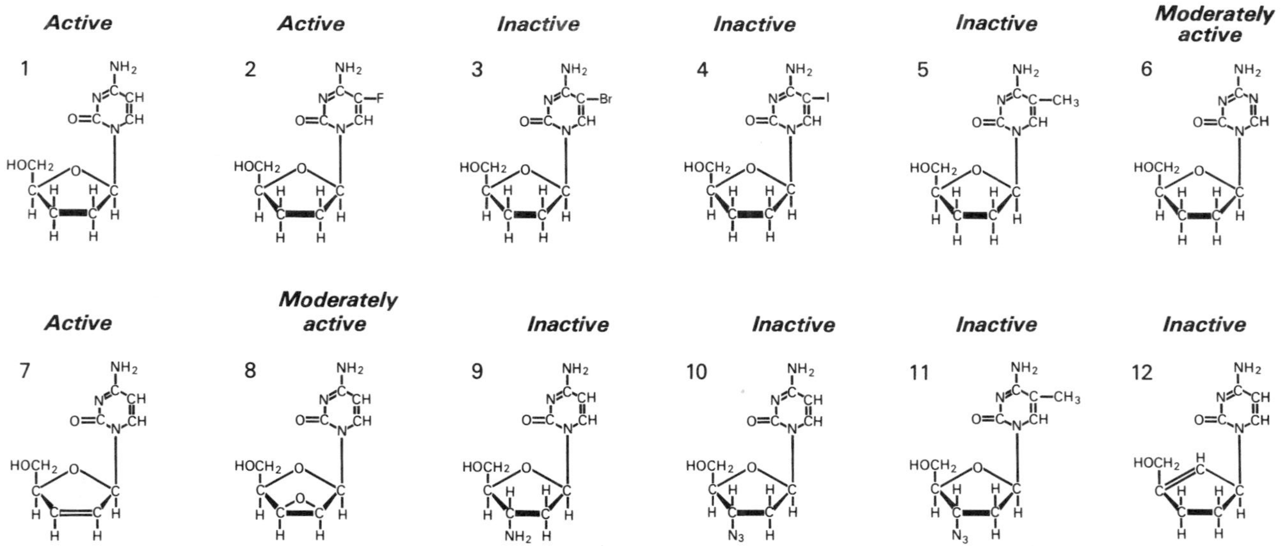

Figure 6. Structures and *in vitro* antiretroviral activity of 2′,3′-dideoxycytidine analogs. Antiretroviral activity of each compound was assessed as described in the legend to Fig. 5. *1,* 2′,3′-dideoxycytidine (ddC); *2,* 5-fluoro-2′,3′-dideoxycytidine; *3,* 5-bromo-2′,3′-dideoxycytidine; *4,* 5-iodo-2′,3′-dideoxycytidine; *5,* 5-methyl-2′,3′-dideoxycytidine; *6,* 5-aza-2′,3′-dideoxycytidine (Mitsuya, Driscoll, and Broder, unpublished); *7,* 2′,3′-didehydro-2′,3′-dideoxycytidine (2′,3′-dideoxycytidine or d4C); *8,* 2′,3′-β-epoxy-2′,3′-dideoxycytidine; *9,* 3′-amino-2′,3′-dideoxycytidine; *10,* 3′-azido-2′,3′-dideoxycytidine; *11,* 3′-azido-2′,3′-dideoxycytidine (AZC); *11,* 3′-azido-5-methyl-2′,3′-dideoxycytidine; *12,* cyclopentenyl-2′,3′-dideoxycytidine (Mitsuya, Driscoll, and Broder, unpublished).

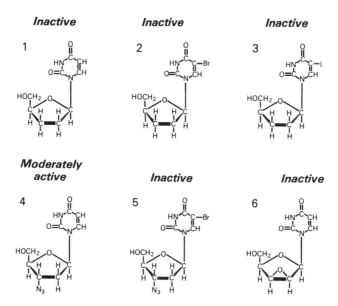

Figure 7. Structures and *in vitro* antiretroviral activity of 2′,3′-dideoxyuridine analogs. Antiretroviral activity of each compound was assessed as described in the legend to Fig. 5. *1*, 2′,3′-dideoxyuridine (ddU); *2*, 5-bromo-2′3′-dideoxyuridine; *3*, 5-iodo-2′,3′-dideoxyuridine; *4*, 3′-azido-2′,3′-dideoxyuridine (AZddU) (Schinazi and Ahn, 1987); *5*, 5-bromo-3′-azido-2′,3′-dideoxyuridine; *6*, 2′,3′,-β-epoxy-2′,3′-dideoxyuridine.

studied and are now known to have higher affinities for HIV reverse transcriptase than for cellular DNA polymerase α (Cheng *et al.*, 1987; Starnes and Cheng 1987; Mitsuya *et al.*, 1988c), although cellular DNA polymerases β and γ appear to be sensitive to the dideoxynucleoside 5′-triphosphates (Cheng *et al.*, 1987; Starnes and Cheng, 1987). These dideoxynucleoside 5′-triphosphates can compete with normal nucleotides for reverse transcriptase binding, and also can be incorporated into the growing DNA chain and bring about termination of the viral DNA chain because a normal 5′ → 3′ phosphodiester linkage cannot be completed (Mitsuya and Broder, 1986, 1987; Mitsuya *et al.*, 1987a, b). It has been shown that at concentrations that are achievable in human cells, dideoxynucleoside 5′-triphosphates can serve as substrates for reverse transcriptase to elongate a retroviral DNA chain by one residue, after which the DNA chain is terminated (Mitsuya *et al.*, 1987a).

The 2′,3′-dideoxy version of thymidine, 2′,3′-dideoxythymidine (ddT), was not a particularly potent compound against HIV in our *in vitro* system (Fig. 5: *1*) (Mitsuya and Broder, 1986). This could be because, at least in part, ddT serves as a relatively poor substrate for human thymidine kinase (K_m ~1000 μM). Of interest, ddT had been shown to be comparatively

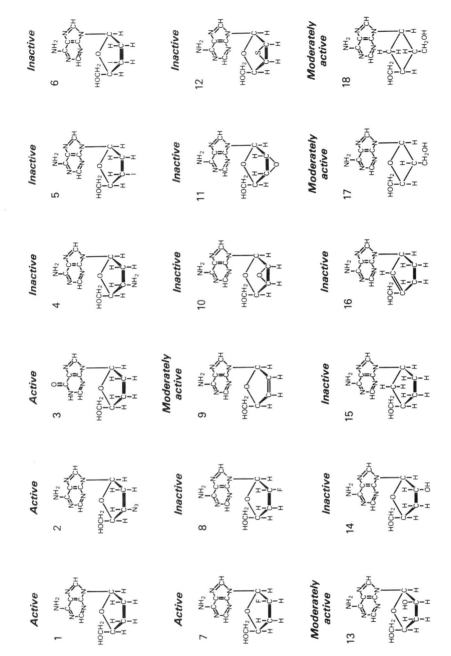

potent in murine cells against murine retroviruses (Furmanski *et al.*, 1980). In 1985, we found that a 2',3'-dideoxythymidine analog, substituted with an azido group ($-N_3$) at the 3'-carbon atom of the 2',3'-dideoxyribose moiety, is a potent *in vitro* inhibitor of HIV replication (Mitsuya *et al.*, 1985) (Fig. 5: **2**). The substitution of an azido group at the 3'-carbon, yielding AZT, produced a compound that was an excellent substrate for thymidine kinase (K_m ~3 μM) and was very potent against divergent strains of HIV *in vitro* (Mitsuya *et al.*, 1985). This 2',3'-dideoxythymidine analog, AZT, also undergoes anabolic phosphorylation to generate AZT 5'-triphosphate, which competes with thymidine 5'-triphosphate (Furman *et al.*, 1986) and functions as a DNA-chain terminator as well. In this sense, AZT parallels the 2',3'-dideoxynucleosides described above. The 3'-azido substitution, however, does not always potentiate the antiretroviral activity of other members of the 2',3'-dideoxynucleoside family (Mitsuya *et al.*, 1987b) (Fig. 5: **8**; Fig. 6: **10**, **11**; Fig. 7: **5**). It should be noted that antiretroviral activity of a nucleoside depends upon multiple factors including penetration into target cells, multistep anabolic phosphorylation, intracellular catabolism, and relative affinities for reverse transcriptase and cellular DNA polymerases. Indeed, substitution of an atom at a certain position can drastically change antiviral activities of a given nucleoside. For instance, 3'-azido substitution in ddC (Fig. 6: **1**), generating AZC (Fig. 6: **10**) results in almost total loss of the *in vitro* antiretroviral activity (Mitsuya *et al.*, 1987b). The 3'-azido substitution of ddA (Fig. 8: **1**), generating AZA (Fig. 8: **2**), potentiates only its toxicity while it does not apparently affect the antiviral activity (Mitsuya *et al.*, 1987b). On the other hand, the 3'-azido substitution in ddG, generating AZG, does not appear to alter its antiviral activity or cell toxicity (Hartman *et al.*, 1987; Baba *et al.*, 1987b). Even simple compounds, which might appear to be similar on first examination (such as the purine analogs ddA and ddG), can show significantly different patterns of biologic activity (*vide infra*).

On the basis of molarity, ddC is one of the most potent dideoxynucleosides as tested in susceptible CD4$^+$ T cell lines such as ATH8, H9, or normal CD4$^+$ T-cell clones (Mitsuya and Broder, 1986). We then asked whether substitutions at the base or the oxacyclopentane moiety in ddC

Figure 8. Structures and *in vitro* antiretroviral activity of 2',3'-dideoxyadenosine analogs. Antiretroviral activity of each compound was assessed as described in the legend to Fig. 5. **1,** 2',3'-dideoxyadenosine (ddA); **2,** 3'-azido-2',3'-dideoxyadenosine (AZA); **3,** 2',3'-dideoxyinosine (ddI); **4,** 3'-amino-2',3'-dideoxyadenosine; **5,** 3'-α-iodo-2',3'dideoxyadenosine; **6,** 3'-β-iodo-2',3'-dideoxyadenosine; **7,** 2'-β-fluoro-2'-3'-dideoxyadenosine; **8,** 2'-α-fluoro-2',3'-dideoxyadenosine **9,** 2',3'-didehydro-2',3'-dideoxyadenosine (2',3'-dideoxyadenosine or d4A); **10,** 2',3'-β-epoxy-2',3'-dideoxyadenosine; **11,** 2',3'-α-epoxy-2', 3'-dideoxyadenosine; **12,** 2',3'-β-episulfide-2',3'-dideoxyadenosine; **13,** 3'-deoxy-2'-arabinofuranosyladenine; **14,** 3'-deoxyadenosine (cordecepin); **15,** 2',3'-dideoxyaristeromycin; **16,** 2',3'-dideoxyneplanocin A (Mitsuya, Driscoll, and Broder, unpublished); **17,** oxetanocin-A; **18,** carbocyclic-oxetanocin-A (cyclobut-A) (Hayashi *et al.*, 1990a). 2',3'-Dideoxyinosine (ddI) (**3**) was added as a reference compound.

could change its antiretroviral activity *in vitro*. When the potent ddC molecule (Fig. 6-1) was substituted with a bromine or iodine substituent, or a methyl group at the 5 position in the base (Fig. 6: *3, 4, 5,* respectively), the antiviral activity was completely abolished, while with a fluorine substituent at the same position (Fig. 6: *2*), its antiviral activity was not affected at all (Kim *et al.*, 1987). Interestingly, when ddC (Fig. 6: *1*) was substituted at the 5 position with a methyl group ($-CH_3$), generating a "hybrid" nucleoside between ddC and AZT (Fig. 6: *11*), the antiretroviral activity was totally lost *in vitro* (Mitsuya and Broder, unpublished).

In many circumstances, a change in the stereochemistry of the substituent converts an active nucleoside to an inert compound. For example, while a substitution of ddT with an azido group, generating AZT (also called *erythro*-AZT) (Fig. 5: *2*) (Mitsuya *et al.*, 1985; Mitsuya and Broder, 1987) potentiates the activity of ddT (Fig. 5: *1*) as discussed earlier, an upward (or β-) 3'-azido substitution of ddT, generating *threo*-AZT (or 3'-azido-2',3'-dideoxy-β-xylofuranosylthymine) (Fig. 5: *3*), completely abrogates the activity of ddT (Mitsuya and Broder, unpublished). Interestingly, Eriksson and his co-workers (1987) have shown that these two AZT isomers apparently bind to different sites of avian retroviral reverse transcriptase.

In other cases, the substitution of another atom for a hydrogen may add a practical advantage to certain antiviral dideoxynucleosides. 2',3'-Dideoxyadenosine and dideoxyinosine (Fig. 8: *1, 3* respectively) are sensitive to low pH (solvolysis) and are decomposed to a base and a sugar moiety, resulting in a total loss of their antiviral activity under acidic conditions (Marquez *et al.*, 1987, 1990). However, the substitution of a β-fluorine atom at the 2' carbon, generating 2'-fluoro-2',3'-dideoxyarabinosyladenine (Fig. 8: *7*) (or -inosine) [also called *threo*-2'-fluoro-2',3'-dideoxyarabinosyladenine (or -inosine)], can produce acid stability, with retention of antiretroviral activity (Marquez *et al.*, 1987, 1990). Downward (or α-) fluorine substitution, generating *erythro* isomers of 2'-fluoro-ddA (Fig. 8: *8*), also produces acid-resistance. But this latter substitution nullifies antiretroviral activity (Marquez *et al.*, 1987, 1990).

V. 2',3'-Dideoxynucleosides as Therapeutics for HIV Infection

A. AZT

1. Clinical Activity of AZT   The synthesis of AZT was first reported in 1964 by Horwitz and his colleagues, and this agent was shown to inhibit C-type murine retrovirus replication *in vitro* by Ostertag and his co-workers in 1973 (Ostertag *et al.*, 1974a, b). These workers, therefore, should be credited with discovering AZT. However, its application for antiretroviral therapy in humans had to await a new perspective. In July 1985, on the basis of the *in vitro* data of antiviral activity described above and results of animal toxicology, AZT was first administered to patients with severe HIV infection in the National Cancer Institute and Duke University Medical Center (Yarchoan *et al.*, 1986). This initial study showed that patients with AIDS or AIDS-related complex had immunologic, virologic, and clinical improvement during the therapy with AZT (Yarchoan *et al.*, 1986). It was also found that patients with HIV-associated dementia had substantial improvement in intellectual function during therapy (Yarchoan *et al.*, 1987). Based on these results, Wellcome Research Laboratories began a multicenter, randomized, placebo-controlled trial of AZT in February 1986.

The phase II clinical data of AZT confirmed the phase I data and demonstrated a reduced mortality in patients receiving AZT, thus ending the placebo-controlled arm in September 1986 (Fischl *et al.*, 1987). The phase II data also showed that patients in the AZT arm had at least a temporary increased in $CD4^+$ lymphocyte counts, a decreased viral load (assessed by serum p24 levels), fewer opportunistic infections, and an average of 0.5 kg weight gain. In addition, some patients with HIV dementia had improved cognitive function (Fischl *et al.*, 1987; Schmitt *et al.*, 1988). In March 1987, AZT was approved as a prescription drug for patients who have episodes of *Pneumocystis carinii* pneumonia or whose $CD4^+$ cell count is below 200/mm^3. Recently, AZT treatment has also been shown to improve and/or delay certain HIV-associated neurologic symptoms in adults and children with HIV infection (Pizzo *et al.*, 1988; Portgies *et al.*, 1989). AZT has also been shown to increase the platelet counts in the patients with HIV infection possibly either by inhibiting reticuloendothelial function, leading to diminished clearance of immunoglobulin-coated platelets, or by inhibiting the HIV infection to megakaryocytes (or both (Hymes *et al.*, 1988). Such AZT effects could be useful in the treatment of thrombocytopenia in individuals with HIV infection (Leaf *et al.*, 1988); however, further studies are necessary.

Several of the toxic effects of AZT have limited its long-term administration. The frequent toxicities include bone marrow suppression, and anemia is its most frequent manifestation (Yarchoan *et al.*, 1986; Richman *et al.*, 1987). In the phase II trial, bone marrow suppression occurred in 45% of the patients who had *P. carinii* pneumonia, which required transfusions or a dose reduction during the first 6 months of the trial (Richman *et al.*, 1987). Other toxic effects of AZT include nausea, vomiting, myalgias, myositis, headaches, liver function abnormalities, and bluish nail pigmentation (Yarchoan *et al.*, 1986, 1989b; Yarchoan and Broder, 1987; Richman *et al.*, 1987). High doses of AZT can also cause anxiety, confusion, and tremor (Yarchoan *et al.*, 1986, 1989b; Yarchoan and Broder, 1987; Richman *et al.*, 1987). The toxic effects of AZT, in particular the bone marrow suppression, might be linked to the inhibition of cellular DNA polymerases β and/or γ (the latter being responsible for mitochondrial DNA replication) (Cheng *et al.*, 1987; Starnes and Chang, 1987; Simpson *et al.*, 1989). Alternatively, AZT may cause a depletion of thymidine triphosphate. However, it is noteworthy that recent data suggest that lower doses of AZT benefit certain subsets of patients with significant reduction of side effects. Various attempts have also been made to reverse the toxic effects of AZT without loss of antiviral activity *in vitro* (Sommadossi *et al.*, 1988). The clinical usefulness of such approaches remains to be determined.

Recently, it has been reported that AZT has tumorigenicity in rodents (Letter to physicians from Burroughs Wellcome Co. dated December 5, 1989). In standard lifetime carcinogenicity bioassay studies in rats and mice receiving high doses of AZT ($\geq$60 mg/kg/day and then $\geq$30 mg/kg/day) for ~22 months, seven vaginal tumors consisting of five nonmetastasizing squamous cell carcinomas and two benign tumors occurred in 60 female mice and a similar finding was observed in rats. Although results from rodent carcinogenicity studies may or may not predict for humans, further close studies are required.

Thus, while AZT can induce a decrease in the morbidity and mortality among patients with HIV infection, its use can be associated with substantial toxicity. Regimens to decrease its toxicity and development of less toxic antiretroviral drugs are obviously urgently required.

2. Emergence of AZT-Insensitive HIV Variants The long-term administration of AZT has posed a new challenge. Larder, Darby, and Richman (1989a) have isolated HIV strains which are insensitive to AZT *in vitro*. Such strains can most readily be obtained from patients who have been under AZT therapy for more than 6 months (Larder *et al.*, 1989). They have recently identified common mutations that occur in the *pol* gene of AZT-insensitive strains (Larder and Kemp, 1989; Richman, 1990). Their studies have suggested that acquisition of AZT-resistance may require accumulation of multiple mutations at specific amino acid residues of the *pol* gene product. Larder and his co-workers (1987) had reported that site-directed mutagenesis in reverse transcriptase could render the enzyme insensitive to AZT triphosphate. However, recombinant infectious viruses carrying the *pol* genes encoding such AZT-insensitive mutant reverse transcriptase have paradoxically shown hypersensitivity to AZT when tested in culture (Larder *et al.*, 1989b). These observations are not easy to explain under one unifying theory. At this time, elucidation of the mechanism of emergence of AZT-insensitive variants appears to be a formidable challenge. However, it should be noted that study of drug-resistant viral strains may make it possible to identify more specific viral targets and to develop more effective therapeutic strategies.

B. ddC

2',3'-Dideoxycytidine exerts its antiretroviral activity *in vitro* at a concentration as low as 10 nM depending on the HIV viral dose used (Mitsuya and Broder, 1986; Mitsuya *et al.*, 1988c). This drug was the second dideoxynucleoside administered to patients with HIV infection (Yarchoan *et al.*, 1988a, b). A short-term phase I clinical trial on ddC has shown that this drug can suppress the replication of HIV *in vivo* and improve clinical syndromes in patients with advanced AIDS or ARC (Yarchoan *et al.*, 1988a, b). A subsequent study has also shown evidence of antiviral efficacy (Merigan *et al.*, 1989). In some patients, however, the immunologic and virologic improvements were temporary (Yarchoan *et al.*, 1988b). Toxic effects of ddC defined in the phase I clinical trial, particularly at high doses, included eruptions, aphthous oral ulcerations, fever, and malaise, which developed between 1 and 4 weeks of therapy. These symptoms usually resolved in 1–2 weeks, even with continuous therapy of ddC. However, a painful sensory-motor peripheral neuropathy, which developed after 8–14 weeks of ddC therapy, was the dose-limiting toxic effect (Yarchoan *et al.*, 1988a, b; Merigan *et al.*, 1989). Since the toxicity profile of ddC is strikingly different from that of AZT, a combination of these two drugs was thought likely to reduce overall toxicity without necessarily lowering antiretroviral activity. To test this possibility, patients with advanced HIV infection underwent a regimen alternating AZT for 1 week and ddC for 1 week (Yarchoan *et al.*, 1988a, b). Preliminary results have shown that the individual toxicities of the drugs were reduced, and the patients, overall, had an average increase of more than 70 $CD4^+$ cells/mm^3 at week 22, and sustained decreases in serum p24 antigen level and a mean weight gain of 5 kg (Yarchoan *et al.*, 1988a). We will discuss the strategy of drug combinations for the therapy of AIDS later.

It is noteworthy that ddC is the most potent antiretroviral dideoxynucleoside we have thus far tested *in vitro* on the basis of molarity (Mitsuya and Broder, 1986). ddC can exert complete suppression of HIV replication at $\frac{1}{2}$–$\frac{1}{5}$ the effective concentration of AZT and $\frac{1}{4}$–$\frac{1}{10}$ the effective concentrations of ddI (another agent being tested in patients with HIV infection: *vide infra*) *in vitro*. Furthermore, the activity of ddC appears to be more durable than that of AZT *in vitro* (Mitsuya and Broder, 1986; Mitsuya *et al.*, 1987a). ddC may have an antiretroviral activity *in vivo* at doses substantially lower than the effective daily dose of AZT. The current large phase II/III study employing a low dose of ddC should define the role in the treatment of HIV infection.

C. ddI

2′,3′-Dideoxyinosine (ddI) is an analog of the naturally occurring purine nucleoside, inosine. A closely related compound, 2′,3′-dideoxyadenosine (ddA) was first synthesized in 1964 by Robins and Robins. These two dideoxypurine nucleosides, ddI and ddA, were first identified in 1985 as equally potent inhibitors of HIV replication *in vitro* (Mitsuya and Broder, 1986). These compounds were also effective against HIV in monocytes/macrophages *in vitro* (Perno *et al.*, 1988). At least in culture, ddI is less toxic than AZT or ddC and its therapeutic index appears to be wider (Mitsuya and Broder, 1986). The toxicity of ddI to hematopoietic precursor cells also seems to be much less than AZT or ddC *in vitro* (Molina and Groopman, 1989; Du *et al.*, 1989; Fine *et al.*, 1990).

The cellular uptake and metabolism of ddI involves a series of complex events. Although ddA was first synthesized more than 25 years ago, ddI had never been explored as a therapeutic agent until its possible applicability for treating retroviral infection was shown in 1985 in our laboratory (Mitsuya and Broder, 1986). Its biochemical pharmacology was unknown. Recently, studies by Johns and his co-workers and Fridland and his co-workers for the first time have clarified the potential mode of anabolism of ddI and ddA (Ahluwalia *et al.*, 1987; Johnson *et al.*, 1988; Johnson and Fridland, 1989). Both ddI and ddA are similar to AZT in that they diffuse into the cell without active transport. However, ddI and ddA may have a passive carrier. ddA can be directly phosphorylated to ddAMP in the cytoplasm by both deoxycytidine kinase and adenosine kinase (Ahluwalia *et al.*, 1987). Alternatively, ddA may be rapidly deaminated by adenosine deaminase to ddI, phosphorylated to ddIMP by cytosolic 5'-nucleotidase (Johnson and Fridland, 1989), and converted to ddAMP by adenylosuccinate synthetase/lyase (Ahluwalia *et al.*, 1987; Johnson *et al.*, 1988; Johnson and Fridland, 1989). The latter route appears to be the predominant pathway in human cells. Accordingly, ddA can be viewed as a prodrug for ddI. Once ddI and ddA are converted to ddA 5'-triphosphate (ddATP) in the cytoplasm, it remains there for a long period of time. Its intracellular half-life is more than 12 hr, and perhaps as long as 24 hr, which differs from AZT and ddC (Ahluwalia *et al.*, 1988). Thus, even if the plasma half-life of ddI is relatively short (about 35 min), ddI may perhaps be administered with less frequent dosing schedules (e.g., every 8–24 hr) (Hoshino *et al.*, 1987; Hayashi *et al.*, 1988) than AZT.

ddI and ddA are distinct from the other three dideoxynucleosides, AZT, ddC, and ddG, in that the antiretroviral activity of ddI/ddA is not reversed by any of the normal 2'-deoxynucleosides alone or in combinations (Mitsuya and Broder, 1987; Mitsuya *et al.*, 1987b). This is not the case with AZT, ddC, and ddG, drugs whose antiretroviral effets are readily nullified by their corresponding normal 2'-deoxynucleosides *in vitro* (Mitsuya *et al.*, 1985, 1987b). The reversal of the antiretroviral effect of AZT, ddC, and ddG by their normal counterparts suggests that these dideoxynucleosides are in competition with normal 2'-deoxynucleosides in one or more metabolic reactions that lead to the triphosphate formation. This feature of ddI/ddA, in addition to the longer intracellular half-life of ddATP, suggests that ddI/ddA could be a less toxic, more effective, and more durable antiviral drug *in vivo* as compared to the other dideoxynucleosides.

ddI was first administered to patients with AIDS or AIDS-related complex in an escalating-dose phase I study at the National Cancer Institute in July 1988 (Yarchoan *et al.*, 1989a, 1990). Patients receiving ddI had improvement in immunologic function and evidence of a decrease of viral load. They had an increase in their CD4 cells ($P < 0.0005$) and CD4/CD8 lymphocyte ratio ($P < 0.01$). In addition, more than an 80% decrease in serum HIV p24 antigen was observed in evaluable patients ($P < 0.05$). Half the patients reported increased energy and appetite on ddI and, overall, the patients gained 1.5 kg during the first 6 weeks of therapy. The drug had a virustatic effect, even in patients whose disease had become refractory to AZT. The most notable adverse effects at the ddI doses used in the phase I study included painful peripheral neuropathy, sporadic pancreatitis, mild headaches, and insomnia. At the highest doses tested, peripheral neuropathy and pancreatitis appeared to be the dose-limiting toxicities. However, doses up to 9.6 mg/kg have been tolerated in some patients for over 2 years without toxicity (Yarchoan et al., 1990).

These data strongly suggest that ddI has antiviral activity against HIV at doses that can be well tolerated. However, phase I studies, by definition, do not prove the efficacy of drugs. At this time, three large multicenter phase II trials of ddI are underway in the United States: the first comparing the clinical efficacy of AZT and ddI, the second testing the effect of AZT and ddI in patients who have been undergoing long-term AZT therapy, and the third comparing three different doses of ddI in patients who do not tolerate AZT therapy. Only these controlled clinical trials will be able to determine the clinical usefulness and safety of ddI.

VI. Carbocyclic and Acyclic Nucleoside Analogs Active against HIV *in Vitro*

To date, a number of nucleosides that do not have a normal oxacyclopentane (pentafuranosyl) sugar moiety have been reported to be active against HIV *in vitro* (Hoshino et al., 1987; Hayashi et al., 1988, 1990a; Pauwels et al., 1988; Vince et al., 1988; DeClerq, 1989; Norbeck et al., 1990). We will discuss the potential of some of these drugs for use in therapy against HIV infection.

A. Carbocyclic Nucleosides

Carbocyclic nucleosides are analogs in which the oxygen of the oxacyclopentane is substituted with a carbon atom (e.g., Fig. 5: *7, 8*). Earlier studies have shown that carbocyclic nucleosides are refractory to cleavage by phosphorylases and hydrolases while retaining some potential for therapeutically useful interaction with enzymes involved in nucleoside metabolism (Bennett et al., 1975). Carbocyclic 2′,3′-didehydro-2′,3′-dideoxyguanosine (designated carbovir), which is one member of the broad family of 2′,3′-dideoxynucleoside analogs, has recently been reported to be active against HIV *in vitro* (Vince et al., 1988). The triphosphate of this compound apparently has a highly specific inhibitory effect on HIV reverse transcriptase with little, if any, effect on the cellular DNA-synthetic enzymes (White et al., 1989). This contrasts with AZT which can substantially suppress the activity of DNA polymerases β and γ (Cheng et al., 1987). Carbovir is scheduled for phase I clinical trial in the United States.

Oxetanocin-A (9-[(2R,3R,4S)-3,4-bis(hydroxymethyl)-2-oxetanyl]-adenine) is an adenine nucleoside analog naturally synthesized by *Bacillus megaterium*, which has an oxytane sugar ring (Fig. 8: *17*) (Shimada *et al.*, 1986). This compound has been shown to be active against HIV *in vitro* by Hoshino and his co-workers (1987). We have recently observed that two carbocyclic oxetanocin analogs, designated cyclobut-A (Fig. 8: *18*) and cyclobut-G, could block the infectivity and replication of HIV-1 and HIV-2 *in vitro* (Hayashi *et al.*, 1990a). It is noteworthy that both cyclobut compounds also demonstrated potent and selective *in vitro* activity against the herpes viruses (HSV-1 and HSV-2), cytomegalovirus (CMV), and varicella zoster virus (VZV) (Norbeck *et al.*, 1990). Dual infection with HIV and CMV has been found in patients with AIDS (Nelson *et al.*, 1988) and appears to correlate with rapid progression of retinal necrosis. (Skolnik *et al.*, 1989). In this regard, cylobut-A and G or combinations of anti-HIV agents and anti-CMV drugs could be useful.

B. Acyclic Nucleosides

Drug-searching efforts now have been extended to acyclic nucleosides, which do not have a ring structure for the sugar moiety. Pauwels, De Clercq, and co-workers have identified an acyclic nucleoside derivative, phosphonylmethoxyethyladenine (PMEA), which had a broad spectrum activity *in vitro* against a number of DNA and RNA viruses including HIV (Fig. 9: *1*) (Pauwels *et al.*, 1988; De Clercq, 1989). The exact mechanism of antiretroviral activity of this acyclic nucleoside has not yet been fully understood. However, taking into consideration its structure and the fact that anabolic phosphorylation of its related analog is likely to be a factor for antiviral activity (Vorruba *et al.*, 1987), it is probable that the inhibition of reverse transcriptase (and possibly DNA-chain termination) is involved in its antiretroviral activity. PMEA is acid-resistant and as potent as ddI/ddA on the basis of molarity in several assay systems (Mitsuya and Broder, unpublished).

We have recently identified two additional antiretroviral acyclic nucleoside analogs, designated adenallene and cytallene (Fig. 9: *2, 3*) (Hayashi *et*

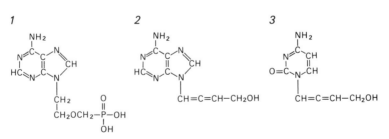

Figure 9. Structures and *in vitro* antiretroviral activity of selected acyclic nucleosides. Antiretroviral activity of each compound was assessed as described in the legend to Fig. 5. *1*, phosphonylmethoxyethyladenine (PMEA); *2*, 9-(4'-hydroxy-1',2'-butadienyl)adenine (adenallene); *3*, 1-(4'-hydroxy-1',2'-butadienyl)cytosine (cytallene).

al., 1988). These acyclic nucleoside analogs do not have an oxacyclopentane moiety but rather a four-carbon chain with two cumulative double bonds and a 4'-hydroxyl group. Both drugs are known to be acid-resistant (Hayashi *et al.*, 1988) and may therefore have a reasonable oral bioavailability. But more research is required.

The exact mechanism(s) of the antiretroviral activity of the acyclic nucleosides described here remains to be determined. However, their unique structure/activity relationships may lead to the discovery of new classes of anti-HIV agents.

VII. Combination of Multiple Antiretroviral Drugs for Therapy against AIDS

As discussed above, there are a reasonable number of agents that have potential usefulness in the treatment of HIV infection. However, many of these drugs may have serious dose-limiting toxicities. A logical extension of current therapeutic approaches would be the use of combinations of agents which have different antiretroviral mechanisms or have different metabolic pathways. Such combination therapy will likely enhance the efficacy and/or reduce the toxic effects of each drug and also could minimize or retard the emergence of drug-resistant HIV variants. Indeed, in the treatment of a variety of microbial and neoplastic diseases, notably in tuberculosis and leukemias-lymphomas, only the judicious application of combination chemotherapies made it possible to successfully cure the patients (DeVita and Schein, 1973). Several groups have reported that combinations of 2',3'-dideoxypyrimidine nucleosides plus 2',3'-dideoxypurine nucleosides (Mitsuya *et al.*, 1987b, 1990), and combinations of 2',3'-dideoxynucleosides plus different types of antiretroviral drugs such as IFN-α (Hartshorn *et al.*, 1987), soluble CD4 (Johnson *et al.*, 1989; Hayashi *et al.*, 1990b), or PMEA (Smith *et al.*, 1989) can produce synergistic (or at least additive) antiretroviral activities against HIV without concomitant increases in toxicities *in vitro*.

Another approach to exploring drug combination is synthesis of nucleotide oligomers. For example, we, in collaboration with Tam and Weigele at the Hoffmann-La Roche Research Institute, have synthesized a heterodimer of AZT and ddC, and a homodimer of ddC (Fig. 10: *1, 2*). These dimers can have higher lipophilicity than the parent compound alone while retaining the antiretroviral activity *in vitro* (Tam *et al.*, 1989). Although the value of acquisition of lipophilicity in antiretroviral dideoxynucleosides is not determined, this feature might have a direct clinical relevance.

However, not all combinations of antiretroviral agents have synergistic or additive effects. One nucleoside analog, ribavirin, which has been reported to be active against HIV *in vitro* (possibly by blocking the capping of viral RNA) (McCormick *et al.*, 1984), is intriguing, because this agent antagonizes the AZT effect by blocking the phosphorylation of AZT *in vitro* (Vogt *et al.*, 1987). In contrast, ribavirin can increase the phosphorylation of dideoxypurine nucleosides such as ddI/ddA and ddG *in vitro* (Baba *et al.*,

1. Active

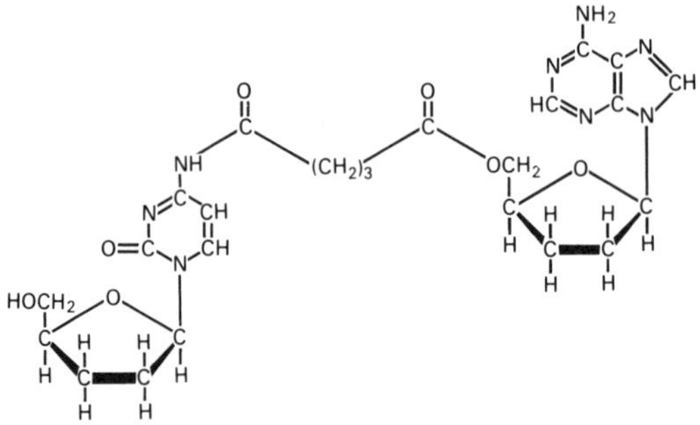

2. Active

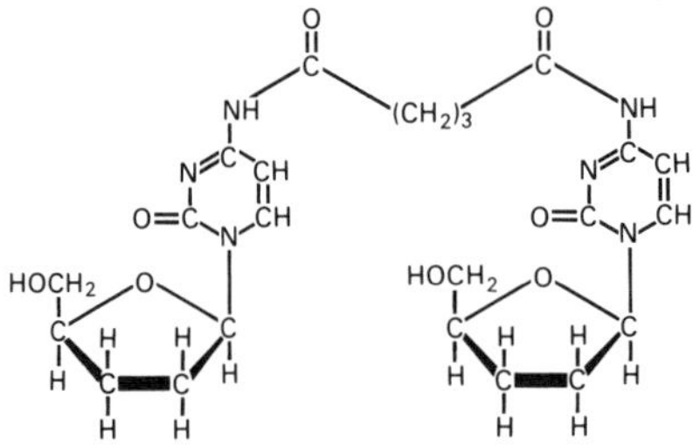

Figure 10. Structures and *in vitro* antiretroviral activity of selected nucleoside dimers. Antiretroviral activity of each compound was assessed as described in the legend to Fig. 5. **1,** heterodimer of 2',3'-dideoxyadenosine (ddA) and 2',3'-dideoxycytidine (ddC); **2,** homodimer of ddC.

1987b). Phase I/II clinical trials with ribavirin as a single agent are underway in the United States. Although its value as a single agent still remains to be determined, coadministration of ribavarin with dideoxypurine nucleosides may theoretically be useful. However, it should be noted that such coadministration should take place only in the setting of a clinical trial.

Combinations of antiretroviral agents with certain drugs, which are not active against HIV but which have activities against other viruses, or modify the metabolism or *in vivo* distribution of antiretroviral agents, could also be useful. For example, acyclovir, an anti-herpes agent, which has little antiretroviral activity against HIV-1 *in vitro*, has been shown to be synergistic against HIV *in vitro* when used in combination with AZT (Mitsuya and Broder, 1987; Mitsuya *et al.*, 1987b). In theory, this regimen could be useful, because the suppression of herpes virus, a frequent opportunistic pathogen in patients with AIDS, may lead to effective inhibition of HIV in individuals with superinfection of HIV and herpes viruses. In a short-term pilot clinical trial, patients with AIDS and AIDS-related complex have tolerated the combination regimen of AZT and acyclovir for more than 30 weeks (Surbone *et al.*, 1988). Another drug, probenecid, a drug for the treatment of hyperuricemia associated with gout, has been shown to block the renal excretion of AZT (Kornhauser *et al.*, 1989), possibly by acting as a competitive inhibitor in the renal tubule (Hedaya and Sawchuk, 1989), thereby decreasing the total body clearance of AZT. In rabbits, probenecid has been shown to increase the AZT concentration in cerebrospinal fluid by a factor of seven as compared to that with AZT alone (Hedaya and Sawchuk, 1989). Thus, coadministration of probenecid with AZT could reduce the daily dose of AZT or allow its dosing interval to be longer, and may have a potential to reduce the cost of any regimen compared with AZT alone. However, it remains to be determined whether the therapeutic AZT levels are consistently achieved in plasma, and whether further unexpected toxic effects might occur, particularly in patients susceptible to AZT toxicities. The routine use of probenecid coadministered with AZT is not recommended until further clinical trials prove the safety and efficacy of this combination. Dipyridamole (DPM), a coronary vasodilator and antithrombotic drug, has also been shown to potentiate the antiviral activity of AZT and ddC against HIV-1 in monocytes/macrophages *in vitro* (Szebeni *et al.*, 1989). DPM apparently inhibits cellular uptake and phosphorylation of thymidine but does not affect those of AZT in monocytes and macrophages. A pharmacologically-oriented clinical trial administering the combination regimen of DPM and AZT is now underway.

Antiretroviral agents can also antagonize the therapeutic effects of other drugs when coadministered. AZT has been shown to reverse the action of pyrimethamine against *Toxoplasma gondii in vitro* (Israelski *et al.*, 1989). The synergism of pyrimethamine and sulfadiazine against *T. gondii* was also antagonized by AZT *in vitro* (Israelski *et al.*, 1989). The clinical significance of these *in vitro* observations has not been determined; however, AZT should be used with caution in patients with toxoplasmic encephalitis receiving pyrimethamine, with careful monitoring for evidence of poor response to or relapse during antitoxoplasmosis therapy.

VIII. Conclusion

Since AIDS was reported as a new clinical entity in 1981, new knowledge has been accumulated with unprecedented speed. Perhaps one can specifically say that the development of antiretroviral therapy against HIV infection has also proceeded with unprecedented speed. In fact, a number of approaches to the therapy of HIV infection from different directions has provided significant hope for HIV-infected individuals. We believe that the major advances in the development of antiviral drugs for therapy of HIV infection will require adherence to the principles of randomized, controlled trials, using careful monitoring and effective trial termination rules by independent review boards. Such trials should provide an effective safeguard against investigator bias and protect patients and society against the possibility of commercial exploitation (Broder, 1989). But if we adhere to the principles of the scientific method and controlled clinical trials, progress against AIDS and HIV infection seems certain.

Acknowledgments

The authors thank Drs. John S. Driscoll, Victor E. Marquez, and Marvin S. Reitz for critical reading of the manuscript.

References

Abrmas, D. I., Kuno, S., Wong, R., and Ueno, R. (1989). *Ann. Intern. Med.* **110**, 183–188.

Ahluwalia, G., Cooney, D. A., Mitsuya, H., Fridland, A., Flora, K. P., Hao, Z., Dalal, M., Broder, S., and D. J. Johns. (1987). *Biochem. Pharmacol.* **36**, 3797–3800.

Ahluwalia, G., Johnson, M. A., Fridland, A., Cooney, D. A., Broder, S., and Johns, D. G. (1988). *Proc. Am. Assoc. Cancer Res.* **29**, 349.

Atkinson, M. R., Deutscher, M. P., Kornberg, A., Russe, A. F., and Moffatt, J. G. (1969). *Biochemistry* **12**, 4897–4903.

Baba, M., Pauwels, R., Balzarini, J., Herdewijn, P., and De Clercq, E. (1987a). *Biochem. Biophys. Res. Commun.* **145**, 1080–1086.

Baba, M., Pauwels, R., Balzarini, J., Herdewijn, P., De Clercq, E., and Desmyer, J. (1987b). *Antimicrob. Ag. Chemother.* **31**, 1613–1617.

Baba, M., Pauwels, R., Balzarini, J., Arnout, J., Desmyter, J., and De Clercq, E. (1988). *Proc. Natl. Acad. Sci. U.S.A.* **85**, 6132–6136.

Balzarini, J., Pauwels, R., Herdewijn, P., De Clercq, E., Cooney, D. A., Kang, G. J., Dalal, M., Johns, D. G., and Broder, S. (1986). *Biochem. Biophys. Res. Commun.* **140**, 735–742.

Balzarini, J., Kang, G. J., Dalal, M., Herdewijn, P., De Clercq, E., Broder, S., and Johns, D. G. (1987a). *Mol. Pharmacol.* **32**, 162–167.

Balzarini, J., Pauwels, R., Baba, M., Robins, M. J., Zou, R., Herdewijn, P., and De Clercq, E. (1987b). *Biochem. Biophys. Res. Commun.* **145**, 269–276.

Balzarini, J., Van Aerschot, A., Herdewijn, P., Rosenberg, I., Holy, A., Pauwels, R., Baba, M., Johns, D. G., and De Clercq, E. (1989). *Biochem. Pharmacol.* **38**, 869–874.

Barré-Sinoussi, F., Chermann, J. C., Rey, F., Nugeyre, T., Charmaret, S., Gruest, J., Dauguet, C., Axler-Blin, C., Vésinet-Brun, F., Rouzioux, C., Rozenbaum, W., and Montagnier, L. (1983). *Science* **220**, 868–871.

Bennett, L. L., Shannon, W. M., Allan, P. W., and Arnett, G. (1975). *Ann. N.Y. Acad. Sci.* **255**, 342–358.

Beral, V., Peterman, T. A., Berkelman, R. L., and Jaffe, H. W. (1990). *Lancet:* **i**, 123–128.

Blundel, T., and Pearl, L. (1989). *Nature (London)* **337**, 596–597.

Bowerman, B., Brown, P. O., Bishop, J. M., and Varmus, H. E. (1989). *Genes & Devel.* **3**, 469–478.

Broder, S. (1989). *Ann. Intern. Med.* **110**, 417–418.

Brown, P. O., Bowerman, B., Varmus, H. E., and Bishop, J. M. (1989). *Proc. Natl. Acad. Sci. USA.* **86**, 2525–2529.

Capon, D. J., Chamow, S. M., Mordenti, J., Marster, S. A., Gregory, T., Mitusya, H., Byrn, R. A., Lucas, C., Wurm, F. M., Groopman, J. E., Broder, S., and Smith, D. H. (1989). *Nature (London)* **337**, 525–531.

Chaudhary, V. K., Mizukami, T., Fuerst, T. R., FitzGerald, D. J., Moss, B., Pastan, I., and Berger, E. A. (1988). *Nature (London)* **335**, 369–372.

Cheng, Y.-C., Dutschman, G. E., Bastow, K. F., Sarngadharan, M. G., and Ting, R. Y. C. (1987). *J. Biol. Chem.* **262**, 2187–2189.

Clapham, P. R., Weber, J. N., Whitby, D., McIntosh, K., Dalgleish, A. G., Maddon, P. J., Deen, K. C., Sweet, R. W., and Weiss, R. W. (1989). *Nature (London)* **337**, 368–370.

Cooley, T. P., Kunches, L. M., Saunders, C. A., Ritter, J. K., Perkins, C. J., McLaren, C., McCaffrey, R. P., and Liebman, H. A. (1990). *N. Engl. J. Med.* **322**, 1340–1345.

Cooney, D. A., Dalal, M., Mitsuya, H., McMahon, J. B., Nadkarni, M., Balzarini, J., Broder, S., and Johns, D. G. (1986). *Biochem. Pharmacol.* **35**, 2065–2068.

Daar, E. S., Li, E. L., Moudgil, T., and Ho, D. D. (1990). *Proc. Natl. Acad. Sci. U.S.A.* **87**, 6574–6579.

Dahlberg, J. E., Mitsuya, H., Broder, S., Blam, S. B., and Aaronson, S. A. (1987). *Proc. Natl. Acad. Sci. USA* **84**, 2469–2473.

Dalgleish, A. G., Beverleym, P. C. L., Clapham, P. R., Crawford, D. H., Greaves, M. F., and Weiss, R. A. (1984). *Nature (London)* **312**, 763–767.

De Clercq, E. (1989). *Antiviral Res.* **12**, 1–20.

Dedera, D., Hu, W., Heyden, N. V., and Ratner, L. (1989). *J. Virol.* **63**, 3205–3208.

Deen, K. C., McDougal, J. S., Inacker, R., Folena-Wasseman, G., Arthos, J., Rosenberg, J., Maddon, P. J., Axel, R., and Sweet, R. W. (1988). *Nature (London)* **331**, 82–84.

DeVita, V. T., and Schein, P. S. (1973). *N. Engl. J. Med.* **288**, 998–1006.

Di Marzo Veronese, F., Copeland, T. D., DeVico, A. L., Rahman, R., Oroszlan, S., Gallo, R. C., and Sarngadharan, M. G. (1986). **231**, 1289–1291.

Dingwall, Ernberg, C. I., Gait, M. J., Green, S. M., and Balerio, R. (1989). *Proc. Natl. Acad. Sci. U.S.A.* **86**, 6925–6929.

Du, D. L., Volpe, D. A., Murphy, M. J., and Grieshaber, C. K. (1989). *Proc. Am. Assoc. Cancer Res.* **30**, 606.

Erickson, J., Neidhart, D. J., VanDrie, J., Kempf, D. J., Wang, X. C., Norbeck, D. W., Plattner, J. J., Rittenhouse, J. W., Turon, M., Wideburg, N., Kohlbrenner, W. E., Simmer, R., Helfrich, R., Paul, D. A., and Knigge, M. (1990). *Science* **249**, 527–533.

Eriksson, B., Vrang, L., Bazin, H., Chattopadhyaya, J., and Öberg, B. (1987). *Antimicrob. Ag. Chemother.* **31**, 600–604.

Feinberg, M. B., Jarrett, R. F., Aldovini, A., Gallo, R. C., and Wong-Staal, F. (1986). *Cell* **46**, 807–817.

Felber, B. K., Hadzopoulou-Cladaras, M., Cladaras, C., Copeland, T., and Pavalakis G. N. (1988). *Proc. Natl. Acad. Sci. U.S.A.* **86**, 1495–1499.

Fields, A. P., Bennarik, D. P., Hess, A., and May, S. (1988). *Nature (London)* **333**, 278–280.

Fine, R. L., Poston, C., Williams, A., Chabner, B. A., Broder, S., and Mitsuya, H. (1990). (in preparation).

Fischl, M. A., Richman, D. D., Grieco, M. H., Gottlieb, M. S., Volberding, P. A., Laskin, O. L., Leedom, J. M., Groopman, J. E., Mildvan, D., Schooley, R. T., Jackson, G. G., Durack, D. T., King, D., and the AZT Collaborative Group (1987). *N. Engl. J. Med.* **317**, 185–191.

Fisher, A. G., Ensoli, B., Ivanoff, L., Chamberlain, M., Pettway, S., Ratner, L., Gallo, R. C., and Wong-Staal, F. (1987). *Science* **237**, 888–893.

Fisher, R. A., Bertonis, J. M., Meier, W., Johnson, V. A., Costopoulos, D. S., Liu, T., Tizzard, R., Walker, B. D., Hirsch, M. S., Schooley, R. T., and Flavell, R. A. (1988) *Nature (London)* **331**, 76–78.

Frankel, A. D., and Pabo, C. O. (1988). *Cell* **55**, 1189–1193.

Frankel, A. D., Bredt, D. S., and Pabo, C. O. (1988a). *Science* **240**, 70–73.

Frankel, A. D., Chen, L., Cotter, R. J., and Pabo, C. O. (1988b). *Proc. Natl. Acad. Sci. U.S.A.* **85**, 6297–6300.

Franza, B. R. Jr., Josephs, S. F., Gilman, M. Z., Ryan, W., and Clarkson, B. (1987). *Nature (London)* **330**, 391.

Fujiwara, T., and Mizuuchi, K. (1988). *Cell* **54**, 497–504.

Furman, P. A., Fyfe, J. A., St. Clair, M. H., Weinhold, K., Rideout, J. L., Freeman, G. A., Lahrman, S. N., Bolognesi, D. O., Broder, S., Mitsuya, H., and Barry, D. W. (1986). *Proc. Natl. Acad. Sci. U.S.A.* **83**, 8333–8337.

Furmanski, P., Bourguignon, G. J., Bolles, C. S., Corombos, J. D., and Das, M. R. (1980). *Cancer Lett.* **8**, 307–315.

Gallo, R. C., Wong-Staal, F., Montagnier, L., Haseltine, W. A., and Yoshida, M. (1988). *Nature (London)* **333**, 504.

Garcia, J. A., Harrich, D., Pearson, L., Mitsuyasu, R., and Graynor, R. B. (1988). *EMBO J.* **7**, 3143–3147.

Gartner, S., Markovits, P., Markovits, D. M., Kaplan, M. H., Gallo, R. C., and Popovic, M. (1986). *Science* **233**, 215–219.

Gatignol, A., Kumar, A., Rabson, A., and Jeang, K. T. (1989). *Proc. Natl. Acad. Sci. U.S.A.* **86**, 7828–7832.

Gottlieb, M. S., Schroff, R., Schanker, M., Weisman, J. D., Peng, T. F., Wolf, R. A., and Saxon, A. (1981). *N. Engl. J. Med.* **305**, 1425–1430.

Green, M., and Loewenstein, P. M. (1988). *Cell* **55**, 1179–1188.

Greengrass, C. W., Hoople, D. W. T., Street, S. D. A., Hamilton, F., Marriott, M. S., Bormer, J., Dalgleish, A. G., Mitsuya, H., and Broder, S. (1989). *J. Med. Chem.* **32**, 618–622.

Groopman, J. E., Gottlieb, M. S., Goodman, J., Mitsuyasu, R. T., Conant, M. A., Prince, H., Fahey, J. L., Derezin, M., Weinstein, W. M., Casavante, C., Rothman, J., Rudnick, S. A., and Volberding, P. A. (1984). *Ann. Intern. Med.* **100**, 671–676.

Gruters, R. A., Neefjes, J. J., Tersmette, M., de Goede, R. E. Y., Tulp, A., Huisman, H. G., Miedema, F., and Ploegh, H. L. (1987). *Nature (London)* **330**, 74–77.

Guy, B., Kieny, M. P., Riviere, Y., Le Peuch, C., Dott, K., Girard, M., Montagnier, L., and Lecocq, J.-P. (1987). *Nature (London)* **330**, 266–269.

Haertle, T., Carrera, C. J., Wasson, D. B., Sowers, L. C., Richman, D. D., and Carson, D. A. (1988). *J. Biol. Chem.* **263**, 5870–5875.

Hamamoto, Y., Nakashima, H., Matsui, T., Matsuda, A., Ueda, T., and Yamamoto, N. (1987). *Antimicrob. Ag. Chemother.* **31**, 907–910.

Hansen, J., Schulze, T., and Moeling, K. (1987). *J. Biol. Chem.* **262**, 12393–12396.

Harada, S., Koyanagi, Y., and Yamamoto, N. (1985). *Science* **229**, 563–566.

Harper, M. E., Kaplan, M. H., Marselle, L. M., Pahwa, S. G., Chayt, K. J., Sarngadharan, M. G., Wong-Staal, F., and Gallo, R. C. (1986). *N. Engl. J. Med.* **315**, 1073–1078.

Hartmann, H., Hunsmann, G., and Eckstein, F. (1987). *Lancet* **i**, 40–41.

Hartman, N. R., Johns, D. J., and Mitsuya, H. (1990). *AIDS Res. Hum. Retroviruses* **6**, 805–812.

Hartshorn, K. L., Vogt, M. W., Chou, T.-C., Blumberg, R. S., Byington, R., Schooley, R. T., and Hirsch, M. S. (1987). *Antimicrob. Ag. Chemother.* **31**, 168–172.

Hayashi, S., Phadtare, S., Zemlicka, J., Matsukura, M., Mitsuya, H., and Broder, S. (1988). *Proc. Natl. Acad. Sci. U.S.A.* **85**, 6127–6131.

Hayashi, Y., Ikuta, M., Fujii, N., Ezawa, K., and Kato, S. (1989). *Arch. Virol.* **105**, 129–135.

Hayashi, S., Norbeck, D. W., Rosenbrook, W., Fine, R. L., Matsukura, M., Plattner, J. J., Broder, S., and Mitsuya, H. (1990a). *Antimicrob. Ag. Chemother.* **34**, 287–294.

Hayashi, S., Fine, R. L., Chou, T.-C., Currens, M. J., Broder, S., and Mitsuya, H. (1990b). *Antimicrob. Ag. Chemother.* **34**, 82–88.

Hedaya, M. A., and Sawchuk, R. J. (1989). *J. Pharm. Sci.* **78**, 716–722.

Hirsch, M. S., and Kaplan, J. C. (1987). *Antimicrob. Chemother.* **31**, 839–843.

Ho, D. D., Hartshorn, K. L., Rota, T. R., Andrews, C. K., Kaplan, J. C., Schooley, R. T., and Hirsch, M. S. (1985). *Lancet* **i**, 602.

Ho, D. D., Rota, T. R., and Hirsch, M. S. (1986). *J. Clin. Invest.* **77**, 1712–1715.

Horwitz, J. P., Chua, J., and Noel, M. (1964). *J. Org. Chem.* **29**, 2076–2078.

Horwitz, J. P., Chua, J., Noel, M., and Donatti, J. T. (1966). *J. Org. Chem.* **32**, 817–818.

Hoshino, H., Shimizu, N., Shimada, N., Takita, T., and Takeuchi, T. (1987). *J. Antibiot.* **40**, 1077–1078.

Hussey, R. E., Richardson, N. E., Kowalski, M., Brown, N. R., Chang, H.-S., Siliciano, R. F., Dorkman, T., Wlaker, B., Sodroski, J., and Reinherz, E. L. (1988). *Nature (London)* **331**, 78–81.

Hymes, K. B., Greene, J. B., and Karpatkin, S. (1988). *N. Engl. J. Med.* **318**, 516–517.

Jacks, T., Power, M. D., Masiarz, F. R., Luciw, P. A., Barr, P. J., and Varmus, H. E. (1988). *Nature (London)* **331**, 280–283.

Jacks, T., Madhani, H. D., Masiarz, F. R., Varmus, H. E. (1989). *Cell* **55**, 447–458.

Israelski, D. M., Tom, C., Remington, J. S. (1989). *Antimicrob. Ag. Chemother.* **33**, 30–34.

Ito, M., Baba, M., Sato, A., Powels, R., De Clercq, E., Shigeta, S., and Yamamoto, N. (1987). *Antiviral Res.* **7**, 361–367.

Johnson, M. A., and Fridland, A. (1989). *Mol. Pharmacol.* **36**, 291–295.

Johnson, M. A., Ahluwalia, G., Connelly, M. C., Cooney, D. A., Broder, S., Johns, D. G., and Fridland, A. (1988). *J. Biol. Chem.* **30**, 15354–15357.

Johnson, V. A., Barlow, M. A., Chou, T.-C., Fisher, R. A., Walker, B. D., Hirsch, M. S., and Schooley, R. T. (1989). *J. Infect. Dis.* **159**, 837–844.

Kappes, J. C., Morrow, C. D., Lee, S.-W., Jameson, B. A., Kent, S. B., Hood, L. E., Shaw, G. M., and Hahn, B. H. (1988). *J. Virol.* **62**, 3501–3505.

Karpas, A., Fleet, G. W. J., Dwek, R. A., Petursson, S., Namgoong, S. K., Ramsden, N. G., Jacob, G. S., and Rademacher, T. W. (1988). *Proc. Natl. Acad. Sci. U.S.A.* **85**, 9229–9233.

Kawakami, K., Schedereit, C., and Roeder, R. G. (1988). *Proc. Natl. Acad. Sci. U.S.A.* **85**, 4700–4704.

Klatzman, Champagne, D. E., Chamaret, S., Gruest, J., Guetard, D., Hercend, T., Gluckman, J.-C., and Montagnier, L. (1984). *Nature (London)* **312**, 767–768.

Kim, C.-H., Marquez, V. E., Broder, S., Mitsuya, H., and Driscoll, J. S. (1987). *J. Med. Chem.* **30**, 862–866.

Koenig, S., Gendelman, H. E., Orenstein, J. M., Dal Canto, M. C., Pezeshpour, G. H., Yungbluth, M., Janotta, F., Aksamit, A., Martin, M. A., and Fauti, A. S. (1986). *Science* **233**, 1089–1093.

Kornhauser, D. M., Petty, B. G., Hendrix, C. W., Woods, A. S., Nerhood, L. J., Bartlet, J. G., and Lietman, P. S. (1989). *Lancet* **ii**, 473–475.

Kovacs, J. A., Deyton, L., Davey, R., Falloon, J., Zunich, K., Lee, K., Metcalf, J. A., Bigley, J. W., Sawyer, L. A., Zoon, K. C., Masur, H., Fauci, A. S., and Lane, H. C. (1989). *Ann. Intern. Med.* **111**, 280–287.

Krown, S. E., Real, F. X., Cunningham-Rundles, S., Myskowski, P. L., Koziner, B., Fein, S., Mittelman, A., Oettgen, H. F., and Safai, B. (1983). *N. Engl. J. Med.* **308**, 1071–1076.

Lambert, J. S., Seidlin, M., Reichman, R. C., Plank, C. S., Laverty, M., Morse, G. D., Knupp, C., McLaren, C., Pettinelli, C., Valentine, F. T., and Dolin, R. (1990). *N. Engl. J. Med.* **322**, 1333–1340.

Lapatto, R., Blundel, T., Hemmings, A., Overington, J., Wildershpin, A., Wood, S., Merson, J. R., Whittle, P. J., Danley, D. E., Geogheegan, K. F., Hawrylik, S. J., Lee, S. E., Scheld, K. G., and Hobart, P. M. (1989). *Nature (London)* **342**, 299–302.

Larder, B. A., and Kemp, S. D. (1989). *Science* **246**, 1155–1158.

Larder, B. A., Purifoy, D. J. M., Powell, K. L., and Darby, G. (1987). *Nature (London)* **327**, 716–717.

Larder, B. A., Darby, G., and Richman, D. D. (1989a). *Science* **243**, 1731–1734.

Larder, B. A., Kemp, S. D., and Purifoy, D. J. M. (1989b). *Proc. Natl. Acad. Sci. U.S.A.* **86**, 4803–4807.

Leaf, A. N., Laubenstein, L. J., Raphael, R., Hochster, H., Baez, L., and Karpatkin, S. (1988). *Ann. Intern. Med.* **109**, 194–197.

Lenardo, M. J., Kuang, A., Gifford, A., and Baltimore, D. (1988). *Proc. Natl. Acad. Sci. U.S.A.* **85**, 8825–8829.

Lever, A., Gottlinger, H., Haseltine, W., and Sodroski, J. (1989). *Virology,* **63**, 4085–4087.

Lin, T.-S., Schinazi, R. F., and Prusoff, W. H. (1987). *Biochem. Pharmacol.* **36**, 2713–2718.

Looney, D. J., Hayashi, S., Nicklas, M., Redfiled, R. R., Broder, S., Wong-Staal, F., and Mitsuya, H. (1990). *J. AIDS* (in press).

Lorentsen, K., Hendrix, C., Collins, J., Eckel, R., Petty, B., and Lietman, P. (1989). *Ann. Intern. Med.* **111**, 561–566.

Luciw, P. A., Cheng-Meyer, C., and Levy, J. A. (1987). *Proc. Natl. Acad. Sci. U.S.A.* **84**, 1434–1438.

Lusso, P., Ensoli, B., Markham, P. D., Ablashi, D. V., Salahuddin, S. Z., Tschachler, E., Wong-Staal, F., Gallo. R. C. (1989). *Nature (London)* **337**, 370–373.

Maddon, P. J., Dalgleish, A. G., McDougal, J. S., Clapham, P. R., Weiss, R. A., and Axel, R. (1986). *Cell* **47**, 333–348.

Maekawa, T., H. Sakura, T. Sudo, and Ishii, S. (1989). *J. Biol. Chem.* **264**, 14591–14593.

Majumdar, C., Abbotts, J., Broder, S., and Wilson, S. H. (1988). *J. Biol. Chem.* **263,** 15657–15665.

Malim, M. H., Hauber, J., Le, S.-Y., Maziel, J. V., and Cullen, B. R. (1989). *Nature (London)* **338,** 254–257.

Marquez, V. E., Tseng, C. K. H., Driscoll, J. S., Mitsuya, H., Broder, S., Roth, J. S., and Kelley, J. A. (1987). *Biochem. Pharmol.* **36,** 2719–2722.

Marquez, V. E., Tseng, C. K. H., Mitsuya, H., Aoki, S., Kelly, J. A., Ford, H., Roth, J. S., Broder, S., and Driscoll, J. S. (1990). *J. Med. Chem.* **33,** 978–985.

Masood, R., Ahluwalia, G. S., Cooney, D. A., Fridland, A., Marquez, V. E., Driscoll, J. S., Hao, Z., Mitsuya, H., Broder, S., and Johns, D. G. (1990). *Mol. Pharmacol.* **37,** 590–596.

Masur, H., Michelis, M. A., Greene, J. B., Onarato, I., Vande Stouwe, R. A., Holzman, R. S., Wormser, G., Brettman, L., Lange, M., Murray, H. W., and Cunningham-Rundles, S. (1981). *N. Engl. J. Med.* **305,** 1431–1438.

Matsukura, M., Shinozuka, K., Zon, G., Mitsuya, H., Reitz, M., Cohen, J. S., and Broder, S. (1987). *Proc. Natl. Acad. Sci. U.S.A.* **84,** 7706–7710.

Matsurka, M., Zon, G., Shinozuka, K., Robert-Gurroff, M., Shimada, T., Stein, C. A., Mitsuya, H., Wong-Staal, F., Cohen, J. S., and Broder, S. (1989). *Proc. Natl. Acad. Sci. U.S.A.* **86,** 4244–4248.

Matsushita, S., Mitsuya, H., Reitz, M. S., and Broder, S. (1987). *J. Clin. Invest.* **80,** 394–400.

Matsushita, S., Robert-Guroff, M., Rusche, J., Koito, A., Hattori, T., Hoshino, H., Javaherian, K., Takatsuki, K., and Putney, S. (1988a). *J. Virol.* **62,** 2107–2115.

Matsushita, S., Putney, S., Maeda, Y., Koito, A., Hattori, T., Robert-Guroff, M., Takatsuki, K. (1988b). *Proceedings of IV International Conference on AIDS, Stockholm, 1988,* p. 234.

McCormick, J. B., Gethcell, J. P., Mitchell, S. W., and Hicks, D. R. (1984). *Lancet* **ii,** 1367–1369.

Meek, T. D., Dayton, B. D., Metcalf, B. W., Dreyer, G. B., Strichler, J. E., Gorniak, J., Rosenber, M., Moore, M. L., Mafard, V. W., and Debouck, C. (1989). *Proc. Natl. Acad. Sci. U.S.A.* **86,** 1841–1845.

Merigan, T. C., Skowron, G., Bozzette, S. A., Richman, D., Uttamchandani, R., Fischl, M., Schooley, R., Hirsch, M., Soo, W., Pettinelli, C., Schaumburg, H., and the ddC Study Group of the AIDS Clinical Trials Group (1989). *Ann. Intern. Med.* **110,** 189–194.

Miller, M., Schneider, J., Sathyanarayana, B. K., Toth, M. V., Marshall, G. R., Clawson, L., Selk, L., Kent, S. B. H., and Wlodawer, A. (1989a). *Science* **246,** 1149–1152.

Miller, M., Jaskólski, M., Rao, J. K. M., Leis, J., and Wlodawer, A. (1989b). *Nature (London)* **337,** 576–579.

Mitsuya, H., and Broder, S. (1986). *Proc. Natl. Acad. Sci. USA.* **83,** 1911–1915.

Mitsuya, H., and Broder, S. (1987). *Nature (London)* **325,** 773–778.

Mitsuya, H., and Broder, S. (1988). *AIDS Res. Hum. Retroviruses* **4,** 107–113.

Mitsuya, H., and Broder, S. (1990). *In* "Retroviruses Biology and Human Disease" (R. C. Gallo and F. Wong-Staal, eds.), pp. 331–358. Marcel Dekker, New York and Basel.

Mitsuya, H., Popovic, M., Yarchoan, R., Matsushita, S., Gallo, R. C., and Broder, S. (1984). *Science* **226,** 172–174.

Mitsuya, H., Weinhold, K. J., Furman, F. A., St. Clair, M. H., Lehrman, S. N., Gallo, R. C., Bolognesi, D., Barry, D. W., and Broder, S. (1985). *Proc. Natl. Acad. Sci. U.S.A.* **82,** 7096–7100.

Mitsuya, H., Jarrett, R. F., Matsukura, M., Di Marzo Veronese, F., Devico, A. L., Sarngadharan, M. G., Johns, D. G., Reitz, M. S., and Broder, S. (1987a). *Proc. Natl. Acad. Sci. U.S.A.* **84,** 2033–2037.

Mitsuya, H., Matsukura, M., and Broder, S. (1987b). *In* "AIDS: Modern Concepts and Therapeutic Challenges" (S. Broder, ed.), pp. 303–333. Marcel-Dekker, New York.

Mitsuya, H., Looney, D. J., Kuno, S., Ueno, R., Wong-Staal, F., and Broder, S. (1988a). *Science* **240,** 646–649.

Mitsuya, H., Matsukura, M., Hayashi, S., Shinozuka, K., Perno, C. F., and Broder, S. (1988b). *Abstracts of IV International Conference on AIDS, Stockholm, 1988,* p. 226.

Mitsuya, H., Dahlberg, J. E., Spigelman, Z., Matsushita, S., Jarrett, R. F., Matsukura, M., Currens, M. J., Aaronoson, S. A., Reitz, M. S., McCaffrey, R. S., and Broder, S. (1988c). *In* "Human Retroviruses, Cancer and AIDS: Approaches to Prevention and Therapy" (D. Bolognesi, ed.), pp. 407–421. Alan R. Liss, New York.

Mitsuya, H., Looney, D. J., Kuno, S., Ueno, R., Wong-Staal, F., and Broder, S. (1989). *In* "Mechanisms of Action and Therapeutic Applications of Biologicals in Cancer and Immune Deficiency Disorders." (J. E. Groopman, D. W. Golde, and C. H. Evans, eds.), pp. 331–341. Alan R. Liss, New York.

Mitsuya, H., Yarchoan, R., Hayashi, S., and Broder, S. (1990). *J. Am. Acad. Dermatol.* **22**, 1282–1294.

Miyoshi, I., Kubonishi, I., Yoshimoto, S., Akagi, T., Ohtsuki, Y., Shiraishi, Y., Nagata, K., and Hinuma, Y. (1981). *Nature (London)* **294**, 770–771.

Molina, J. M., and Groopman, J. E. (1989). *N. Engl. J. Med.* **23**, 1478.

Montagnier, L., Gruest, J., Chamaret, S., Dauguet, C., Axler, C., Guétard, D., Nugeyre, M. T., Barré-Sinoussi, F., Chermann, J.-C., Brunet, J. B., Klatzman, D., and Gluckman, J. C. (1984). *Science* **225**, 63–66.

Mosca, J. D., Bednarik, D. P., Raj, N. B. K., Rosen, C. A., Sodroski, J. G., Haseltine, W. A., and Pitha, P. M. (1987). *Nature (London)* **325**, 67–70.

Nabel, G., and Baltimore, D. (1987). *Nature (London)* **326**, 711–713.

Nabel, G. J., Rice, S. A., Knipe, F. M., and Baltimore, D. (1988). *Science* **239**, 1299–1302.

Nakamura, S., Salahuddin, S. Z., Biberfeld, P., Ensoli, B., Markam, P. D., Wong-Staal, F., and Gallo, R. C. (1988). *Science* **242**, 426–430.

Nakashima, H., Kido, Y., Kobayashi, N., Motoki, Y., Neushul, M., and Yamamoto, N. (1987). *Antimicrob. Ag. Chemother.* **31**, 1524–1528.

Navia, A., Fitzgerald, P. M. D., McKeever, M. M., Leu, R.-T., Heimbach, J. C., Herber, W. K., Sigal, I. S., Daeke, P. L., and Springer, J. P. (1989). *Nature (London)* **337**, 615–620.

Nelson, J. A., Reynolds-Kohler, C., Oldstone, M. B. A., Willey, C. A. (1988). *Virology* **165**, 286–290.

Niederman, T. M. J., Thielan, B., and Ratner, L. (1989). *Proc. Natl. Acad. Sci. U.S.A.* **86**, 1128–1132.

Norbeck, D. W., Kern, E., Hayashi, S., Rosenbrook, W., Sham, H., Herrin, T., Plattner, J. J., Erickson, J., Clement, J., Swanson, R., Shipkowitz, N., Hardy, D., Marsh, K., Arnett, G., Shannon, W., Broder, S., and Mitsuya, H. (1990). *J. Med. Chem.* **33**, 1281–1285.

Ogawa, K., Shibata, R., Kiyomasu, T., Higuchi, I., Kishida, Y., Ishimoto, A., and Adachi, A. (1989). *J. Virol.* **63**, 4110–4114.

Ono, K., Ogasawara, M., and Matsukage, A. (1979). *Biochem. Biophys. Res. Commun.* **88**, 1255–1262.

Ostertag, W., Cole, T., Crozier, T., Gaedicke, G., Kind, J., Kluge, N., Krieg, J. C., Roesler, G., Steinheider, G., Weismann, B. J., and Dube, S. K. (1974a). *In* "Differentiation and Control of Malignancy of Tumor Cells" (W. Nakahara, T. R. Ono, T. Sugimura, and H. Sugano, eds.), pp. 485–513. University Park Press, Baltimore, London, Tokyo.

Ostertag, W., Roesler, G., Krieg, C. J., Kind, J., Cole, T., Crozier, T., Gaedicke, G., Steinheider, G., Kludge, N., and Dube, S. (1974b). *Proc. Natl. Acad. Sci. U.S.A.* **71**, 4980–4985.

Pal, R., Gallo, R. C., and Sarngadharan, M. G. (1988). *Proc. Natl. Acad. Sci. U.S.A.* **85**, 9283–9286.

Pauwels, R., Balzarini, J., Schols, D., Baba, M., Desmyter, J., Rosenberg, I., Holy, A., and De Clercq, E. (1988). *Antimicrob. Ag. Chemother.* **32**, 1025–1030.

Perno, C. F., Yarchoan, R., Cooney, D. A., Hartman, N. R., Gartner, S., Popovic, M., Hao, Z., Gerard, T. L., Wilson, Y. A., Johns, D. G., and Broder, S. (1988). *J. Exp. Med.* **168**, 1111–1125.

Perno, C.-F., Yarchoan, R., Cooney, D. A., Hartman, N. R., Webbs, D. S. A., Hao, Z., Mitsuya, H., Johns, D. G., and Broder, S. (1989). *J. Exp. Med.* **169**, 933–951.

Peterson, A. and Seed, B. (1988). *Cell* **54**, 65–72.

Pitha, P. M., Wivel, N. A., Fernie, B. F., and Harper, H. P. (1979). *J. Gen. Virol.* **42**, 467.

Pizzo, P. A., Eddy, J., Faloon, J., Balis, F., Murphy, R. F., Moss, H., Wolters, P., Browers, P., Jarosinski, P., Rubin, M., Broder, S., Yarchoan, R., Brunetti, A., Maha, M., Nusinoff-Lehrman, S., and Poplack, D. G. (1988). *N. Engl. J. Med.* **319**, 889–896.

Pluda, J., Yarchoan, R., McAtee, N., Thomas, R. V., Oette, D., and Broder, S. (1989). *Abstracts of the V International Conference on AIDS, Montreal, 1989*, p. 406.

Poli, G., Orenstein, J. M., Kinter, A., Folks, T. M., and Fauci, A. S. (1989). *Science* **244**, 575–577.

Popovic, M., Sarngadharan, M. G., Read, E., and Gallo, R. C. (1984). *Science* **224**, 497–500.

Portgies, P., De Gans, J., Lange, J. M. A., Derix, M. M. A., Speelman, H., Bakker, M., Danner, S. A., and Goudsmit, J. (1989). *Brit. Med. J.* **299**, 819–821.

Preston, B. D., Poiesz, B. J., and Loeb, L. A. (1988). *Science* **242**, 1168–1171.

Ratner, Haseltine, L. W., Patarca, R., Livak, K. J., Starcich, B., Josephs, S. F., Doran, E. R., Rafalski, J. A., Whitehorn, E. A., Baumeister, K., Ivanoff, L., Petteway, Jr., S. R., Pearson, M. L., Lautenberger, J. A., Papas, T. S., Ghrayeb, J., Chang, N. T., Gallo, R. C., and Wong-Staal, F. (1985a). *Nature (London)* **313**, 277–284.

Ratner, L., Starcich, B., Joseph, S. F., Hahn, B. H., Reddy, E. P., Liva, K. L., Petteway, Jr., R., Pearson, M. L., Haseltine, W. A., Arya, S. K., and Wong-Staal, F. (1985b). *Nucl. Acids. Res.* **13**, 8219–8229.

Reitz, M. S., Wu, A. M., and Gallo, R. C. (1977). *Int. J. Cancer* **20**, 67–74.

Richman, D. D. (1990). *Rev. Infect. Dis.* (in press).

Richman, D. D., Fischl, M. A., Grieco, M. H., Gottlieb, M. S., Volberding, P. A., Laskin, O. L., Leedom, J. M., Groopman, J. E., Mildvan, D., Hirsch, M. S., Jackson, G. G., Durack, D. T., Nusinoff-Lehrman, S., and the AZT Collaborative Working Group. (1987). *N. Engl. J. Med.* **317**, 192–197.

Robert-Guroff, M., Weiss, S. H., Giron, J. A., Jennings, A. M., Ginzburg, H. M., Morgolis, I. B., Blattner, W. A., and Gallo, R. C. (1986). *JAMA* **255**, 3133–3137.

Roberts, J. D., Bebenek, K., and Kunkel, T. A. (1988). *Science* **242**, 1171–1173.

Roberts, N. A., Martin, J. A., Kinchington, D., Broadhurst, A. V., Craing, J. C., Duncan, I. B., Galpin, S. A., Handa, B. K., Kay, J., Kröhn, A., Lambert, R. W., Merrett, J. H., Mills, J. S., Parkes, K. E. B., Redshaw, S., Ritchie, A. J., Taylor, D. L., Thomas, G. J., and Machin, P. J. (1990). *Science* **248**, 358–361.

Robins, M. J., and Robins, R. K. (1964). *J. Am. Chem. Soc.* **86**, 3585–3586.

Rosen, C. A., Sodroski, J. G., and Haseltine, W. A. (1985). *Cell* **41**, 813–823.

Rosen, C. A., Terwilliger, E., Dayton, A., Sodroski, J. G., and Haseltine, W. A. (1988). *Proc. Natl. Acad. Sci. U.S.A.* **85**, 2071–2075.

Ruben, S., Perkins, A., Purcell, R., Joung, K., Sia, R., Burghoff, R., Haseltine, W. A., Rosen, C. A. (1989). *J. Virol.* **63**, 1–8.

Ruprecht, R. M., O'Brien, L. G., Rossoni, L. D., and Nusinoff-Lehrman, S. (1986). *Nature (London)* **323**, 467–469.

Salahuddin, S. Z., Ablashi, D. V., Markham, P. D., Josephs, S. F., Sturzenegger, S., Kaplan, M., Halligan, G., Biberfeld, P., Wong-Staal, F., Kramarsky, B., and Gallo, R. C. (1986). *Science* **234**, 596–601.

Salahuddin, S. Z., Nakamura, S., Biberfeld, P., Kaplan, M. H., Markham, P. D., Larsson, L., and Gallo, R. C. (1988). *Science* **242**, 430–433.

Samuel, K. P., Seth, A, Konopka, A., Lautenberger, J. A., and Papas, T. S. (1987). *FEBS Lett.* **218**, 81–86.

Sanger, F., Nicklen, S., and Coulson, A. R. (1977). *Proc. Natl. Acad. Sci. U.S.A.* **75**, 5463–5467.

Schinazi, R. F., and Ahn, S.-K. (1987). *J. Clin. Biochem. Suppl.* **11D**, 74.

Schmitt, R. A., Bigley, J. W., McKinnis, R., Logue, P. E., Evans, R. W., Drucker, J. L., and the AZT Collaborative Working Group. (1988). *N. Engl. J. Med.* **319**, 1573–1578.

Seiki, M., Hattori, S., Hirayama, Y., and Yoshida, M. (1983). *Proc. Natl. Acad. Sci. U.S.A.* **80**, 3618.

Sharp, A. H., Jaenisch, R., and Ruprecht, R. M. (1987). *Science* **236**, 1671–1674.

Shimada, N., Hasegawa, S., Harada, T., Tomisawa, T., Fujii, A., and Takita, T. (1986). *J. Antibiot.* **39**, 1623–1625.

Siegal, F. P., Lopez, C., Hammer, G. S., Brown, A. E., Kornfeld, S. J., Gold, J., Hassett, J., Hirshman, S. Z., Cunningham-Rundles, C., Adelsberg, B. R., Parham, D. M., Siegal, M., Cunningham-Rundles, S., and Armstrong, D. (1981). *N. Engl. J. Med.* **305**, 1439–1444.

Simpson, M. V., Chin, C. D., Keilbaugh, S. A., Lin, T.-S., and Prusoff, W. H. (1989). *Biochem. Pharmacol.* **38**, 1033–1036.

Siomi, H., Sha, H., Nam, S. H., Nosaka, T., Maki, M. and Hatanaka, M. (1988). *Cell* **55**, 197–209.

Skolnik, P. R., Kosloff, B. R., Hirsch, M. S. (1988). *J. Infect. Dis.* **157**, 508–514.

Skolnik, P. R., Pomerantz, R. J., de la Monte, S., Lee, S. F., Hsing, G. D., Foos, R. Y., Cowan, G. M., Kosloff, B. R., Hirsch, M. S., and Pepose, J. S. (1989). *Am. J. Ophthalmol.* **107**, 361–372.

Smith, C. C., Aurelian, L., Reddy, M. P., Miller, P., and Ts'o, P. O. P. (1986). *Proc. Natl. Acad. Sci. U.S.A.* **83**, 2787–2791.

Smith, D. H., Byrn, R. A., Marsters, S. A., Gregory, T., Groopman, J. E., and Capon, D. J. (1987). *Science* **238**, 1704–1707.

Smith, M. S., Brian, E. L., De Clercq, E., and Pagano, J. S. (1989). *Chemother.* **33**, 1482–1486.

Smith, T. J., Kremer, M. J., and Luo, M. (1986). *Science* **233**, 1286–1293.

Sodroski, J., Rosen, C., Wong-Staal, F., Salahuddin, S. Z., Popovic, M., Arya, S., Gallo, R. C., and Haseltine, W. A. (1985). *Science* **227**, 177–181.

Sodroski, J. G., Goh, W. C., Rosen, C., Dayton, A., Terwilliger, E., and Haseltine, W. A. (1986). *Nature (London)* **321**, 412–417.

Sommadossi, J. P., Carlisle, R., Schinazi, R. F., and Zhou, Z. (1988). *Antimicrob. Ag. Chemother.* **32**, 997–1001.

Starnes, M., and Cheng, Y.-C. (1987). *J. Biol. Chem.* **262**, 988–991.

Strebel, K., Klimkait, T., Maldarelli, F., Martin, M. A. (1989). *J. Virol.* **63**, 3784–3791.

Surbone, A., Yarchoan, R., McAtee, N., Blum, M. R., Maha, M., Allain, J. P., Thomas, R. V., Mitsuya, H., Nusinoff Lehrman, S., Leuther, M., Pluda, N. M., Jacobbsen, F. K., Kessler, H. A., Myers, C. E., and Broder, S. (1988). *Ann. Intern. Med.* **108**, 534–540.

Szebeni, J., Wahl, S. M., Popovic, M., Wahl, L. M., Gartner, S., Fine, R. L., Skaleric, U., Friedman, R. M., and Weinstein, J. N. (1989). *Proc. Natl. Acad. Sci. U.S.A.* **86**, 3842–3846.

Tam, S., Weigele, Broder, S., and Mitsuya, H. (1989). United States Patent Number 4,837,311.

Tateno, M., Gonzalez-Scarano, F., and Levy, J. A. (1989). *Proc. Natl. Acad. Sci. U.S.A.* **86**, 4287–4290.

Terwilliger, E. F., Cohen, E. A., Lu, Y., Sodroski, J. G., and Haseltine, W. A. (1988). *Proc. Natl. Acad. Sci. U.S.A.* **86**, 5163–5167.

Toji, L., and Cohen, S. S. (1970). *J. Bacteriol.* **103**, 323–328.

Traunecker, A., Lüke, W., and Karjalaine, K. (1988). *Nature (London)* **331**, 84–86.

Traunecker, A., Schneider, J., Kefer, H., and Karjalaine, K. (1989). *Nature (London)* **339**, 68–70.

Travares, L., Roneker, C., Johnston, K., Lehrman, S. N., and de Noronha, F. (1987). *Cancer Res.* **47**, 3190–3194.

Tsai, C.-C., Follis, K. E., and Benveniste, R. E. (1988). *AIDS Res. Hum. Retroviruses* **4**, 359–368.

Ueno, R., and Kuno, S. (1987). *Lancet* **i**, 1379–1380.

Vince, R., Hua, M., Browne, J., Daluge, S., Lee, F., Shannon, W. M., Lavelle, G. C., Qualls, J., Weislow, O. S., Kiser, R., Canonico, P. G., Schultz, R. H., Narayanan, V. L., Mayo, J. G., Shoemaker, R. H. and Boyd, M. R. (1988). *Biochem. Biophys. Res. Commun.* **156**, 1046–1053.

Vogel, J., Hinrichs, S. H., Reynold, R. K., Luciw, P. A., and Jay, G. (1988). *Nature (London)* **335**, 606–611.

Vogt, M. W., Hartshorn, K. L., Furman, P. A., Chou, T.-C., Fyfe, J. A., Colemsn, L. A., Crumpacker, C., Schooley, R. T., and Hirsch, M. S. (1987). *Science* **235**, 1376–1379.

Votruba, I., Bernaerts, R., Sakuma, T., De Clercq, E., Rosenburg, I., and Holy, A. (1987). *Mol. Pharmacol.* **32,** 524–529.

Waqar, M. A., Evans, M. J., Manly, K. F., Hughes, R. G., and Huberman, J. A. (1984). *J. Cell. Physiol.* **121,** 402–408.

Wain-Hobson, S., Danos, P. O., Cole, S., and Alizon, M. (1985). *Cell* **40,** 9–17.

Walker, B. D., Kowalski, M., Goh, W. C., Kozarsky, K., Krieger, M., Rosen, C., Rohrschneider, L., Haseltine, W. A., and Sodroski, J. (1987). *Proc. Natl. Acad. Sci. U.S.A.* **84,** 8120–8124.

Webb, T. R., Mitsuya, H. and Broder, S. (1988). *J. Med. Chem.* **31,** 1475–1479.

Weber, I. T., Miller, M., Jaslóski, M., Leis, J., Skalka, A. M., and Wlodawer, A. (1989). *Science* **243,** 928–931.

Weiss, R., Clapham, P. R., Cheingsong-Popov, R., Dalgleish, A. G., Carne, C. A., Weller, I. V. D., and Tedder, R. S. (1985). *Nature (London)* **316,** 69–72.

Weiss, R., Clapham, P., Weber, J., Dalgleish, A., Lasky, L., and Berman, P. (1986). *Nature (London)* **324,** 572–575.

White, E. L., Parker, W. B., Macy, L. J., Shaddix, S. C., McCaleb, G., Secrist, J. A., Vince, R., and Shannon, M. W. (1989). *Biochem. Biophys. Res. Commun.* **161,** 393–398.

White, J. M. and Littman, D. R. (1989). *Cell* **56,** 725–728.

Wong-Staal, F., and Gallo, R. C. (1985). *Nature (London)* **317,** 395–403.

Wu, F., Garcia, J., Harrich, D., and Gaynor, R. G. (1988). *EMBO J.* **7,** 2117–2129.

Yano, O., Kanellopoulos, J., Kieran, M., Lebail, O., Israel, A., and Mourilsky, P. (1987). *EMBO J.* **6,** 3317–3324.

Yarchoan, R., and Broder, S. (1987). *N. Engl. J. Med.* **316,** 557–564.

Yarchoan, R., Klecker, R., Weinhold, K. J., Markham, P. D., Lyerly, H. K., Durack, D. T., Gelmann, E., Lehrman, S. N., Blum, R. M., Barry, D. W., Shearer, G. M., Fischl, M. A., Mitsuya, H., Gallo, R. C., Collins, J. M., Bolognesi, D. P., Myers, C. E., and Broder, S. (1986). *Lancet* **i,** 575–580.

Yarchoan, R., Berg, G., Brouwers, P., Fischl, M. A., Spitzer, A. R., Wichman, A., Grafman, J., Thomas, R. V., Safai, B., Brunetti, A., Perno, C. F., Schmidt, P. J., Larson, S. M., Myers, C. E., and Broder, S. (1987). *Lancet* **i,** 132–135.

Yarchoan, R., Perno, C.-F., Thomas, R. V., Klecker, R. W., Allain, J. P., Wills, J., McAtee, N., Fischl, M. A., Dubinski, R., Mcneely, M. C., Mitsuya, H., Pluda, J. M., Lawley, T. J., Leuther, M., Safai, B., Collins, J. M., Myers, C. E., and Broder, S. (1988a). *Lancet* **i,** 76–81.

Yarchoan, R., Thomas, R. V., Pluda, J., McAtee, N., Perno, C. F., Myers, C. E., and Broder, S. (1988b). *Clin. Res.* **36,** 450A.

Yarchoan, R., Mitsuya, H., Thomas, R. V., Pluda, J. M., Hartman, N. R., Perno, C. F., Marczyk, K. S., Allin, J.-P., Johns, D. G., and Broder, S. (1989a). *Science* **245,** 412–415.

Yarchoan, R., Mitsuya, H., Meyer, C. E., and Broder, S. (1989b). *N. Engl. J. Med.* **321,** 726–738.

Yarchoan, R., Thomas, R. V., Pluda, J. M., Perno, C. F., Mitsuya, H., Marczyk, K. S., Sherwin, S. A., and Broder, S. (1989c). *Proceedings of V International Conference on AIDS, Montreal, 1989,* p. 564, MCP 137.

Yarchoan, R., Mitsuya, H., Pluda, J. M., Marczyk, K. S., Thomas, R. V., Hartman, N. R., Browers, P., Perno, C.-F., Allain, J.-P., Johns, D. G., and Broder, S. (1990). *Rev. Inf. Dis.* (in press).

Zagury, D., Bernard, J., Leonard, R., Cheynier, R., Feldman, M., Sarin, P. S., and Gallo, R. C. (1986). *Science* **231,** 850–853.

Zamecnik, P. C., and Stephenson, M. L. (1978). *Proc. Natl. Acad. Sci. U.S.A.* **75,** 280–284.

Zamecnik, P. C., Goodchild, J., Taguchi, Y., and Sarin, P. S. (1986). *Proc. Natl. Acad. Sci. U.S.A.* **83,** 4143–4146.

Zapp, M. L., and Green, M. R. (1989). *Nature (London)* **342,** 714–716.

CD4-PE40—A Chimeric Toxin Active against Human Immunodeficiency Virus (HIV)-Infected Cells

Vijay K. Chaudhary, Bernard Moss, Edward A. Berger,
David J. FitzGerald, and Ira Pastan

I. Introduction
II. Construction and Evaluation of CD4-PE40
III. Clinical Considerations
    References

I. Introduction

The acquired immunodeficiency syndrome (AIDS) has rapidly developed into one of the major public health problems. As detailed in other chapters in this volume, an enormous amount of information has been obtained about the life cycle of the virus (HIV) that causes AIDS. Utilizing this information, it has been possible to develop strategies for preventing or arresting viral infection and spread. Our laboratory has been interested in developing new cytotoxic agents that selectively kill cells expressing specific surface proteins. With the realization that HIV utilizes CD4 on the surface of T cells as its receptor, and later in its cycle expresses gp120 (envelope glycoprotein) on the cell surface, it has been possible to design therapies to kill HIV-infected cells and thereby interrupt the life cycle of the virus. To accomplish this task, we have designed a novel toxin molecule composed of a portion of CD4 and a portion of *Pseudomonas* exotoxin (PE). This molecule, termed CD4-PE40, binds to viral gp120 present on the surface of HIV-infected cells, is internalized, and leads to the death of the infected cells (Chaudhary *et al.*, 1988; Berger *et al.*, 1989, 1990), presumably by mechanisms analogous to those for native *Pseudomonas* exotoxin and derivatives (Pastan and FitzGerald, 1989). This is the first example of a virally infected cell being killed by a targeted toxin.

Cytotoxic agents that kill HIV-infected cells have also been created by coupling ricin A chain to either CD4 (Till *et al.*, 1988) or monoclonal antibodies that react with HIV-infected cells (Till *et al.*, 1989; Pincus *et al.*, 1989). Like CD4-PE40 these agents must be internalized by endocytosis in order for the ricin A chain to kill the HIV-infected cell.

II. Construction and Evaluation of CD4-PE40

Pseudomonas exotoxin is a 66,000-molecular-weight enzyme that is secreted by *Pseudomonas aeruginosa*. It kills cells by catalyzing the ADP ribosylation of elongation factor 2 which leads to its irreversible inactivation. The inability of these cells to make new protein leads to cell death. Several years ago, the three-dimensional structure of PE was solved by McKay and colleagues who showed that the toxin was made up of three major structural domains (Allured *et al.*, 1986). In order to determine the function of each of these domains, Hwang *et al.* (1987) isolated a clone of the PE gene and expressed both the full length gene and various portions of the gene in *Escherichia coli*. This study revealed that domain I of PE (amino acids 1–253) contains the cell binding region which enables the toxin to bind to many different types of cell. The study also showed that domain II of PE was responsible for translocation of the toxin across an intracellular membrane into the cytosol. Finally, domain III of PE was shown to contain the enzymatic activity responsible for the ADP ribosylation of elongation factor 2. On the basis of these findings, it was possible to remove domain I and create a 40,000-molecular-weight molecule (PE40) that had the translocating and ADP ribosylation functions of PE, but could not bind to cells and therefore could not kill them. PE40 could be rendered cytotoxic by attaching cell recognition molecules to it, in the place of domain I. A summary of the various chimeric toxin molecules that have been made using this approach, including CD4-PE40, is shown in Table I.

Table I
Chimeric Toxin Molecules

Chimeric toxin	Target protein	Cell types
TGFα-PE40[a]	EGF receptor	Various carcinomas
IL-2-PE40	IL-2 receptor	Activated T cells
Anti-Tac(Fv)-PE40	IL-2 receptor	Activated T cells
IL-4-PE40	IL-4 receptor	Some monocytes
IL-6-PE40	IL-6 receptor	Myelomas, hepatomas
CD4-PE40	gp120	HIV-infected cells

[a]TGFα, transforming growth factor, alpha; IL2, interleukin 2; Anti-Tac, murine monoclonal antibody that binds the p55 subunit of the IL2 receptor; IL4, interleukin 4; IL6, interleukin 6; EGF, epidermal growth factor.

A variety of bacterial expression systems have been evaluated to determine an efficient method of producing CD4-PE40 and other chimeric toxins in *E. coli*. A very convenient and efficient one utilizes a T7 expression system developed by Studier and Moffatt (1986). In this expression system a plasmid has been constructed which contains a T7 promoter, an appropriate Shine-Delgarno sequence, and a cloning site into which a gene such as CD4-PE40 can be inserted. To transcribe the plasmid and initiate synthesis of the chimeric toxin, the plasmid is placed into the BL21 (λDE3) strain of *E. coli*. The host bacterium has a T7 polymerase gene under the control of the *lac* promoter in its chromosome. To initiate synthesis of the recombinant toxin, an inducer of the *lac* operon such as isopropylthiogalactopyranoside (IPTG) is added. The cells then make large amounts of the recombinant protein which usually accumulates in an insoluble mass termed an inclusion body. The expression vector containing CD4-PE40 is shown in Fig. 1.

Following the harvest and lysis of the bacteria, the inclusion bodies are isolated, dissolved in guanidine HCl, and the recombinant protein renatured and purified by column chromatography. An outline of the steps used in the purification is shown in Fig. 2. Typically, 1–2 mg of chimeric toxin can be produced from 1 l of cells induced at an optical density $(OD)_{650}$ of 0.6. Examples of the purity of CD4-PE40 at various steps in the purification scheme are shown in Fig. 3.

Initially, the purified chimeric toxin was examined for its ability to bind to gp120 in solution and to bind to cells expressing gp120. The chimeric toxin fulfilled both these criteria (Chaudhary *et al.*, 1988). However, the most important and final test was to examine its activity on HIV-infected cells expressing gp120 on their surface. As shown in Table II, CD4-PE40 selectively killed T cells infected with HIV that expressed gp120 on their surface.

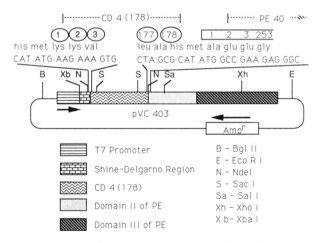

Figure 1. Expression vector for CD4-PE40.

EXPRESSION/ EXTRACTION/PURIFICATION
(All operations performed at 4 C unless otherwise stated)

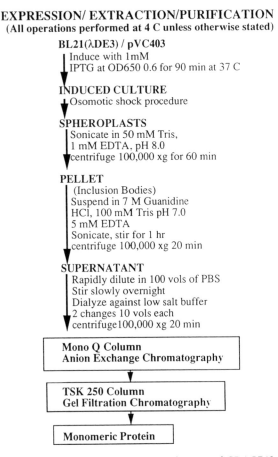

Figure 2. Flow diagram for purification of CD4-PE40.

Control cells not infected by the virus were not killed. In these studies, cell killing was conveniently assessed by measuring the incorporation of [³H]thymidine into DNA. This is an easy and sensitive method of demonstrating the cytotoxic effect of this molecule. CD4-PE40 has been tested on a variety of cells expressing gp120 (Chaudhary *et al.*, 1988; Berger *et al.*, 1989, 1990). These include 8E5, H9, CHO, and CV1 cells. In all these systems, CD4-PE40 is an effective cytotoxic agent. Furthermore, it has been shown that CD4-PE40 blocks HIV spread in mixtures of infected and uninfected cells (Berger *et al.*, 1989).

The normal function of membrane-associated CD4 is to participate in the interaction of T cells with antigen-presenting cells containing the major histocompatibiilty complex (MHC) Class II protein on their surface. By itself,

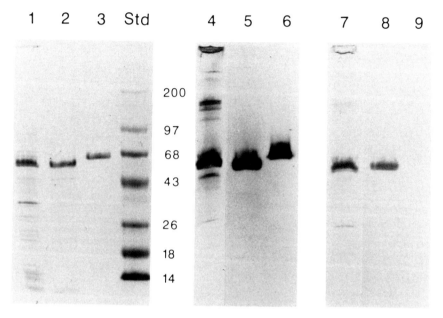

Figure 3. Characterization of CD4(178)-PE40 by SDS-PAGE of samples at various stages of purification. Gels were either stained with Coomassie Blue (lanes 1–3) or immunoblotted with polyclonal antibodies to PE (lanes 4–6) or CD4 (lanes 7–9). Lanes 1, 4, and 7, spheroplasts; lanes 2, 5, and 8, peak fraction 19 from Mono Q column; lanes 3, 6, and 9, authentic PE, 100 ng. Protein standards of known M_r are shown. M_r of authentic PE is 66K.

Table II

Cytotoxicity of CD4-PE40 and Other PE Derivatives on HIV-Infected and Uninfected H9 Cells[a]

	ID_{50} (ng/ml)	
Protein	H9/HIV-1	H9
CD4(178)-PE40	<1	>2500
PE	370	290
PE40	>2500	>2500
HB21-PE40	<1	<1

[a]All the proteins were tested in a 4-day assay and incorporation of [3H]thymidine was measured. ID_{50} is the concentration of protein required to inhibit [3H]thymidine incorporation by 50% as compared to control where no toxin was added. HB21, murine monoclonal antibody that binds the human transferrin receptor.

Table III

Effect of CD4(178)-PE40 on Protein Synthesis in Human Cell Lines
Expressing MHC Class II Molecules[a]

Cell line	gp120	MHC Class II	Protein synthesis (% of control) CD4-PE40	PE40
A3.01	−	−	97	95
8E5	+	−	4	98
Raji	−	+	107	115
BHM23	−	+	106	103
I937	−	+	91	81

[a]The CD4(178)-PE40 and PE40 were added at final concentration of 17 nM
and 25 nM, respectively for 23 hr. The cells were labeled with [^{35}S]methionine for
1 hr and radioactivity in the trichloroacetate (TCA) precipitates of cells was measured. Results are expressed as the % of radioactivity incorporated into the experimental samples as compared to controls where no toxin was added.

soluble CD4 has a very weak affinity for MHC Class II. To determine if CD4-
PE40 would be toxic to cells bearing MHC Class II, we performed experiments in which CD4-PE40 was incubated with cells with or without MHC
class II on their surface (Table III). We could not demonstrate any selective
toxicity on MHC Class II-bearing cells by CD4-PE40. This is probably because the binding of CD4-PE40 to MHC Class II is very weak and also
because MHC Class II may be slowly internalized. CD4-PE40 has also been
tested against a variety of other uninfected cell lines and is not cytotoxic to
any of these even at a concentration 1000 times higher than that capable of
killing HIV-infected cells.

One of the unusual features of the human immunodeficiency virus is its
ability to tolerate changes in the amino acid sequence of its proteins resulting
from its high mutation frequency. One exception is the region of gp120 that
binds to CD4 which appears to be invariant, because changes that abolish
binding to CD4 could lead to loss of viral infectivity. To study the killing of
cells expressing different forms of gp120, a vaccinia virus expression system
was used. The envelope glycoproteins from different viral isolates were
expressed in CV1 cells which were then exposed to CD4-PE40 and control
proteins. As shown in Table IV, CD4-PE40 was able to kill cells expressing
the envelope glycoprotein from divergent isolates of HIV-1, as well as
HIV-2 and simian immunodeficiency virus.

Because of the ability of CD4-PE40 to kill HIV-infected cells, its low
cytotoxicity against cells not infected with the virus, and its ability to kill
cells infected with all strains of the virus, this agent warrants animal testing
and possibly clinical evaluation in patients.

Table IV
Effect of CD4-PE40 on Cells Infected with
Recombinant Vaccinia Viruses Encoding Different
Envelope Glycoproteins[a]

	Protein synthesis (% of control)	
Envelope glycoprotein	CD4-PE40	PE40
None	91	108
HIV-1 (HTLV-IIIb)	15	90
HIV-1 (HTLV-IIIRF)	7	122
HIV-2	7	92
SIVmac	11	122

[a]CV1 cells were infected with vaccinia viruses for 14 hr and then toxins were added to a final concentration of 10 nM and incubations were continued for an additional 4 hr. After a 1-hr pulse with [^{35}S]methionine, the cells were treated as described in Table III. HTLV, human T-cell leukemia virus.

III. Clinical Considerations

It is believed that in patients with active AIDS the number of cells producing virus is relatively low and that products of either the virus or the infected cells produce the immune-deficient state. If one could kill cells that are actively producing HIV, it is possible that the severe immunodeficiency state that makes AIDS patients susceptible to infection could be ameliorated In order for chimeric toxins to kill cells, there must be a minimum number of receptor molecules present on the cell surface. The number of receptor molecules needed varies among different receptor systems and a number of factors are involved. One is how efficiently the receptor is internalized from the cell surface into an endocytic vesicle where the toxin molecule can translocate into the cytosol. Because translocation is not an efficient process, hundreds or thousands of toxin molecules must be delivered from the surface into endocytic vesicles in order to achieve a cytotoxic affect. This means that HIV-infected cells that are producing very little virus and therefore have very little gp120 on their surface are not likely to be good targets for CD4-PE40. Despite this reservation, it is likely that CD4-PE40 could produce a beneficial effect in AIDS patients by killing those cells that are producing substantial amounts of virus. In addition to killing virus-producing cells, CD4-PE40, like CD4 itself, may also act by interfering with the spread of virus from one cell to another by binding to the free virus and preventing its interaction with cellular CD4.

Currently AZT and related drugs are widely used in the treatment of AIDS. These drugs act by inhibiting viral replication but have no effect after the virus has infected a cell. In contrast, the mechanism of action of CD4-PE40 is to kill infected cells; CD4-PE40 does not have a direct effect on viral replication. Because the mechanism of action of these agents is different, a combination of the two agents should have a synergistic effect on inhibiting the spread of the virus to uninfected cells. Recent experiments have shown that this prediction is correct (Ashorn *et al.*, 1990).

Initial studies in mice have indicated that CD4-PE40 has a half-life of between 20 and 40 min (unpublished data). The molecular weight of the molecule is about 60,000 which suggests it should not be rapidly cleared by filtration in the kidney. Currently the mechanism by which CD4-PE40 is removed from the circulation is not known. Because of the short half-life, therapy with CD4-PE40 should probably be carried out by continuous intravenous infusion to maintain an adequate blood level for extended periods. Bolus injection would lead to effective blood levels for only a few hours. New forms of CD4-PE are being developed which may have a longer survival time in the blood and therefore be more clinically useful.

Chimeric toxins, which are composed of foreign proteins, provoke an immune response in normal animals and people. Therefore, the window for initial treatment may be as short as 7–10 days. After that, neutralizing antibodies to PE may develop. Because of the severe immunosuppression evident in patients with AIDS, it is not known whether these patients will actively produce antibodies to PE. In patients with cancer receiving immunotoxins, it is anticipated that an immunosuppressive agent such as cyclosporin or 15-deoxyspergualin (Makind *et al.*, 1987) may have to be administered in conjunction with the immunotoxin to suppress the neutralizing antibody response. Whether such a strategy will be useful or necessary in HIV-infected patients remains to be determined.

References

Allured, V. S., Collier, R. J., Carroll, S. F., and McKay, D. B. (1986). *Proc. Natl. Acad. Sci. U.S.A.* **83**, 1320–1324.

Ashorn, P., Moss, B., Weinstein, J. N., Chaudhary, V. K., FitzGerald, D. J., Pastan, I., and Berger, E. A. (1990). *Proc. Natl. Acad. Sci. U.S.A.* **87**, 8889–8893.

Berger, E. A., Clouse, K. A., Chaudhary, V. K., Chakrabarti, S., FitzGerald, D. J., Pastan, I., and Moss, B. (1989). *Proc. Natl. Acad. Sci. U.S.A.* **86**, 9539–9543.

Berger, E. A., Chaudhary, V. K., Clouse, K. A., and Jaraquemada, D. (1990). *AIDS Res. Hum. Retroviruses* **6**, 795–804.

Chaudhary, V. K., Mizukami, T., Fuerst, T. R., FitzGerald, D. J., Moss, B., Pastan, I., and Berger, E. A. (1988). *Nature (London)* **335**, 369–372.

Hwang, J. FitzGerald, D. J., Adhya, S., and Pastan, I. (1987). *Cell* **48**, 129–136.

Makind, M., Fujiwara, M., Watanabe, H., Aoyagi, T., and Umezawa, H. (1987). *Immunopharmacology* **14**, 115–121.

Pastan, I., and FitzGerald, D. (1989). *J. Biol. Chem.* **264**, 15157–15160.

Pincus, S. H., Wehrly, K., and Chesebro, B. (1989). *J. Immunol.* **142,** 3070–3075.

Studier, F. W., and Moffatt, B. A. (1986). *J. Mol. Biol.* **189,** 113–130.

Till, M., Ghetie, V., Gregory, T., Patzer, E., Porter, J. P., Uhr, J. W., Capon, D. J., and Vitetta, E. S. (1988). *Science* **242,** 1166–1168.

Till, M. A., Zolla-Pazner, S., Gorny, M. K., Patton, J. S., Uhr, J. W., and Vitetta, E. S. (1989). *Proc. Natl. Acad. Sci. U.S.A.* **86,** 1987–1991.

Vaccines against Acquired Immunodeficiency Syndrome (AIDS)

Dani P. Bolognesi

I. Introduction

Society has benefited greatly from vaccines that prevent diseases caused by microorganisms as vaccination has proved to be the simplest and most effective means to control outbreaks of disease in both animals and man. However, the science of vaccinology has been empirical for the most part, being based largely on trial and error using various formulations of the disease-causing organism itself. Otherwise stated, the underlying principles of vaccination are poorly understood. This is in spite of the resounding triumphs that have been achieved with vaccines most notably against smallpox and polio but also against yellow fever, measles, mumps, and rubella (for review, see Hilleman, 1985).

It is important to note, however, that even in the best of circumstances, vaccination carries elements of risk. Adverse effects of vaccination with preparations including whole organisms (live, attenuated, or killed) relate primarily to unanticipated active replication of the pathogen in the vaccine. The experience with live polio virus vaccines illustrates this best where imperfections in vaccine design as well as vaccine manufacture have led to infection and inadvertent spread of the virus to nonvaccinated individuals. A second category includes injurious effects resulting from undesirable or inappropriate immune responses. To a large extent, such consequences have arisen when components of the pathogen necessary for induction of protective immunity have been eliminated or destroyed during formulation of the vaccine, resulting in either inconsequential or deleterious immune responses. Indeed, certain components of a pathogen can induce immune responses that facilitate infection of target cells which bear surface receptors for antibodies or antibody ligands (e.g., complement). If such responses prevail, not only is protection not achieved but disease is exacerbated upon exposure of the vaccinee to the pathogen. A classical example of this is the situation with flaviviruses such as Dengue (Porterfield, 1982).

These mishaps and concerns have resulted in liability claims that have plagued the vaccine industry over the years. However, they have also had a beneficial effect in that they have brought emphasis to the need for development of better defined, more efficacious, and safer vaccines. Indeed, contemporary vaccines have relied increasingly on modern biotechnology of which the prototype is the recombinant hepatitis B vaccine (Zajac *et al.*, 1986). For various reasons outlined in this review, a vaccine against the AIDS virus will depend heavily on the depth of knowledge of HIV/host interactions at the molecular level and the use of technological advances to produce an efficacious and safe product.

II. General Features of HIV Pertinent to Vaccine Design

HIV-1 has developed mechanisms to establish a persistent infection in man which are far more elaborate and difficult to counter than those of viruses against which successful vaccines exist. First and foremost, like other members of the retrovirus family it is able to integrate its genetic information into the genome of its target cells. The survival of the virus is thus directly linked to the survival of the cell. Although HIV-1 can rapidly kill its T-cell targets, others (e.g., monocyte/macrophages) can tolerate virus replication for much longer periods and serve as reservoirs for production of infectious virus.

As part of this framework, the virus can also establish a latent infection, that is, the silent integration of its genome into that of the target cell (i.e., no synthesis of virus or viral gene products). Latency is a well-known property of viruses of the herpes family against which vaccines are still being sought. Latently infected cells are invisible to the immune system, unless signals are applied which are able to activate the expression of viral genes. For HIV, these signals can be immune activators such as phytohemagglutinin (PHA), other viruses, and/or host factors that include elements that regulate immune function (for review see Rosenberg and Fauci, 1989). From this alone, one can glean how intimately virus and target cell functions are likely to be intertwined, providing considerable selective advantage to the survival of the pathogen.

There are other features of HIV-1 which merit some attention. One of these relates to the natural transmission of virus in both free and cell-associated form. Moreover, virus can pass from cell to cell by a variety of mechanisms that can escape immune defenses. One of these has been brought to light by recent studies in monocyte/macrophages which indicate that large concentrations of virus can exist in intracellular vesicles (Orenstein et al., 1988). In certain subpopulations of these cells, virus synthesis actually occurs within such structures rather than at the cell surface. Through phagocytosis, macrophages could also sequester HIV-1 within vacuoles and disseminate the virus to other cells by a "trojan horse" mechanism, similar to what occurs with lentiretroviruses such as visna (Peluso et al., 1985). Moreover, if macrophages carrying sacks filled with virus are ruptured, large quantities of HIV-1 would be liberated and further propagated. Thus, immune attack on such cells may bear negative consequences.

HIV-1 may also be able to escape immune attack as a result of variation within its envelope gene (Starcich et al., 1986). Thus, large numbers of variants of this virus exist in the population. The variation affects certain important targets for immune attack, most notably the major site to which virus-neutralizing antibodies are directed (Palker et al., 1988; Rusche et al., 1988). A single change in the amino acid sequence of certain segments of this hypervariable region can alter susceptibility of the virus to neutralization (Looney et al., 1988). Thus, it is tempting to speculate that under pressure of the immune response, virus variants emerge which can escape immune attack. This is not unlike the properties of influenza viruses, but is actually more acute since HIV-1 replicates as a swarm of viruses rather than the discrete variant influenza strains which emerge seasonally.

The very fact that macrophages can be infected by HIV-1 raises the question of enhancing antibodies. Thus, antibodies would bind the virus and direct it to the macrophage surface by attachment to Fc receptors present on such cells. Virus infection could then occur through any of several possible mechanisms. Indeed, antibody potentiation of HIV-1 infection in vitro has been reported for Fc-receptor-bearing cells (Takeda et al., 1988) but its role in vivo has not been established.

From this cursory look alone it is evident that HIV is a formidable adversary. It attacks the immune system upon which a vaccine depends in order to be effective. While doing so, it is able to hide from immune defenses by establishing latent infections and by developing covert mechanisms of transmission. Such effects are punctuated by HIV's ability to infect sanctuaries such as the central nervous system (for review see Price et al., 1988). Another potential mechanism of immune escape results from the extensive diversity of the virus; and the large number of variants within the population presents a major obstacle for design of a protective immunogen. These features have led many to speculate that no degree of infection by HIV can be allowed if a vaccine is to be successful. If this is indeed the case, an issue that will be dealt with later, it would place demands on a vaccine for HIV far above what has ever been required for other pathogens where complete blockade of infection is not a requirement for protection, which is limited to prevention of disease.

III. General Features of the Immune Responses to HIV during Natural Infection

Individuals infected with HIV mount a vigorous immune response against the virus involving both humoral and cellular arms. In spite of this there is no epidemiological evidence to date that this immune response is capable of suppressing the infection and influencing the course of disease although many models favor such an interpretation (Bolognesi, 1989a). This represents a major bottleneck to a rational approach for vaccine design since it raises the issue of whether or not an immunological clearance mechanism exists against HIV, or, more to the point, whether it is possible to induce one through vaccination or postexposure prophylaxis.

Given the mechanisms that HIV has evolved to escape immune attack, one can take the position that an immune response of full force is present but simply insufficient to deal with the unique features of this particular pathogen. A counterview would be that the immune response is somehow defective, perhaps due to the ability of the virus to induce responses that exacerbate infection and disease progression. Analysis of the natural immune response to HIV indeed reveals that immunity to HIV represents a mix of counteractive features. This being the case, one could view the natural course of infection as representing a balance where the positive responses are in control during the early phases of infection but that these are eventually overwhelmed by the destructive effects of the virus on the immune system and ultimately dominated by the negative responses. This being the case, it follows that elimination of the negative features might result in better control of the virus. It is thus worthwhile to define the various epitopes that elicit both classes of response such that design of immunogens and immunotherapeutic regimens can be safer and more effective. By way of example, the properties of one of the most intensely studied gene products of the virus, namely its envelope glycoproteins, will be reviewed next in this context.

IV. Properties of the HIV Envelope Important for Vaccine Design

The products of the *env* gene are two glycoproteins, designated gp120 and gp41, which represent the outer knob and viral membrane anchor respec-

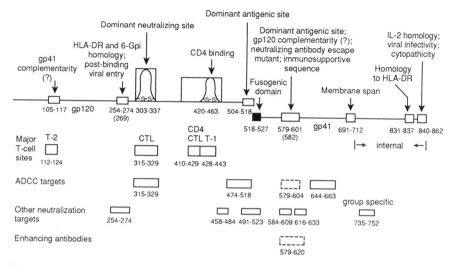

Figure 1. Composite of selected sites on the HIV envelope discussed in this report. Additional regions of the envelope that are critical for various viral functions have been identified by insertional deletion mutagenesis and are reviewed in Thomas *et al.* (1988).

tively. Both are essential building blocks of the virus and play major roles in its life cycle, particularly during the early stages of infection. The envelope very likely determines at least in part, the host range and tissue specificity of HIV and participates in pathogenic processes mediated by the virus. Because of its strategic location on the outer surface of the virion and the infected cell, it represents an optimal target for immune attack and thus a prime candidate for development of vaccine and therapeutic strategies.

Studies from a number of laboratories have gone to great lengths to define the functional and immunogenic regions of the HIV envelope, a summary of which is outlined and depicted in Fig. 1. The salient features that relate to vaccine development are represented by regions defining epitopes responsible for antibodies that neutralize virus, mediate antibody-dependent cell cytoxicity (ADCC), and generate cytotoxic lymphocytes. Equally important are domains that induce undesirable responses such as enhancing antibodies or antibodies that suppress immune function. Finally, note should be taken of properties of the molecules that themselves lead directly to undesirable effects. Selected examples of each will be given.

V. Neutralizing Epitopes

A prominent neutralization site on gp120 resides within the third hypervariable region (V3) of the molecule (Putney *et al.*, 1986). Reports from a number of laboratories (Rusche *et al.*, 1987, 1988; Goudsmit *et al.*, 1988; Palker *et al.*, 1988) identify the portion of gp120 as dominant for development of high-

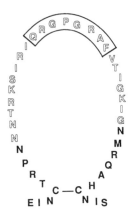

Figure 2. Depicted is the loop sequence of the human T-cell leukemia virus (HTLV)-IIIB strain of HIV which defines the principle type-specific neutralizing determinant. The eight amino acid peptide at the tip of the loop is sufficient to elicit neutralizing and fusion-inhibiting antibodies. However, a consensus sequence for this region (IHIGPGRAF) identi-fied by examining over 100 field isolates differs from the IIIB isolate in that the QR insert is not present. The IHIGPGRAF sequence is, however, representative of the MN and other isolates from the Gallo laboratory. The amino acids at the two sides of the loop represent the regions where the variability is most extensive. By contrast the sequences at the base of the loop are more conserved as are the cysteines and the flanking glycosylation sites.

titered, type-specific neutralizing antibodies against HIV-1. It is thought to exist as a loop formed by two disulfide-linked cysteine residues (Fig. 2; Leonard *et al.*, 1990). The two cysteines spanning this region, residue num-bers 303 and 337, are themselves highly conserved. Thus, although the amino acid sequence within this loop is variable, the loop itself is present in gp120 from most if not all HIV-1 isolates. Several neutralizing monoclonal antibodies have been developed which map to various regions of this hyper-variable loop (Matsushita *et al.*, 1988, Skinner *et al.*, 1988a; Thomas *et al.*, 1988).

In addition to the conservation of the cysteines, several pieces of infor-mation indicate that the loop is also an important functional region of the virus. Recent evidence indicates that deletion of this loop generates a virus that is no longer infectious (S. Putney, D. Looney, and F. Wong-Staal, personal communication). Moreover, certain mutations within the loop pro-duce a virus with substantially reduced infectivity (T. Matthews, S. Pette-way, D. Looney *et al.*, in preparation).

The hypervariability of this region suggests that HIV may be able to mutate away from neutralizing antibodies directed to the loop. A single change in its amino acid sequence can alter susceptibility of the virus to neutralization (Looney *et al.*, 1988). Recent studies have demonstrated that virus variants can be generated *in vitro* under the selective pressure of monoclonal antibodies to the loop (McKeating *et al.*, 1988). The results indicate that neutralization escape mutants do indeed demonstrate changes within the loop but interestingly also in regions outside the loop. Moreover, studies by Nara and colleagues (1989) in chimpanzees infected with a laboratory HIV isolate indicate that the same phenomenon can occur *in vivo*. The related regions outside of the loop may be important for the proper exposure of the loop in the native configuration of gp120.

Antibodies to this region neutralize virus infectivity by interfering with a postbinding step in virus infection (Linsley *et al.*, 1988; Skinner *et al.*, 1988b). This is probably linked to the fusion process because these antibodies are very effective in blocking cell/cell syncytium formation (Skinner *et al.*, 1988b). A plausible mechanism for this might be that the binding of gp120 to CD4 serves as a trigger for releasing the fusogenic domain within gp41 (Gallagher, 1987; Gonzalez-Scarano *et al.*, 1987) which anchors the respective membranes. The hypervariable loop may itself be associated with a critical contact region of gp120 to gp41 and antibodies to the loop may prevent the process of dissociation. Parenthetically such antibodies need not have to compete with the high-affinity binding that occurs between gp120 and CD4 in order to block virus infection and cell fusion.

A logical candidate for antibodies that would neutralize virus by blocking the fusogenic step would be the transmembrane component of the envelope, gp41. Various studies point to a conserved site on gp41 (735–752) (Fig. 1) as a region that can induce antibodies that neutralize HIV and block fusion in a broad fashion (Chahn *et al.*, 1986, Dalgleish *et al.*, 1988; Thomas *et al.*, 1988). Curiously, most models of how gp41 is situated in the virus outer membrane indicate that this region is probably internal (Gallaher *et al.*, 1989). Yet monoclonal antibodies to this epitope stain the surface of virus-infected cells (Dalgleish *et al.*, 1988), a paradox which remains unexplained. Finally, studies by Jeffrey Almond and colleagues (Evans *et al.*, 1989) demonstrate that when this region is inserted as a surface component of polio virus, it is not only able to induce similar antibody reactivities but is also capable of removing group common neutralizing and fusion-inhibiting antibodies from some human sera.

In sum, epitopes on both gp120 and gp41 are targets of neutralizing and fusion-inhibiting antibodies. One immunodominant site on gp120 is a hypervariable region while a second in gp41 is conserved. Recent studies with field isolates suggest that virus clusters resembling the MN prototypic isolate (Sargo *et al.*, 1988) exist in the population which also exhibit common neutralization loops (Putney *et al.*, 1989), making this an additional stimulus for broadly reactive immune responses. The identity of the full range broad reactivity which is/are responsible for HIV neutralization and fusion inhibition in human sera remain to be elucidated, particularly the targets for antibodies to gp120 which interfere with binding of the virus to its receptor alluded to earlier. Moreover, very little information is available on epitopes that depend upon conformation of the envelope glycoproteins.

VI. Epitopes Associated with Cellular Immunity

To date, various regions of gp120 are known to be recognized by T cells (Figure 1). Some of the best studied are in (a) a conserved region toward the N-terminus of gp120 (Cease *et al.*, 1987), (b) within the hypervariable neutralization loop (Takahashi *et al.*, 1988), and (c) within the conserved binding region of gp120 to CD4 (Cease *et al.*, 1987) as well as an adjacent region towards the N-terminus (Siliciano *et al.*, 1988). There are also T-cell epitopes on gp41 (Ahearne *et al.*, 1988; Schrier *et al.*, 1988; Clerici *et al.*, 1989). While some of the studies were conducted in mice, evidence that human T cells recognize these epitopes has also been reported (Berzofsky *et al.*, 1988; Lanzavecchia *et al.*, 1988; Siliciano *et al.*, 1988; Clerici *et al.*, 1989). Many more epitopes are certain to be discovered and particular emphasis is being placed on those that are targets for cytotoxic lymphocytes.

One of the epitopes that serves as a target for cytotoxic T lymphocytes (CTL) is present within, the hypervariable neutralization loop in gp120 (Takahashi *et al.*, 1988). Variable T-cell epitopes may be part of the strategy of HIV to escape immune destruction, much like the targets for neutralizing antibodies. This could occur through mutations that reflect recognition by the T-cell receptor for that particular site (Takahashi *et al.*, 1989a), but also at sites that are critical for association with Class I or Class II major histocompatibility complexes (MHC). Without the latter, a virus would become invisible to the immune system. However, there are likely to be a sufficient number of T-cell epitopes in the virus to overcome allotype restriction and this would apply only to those that are immunodominant, if such indeed exist.

A distinct but related issue relates to regions that are targets for antibodies that mediate cellular cytotoxicity (ADCC). Although, fine mapping studies have not been done, it appears that conserved regions of the envelope, situated mainly on gp120, may be primary (Lyerly *et al.*, 1987). One such domain is situated at the C-terminus of gp120 (Lyerly *et al.*, 1988). More recently, evidence that variable regions are also targets has been obtained (D. Tyler, and K. Weinhold, in preparation) and discriminatory sites found in virus isolates from a given patient have been identified (Ljundggren *et al.*, 1989). Finally, it is now evident through use of human monoclonal antibodies directed toward gp41, that various sites within the transmembrane glycoprotein are also targets for ADCC (Gorny *et al.*, 1989; D. Tyler, in preparation).

To date no solid evidence has been obtained that antibodies to the HIV envelope are able to direct complement-dependent lysis of virus-infected cells; in fact, there is evidence to the contrary (Lyerly *et al.*, 1987; Nara *et al.*, 1987). On the other hand, antibodies that associate with complement have been reported to facilitate infection of target cells bearing complement receptors (see below).

VII. Antibodies That Enhance Virus Infectivity

Antibodies that are able to bind to the surface of cells through sites such as Fc or complement binding regions conceivably could facilitate infection by concentrating the virions to the surface of such cells. Such antibodies are presumably of the nonneutralizing class. Infections with flaviviruses can be exacerbated by the presence of enhancing antibodies but controlled by neutralizing antibodies depending on which are prevalent (Halstead, 1988).

Because HIV is able to infect cells of the monocyte/macrophage lineage which bear Fc receptors on their surface, investigators have sought to determine if the phenomenon of enhancement is operative with this virus. Studies by Ennis and colleagues (Takada *et al.*, 1988) do indeed demonstrate some enhancement of virus infectivity but at the same time, it is an effect that is much less pronounced than that observed with flaviviruses (Halstead, 1988). Others have also obtained evidence of this form of enhancement (Homsy *et al.*, 1989) and for which the Fc portion of the antibody is required. However, there is disagreement as to whether the subsequent steps in virus infection involve the CD4 receptor (for review, see Bolognesi, 1989c). Several investigators have shown an absolute requirement while others report independence of CD4. The former are consistent with a model where the virus would dissociate from the antibody given the higher affinity of gp120 for CD4 and proceed with the traditional mechanisms of binding, anchorage, fusion, and entry. The latter would support models where there may be secondary mechanisms of virus entry independent of CD4. Resolution of this puzzle awaits further experimentation.

A second mechanism of viral enhancement requires the presence of certain components of complement. Antibodies that bind complement and virus could likewise focus HIV to the surface of cells bearing complement receptors. Studies by Mitchell and colleagues (Robinson *et al.*, 1988, 1989a); using a cell line that is apparently rich in receptors for complement have reported this form of enhancement. Moreover, a region of gp41 has been identified as one of the targets for complement-mediated enhancement (Robinson *et al.*, 1990). Numerous other cell lines have been examined that do not reveal complement dependent enhancement, presumably due to the absence of sufficient numbers of receptors. Neither of the two mechanisms have been directly implicated *in vivo* either in man or in animal models although suggestive evidence exists that complement-mediated enhancement may have been responsible for failure of passive immunoprophylaxis trials in chimps (Robinson *et al.*, 1989b).

VIII. Epitopes That Mimic Products of Normal Cellular Genes

Molecular mimicry is recognized as an important process in pathogenesis and immune suppression accompanying virus infections (Oldstone, 1987). Viruses bearing structures analogous to those present on the surface of normal cells can present such regions to the immune system in a manner such that they are recognized as foreign antigens and thereby elicit immune responses that attack normal cells. Alternatively these regions may represent growth-factor-like elements which could influence a variety of normal cellular functions. HIV displays examples of each of these as well as other mechanisms by which it can cause the destruction or impairment of normal cells. One of the most important regions in this regard is the homology to HLA-DR (Golding *et al.*, 1988). Anti-DR activities are present in a substantial proportion of HIV-infected individuals which can impair normal immune function *in vitro* (Golding *et al.*, 1988, 1989). This is clearly a region to be avoided in vaccine design.

IX. Immunosuppressive Effects of the HIV Envelope

Based primarily on *in vitro* studies, HIV has been to suggested to evolve some other unique ways to impair the immune system. By releasing its exterior envelope glycoprotein, which apparently occurs readily from virus or infected cells (Gelderblom *et al.*, 1985), a molecule is generated which actively binds to CD4-bearing normal lymphocytes. This event now targets these cells for immune attack by both antibodies which mediate ADCC and cytotoxic T cells which recognize processed forms of gp120 after its internalization by CD4$^+$ lymphocytes (see above). Both events can result in destruction of normal CD4$^+$ T cells without the necessity for HIV infection (noninfectious lympholysis). To what extent this occurs *in vivo* depends on the level of gp120 synthesis, secretion, and shedding. To date, free gp120 has not been measured in the circulation but this is not surprising given its powerful affinity for CD4. Adding another twist, it has been suggested that gp120/CD4 complexes might give rise to new epitopes on CD4 which elicit anti-CD4 antibodies in a fraction of HIV-infected patients (Kowalski *et al.*, 1989).

The gp120 molecule can also suppress immune function by virtue of its ability to bind to CD4 (Weinhold *et al.*, 1989). Recent studies demonstrate that this might occur as a result of its ability to occlude the binding of CD4 to Class II MHC, an event that is necessary in antigen recognition (Clayton *et al.*, 1989). Indeed the evidence suggests that the binding site of CD4 to gp120 is contained within the binding domain of CD4 to MHC Class II (Clayton *et al.*, 1989).

Finally, immunosuppressive sequences may exist in the envelope itself. Regions homologous to those present in the transmembrane glycoprotein of animal and human type C retroviruses known to be suppressive for macrophage function (Cianciolo et al., 1980) are also present in HIV gp41 (Table I) (Cianciolo et al., 1985). In addition it has been reported that disappearance of antibodies to this region correlated with disease progression (Klasse et al., 1988).

X. Recent Progress in Vaccine Development

The pessimism shadowing the development of a vaccine against HIV is showing some signs of receding due to several recent successes of experimental vaccines in animal models (Bolognesi, 1989b; for review).

The most striking results are the animal model studies with simian immunodeficiency virus (SIV) by Desrosiers et al. (1989), Mickey Murphy-Corb (personal communication), and Sutjipto et al. (1989) as well as with equine infectious anemia virus (EIAV) by Ron Montelaro (personal communication). In each example whole killed virus vaccines were able to delay the onset of disease for significant periods of time possibly permanently as the experiments may eventually demonstrate. Notably, protection against the respective diseases caused by experimental challenge with SIV and EIAV could be achieved even in cases where actual infection by the virus was not prevented. The importance of this development is underscored because it suggests that the same general rules apply to HIV vaccines as have existed for other viruses, namely that it may not be necessary to block infection completely in order to vaccinate successfully. If some degree of infection with HIV can indeed be tolerated, a vaccine against the virus is much more within reach. It also suggests that some of the prior HIV vaccine trials in chimpanzees with HIV (which were considered failures because infection was not prevented) might have had a more favorable outcome had a disease end point been available.

The above issue notwithstanding, it is becoming more and more evident that a significant number of the animals in the SIV and EIAV vaccine trials were also able to resist infection altogether (Desrosiers et al., 1989; Murphey-Corb et al., 1989; R. Montelaro et al., personal communication). Quite recently similar results have emerged from HIV vaccine trials in chimps (M. Girard, 1989, Berman et al., 1990). In these latter studies, viral subunits and peptides (alone and in various combinations) were used to induce protective immune responses, indicating that approaches other than killed virus vaccines can be successful.

While these animal models are of extreme importance for providing concepts and strategies for vaccine development, it must be recognized that they fall in the category of feasibility studies as opposed to true vaccine trials. In order to progress toward practical vaccine candidates, a number of criteria need to be fulfilled, including the ability to (1) prevent infection by widely diverse virus strains, (2) block infection when the challenge is cell-associated virus, and (3) induce immunity which would prevent transmission across a mucosal surface. Moreover, the vaccine must be capable of inducing long-term immunity, preferably with a minimum number of immunizations. One can look toward the SIV system as a model for devising ways to overcome these barriers and to approach the definition of what represents protective immunity.

XI. Discussion

The morbidity and mortality associated with infection with HIV-1 together with the uncertainty of vaccine efficacy against this virus generally precludes the use of conventional vaccine strategies employing live-attenuated or even killed preparations which have been successful with other viruses. (A notable exception would be in a postexposure setting such as has been practiced with rabies virus.) However, the potential adverse effects of vaccination against HIV-1 may even exceed those documented for other viruses. What then is a rational approach to vaccine design which embodies the features of efficacy and safety in a practical formula?

On the basis of the above information relative to the HIV envelope, the emphasis should be directed toward constructing immunogens containing the different categories of epitopes that induce protective immunity against the virus and its infected target cells. By inference such preparations should exclude the regions of the virus that elicit undesirable responses. Thus, domains representing targets for neutralizing antibodies, ADCC, and cytotoxic lymphocytes should continue to be refined and studies along these lines are being vigorously pursued in many laboratories. All regions responsible for molecular mimicry with host products, immunosuppressive sequences, or regions of HIV that bind to normal cell surface molecules (and might induce anti-idiotype antibodies) need to be identified.

With such information in hand, viral genes can be essentially reconstructed by deletion of unwanted sequences and ligation of the portions to be included. They can then be inserted into various expression systems or chemically synthesized to produce the desired products. One can thus envision combinations of T and B cell epitopes in linear arrays which would produce a safe and effective immunogen (Fig. 3).

It should be emphasized that while the envelope of HIV may well be a primary target of immune attack, other viral components are likely to be equally important. The phenomenon of antigen processing and presentation on the cell surface in association with MHC makes other HIV gene products (structural or regulatory) potential targets for immune attack as has been amply documented for other viral systems (Townsend *et al.*, 1985). Indeed T-cell epitopes also are present on internal antigens such as p24 (Nixon *et*

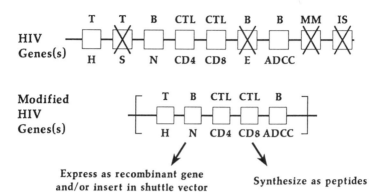

Figure 3. Selection of epitopes that induce protective responses and exclusion of undesired regions. As domains of HIV gene products that induce defined immune responses are identified, some rational selection can be made to devise an improved immunogen through recombinant DNA technology or by direct chemical synthesis. T/H, T-cell epitopes required for help; T/S, T-cell epitopes that give rise to suppressor cells; B/N, B-cell epitopes responsible for neutralizing Ab; B/ADCC, B-cell epitopes for ADCC; B/E, B-cell epitopes that elicit enhancing Ab; CTL/CD4, epitopes for CD4+/CTL (Class II MHC dependent); CTL/CD8, epitopes for CD8+/CTL (class I MHC dependent); MM, regions that mimic normal cell surface antigens (molecular mimicry); IS, regions that are directly immunosuppressive. Various arrays could be configured in the restructured gene or peptides including multiple representation of the indicated regions.

al., 1988), and products of the *pol* (Walker *et al.*, 1988) and *nef* (Riviere *et al.*, 1989). Depending on the level of expression on the infected target cell, it may also be that certain epitopes are more dominant than others and better suited as targets for immune attack. Thus cocktails of epitopes from products of envelope, core, and regulatory genes may be needed to induce a protective response against the virus. Combinations of T-helper cell sites and B-cell neutralization epitopes can induce high titers of neutralizing antibodies as well as T-cell response to more than one HIV isolate (Palker *et al.*, 1989).

Given the unprecedented obstacles that HIV presents toward development of a vaccine, it can be anticipated that multiple arms of the immune system will need to be activated in order to achieve a protective immune response against infection. Recent advances in our understanding of how antigens are processed and presented to the immune system are noteworthy in this regard. Until recently, it was generally accepted that antigen processing followed two separate pathways with distinct outcomes in relation to antibody synthesis and generation of classical cytotoxic lymphocytes (CD8+ CTL). It is further suggested that entry into these pathways is dependent upon whether the antigen is administered from without or synthesized within the antigen-presenting cell (for review see Bolognesi, 1990). However, as

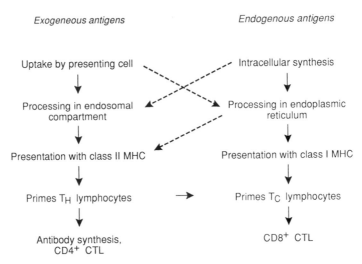

Exogeneous antigens   *Endogenous antigens*

Uptake by presenting cell → Intracellular synthesis
↓ ↓
Processing in endosomal compartment → Processing in endoplasmic reticulum
↓ ↓
Presentation with class II MHC → Presentation with class I MHC
↓ ↓
Primes T$_H$ lymphocytes → Primes T$_C$ lymphocytes
↓ ↓
Antibody synthesis, CD4$^+$ CTL   CD8$^+$ CTL

Figure 4. Solid arrows indicate current models and may represent primary pathways. Alternative pathways (dashed arrows) have recently emerged, the prevalence and efficiency of which may depend on antigen formulation or the nature of the antigen itself. T$_H$, helper T cell; T$_C$ cytotoxic T cell; CTL, cytotoxic T lymphocytes.

illustrated in Fig. 4, a number of crossover points between these two pathways have recently been established. This is welcome news for vaccine developers, particularly those whose focus is on approaches involving subunits or peptides. Until these developments, one could theoretically rely only on replicating entities in order to elicit CD8$^+$ CTL. The demonstration that subunit and peptide immunogens can be delivered in a manner such that they are presented within the context of both class I and class II MHC and thereby are able to stimulate a full range of immunity guarantees that such approaches will now receive considerable emphasis. The prospect of a subunit or peptide immunogens consisting only of desired epitopes which can also be mass-produced using modern biotechnology is an attractive approach for vaccine developers.

These and other encouraging steps aside, the challenges standing before the development of an HIV vaccine are prodigious. Much more has to be learned about approaches which guarantee sustained protective immunity, including the realm of new and more effective adjuvants. Along with this is the need for a better appreciation of how to induce mucosal immunity. However, by far the most formidable barrier will be the evaluation of vaccine efficacy in humans, particularly as this relates to identifying suitable populations for testing and determination of measurable prognostic end points for both infection and disease onset. As research produces more compelling vaccine candidates, it is hoped that such obstacles will also be overcome.

References

Ahearne, P. M., Matthews, T. J., Lyerly H. K., White, G. C., Bolognesi, D. P., and Weinhold, K. J. (1988). *AIDS Res. Hum. Retroviruses* **4**, 259–267.

Berman, P. W., Gregory, T. J., Riddle, L., Nakamura, G. R., Champe, M. A., Porter, J. P., Wurm, F. M., Hershberg, R. D., Cobb, E. K., and Eichberg, J. W. (1990). *Nature (London)* **345**, 622–625.

Bolognesi, D. P. (1989a). *JAMA* **261**, 3007–3013.

Bolognesi, D. P. (1989b). *Science* **246**, 1233–1234.

Bolognesi, D. P. (1989c). *Nature (London)* **340**, 431–432.

Bolognesi, D. P. (1990). *Nature (London)* **344**, 818–819.

Cease, K. B., Margalit, H., Cornette, J. L., Putney, S. D., Robey, W. G., Ouyang, C., Streicher, H. Z., Fischinger, P. J., Gallo, R. C., DeLisi, C., and Berzofsky, J. A. (1987). *Proc. Natl. Acad. Sci. U.S.A.* **84**, 4249–4253.

Chahn, T. L., Dreesman, G. R., and Kanda, P. (1986). *Eur. Mol. Biol. Org. J.* **5**, 3065–3071.

Cianciolo, G. J., Matthews, T. J., Bolognesi, D. P., and Snyderman, R. (1980). *J. Immunol.* **124**, 2900–2905.

Cianciolo, G. J., Copeland, T. D., Oroszlan, S., and Snyderman, R. L. (1985). *Proc. Natl. Acad. Sci. U.S.A.* **230**, 453–455.

Clayton, L. K., Sieh, M., Pious, D. A., and Reinherz, E. L. (1989). *Nature (London)* **339**, 548–551.

Clerici, M., Stocks, N. I., Zajac, R. A., Boswell, R. N., Bernstein, D. C., Mann, D. L., Shearer, G. M., and Berzofsky, J. A. (1989). *Nature (London)* **339**, 383–385.

Dalgleish, A. G., Chanh, T. C., Kennedy. R. C., Kanda, P., Clapham, P. R., and Weiss, R. A. (1988). *Virology* **165**, 209–215.

Desrosiers, R. C., Wyand, M. S., Kodama, T., Ringler, D. J., Arthur, L. O., Sehgal, P. K., Letvin, N. L., King, N. W., and Daniel, M. D. (1989). *Proc. Natl. Acad. Sci. U.S.A.* **86**, 6353–6357.

Evans, D. F., McKeating, J., Meredith, J. M., Burke, K. L., Katrak, K., John, A., Ferguson, M., Minor, P. D., Weiss, R. A., and Almond, J. W. (1989). Nature (*London*) **339**, 385–388.

Gallaher, W. R. (1987). *Cell* **50**, 327–328.

Gallaher, W. R., Ball, J. M., Garry, R. F., Griffin, M. C., and Montelaro, R. C. (1989). *AIDS Res. Hum. Retroviruses* **5**, 431–440.

Gelderblom, H. R., Reupke, H., and Pauli, G. (1985). *Lancet* **i**, 1016–1017.

Girard, M. P. (1989). 4th Colloque Des Cent Gardes October 26–28.

Golding, H., Robey, F. A., Gates, F. T., Linder, W., Beining, P. R., Hoffman, T., and Golding, B. (1988). *J. Exp. Med.* **167**, 914–923.

Golding, H., Shearer, G. M., Hillman, K., Lucas, P., Manischewitz, J., Zajac, R. A., Clerici, M., Gress, R. E., Boswell, R. N., and Golding, B. (1989). *J. Clin. Invest.* **83(4)**, 1430–1435.

Gonzalez-Scarano, F., Wazham, M. N., Ross, A. M., and Hoxie, J. A. (1987). *AIDS Res. Hum. Retroviruses* **3(3)**, 245–252.

Gorny, M. K., Gianakakos, V., Sharpe, S., and Zolla-Pazner, S. (1989). *Proc. Natl. Acad. Sci. U.S.A.* **86**, 1624–1628.

Goudsmit, J., Debouck, C., Meloen, R. H., Smit, L., Bakker, M., Asher, D. M., Wolff, A. V., Gibbs, C. J., and Gajdusek, D. C. (1988). *Proc. Natl. Acad. Sci. U.S.A.* **85**, 4478–4482.

Halstead, S. B. (1988). *Science* **239**, 476–481.

Hilleman, M. R. (1985). *J. Infect. Dis.* **151**, 407–419.

Homsy, J., Meyer, M., Tateno, M., Clarkson, S., and Levy, J. A. (1989). *Science* **244**, 1357–1360.

Klasse, P. J., Pipkorn, R., and Blomberg, J. (1988). *Proc. Natl. Acad. Sci. U.S.A.* **85**, 5225–5229.

Kowalski, M., Ardman, B., Basiripour, L., Lu, Y., Blohm, D., Haseltine, W., and Sodroski, J. (1989). *Proc. Natl. Acad. Sci. U.S.A.* **86**, 3346–3350.

Lanzavecchia, A., Roosnek, E., Gregory, T., Berman, P., and Agrignani, S. (1988). *Nature (London)* **334**, 530–532.

Leonard, C. K., Spellman, M. W., Riddle, L., Harris, R. J., Thomas, J. N., and Gregory, T. (1990). *J. Biol. Chem.* **265**, 10373–10382.

Linsley, P. S., Ledbetter, J. A., Kinney-Thomas, E., and Hu, S. (1988). *J. Virol.* **62**, 3695–3702.

Ljunggren, K., Biberfeld, G., Jondal, M., and Fenyo, E. (1989). *J. Virology* **63**, 3376–3381.

Looney, D. J., Fisher, A. G., Putney, S. D., Rusche, J. R., Redfielf, R. R., Burke, D. S., Gallo, R. C., and Wong-Staal, F. (1988). *Science* **241**, 357–359.

Lyerly, H. K., Matthews, T. J., Langlois, A. J., Bolognesi, D. P., and Weinhold, K. J. (1987). *Proc. Natl. Acad. Sci. U.S.A.* **84**, 4601–4605.

Lyerly, H. K., Reed, D. L., Matthews, T. J., Langlois, A. J., Ahearne, P. M., Petteway, S. R., Bolognesi, D. P., and Weinhold, K. J. (1988). *AIDS Res. Hum. Retroviruses* **3(4)**, 409–422.

Matsushita, S., Robert-Guroff, M., Rusche, J., Kioto, A., Hatori, T., Hoshino, H., Javaherian, K., Takatsuki, K., and Putney, S. D. (1988). *J. Virol.* **62**, 2107–2114.

McKeating, J. A., Gow, J., Goudsmit, J., Mulder, C., McClure, J., and Weiss, R. (1988). *In* "Retroviruses of Human AIDS and 4ᶜ Colloque des "Cent Gardes. Edited by M. Girard and L. Valette. Related Animal Diseases" pp. 159–164.

Murphey-Corb, M., Martin, L. N., Davison-Fairburn, B., Montelaro, R. C., Miller, M., West, M., Ohkawa, S., Baskin, G. B., Zhang, J., Putney, S. D., Allison, A. C., and Eppstein, D. A. (1989). *Science* **246**, 1293–1297.

Nara, P. L., Robey, W. G., Gonda, M. A., Cater, S. G., and Fischinger, P. J. (1987). *Proc. Natl. Acad. Sci. U.S.A.* **84**, 3797–3801.

Nara, P., Dunlop, N., Waters, D., Smit, L., Goudsmit, J., and Gallo, R. (1989). *Abstracts of V International Conference on AIDS, Montreal, 1989,* p. 518, #T.C.O.22.

Nixon, D. F., Townsend, A. R. M., Elvin, J. G., Rizza, C. R., Gallwey, J., and McMichael, A. J. (1988). *Nature (London)* **336**, 484–487.

Oldstone, M. B. A. (1987). *Cell,* **50**, 819–820.

Orenstein, J. M., Meltzer, M. S., Phillips, T., and Gendelman, H. E. (1988). *J. Virol.* **62**, 2578–2586.

Palker, T. J., Clark, M. E., Langlois, A. J., Matthews, T. J., Weinhold, K. J., Randall, R. R., Bolognesi, D. P., and Haynes, B. F. (1988). *Proc. Natl. Acad. Sci. U.S.A.* **85**, 1–5.

Palker, T. J., Matthews, T. J., Langlois, A., Tanner, M. E., Martin, M. E., Scearce, R. M., Kim, J. E., Berzofsky, J. A., Bolognesi, D. P., and Haynes, B. F. (1989). *J. Immunol.* **142**, 3612–3619.

Peluso, R., Haase, A., Stowring, L., Edwards, M., and Ventura, P. (1985). *Virology* **147**, 231–236.

Porterfield, J. S. (1982). *J. S. J. Hyg., Camb.* **89**, 355–364.

Price, R. W., Brew, B., Sidtis, J., Rosenblum, M., Scheck, A. C., and Cleary, P. (1988). *Science* **239**, 586–592.

Putney, S. D., Matthews, T. J., Robey, W. G., Lymm, D. L., Robert-Guroff, M., Mueller, W. T., Langlois, A. J., Ghrayeb, J., Petteway, Jr., S. R., Weinhold, K. J., Fischinger, P. J., Wong-Staal, F., Gallo, R. C., and Bolognesi, D. P. (1986). *Science* **234**, 1392–1395.

Putney, S. D., LaRosa, G., Javaherian, K., Emini, E., Bolognesi, D., and Matthews, T. (1989). *Abstracts of V International Conference on AIDS, Montreal 1989,* p. 527, #W.C.O.18.

Riviere, Y., Tanneau-Salvadori, F., Regnault, A., Lopez, O., Sansonetti, P., Guy, B., Kieny, M. P., Fournel, J. J., and Montagnier, L. (1989). *J. Virol.* **63**, 2270–2277.

Robinson, Jr., W. E., Montefiori, D. C., and Mitchell, W. M. (1988). *Lancet* **i**, 790–794.

Robinson, Jr., W. E., Montefiori, D. C., Gillespie, D. H., and Mitchell, W. M. (1989a). *J. AIDS* **2**, 33–42.

Robinson, Jr., W. E., Montefiori, D. C., Mitchell, W. M., Prince, A. M., Alter, H. J., Drees-man, G. R., and Eichberg, J. W. (1989b). *Proc. Natl. Acad. Sci. U.S.A.* **86,** 4710–4714.

Robinson, Jr., W. E. Kawamure, T., Gorny, M. K., Lake, D., Xu, J. Y., Matsumoto, Y., Sugano, T., Masuho, Y., Mitchell, W. M., Hersh, E., and Zolla-Pazner, S., (1990). *Proc. Natl. Acad. Sci. U.S.A.* **87,** 3185–3189.

Rosenberg, Z. F., and Fauci, A. S. (1989). *AIDS Res. Hum. Retroviruses,* **5,** 1–4.

Rusche, J. R., Lynn, D. L., Robert-Guroff, M., Langlois, A. J., Lyerly, H. K., Carson, J., Krohn, K., Ranki, A., Gallo, R. C., Bolognesi, D. P., Putney, S. D., and Matthes, T. J. (1987). *Proc. Natl. Acad. Sci. U.S.A.* **84,** 6924–6928.

Rusche, J. R., Javaherian, K., McDanal, C., Petro, J., Lynn, D. L., Grimaili, R., Langlois, A. J., Gallo, R. C., Arthur, L. O., Fischinger, P. J., Bolognesi, D. P., Putney, S. D., and Matthews, T. J. (1988). *Proc. Natl. Acad. Sci. U.S.A.* **85,** 3198–3202.

Sargo, C., Guo, H. G., Franchini, G., Aldovini, A., Collalti, E., Farrell, K., Wong-Staal, F., Gallo, R. C., and Reitz, M. S. (1988). *Virology* **164,** 531.

Schrier, R. D., Gnann, J. W., Langlois, A. J., Shriver, K., Nelson, J. A., and Oldstone, M. B. A. (1988). *J. Virol.* **62,** 2531–2536.

Siliciano, R. F., Lawton, T., Knall, C., Karr, R. W., Berman, P., Gregory, T., and Reinherz, E. L. (1988). *Cell* **54,** 561–575.

Skinner, M. A., Ting, R., Langlois, A. J., Weinhold, K. J., Lyerly, H. K., Javaherian, K., and Matthews, T. J. (1988a). *AIDS Res. Hum. Retroviruses* **4,** 187–197.

Skinner, M. A., Langlois, A. J., McDanal, C. B., Bolognesi, D. P., and Matthews, T. J. (1988b). *J. Virol.* **62,** 4195–4200.

Starcich, B. R., Hahn, B. H., Shaw, G. M., McNeely, P. D., Modrow, S., Wolf, H., Parks, E. S., Parks, W. P., Josephs, S. F., Gallo, R. C., and Wong-Staal, F. (1986). *Cell* **45,** 637–648.

Sutjipto, S., Pendersen, N., Gardner, M., Hanson, C., Miller, C., Gettie, A., Jennings, M., Higgins, J., and Marx, P. (1989). *Abstracts of V International Conference on Aids, Montreal, 1989,* C.Th.C.O.45

Takahashi, H., Cohen, J., Hosmalin, A., Cease, K. B., Houghten, R., Cornette, J. L., DeLisi, C., Moss, B., Germain, R. N., and Berzofsky, J. A. (1988). *Proc. Natl. Acad. Sci. U.S.A.* **85,** 3105–3109.

Takahashi, H., Merli, S., Putney, S. D., Houghten, R., Moss, B., Germain, R. N., and Ber-zofsky, J. A. (1989a). *Science* **246,** 118–121.

Takahashi, H., Houghton, R., Putney, S. D., Margulies, D. H., Moss, B., Germain, R. N., and Berzofksy, J. A. (1989b). *J. Exp. Med.* **170,** 2023–2035.

Takeda, A., Tuazon, C. U., and Ennis, F. A. (1988). *Science* **242,** 580–583.

Thomas, E. K., Weber, J. N., McClure, J., Clapham, M. C., Singhal, M. C., Shriver, M. K., and Weiss, R. A. (1988). *AIDS* **2,** 25–29.

Townsend, A. R. M., Gotch, F. M., and Davey, J. (1985) *Cell* **42,** 457–467.

Walker, B. D., Flexner, C., Paradis, T. J., Fuller, T. C., Hirsch, M. S., Schooley, R. T., and Moss, B. (1988). *Science* **240,** 64–66.

Weinhold, K. J., Lyerly, H. K., Stanley, S. D., Austin, A. A., Matthews, T. J., and Bolognesi, D. P. (1989). *J. Immunol.* **142,** 3091–3097.

Zajac,. B. A., West, D. J., McAleer, W. J., and Scolmick, E. M. (1986). *J. Infect. Dis.* **13,** 39–45.

Index